Flexible Towpregs and Their Thermoplastic Composites

Flexible Towpregs and Their Thermoplastic Composites

Edited by
R. Alagirusamy

CRC Press
Taylor & Francis Group
Boca Raton London New York

CRC Press is an imprint of the
Taylor & Francis Group, an **informa** business

First edition published 2023
by CRC Press
6000 Broken Sound Parkway NW, Suite 300, Boca Raton, FL 33487-2742

and by CRC Press
4 Park Square, Milton Park, Abingdon, Oxon, OX14 4RN

© 2023 Taylor & Francis Group, LLC

CRC Press is an imprint of Taylor & Francis Group, LLC

ISBN: 978-0-367-46912-2 (hbk)
ISBN: 978-0-367-50380-2 (pbk)
ISBN: 978-1-003-04971-5 (ebk)

DOI: 10.1201/9781003049715

Typeset in Times
by KnowledgeWorks Global Ltd.

Contents

Preface

Thermoplastic polymers present a formidable alternative to the traditional matrix materials due to shorter processing cycles, effortless handling, indefinite shelf life at room temperature, and superior energy absorption characteristics. In spite of the virtues offered, high melt viscosity of these materials remains as a hurdle to obtain void free high-quality composites with similar production methods employed for thermoset matrices. These overwhelmingly high viscosity, bars the impregnation of the resin into woven, knitted and braided textile preforms to displace the air as well as produce the thoroughly impregnated, well consolidated, void free composite structures. This necessitates development of flexible towpregs, also known as hybrid yarns, which can undergo textile preforming operations and result in good quality void free composites.

In this book, various methods developed for production of flexible towpregs, namely commingling, powder coating, friction spinning, and other techniques, are discussed at length. Various studies conducted on properties of composites produced with these techniques are analysed. Attempts at modelling consolidation behaviours of these towpregs are explored at length. Applications of the thermoplastic matrix composites in various areas are highlighted. Special emphasis is given on application of these composites in electromagnetic shielding. This book will be highly helpful to students, researchers, and industry experts working in the area of thermoplastic composites.

Editor

R. Alagirusamy, PhD, is a Professor in the Department of Textile and Fibre Engineering at Indian Institute of Technology, Delhi (IIT Delhi), India. He earned a PhD at the Georgia Institute of Technology, Atlanta, Georgia, USA, in the field of textile preforming for composite applications using thermoplastic powder coated carbon towpregs. He has published more than 100 research articles in international journals and conferences. He has also authored and edited books, and contributed book chapters and monographs in the area of textile structures used in composite applications. His research areas of interest are textile performing for composites, hybrid yarn development for thermoplastic composites, natural fibre composites, and structure property relations in spun yarns.

Contributors

R. Alagirusamy
Department of Textile and Fibre
 Engineering
Indian Institute of Technology Delhi
New Delhi, India

V. Balakumaran
Centre for Biomedical Engineering
Indian Institute of Technology Delhi
New Delhi, India

Mahadev Bar
Laboratoire Génie de Production
INP-ENIT
Université de Toulouse
Tarbes, France

Nabo Kumar Barman
Department of Textile Engineering
Faculty of Technology and Engineering
The Maharaja Sayajirao University of
 Baroda
Vadodara, Gujarat, India

S.S. Bhattacharya
Department of Textile Engineering
Faculty of Technology and Engineering
The Maharaja Sayajirao University of
 Baroda
Vadodara, Gujarat, India

Apurba Das
Department of Textile and Fibre
 Engineering
Indian Institute of Technology Delhi
New Delhi, India

Vijay Goud
Process Engineering
3B the Fibreglass Company
Colvale, Goa, India

Ganesh Jogur
Department of Textile and Fibre
 Engineering
Indian Institute of Technology Delhi
New Delhi, India

Dinesh Kalyanasundaram
Center for Biomedical Engineering
Indian Institute of Technology Delhi
New Delhi, India

Ashraf Nawaz Khan
Department of Textile and Fibre
 Engineering
Indian Institute of Technology Delhi
New Delhi, India

J. Krishnasamy
Department of Textile Technology
PSG College of Technology
Peelamedu, Coimbatore, India

Puneet Mahajan
Department of Applied Mechanics
Indian Institute of Technology Delhi
Delhi, India

Vinayak Ogale
Research and Development
Saint-Gobain India Private Ltd.
Chennai, India

Pierre Ouagne
Laboratoire Génie de Production
INP-ENIT
Université de Toulouse
Tarbes, France

Naveen Padaki
Central Silk Technological Research
    Institute
Central Silk Board
Ministry of Textiles, Government of
    India
Bengaluru, India

G. Thilagavathi
Department of Textile Technology
PSG College of Technology
Peelamedu, Coimbatore, India

1 Introduction to Thermoplastic Composites

R. Alagirusamy[a] and Mahadev Bar[b]
[a]Indian Institute of Technology Delhi
New Delhi, India
[b]Université de Toulouse (INP-ENIT)
Tarbes, France

CONTENTS

DOI: 10.1201/9781003049715-1

1.1   INTRODUCTION

Starting from the housing of early civilization, composite materials have played an important role in the progress of human beings and are still contributing [Chang and Lees, 1988]. Composite materials are a kind of material which is made of two or more components with different properties. These components do not dissolve into or react with each other, and inside the material structure they keep their individual identity by separating themselves with a prominent interphase. Out of these two components one is called a matrix (the continuous phase), while the other is known as reinforcement/filler. The characteristics of the matrix determine the shape, surface appearance, environmental tolerance, stress transfer and the durability of a composite system while the filler/reinforcement determines its mechanical properties. Wood, bone etc. are some examples of composite structures that exist in the nature. The first human-made composite was a brick made up of mud and straw, evident around 1500 BC, made by the Mesopotamian settlers (Johnson, 2018). Composite materials are strong and light in weight. This unique characteristic of the composite materials makes them very handy at the time of war. During World War II, many composites were developed and moved from the laboratory to actual production. Polymer composites are the result of this need and development.

Along with polymers, metals and ceramics are also used as a composite matrix. Among these three matrix systems, the polymer matrices are the most popular for composite manufacturing. The polymer matrix composites are manufactured by reinforcing polymers with particle filers or with textile grade fibrous materials. Presently, polymer composite materials are gaining interest as the demand for lightweight, high strength materials are increasing among the consumers (Bar, Das, and Alagirusamy 2018; John and Thomas, 2008). Polymer composites not only offer high specific mechanical properties but also exhibit some exceptional properties such as high durability and damping property, resistance to corrosion, wear, impact, fire etc. Due to these diverse material properties the polymer composite materials have found application in automobile, biomedical, construction, marine, packaging, sports and other areas.

The rate of consumption of the polymer matrix composite materials in these areas is rising exponentially every year (Bar et al. 2020; John and Thomas 2008). This increasing consumption of durable polymer composites has led to the concern of plastic waste accumulation in the environment. Based on their thermal characteristics polymer matrices are of two types, namely thermoset and thermoplastic. At present, thermoset polymers are mostly used as matrices for composite manufacturing. The thermoset polymers are non-recyclable while the thermoplastic polymers are recyclable. Hence, considering the problem of non-recyclable solid waste disposals the composite researchers and manufacturers have shifted their interest from thermoset polymers to thermoplastic polymers (Fortea-Verdejo et al. 2017). This chapter seeks to provide an overview of the science and technology related to thermoplastic polymers and their composites.

1.2   THERMOSET AND THERMOPLASTIC POLYMERS

Based on their thermal characteristics, polymers are of two types, namely thermoset and thermoplastic polymers. Thermoset polymers are those that are irreversibly hardened from their viscous liquid state by the curing process. The curing process involves a chemical reaction called cross-linking. The thermoset polymers in their uncured state have short and unlinked molecules known as monomers. Then with the addition of a second phase known as cross-linker and/or at suitable conditions, the curing reaction is initiated. This curing process is induced by heat or suitable radiation and is influenced by other external conditions such as pressure, catalyst etc. On curing, the thermoset polymers form a three-dimensional cross-linked network, which is irreversible. A schematic diagram explaining the curing of thermoset polymers is shown in Figure 1.1. If heat is applied to the cured thermoset resin, it will eventually degrade thermally without melting. Epoxy, vinyl ester, unsaturated polyester, phenolic resin etc. are the examples of some thermoset resins. At present, thermoset polymers are mostly used as matrices for composite reinforcement. It is mainly due to ease of processing, ease to wet the reinforcing filler or fibres, design flexibility etc. Fibre reinforced thermoset composites have found utilization in the areas of aerospace, automobile, construction, sports, packaging etc.

However, the thermoset polymers as a composite matrix forming component have some drawbacks, which are as follows (Faruk et al. 2012):

- Thermoset polymers are non-biodegradable and non-recyclable.
- They are sticky, which contaminates the machine parts.
- They have shorter self-life.

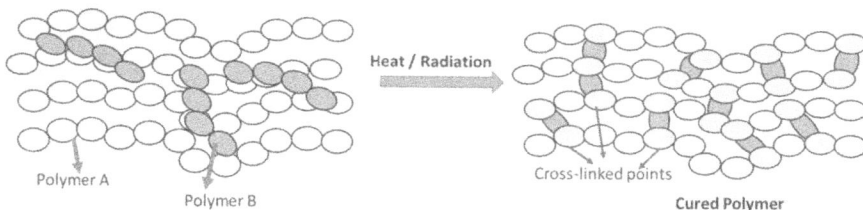

FIGURE 1.1   Curing of the thermoset polymer.

- It needs longer curing cycle.
- Thermoset polymers are prone to trap the air bubble, which ultimately leads to development of a composite with a high amount of voids and a composite with inferior mechanical properties.

Considering the drawbacks of the thermoset resins, the interest of the composite researcher and the manufacturer is shifting from the thermoset to thermoplastic polymers. Thermoplastics are melt-processable polymers. When a thermoplastic polymer is subjected to a heat source, it melts above its melting point. Then on the removal of the heat source, it again solidifies when its temperature goes below the melting point. This process is repeatable. However, there is a practical limit to the number of this recycling above which the material properties will begin to suffer. Figure 1.2 shows a differential scanning calorimetry (DSC) graph of a thermoplastic polymer (polypropylene). During DSC, the polymer initially is subjected to a heating cycle where it melts at a temperature of 164°C and then on cooling, the molten polymer solidifies. The solidifying temperature or the crystallization point (during cooling) is lower than the melting temperature point (Bar, Das, and Alagirusamy 2019).

Thermoplastic polymers as a composite matrix have many advantages which are mentioned below (John and Thomas 2008).

i. Enhanced mechanical properties, notably in terms of impact and abrasion
ii. Enhanced environmental, moisture and corrosion resistance
iii. Unlimited shelf life of raw materials
iv. Product forms and fabrication processes can be tailored to meet the needs of the application
v. Environmentally benign because there are no exothermic reactions, toxic or solvent emissions

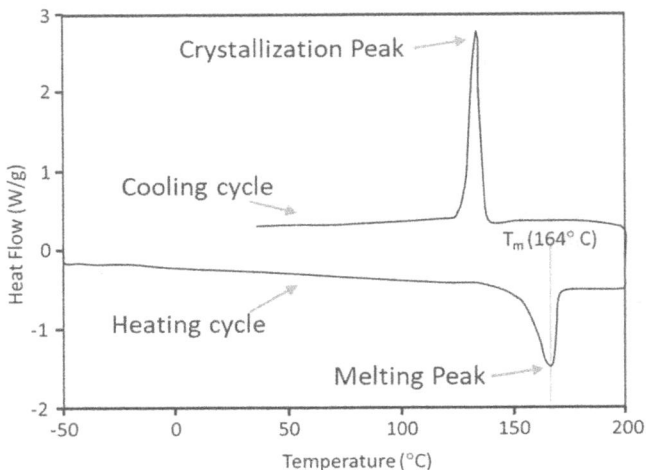

FIGURE 1.2 Differential scanning calorimetry (DSC) graph of a thermoplastic polymer (polypropylene).

vi. Readily adaptable to manufacturing for low as well as high volumes
vii. Low tooling costs and rapid cycle times
viii. Ability to be recycled
ix. Improved assembly and joining methods
x. Ability to mould large integrated complex mouldings in one operation
xi. Potential for functional tailoring of the fibre

1.3 THERMOPLASTIC POLYMERS FOR STRUCTURAL COMPOSITE APPLICATION

Since the mid-1980s, thermoplastic polymers have gained considerable attention as matrix materials for fibre reinforced polymer composites. The advantages of thermoplastic polymers make them attractive as a composite matrix over thermoset polymers (John and Thomas 2008). There are plenty of thermoplastic resins available, which can be segregated in two groups, namely high-performance thermoplastics and engineering thermoplastics. Based on their degradability characteristics, thermoplastic polymers are further classified as biodegradable thermoplastic polymer and non-biodegradable thermoplastic polymer. Thermoplastic polymers which are widely used as a composite matrix are discussed below.

1.3.1 POLYETHYLENE

Polyethylene (PE) is one of the most versatile and widely used thermoplastics in the world because of its excellent properties such as near-zero moisture absorption, excellent chemical inertness, high toughness, low coefficient of friction, ease of processing and unusual electrical properties. Polyethylene is produced from ethylene (CH_2) monomers and is available in numerous forms, such as low-density polyethylene (LDPE), linear low-density polyethylene (LLDPE), high-density polyethylene (HDPE), and ultrahigh molecular weight polyethylene (UHMWPE). These deviations are based on branching in the molecular chain and density (Gopanna et al. 2019). Figure 1.3 shows the schematics of the molecule alignment in different forms of PE.

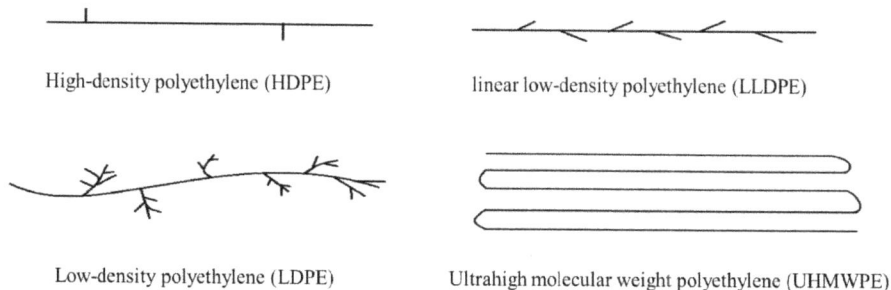

High-density polyethylene (HDPE)

linear low-density polyethylene (LLDPE)

Low-density polyethylene (LDPE)

Ultrahigh molecular weight polyethylene (UHMWPE)

FIGURE 1.3 Schematics of the molecule alignment in different forms of PE.

The molecular weight of PE has a great influence on the properties of PE. Thus, it is important to know the structure and properties of PE to control and modify the properties of PE composites for different applications. The LDPE has a branched molecular structure, which prevents the close packing of the molecules and ultimately resulted in low crystallinity. LDPE is flexible and has low tensile and compressive strength. LDPE is generally used in food packaging materials, plastic film and rigid container applications. On other hand, UHMWPE has no branch in the molecular chain. Thus, it has the highest density and crystallinity percentage among all polymers. UHMWPE is available in powder and in film form. It has very good resistance to wear and impact properties and because of that it finds wide industrial applications (Pleşa et al. 2019).

1.3.2 POLYPROPYLENE

Polypropylene (PP) is one of the most studied thermoplastic resins used for composite manufacturing. It is produced by addition polymerization of the propylene monomer. Propylene has one additional CH_3 group in its structure as compared to ethylene. The arrangement of this CH_3 group in the molecular chain is very important as it determines various properties of the polypropylene. Based on the special conformation of the CH_3 group in the macromolecule, PP are of three types, namely atactic PP, syndiotactic PP and isotactic PP. The chemical structure of propylene monomers and different types of PP are shown in Figure 1.4.

Among these three categories the isotactic PP has the highest melting point due to its high degree of crystallinity. This isotactic PP is mainly used as a matrix for composites and other industrial applications. The molecular weight and molecular weight distribution of any polymer affect its properties, especially the rheological and mechanical properties. PP has a very high molecular weight (M_w = 220,000 – 700,000) and high molecular weight distribution (M_w/M_n = 5.6 – 11.9). PP has a density of 0.906 g/cm^3, which is lower than PE, but its glass transition temperature and melting temperature are higher than that of PE. Polypropylene has high toughness, good di-electrical properties and high chemical resistance, but it is susceptible to degradation by UV radiation. PP is available in different shapes such as chips, film, fibres, tapes etc. and it can be processed by compression moulding, injection moulding, blow moulding, extrusion and thermoforming techniques (Pleşa et al. 2019).

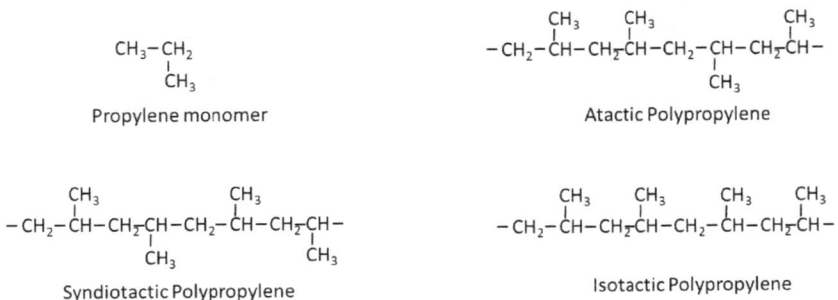

FIGURE 1.4   Propylene monomer and typical structures of PP.

1.3.3 POLYSTYRENE

Polystyrene (PS) is considered as a commodity plastic having a wide range of commercial application opportunities. Switches, shells of televisions, computers etc. are some examples of PS commodities that we use on a daily basis. This widespread use and popularity of PS makes it an obvious candidate for fibre reinforced polymer composite research. PS is produced by the free radical polymerization of the styrene monomers. Figure 1.5 shows the different steps of PS polymerizations.

PS has a similar structure of PP except that the –CH₃ side group of the PP is replaced by a benzene ring. Like PP, PS chains can also exit in three different configurations, i.e. in isotactic, atactic and syndiotactic configurations. The physical and chemical characteristics of each PS configuration differ from the other. Among these three configurations of PS, the atactic PS is the most commercially used PS. Atactic PS is amorphous in nature, which makes it transparent. It has a glass transition temperature of about 100°C and exists in the glassy state at room temperature. Although it is considered a brittle polymer, it can sustain some yielding by crazing (Marshall, Culver, and Williams 1973).

1.3.4 POLYETHYLENE TEREPHTHALATE

Polyethylene terephthalate (PET) is one of the most diffused thermoplastic polymers available in the market. PET belongs to the polyester family, which is a wide category of polymers characterized by having esters functionalities within the macromolecular main chains. Among polymers, PET occupied 7.7% of the European market and it is mainly used for the production of packaging containers and bottles, textile fibres, films, composites etc. The traditional production of PET relies on direct esterification of terephthalic acid and ethylene glycol or on transesterification reaction of dimethyl terephthalate and ethylene glycol. The process initially produces an intermediate compound named bis(2-hydroxyethyl) terephthalate which further produces PET polymer through polycondensation reaction. Figure 1.6 shows the schematic representation of the PET polymerization mechanism.

Initiation

Propagation

Termination

FIGURE 1.5 Free-radical polymerization of styrene monomers.

Esterification

Transesterification

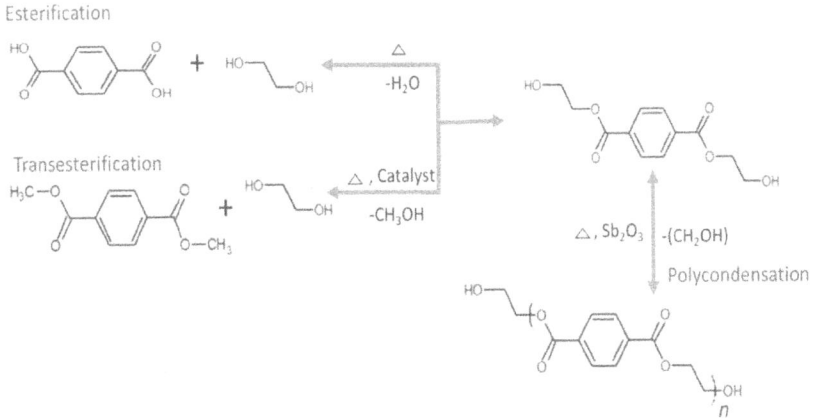

FIGURE 1.6 PET polymerization mechanism.

The density of PET is more than 1.00and it depends on the crystalline percentage of the PET. The amorphous PET has a density in the range between 1.29 g/cm^3 and 1.39 g/cm^3 and it transparent while the semi-crystalline PET is opaque and its density varies between 1.37 g/cm^3 and 1.40 g/cm^3. The PET main chain has aromatic rings and polar groups, which resulted in higher thermal stability of PET. The melting temperature of semi-crystalline PET is around 265°C while its glass transition temperature is around 80°C (Nisticò 2020).

1.3.5 POLYAMIDE

Polyamides (PA) are a type of polymer containing an amide functional group in its main chain. This amide functional group results from the polycondensation of an amine and a carboxylic acid functional group. Based on their skeletal composition, polyamides are of two types, namely aliphatic polyamide and aromatic polyamide. During composite manufacturing aliphatic polyamides are generally considered as a matrix while aromatic polyamides are used as a reinforcement. Kevlar and Nomex from DuPont are some examples of aromatic polyamide while PA 6 and PA 6,6 are some examples of aliphatic polyamide. PA 6 is produced by the polymerizing ε-caprolactam through the ring opening reaction mechanism. PA 6,6 is produced by the poly-condensation reaction of hexamethylene-di-amine and adipic acid. A schematic representation of a PA 6,6 polymerization mechanism is shown in Figure 1.7. Polyamide resins are known as

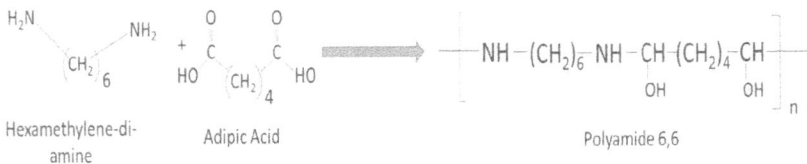

Hexamethylene-di-amine   Adipic Acid   Polyamide 6,6

FIGURE 1.7 Polymerization mechanism of PA 6,6.

engineering thermoplastics, mainly due to their outstanding thermal and mechanical characteristics. Polyamides are available in fibres, films and in polymer chip forms and are generally used in fabrics, carpets and in other tough good applications. The significance of polyamide is limited by moisture absorption, notch sensitivity, poor dimensional stability and low impact strength. These drawbacks are overcome by reinforcing polyamides with different fibrous materials or with various inorganic particles (Kausar and Anwar 2018).

1.3.6 POLY (AMIDE-IMIDE)

Poly(amide-imide) (PAI) is a copolymer of polyamide and polyimide. PAI's can be produced by the condensation of aromatic diamine. The chemical structure of PAI is shown in Figure 1.8. PAI's have excellent mechanical, thermal and oxidative features, which made them appropriate for numerous applications including transportation, industrial processes, electrical equipment etc. PAI is a potential condensation polymer with high heat resistance. PAI has been extensively used in magnetic wire coatings in automotive applications. The composites of polyimide with nanofillers have outstanding thermal, mechanical and electrical features (Kausar and Anwar 2018).

1.3.7 POLYETHER ETHER KETONE

PEEK (poly ether ether ketone) is the foremost member of the polyether ketone family. PEEK is an aromatic, semi-crystalline, thermoplastic polymer. It is commercialized for industrial purposes such as aircraft, turbine blades, piston parts, cable insulation, etc. Chemically, PEEK has a polymer repeat unit of one ketone and two ether groups. This provides a linear, fully aromatic, highly stable structure containing only carbon, hydrogen and oxygen atoms. Because of this unique structure, PEEK poses excellent thermal stability, heat and chemical resistance, and very good mechanical properties. PEEK polymers are synthesized by the step-growth polymerization i.e. through dialkylation of the bisphenolate salts. The synthesis of PEEK polymer through the reaction of disodium salt of hydroquinone with 4, 4-difluorobenzophenone is shown in Figure 1.9. PEEK has a glass transition temperature of around 143°C and it melts around 343°C. PEEK can be processed using compression moulding, injection moulding or extrusion method to manufacture various composite parts.

Salazkin and Shaposhnikova 2020

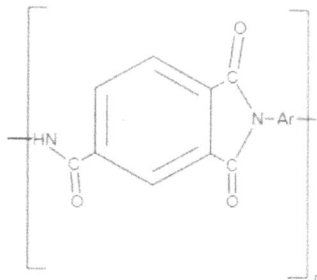

FIGURE 1.8 Chemical structure of poly(amide-imide).

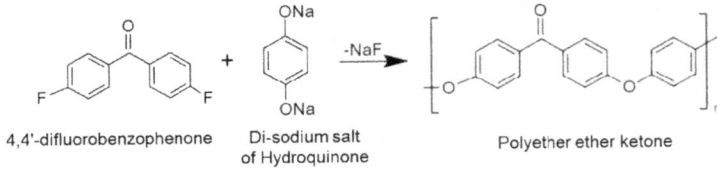

FIGURE 1.9 Chemical synthesis of PEEK polymer.

1.3.8 Poly (Lactic Acid)

Poly (lactic acid) (PLA) is a bio-based, biocompatible aliphatic thermoplastic polyester. Over the last 20 years, PLA has drawn a considerable interest of the composite researchers and manufacturers due to its good processability and properties compared with other biodegradable polymers. PLA is derived from resources such as sugarcane, corn starch, and other renewable biomass products and wastes. PLA can be synthesized through either ring-opening polymerization of cyclic lactide dimers or through direct condensation polymerization of lactic acid monomer. Figure 1.10 shows the PLA synthesis mechanism through the ring-opening mechanism and through the direct condensation mechanism. In general, the ring-opening approach is used to produce high molecular weight, solvent-free PLA. Compared to its competitive bio-based polymers, PLA exhibits high strength and modulus, good clarity and barrier properties which enable PLA to be a suitable candidate in commodity and engineering applications. Some of these applications include fibres and textiles, film and packaging, construction, automotive products etc. PLA melts around 150°C while its glass transition temperature varies between 55°C and 60°C. Although the use of PLA is increasing rapidly, it has some drawbacks which restrict its usage in different applications. These limitations are mainly due to PLA's inherently slow crystallization rate, low melt strength, brittleness, low toughness, low service temperature etc. PLA often blend with other polymers to overcome the above drawbacks (Hamad et al. 2018; Nofar et al. 2019).

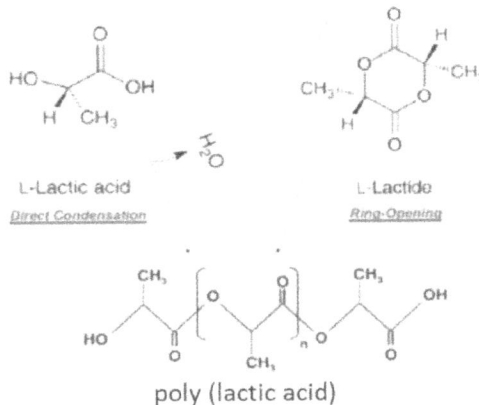

FIGURE 1.10 Synthesis of poly (lactic acid). (Reproduced with permission from Nofar et al. 2019.)

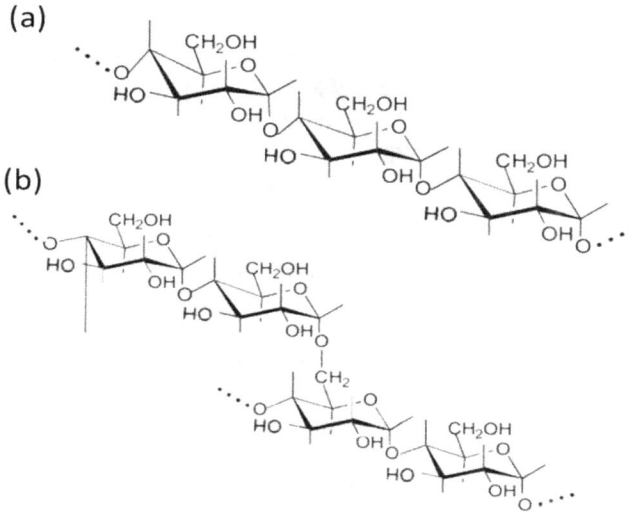

FIGURE 1.11   Chemical structure of amylose and amylopectin. (Reproduced with permission from Wang et al. 2003.)

1.3.9   THERMOPLASTIC STARCH

Starch is a biodegradable and biocompatible polysaccharide derived from renewable resources. It mainly consists of two dissimilar macromolecules named amylose and amylopectin. The amylose is a straight-chain polymer while amylopectin is a branch polymer. Figure 1.11 shows the chemical structure of amylose and amylopectin. Starch is obtained from various plant resources such as wheat, maize, rice, potatoes, peas etc. where its composition, i.e. the amylose and amylopectin content, varied from one source to another. The decomposition temperature of native starch is below its melting point. Hence on heating, the native starch tends to degrade thermally before melting. However, native starch can be transformed into a thermoplastic starch by adding a suitable plasticizer. A plasticizer enhances the mobility of the starch molecules and brings down the melting temperature of the starch below its degradation temperature. Glycerol, glycol, sorbitol, sugar and ethanolamine are some examples of plasticizers which are generally mixed with native starch to produce thermoplastic starch (Wang, Yang, and Wang 2003).

1.4   REINFORCEMENTS FOR THERMOPLASTIC POLYMER COMPOSITES

Reinforcements are the important phase of any composite system. They determine the mechanical properties of the resultant composite. Based on their physical appearances, reinforcing materials for thermoplastic composites are of two types, namely particle-sized reinforcements and fibrous reinforcements.

1.4.1 Particle Reinforcements

Particles are widely used as reinforcing material for thermoplastic polymers due to their wide availability at a reasonably low piece. Compared to fibrous reinforcements, particles have lower efficiency in terms of strengthening polymers. However, polymers can be reinforced to a higher concentration level with particles compared to fibres. Based on strengthening mechanisms, particles reinforced are of two types: dispersion-strengthened composites and particulate composites. Particles for dispersion strengthened composites are comparatively smaller in dimensions, generally between 0.01 μm and 1 μm. In case of dispersion strengthened composites, the matrix bears the major portion of an applied load, while the dispersoids hinder/impede the motion of dislocations. Particulate composites are the other class of particle-reinforced composites that contain large amounts of comparatively coarse particles. These composites are designed to produce unusual combinations of properties rather than to improve the strength. Thermoplastic polymers are often reinforced with various particulate materials to enhance their toughness, abrasion resistance and other properties (Chawla 1998; Hull and Clyne 1996).

1.4.2 Fibrous Reinforcements

Fibrous materials as composite reinforcing components have gained much attention from composite researchers and manufacturers due to their light weight and good mechanical properties. To provide flexibility in cost, composite properties and the process requirements, fibrous materials can be combined with the matrices in various from. Different fibrous architectures or the textile structures used as composite reinforcements are reported in Figure 1.12. The advantages and disadvantages of these structures are discussed below.

1.4.2.1 Short Fibre

Most of the studies on fibre reinforced polymer composites are carried out using short fibre as reinforcement. This is mainly due to abundant availability, easy handling and processability of the short staple fibres. Short fibre reinforced polymer composites can be manufactured by different methods such as compression moulding, vacuum

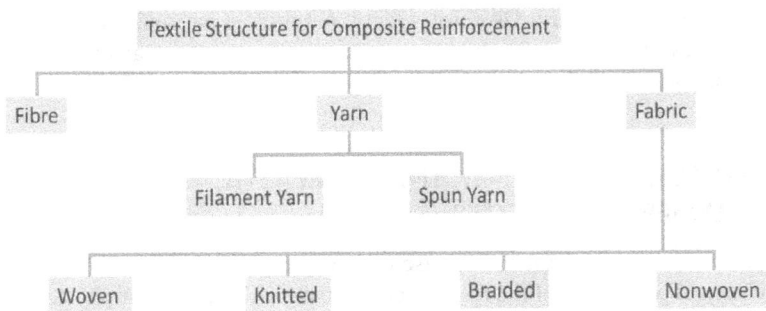

FIGURE 1.12 Textile structures for polymeric composite reinforcement.

bagging, injection moulding etc. Depending on their origin, fibres are of two types, namely natural fibres and synthetic fibres.

It has been observed that compared to virgin polymers, both the synthetic as well as natural fibre reinforced polymers have improved mechanical properties. In general, the synthetic fibres such as glass, carbon, aramid etc. are considered for high-end applications while the natural fibres are considered for low and medium load bearing applications. At present glass fibres are mostly used for composite manufacturing, but with increasing environmental awareness the natural fibres are replacing the glass fibres as composite reinforcements wherever it is possible. Irrespective of fibre types, the main challenge with the short fibre composites is to control the fibre orientation in the composite structure. The random or uncontrolled fibre orientation in the composite structure diminishes the ultimate fibre strength utilization to the final composite (Rowell 2008). In order to control the fibre orientation in the composite structure, short fibres are often transformed in some suitable textile structure such as yarn, woven fabric, knitted fabric etc.

1.4.2.2 Yarn

Yarn is another form of textile structure which is widely used for the composite reinforcing purpose. Textile yarns are of two types: filament yarns and spun yarns. Synthetic yarns used for composite reinforcing purpose generally have a filament structure, while natural fibres (except silk) have short fibre length and are twisted together to produce a continuous yarn. Yarn reinforced polymer composites are fabricated by filament winding method, which is very useful for making hollow, mostly circular or oval shaped composites such as pipes, tanks etc. (Lehtiniemi et al. 2011). Twisted yarn as a reinforcing component can control the fibre orientation in the composite structure, but the fibre obliquity because of twist deteriorates the ultimate fibre strength utilization. Moreover, twist inhibits the uniform fibre-resin distribution in the composite structure. This problem is very prominent in the case of thermoplastic polymer composites. To overcome this problem, hybrid yarns during composite manufacturing are introduced. Different types of yarn used for composite manufacturing are shown in Figure 1.13.

1.4.2.3 Fabric

Textile fabrics are the most preferred textile structures for composite reinforcement because of their easy handling. Based on manufacturing techniques, textile fabrics are of four types, namely nonwoven, knitted, braided and woven fabrics. The advantages and disadvantages of different fabric structures as composite reinforcing components are discussed here.

1.4.2.3.1 Nonwoven

Nonwovens are a type of textile fabric structure produced by the bonding, interlocking or by both, consolidated with mechanical, thermal, chemical or solvent means, with the exception of papers and fabrics made by weaving, knitting and tufting. Nonwovens are produced directly from the fibres without converting them in yarn form. Nonwoven technology is the easiest and fastest method of fabric manufacturing

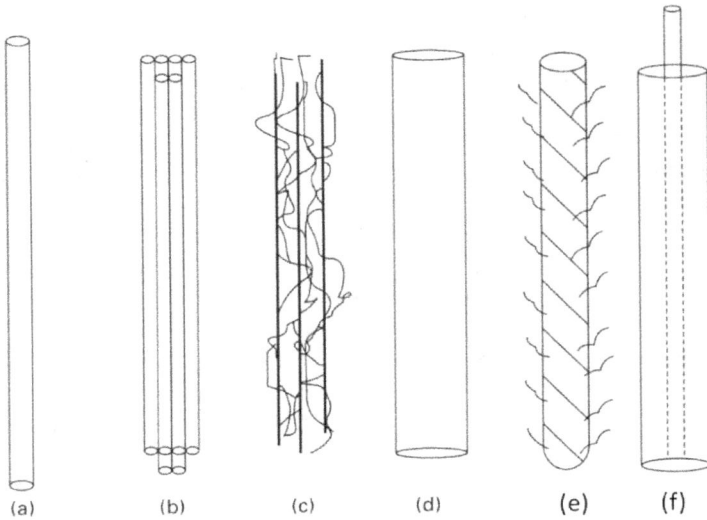

FIGURE 1.13 Yarn structures used for composite manufacturing. (a) Monofilament yarn, (b) multi-filament yarn, (c) commingled yarn, (d) tape yarn, (e) spun yarn, and (f) core-sheath structured yarn.

and is suitable for processing all types of fibres. The manufacturing of nonwoven fabric generally has two important steps: preparation of fibre-web and the bonding of the fibres to impart strength in the web. Drylaid, wetlaid, spun-melt etc. are some examples of web formation techniques, while needle punching, hydro-entanglement, stitch bonding, thermal bonding etc. are some examples of web bonding techniques. Nonwovens are soft, porous and voluminous in nature and are widely used as filtration and hygiene materials. At present, nonwovens are evolving as a material for composite reinforcement. Nonwoven structure-based fibre reinforced polymers exhibit better mechanical properties than that of the virgin polymer. However, nonwovens are anisotropic owing to random orientation of the fibres. This results in lower mechanical performance of the resultant composite (Das et al. 2012).

1.4.2.3.2 *Knitted Fabric*

Knitting is the second-largest technique of fabric formation. The continuous yarns are converted into a fabric by interloping through the knitting technique. This loop structure brings flexibility, stretchability, comfort and shape retention property to the knitted fabric. Depending on the direction of yarn feeding and fabric formation, knitted fabrics are of two types, warp knitted fabrics and weft knitted fabrics. A schematic diagram of different knitted fabric architectures is shown in Figure 1.14. Weft knitting is a more popular and diverse technique than warp knitting. It is mainly due to versatility, low capital cost, easy pattern changing facilities and high production rate of the weft knitting process. In general, a single thread is used for manufacturing weft knitted fabric while a yarn beam is used for the production of warp-knitted fabric.

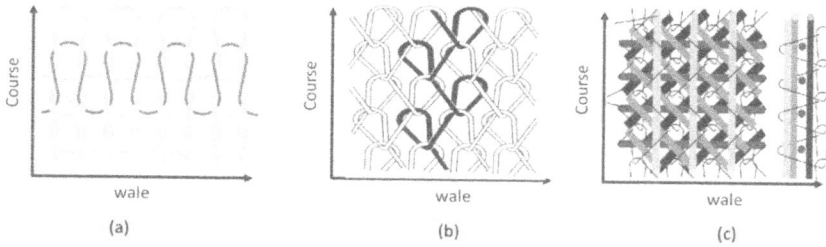

FIGURE 1.14 Schematic diagram of (a) weft knitted fabric, (b) warp knitted fabric, and (c) multi axial 3D knitted fabric. (Adapted from https://en.wikipedia.org/wiki/Knitted_fabric.)

Knitted structures are generally not preferred for composite reinforcement because of their high extensibility and poor stability. To enhance the composite reinforcing ability of the knitted fabrics, float stiches are incorporated into the basic knit architectures (Rudd, Owen, and Middleton 1990). At present, knitted structures are getting interest because of growing awareness of their formability and novel 3D net-shape structure (Alagirusamy et al. 2006). Multi-axial warp knitted fabrics with no crimp are the latest addition to the knitted architectures for composite reinforcement. In this kind of knitted structure, the knitted loops are present only to hold the un-crimped yarns. Hence, the mechanical properties of the non-crimp multi-axial warp knitted fabric composites are expected to be better than that of conventional knitted fabric composites.

1.4.2.3.3 Woven Fabric

Weaving is the oldest and the most common technique that is used for fabric formation. Woven fabrics are manufactured by interlacing two sets of yarns perpendicular to each other, known as warp and weft. Woven structures are the most preferred fabric structures for polymer composite reinforcement. There are mainly three types of woven structures that are used for composite reinforcement namely unidirectional (UD) fabric, bi-dimensional (2D) fabric and three-dimensional (3D) fabric. Schematics of different woven fabric structures are shown in Figure 1.15.

In a UD-fabric structure, the majority of yarns run in one direction while very small amount of yarn or filament runs in the other direction to hold the fabric structure. UD fabrics are very popular as a composite reinforcing material due to their

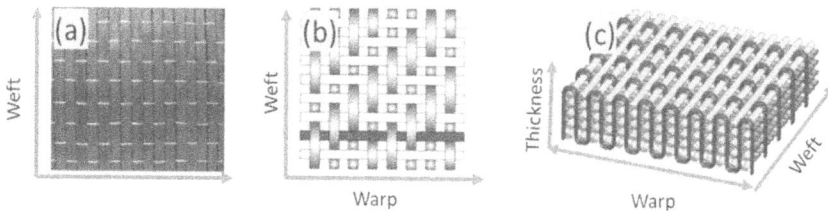

FIGURE 1.15 (a) UD-woven fabric, (b) 2D-woven fabric, and (c) 3D-woven fabric.

crimp-less structure. However, the impact properties of a UD-fabric reinforced composite are very poor. In case of 2D woven fabric, both the warp and weft yarns contribute to the fabric cover. Depending on the yarn interlacement order, 2D-woven fabrics are of different types and among these, plain, twill and sateen/satin structures are widely used as reinforcement for polymer composites (Misnon et al. 2014). 2D laminated composites are monoclinic in nature, i.e., they show very good in-plane properties but poor out-plane properties. To overcome these drawbacks of 2D-woven structures, 3D fabrics are introduced as composite reinforcement. As the name suggests, the 3D fabrics have the third dimension, i.e., along the length and width, the thickness is significant in case of a 3D woven fabric. In a 3D fabric structure, the constituent yarns are disposed in three mutually perpendicular planes. At present, 3D woven structures are getting more interest in polymer composite reinforcement for their higher inter-lamination strength, damage tolerance and impact properties than their 2D counterpart (Bandaru, Vetiyatil, and Ahmad 2015). Traditional 2D fabrics can only be processed into relatively simple and slightly curved shape structures while 3D weaving techniques can produce complex geometries, which is an added advantage for the 3D preforms.

1.4.2.4 Braided Fabrics

Braid is a minor but distinctive form of textile fabric which is manufactured by the braiding process. In braiding process, textile fabrics are manufactured through intertwining or interlacing three or more yarns in a diagonally overlapping manner. Interlacing of yarns in a braided fabric takes place in such a way that no two yarns are twisted around one another. Plaiting of human hair is a very good example of braiding. Braided textiles have been used for centuries for applications such as ropes, cordages and laces. Other uses include parachute cords, cord shock absorbers, fishing lines, wicking, packing, clothes lines and recently as reinforcements in composite materials. Braiding is one of the most cost-effective fabric manufacturing processes as it involves least preparatory processes. Braided fabrics can be classified based on various methods such as yarn intertwining, orientation of yarns in the braided structure, shape of the braided fabric and purpose of its use. Regular braid, diamond braid, basket braid, hercules braid, biaxial braid and triaxial braid are some examples of different braid structure (Rana and Fangueiro 2015).

1.5 MANUFACTURING OF THERMOPLASTIC COMPOSITES

The manufacturing of thermoplastic composites is greatly influenced by the fibre length scale, i.e., short fibre/particle reinforced thermoplastic composites, long fibre reinforced thermoplastic composites and continuous yarn/textile structure reinforced thermoplastic composites. In general, the injection moulding method is suitable for producing short fibre (~ 3mm) reinforced thermoplastic composites; extrusion, compression and sheet extrusion methods are suitable for long fibre (3–25 mm) reinforced thermoplastic composites while pultrusion, thermo stamping and thermoforming methods are suitable for processing textile yarn or fabric reinforced polymer composites. The techniques used for thermoplastic composite manufacturing, their advantages and disadvantages are discussed below.

1.5.1 MELT MIXING

In this process, thermoplastic polymer is initially heated to its melting temperature in a turbulent mixture and then reinforcing fibres are added to mix uniformly. The composite mixture is then formed into the desired shape to produce the final product. The melt mixing process exhibits an excellent mixing homogeneity of the fibre and the polymer matrices. However, melt mixing has some drawbacks. It is a batch process and suitable for processing very short fibre only. Secondly, this process cannot control the fibre orientation in the composite structure (Michaeli et al. 1995).

1.5.2 EXTRUSION

Extrusion is one of the most effective methods for compounding staple fibres with thermoplastic polymers. In this process, thermoplastic polymers and short reinforcing fibres are fed together and drawn through a heated extrusion barrel using either a single screw extruder or two co-rotating screw (twin screw) extruders. The polymer is melted and mixed with the fibres to form a composite melt, which is then drawn forward through the extruder barrel and further mixed and compressed to improve the melt homogeneity. However, the main problem of extrusion moulding is extensive fibre breakage during processing due to inter-fibre friction, friction between fibres and polymer and the friction between fibres and the extruder (Wielage et al. 2003). Like melt mixing, controlling reinforcing fibre orientation in the composite structure is not possible in this process.

1.5.3 INJECTION MOULDING

Injection moulding is a process of manufacturing thermoplastic products in finished form. It is an ideal moulding process for high volume production. This process used for manufacturing various products which are used in our day to day activities, such as bottle caps, chairs, tables, mechanical parts etc. Generally, the polymer chips or the fibre filled polymer chips are used as starting material for the injection moulding process. The process of injection moulding comprises of three basic steps which are filling, packing-holding and cooling respectively. The whole injection moulding process generally takes around 20–60 s time. A schematic diagram of the injection moulding process is shown in Figure 1.16. The injection moulding process can easily produce the complex shapes which are difficult to fabricate by any other means. However, this process involves high capital investment and is not economical for fabricating prototype parts. The materials for injection moulding must be capable of fluid like flow which restricts the use of high amount of fibre and long fibre as reinforcement. The control over the reinforcing fibre orientation is not possible in this process (Debnath and Singh 2017; Krzysik et al. 1990).

1.5.4 COMPRESSION MOULDING

The compression moulding operation begins with the placement of a stack of alternating fibre-mat and thermoplastic polymer sheets onto the bottom half of a preheated

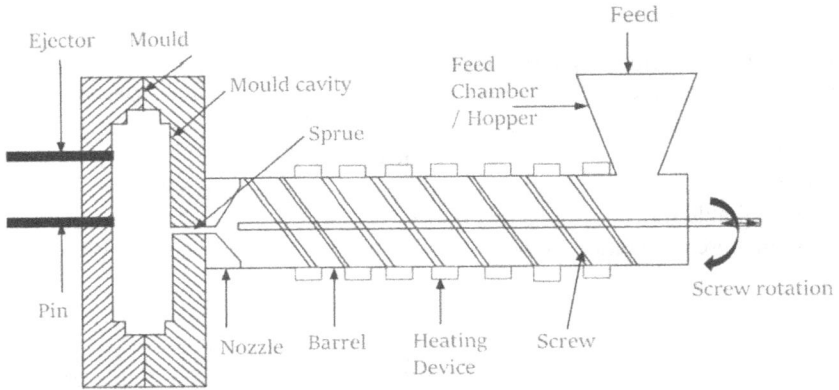

FIGURE 1.16   Schematic diagram of injection moulding machine.

mould cavity (as shown in Figure 1.17). The top half of the mould is lowered at a constant rate until the desired processing pressure is reached, thus melting the polymeric matrix, resulting in consolidation of the composite. After consolidation, the whole set-up is kept at that particular processing temperature and pressure for a specific time, which is known as curing cycle. After curing, the materials are cooled down below the melting temperature of the thermoplastic polymer and are removed from the mould (Debnath and Singh 2017). A schematic diagram explaining compression moulding is shown in Figure 1.17. Compression moulding technique is very popular for producing large scale, relatively simple composite parts with superior mechanical properties. It also capable of manufacturing parts with complex geometry in a short period. In compression moulding, fibrous material can be reinforced in different forms such as in short staple form, continuous fabric etc. Effects such as non-uniform thickness, ribs, bosses, flanges, holes and shoulders can be incorporated during the compression moulding process. It eliminates many secondary finishing operations such as drilling, forming, welding etc. (Wakeman and Rudd 2000).

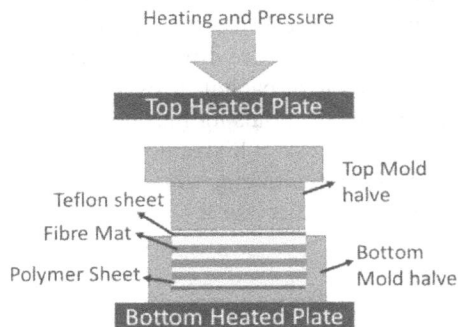

FIGURE 1.17   Working principle of compression moulding.

1.5.5 FILM STACKING

In the film stacking process, layers of matrix components and reinforcement components are stacked in an alternative way and then subjected to heat and pressure. The pressure must be sufficient enough to force the thermoplastic polymeric melt to flow into the reinforcement. The most important factor affecting the melt impregnation process is the rate at which the molten polymer penetrates the fibrous preform. Depending on the fibre architecture, the polymer must flow through and around individual fibres to produce a void-free composite. Film stacking method is considered very effective in producing both small and large-scale thermoplastic composite structures due to its low cost and excellent processability (Kim et al. 2003).

1.5.6 PULTRUSION

Pultrusion is a cost-effective, continuous process of manufacturing composite parts in which the reinforcing fibres are impregnated with the matrix and are pulled through a die to form composites having constant cross-section. A variety of constant cross-section profiles, such as rod, bar, channel, hat, wedge, angle and rectangular lobe can be produced through pultrusion (D'oria, Bourgin, and Coincenot 1995). Figure 1.18 shows a schematic diagram of the thermoplastic pultrusion process.

The pultrusion process with thermoplastics has two main steps, first, impregnation of the reinforcing fibre with the molten thermoplastic polymer/matrix, then shaping the composite under pressure and finally cooling it to preserve the new shape. The pultrusion with thermoplastic resins can be done in different ways using different material forms and curing options (Vaidya and Chawla 2008). Various approaches for thermoplastic pultrusion are summarized in Figure 1.19. However, the main problem of pultrusion with thermoplastics associates with their high melt viscosity, which makes the impregnation of the reinforcements with polymer melts difficult. This leads to poor fibre wetting and thus poor material property.

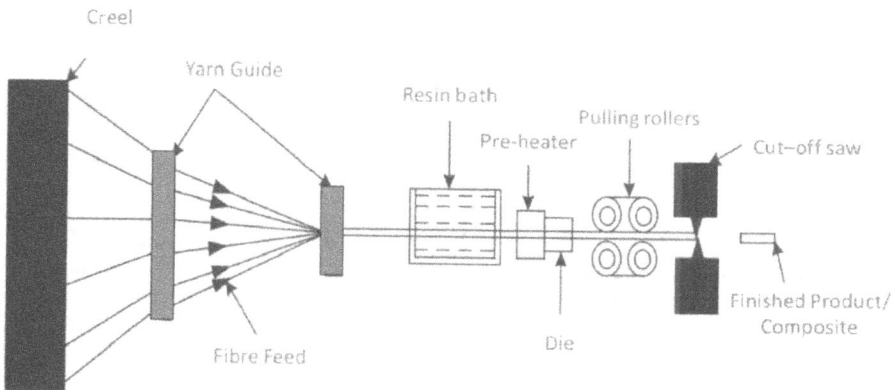

FIGURE 1.18 Schematic diagram of the pultrusion process.

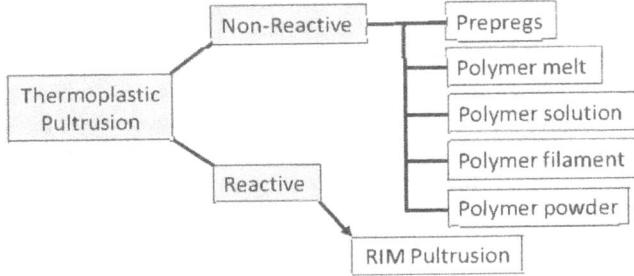

FIGURE 1.19   Various approaches of thermoplastic pultrusion.

1.5.7   Automated Tow Placement

In the automated tow placement process, tows having thermoplastic polymer and reinforcing fibres are consolidated in situ under heat and pressure and therefore avoid costly post-operations, such as autoclave or hot press consolidation. Figure 1.20 shows a schematic diagram of the automated tow placement process.

The point where the tow and the surface of the substrate meet can be heated above the melting temperature by various means such as using ultrasonic heating, infrared lamps, laser, microwave or inductive energy etc. Once the proper molten zone is created by the heating system, adequate pressure is applied via fibre tension and/or roller compaction to compress the uneven melting surfaces and to squeeze the molten matrix to fill the gaps. As a result, intimate contact and diffusion bonding at the interface are established. The automated tow or tape laying process is suitable for producing composites with large surfaces and moderate curvatures such as aircraft wings, large tanks, fly-wheels etc. (Vaidya and Chawla 2008).

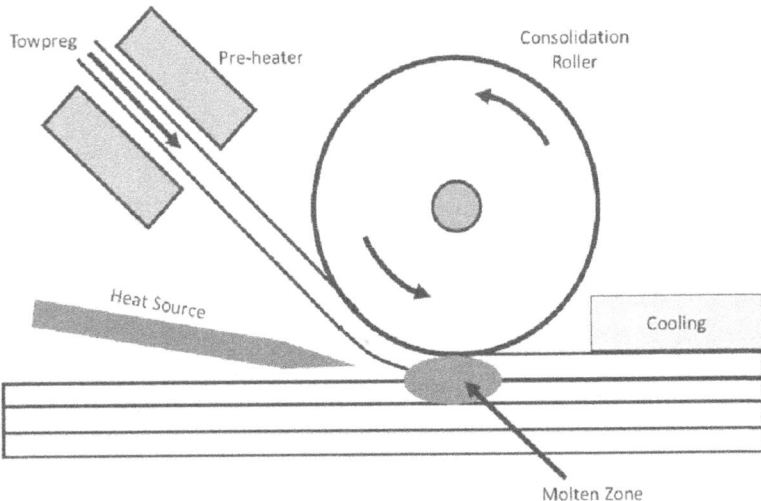

FIGURE 1.20   Schematic diagram of the automated tow placement process.

1.5.8 THERMOFORMING

Thermoforming is a process which is mainly used to produce large size composite components with varying wall thicknesses. This process involves low moulding pressures, generally less than 50 psi. This process has two main steps, firstly heating the plastic or composite sheet and forming the sheet over a male mould or into a female mould. This process deforms a composite sheet into a curvilinear shape with the help of tools or moulds. Various configurations can be used in the thermoforming process such as vacuum forming, matched mould forming, plug assisted forming, vacuum snap back forming and air slip forming. In conventional vacuum forming process, a vacuum is created between a female mould and a heated thermoplastic composite sheet, which is then forced to conform to the mould walls. The formability of the preform plays an important role in this process. The preform/composites containing fabric allow less expansion in the yarn directions and large stretching capability in the ±45 directions, due to the angle reduction between crossing yarns under heated conditions (Breuer and Neitzel 1996).

1.5.9 LIQUID MOULDING

Thermoplastic polymers have high melt viscosity which makes the impregnation of the reinforcements with polymer melts difficult. This leads to poor fibre wetting and thus poor material property of the resultant composites. To overcome the above problem, a substantial interest in liquid moulding of thermoplastic resin has been observed. The liquid or reactive processing of thermoplastics has a lot in common with the processing of thermoset polymers. For instance, the polymerisation rate increases with increasing processing temperature in both cases. Liquid moulding of thermoplastic polymers is generally carried out using anionic polyamide 12, cyclic polybutylene terephthalate and anionic polyamide 6. A liquid moulding process for anionic polyamide 6 composite manufacturing was investigated in the early 1980s, and by mid-1980s this process had started losing the interest of the researchers. Renewed interest in the capabilities of nylon 6 and other thermoplastics for liquid moulding of composites has recently been driven by environmental regulations (especially in Europe), economics and improved performance requirements of components and structures (Vaidya and Chawla 2008).

1.6 FACTORS INFLUENCING THE FIBRE REINFORCED THERMOPLASTIC COMPOSITE PROPERTIES

The problem with pure thermoplastic polymers for structural or semi-structural application are their relatively low stiffness and strength. Incorporation of fibrous materials into the thermoplastic matrix develops a thermoplastic composite that generally has superior mechanical properties to that of virgin thermoplastics. The mechanical properties of the fibre reinforced polymer composites determined by the rule of mixture equation.

$$E_C = V_f E_f + \left(1 - V_f\right) E_m \tag{1.1}$$

Where E_C, E_f, E_m are the modulus of composite, fibre and matrix, respectively, and V_f is the volume fraction of fibre in the composite. Along with the fibres and matrix properties, there are many other factors that influence the thermoplastic composite (reinforced with short fibres) properties such as fibre volume fraction, length of reinforcing fibre, fibre orientation, void content, fibre matrix distribution and the behaviour of fibre matrix interphase. The influences of the above-mentioned factors on thermoplastic composite properties are discussed below.

1.6.1   FIBRE VOLUME FRACTION

Fibrous reinforcements are the main load bearing component of any fibre reinforced polymer composites system. Thus, the fraction of fibre present in the composite structure plays an important role in demining the composite's mechanical properties. The volume fraction of fibre in the composite (v) is expressed in the following ways:

$$v = \frac{V_f}{V_c} \tag{1.2}$$

$$V_f + V_m = V_C \tag{1.3}$$

Where, V_f, V_m, V_C are the volume of fibre, matrix and composite respectively. In practice, fibre weight is generally considered to express the fibre content in the composite. The weight fraction of fibre is converted to fibre volume fraction using the following equation:

$$v = \frac{W_f / \rho_f}{W_f / \rho_f + W_m / \rho_m} \tag{1.4}$$

Where W_f, W_m denotes the weight and ρ_f, ρ_m denotes the density of fibre and matrix respectively. In general, the mechanical properties of a composite increases with increasing fibre volume fraction. However, there is a critical fibre volume fraction below in which the increase in fibre content shows a negative impact on the composite's mechanical properties. This critical volume fraction depends on matrix stress at the composite failure (σ_M^{CF}), ultimate matrix stress (σ_M^u) and on the contribution of fibre strength to the composite ($\lambda\sigma_F^u$). The critical volume fraction is expressed by the following equation (Piggott 2002):

$$v^{crit} = \frac{\sigma_M^u - \sigma_M^{CF})}{\lambda\sigma_F^u - \sigma_M^{CF}} \tag{1.5}$$

On the other hand, at high fibre volume fraction composite strength decreases with increasing fibre volume fraction. It is mainly due to insufficient wetting of the reinforcing fibres at high fibre volume fraction.

1.6.2 LENGTH OF REINFORCING FIBRE

In case of long fibre reinforced composite (where the fibre length is infinitely long) a uniform stress transfer is predicted along the length of the fibre. The modulus of the long fibre reinforced composites is expressed by the rule of mixture equation that means it is the weighted mean between the moduli of the fibre and matrix components. However, the fibre length plays an important role in determining the mechanical properties of the short fibre reinforced polymer composites. In case of short fibre reinforced polymer composites, stress is uniformly transferred along the length of the fibres while at the fibre ends the redistribution of stress takes place. The transfer of stress from one fibre end to another end takes place through high shear stress development in the matrix. Thus, the longitudinal stress increases from zero at the fibre end to a maximum value over a length of the fibre from both ends (as shown in Figure 1.21).

The region at the fibre end over where the longitudinal stress increases from zero to a maximum value $(\sigma_f)_{max}$ is called load transfer length (l_t). If the reinforced fibre length is shorter than l_t, then the load transfer is not sufficient. The length from the fibre end at which the fibre stress transferred from the matrix reaches to its maximum is called critical fibre length (l_c). It is expressed by the following equation:

$$l_c = r_f \left(\sigma_f\right)_{max} / \tau \qquad (1.6)$$

Where, r_f is the fibre radius and τ is the interfacial shear stress between fibre and matrix. Breaking of reinforcing fibres takes place during the manufacturing of short fibre reinforced thermoplastic composites via extrusion moulding and injection moulding. The breakage of fibre in the above processes depends on processing conditions, viscosity of the matrix, fibre content, fibre contact with the surface of the equipment etc.

1.6.3 FIBRE ORIENTATION

The mechanical properties of the fibre reinforced polymer composites can be determined by the rule of mixture (shown in Equation 1.1). This is the most simplified micro-mechanical model to predict tensile modulus of the composite. In the conventional rule of the mixture, it is assumed that the reinforcing fibres are continuous in length and are unidirectionally oriented along the loading direction of the composite.

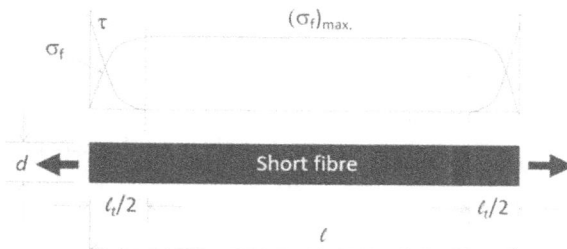

FIGURE 1.21 Stress transfer in short fibre reinforced composite.

Secondly, the model assumes the reinforcing fibres are homogeneous in terms of their mechanical properties and regularly distributed in the matrix without the formation of agglomerates. Short fibre reinforced thermoplastic composites do not meet the above conditions. Thus, the rule of the mixture model was modified by incorporating four other factors to meet the conditions of short fibre composites (Virk, Hall, and Summerscales 2012).

$$E_c = \kappa \eta_l \eta_d \eta_o V_f E_f + \left(1 - V_f\right) E_m \qquad (1.7)$$

Where, κ is the fibre area correction factor for non-circular fibres, η_l and η_d are the fibre length and fibre diameter distribution factor respectively. η_o is the fibre orientation factor, which is quantified by the following equation (Krenchel 1964).

$$\eta_o = \sum_n a_n cos^4 \theta_n \qquad (1.8)$$

Where, a_n is the proportion of fibres, which are oriented in an angle θ_n with respect to the direction of the applied load.

1.6.4 VOID CONTENT

It is often observed that the density measured experimentally does not agree with the theoretical density of the composites. This deviation in theoretical and experimental density of the composite is attributed to its void content. During composite manufacturing if some air gets trapped in the matrix or if some portion of the reinforcing fibres didn't wet sufficiently by the matrix, it creates a void space in the composite structure. The void content of the composite samples is determined using the following equations (Fernández et al. 2016).

$$\%v_f = \%m_f \times \frac{\sigma_c}{\sigma_f} \qquad (1.9)$$

$$\%v_m = \%m_m \times \frac{\sigma_c}{\sigma_m} \qquad (1.10)$$

$$\%v_p = 1 - \left(\%v_f + \%v_m\right) \qquad (1.11)$$

Where, v_f, v_m and v_p are the volume fraction of fibre matrix and void, respectively. σ_c, σ_m and σ_f are the densities of the composites, matrix and fibre, respectively. m_f, m_m are the mass fractions of fibre and matrix, respectively. Presence of void in the composite structure gives a negative impact on their mechanical properties.

1.6.5 FIBRE-MATRIX DISTRIBUTION

Fibre-matrix distribution in the composite structure plays an important role in determining the mechanical properties of the composite. In rule of mixture model, it is

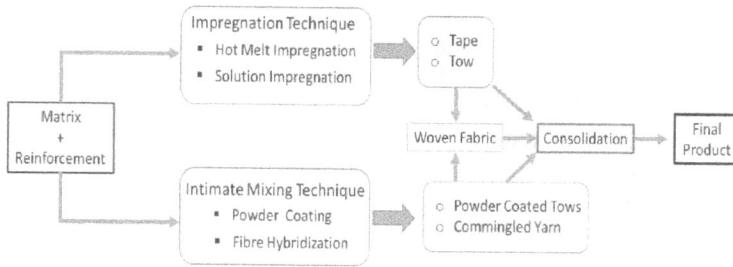

FIGURE 1.22   Towpregs for thermoplastic composite manufacturing.

assumed that the reinforcing fibres are uniformly distributed in the matrix. However, in practice it is not the case. During thermoplastic polymer composite manufacturing, it is often observed that the thermoplastic polymers are not able to impregnate the reinforcing fibre bundles uniformly due to their high melt viscosity. This leads to the development of a composite with a lot of voids and non-uniform fibre resin distribution. Inhomogeneous composite structures have some fibre rich regions and some matrix rich regions. When such composites are subjected to a tensile load, a stress concentration occurs in the mechanically inferior matrix rich region, which ultimately initiates a failure point. To improve the fibre-resin distribution in the thermoplastic composite structure, hybrid towpregs are used during composite manufacturing. Figure 1.22 summarizes various types of towpregs used for thermoplastic composite manufacturing (Virk, Hall, and Summerscales 2012). Manufacturing of different towpreg structures, their properties and influence on thermoplastic composite properties are discussed in other chapters of this book.

1.6.6   Fibre Matrix Interphase

Composite interface plays an important role in determining the mechanical properties of the composite as the stress transfer from the matrix to fibre takes place through this phase. Interfacial bonding between fibre and matrix can occur through four mechanisms which are

    i. Mechanical bonding
    ii. Electrostatic bonding
   iii. Chemical bonding
   iv. Reaction or inter-diffusion.

In a strong fibre-polymer interface, more than one force acts at the same time (Matthews and Rawlings 1999). A composite generally has poor interface if the hydrophilic nature of its constituent is not similar, for instance, the natural fibre reinforced thermoplastic composite. The natural fibres are hydrophilic and the thermoplastics are hydrophobic. This lead to the development of a composite with poor fibre-matrix interface. The interfacial strength of a natural fibre composite system can be improved either by modifying the fibre surface or by modifying the matrix or by modifying both. Among these methods, fibre surface modification is more effective than the matrix modification approach. Fibre surface modification to improve the fibre-matrix interaction can be largely divided into chemical approaches, physical approaches and biological approaches or enzyme treatment. These surface modification techniques modify the natural fibre surfaces mainly in two ways. Firstly, by roughening the natural fibre surfaces and secondly by improving the hydrophobicity of the natural fibre surfaces. The advantages and disadvantages of all three approaches and their effects on composite properties are discussed below.

1.6.6.1 Chemical Approaches

Surface modification of natural fibre through chemical treatment is a very effective way of improving fibre-matrix interaction. There are several chemical treatment techniques available that effectively modify the natural fibre surfaces. Some well-known methods used for the surface modification of natural fibres are discussed below.

1.6.6.1.1 Alkali Treatment

Surface modification of natural fibres through alkali or sodium hydroxide treatment to improve the fibre-matrix interaction is a very well-known approach. Alkali treatment of natural fibre removes the fat, wax and dust particles from the fibre skin, which eliminates the micro-voids from the fibre surface and makes it more uniform, which leads to better fibre-matrix adhesion. Sodium hydroxide also removes some portion of lignin and hemicellulose from the fibre surface. Removal of fats, waxes, lignin and hemicellulose also reduces the effective fibre diameter or increases the aspect (length/diameter) ratio of fibre, which provides more surface area for adhesion (Leonard and Martin 2002).

1.6.6.1.2 Silane Treatment

Silanes are the silicon-based multi-functional compounds that hydrolyse in the presence of water and form a silanol compound. On its application to the natural fibre, one end of the silanol compound reacts with the hydroxyl group of natural fibres through condensation reaction (covalent bonding) and forms a coated layer around the natural fibres. During polymer composite formation, the other end of the silane compound adheres to the polymer matrix and ultimately creates a robust fibre/matrix interface. The silane compound also improves the fibre matrix adhesion by restricting the fibre swelling in the polymer matrix and by mechanical interlocking (Xie et al. 2010).

1.6.6.1.3 Acetylation

Acetylation is known as the esterification method of improving plasticity of the natural fibres. During acetylation reaction, the acetyl groups (CH_3CO) react with the hydroxyl groups of natural fibres and enhance their hydrophobicity. Acetylation of natural fibres also removes the micro-voids from the fibre surface and makes the surface rough, leading to better mechanical bonding with the matrix (Mishra et al. 2004). Acetylation of the natural fibres with and without the presence of acid is shown in Figure 1.23.

Acetylation with acid catalyst

$$\text{Fibre-OH} + CH_3COOH \xrightarrow[\text{Conc. } H_2SO_4]{(CH_3CO)_2O} \text{Fibre-O-} \overset{\overset{\displaystyle O}{\|}}{C}\text{-CH}_3$$

Acetylation without acid catalyst

$$\text{Fibre-OH} + CH_3\text{-}\overset{\overset{\displaystyle O}{\|}}{C}\text{-O-}\overset{\overset{\displaystyle O}{\|}}{C}\text{-CH}_3 \longrightarrow \text{Fibre-O-} \overset{\overset{\displaystyle O}{\|}}{C}\text{-CH}_3 + CH_3COOH$$

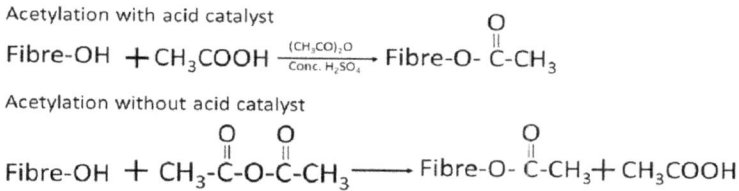

FIGURE 1.23 Acetylation of natural fibres with and without acid treatment.

1.6.6.1.4 Malleated Coupling Agents

Malleated coupling agents are the most promising coupling agents used for natural fibre surface modification. At one end, it has functional hydroxyl groups that react with the polar hydrophilic groups of the natural fibres while on other side it has a long hydrocarbon chain that entangles with the polymeric chains of matrices. In this way, malleated coupling agents form very strong bridge interphase between the natural fibre and polymer matrix through covalent bond formation and mechanical interlocking (Mohanty et al. 2004). The reaction mechanism of a malleated coupling agent is shown in Figure 1.24.

1.6.6.2 Physical Approaches

Surface modification of natural fibre through chemical means successfully improves the interfacial bonding strength of a natural fibre composite, but there are unresolved pollution problems related to the disposal of excess chemicals after the treatment. Moreover, the cost of these chemicals is very high and most of them are hazardous to health (Faruk et al. 2012). Physical approaches of modifying natural fibre surfaces do not have the problems mentioned above as they only change the surface tropology of a fibre and do not involve any hazardous chemicals. Hence, physical means of modifying natural fibre surfaces are gaining attention of the researcher community, although it involves sophisticated machinery and needs a lot of energy which enhances the material processing cost. The physical approaches include plasma treatment, corona discharge, electronic beam radiation, IR treatment, fibre beating etc., which are discussed briefly here.

1.6.6.2.1 Plasma Treatment

Plasma is known as forth state of matter. It can be defined as partially ionized gases that have a collective ionized behaviour. The main advantage of the plasma treatment

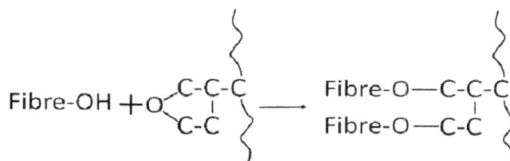

FIGURE 1.24 Reaction mechanism of a malleated coupling agent with natural fibre.

is that it confines to the fibre surface only. It does not change the bulk properties of the material. The proper selection of starting compounds and external plasma parameters such as pressure, power and treatment time can create desired compounds on the fibre surface. Yuan, Jayaraman, and Bhattacharyya (2004) have subjected wood fibre to the cold-plasma and Ar-Plasma treatment and have produced wood-PP composite using the same. It is observed that the plasma treatment enhances the hydrophobicity and roughness of the wood fibre, which significantly improves the tensile and flexural properties of the resultant composite.

1.6.6.2.2   Corona Treatment
Corona is defined as a luminous, audible discharge that occurs due to inhomogeneous electrode geometries such as point electrode and plane. Compared to plasma, the corona discharges are relatively low power electrical discharges that take place at atmospheric pressure. The corona discharge brings the chemical and physical changes of fibres including increasing surface polarity and roughness of the fibre. However, corona discharge is not so effective on three-dimensional surfaces such as textile fabric (Faruk et al. 2012). Belgacem, Bataille, and Sapieha (1994) have studied the effect of corona discharge on cellulose-PP composites. It is observed that the composite strength improves when corona discharge pre-treatment modifies either one or both components.

1.6.6.2.3   Other Physical Approaches
Besides the plasma and corona treatment, there are a few more physical means of modifying natural fibre surfaces to enhance the fibre/matrix bonding. Electron beam radiation is one of them. It improves the interfacial bonding between natural fibres and thermoplastic polymers by producing free radicals on their surfaces. These free radicals encourage cross-linking and ultimately develop a very strong fibre/matrix interface (Huber, Biedermann, and Muessig 2010). The next physical approach of modifying natural fibre surfaces is the treatment with ultrasound. Ultrasound is defined as a very high frequency of sound, above 20 kHz, generally used for medical and diagnostic purposes. Ultrasound treatment of natural fibre cleans the fibre surface and makes it rough. As a result, it improves the fibre matrix interaction (Laine and Goring 1977).

1.6.6.3   Biological Approaches
Surface modification of natural fibres through biological means is another way of improving fibre-matrix interaction. Compared to chemical and physical approaches of natural fibre surface modification, the biological approach is entirely new and advantageous. In biological approaches, natural fibres are treated with enzymes or fungi. The biological treatment is environment-friendly and has a focused performance. In general, the enzyme treatment removes hemicellulose and lignin from the fibre and makes the surface rough. In addition, it also creates holes on the fibre surface, which helps in better interlocking with the matrix. Bledzki et al. (2010) have studied mechanical properties of PP composite reinforced with enzyme treated

abaca fibre. It is observed that the enzyme treatment enhances the surface roughness of the abaca fibres. As a result, up to 45% improvement in composite tensile strength is observed. Besides tensile behaviour, the moisture absorption of the composite samples also reduces by 20–45% due to the enzymatic treatment.

1.7 SUMMARY

Over the past few decades, thermoplastic composites have gained a significant interest in the consumer industry. They have found wide applications in the areas of automobile, aerospace, construction, defence, household, transportation, packaging, sports etc. This wide application of the thermoplastic composites is mainly attributed to their processing advantages, recyclability, low cost, low relative density, high specific mechanical properties and also attributed to environmental legislative pressures and the public's growing awareness on sustainability. However, the thermoplastic polymers have some drawbacks originating from their high melt viscosity. The high melt viscosity of the thermoplastics hinders the resin impregnation into the reinforcing textile structures. This leads to non-uniform fibre-resin distribution in the composite structure and develops a composite full of voids and inferior mechanical properties. Significant research is currently underway around the world to address and overcome the obstacles mentioned above. This effort to develop thermoplastic composites with improved performance for global applications is an ongoing process.

REFERENCES

Alagirusamy, R., R. Fangueiro, V. Ogale, and N. Padaki. 2006. Hybrid yarns and textile preforming for thermoplastic composites. *Textile Progress* 38(4): 1–71. doi.org/10.1533/tepr.2006.0004

Bandaru, A. K., L. Vetiyatil, and S. Ahmad. 2015. The effect of hybridization on the ballistic impact behaviour of hybrid composite armors. *Composite Part B: Engineering* 76: 300–319. doi.org/10.1016/j.compositesb.2015.03.012

Bar, M., A. Das, and R. Alagirusamy. 2018. Effect of interface on composites made from DREF spun hybrid yarn with low twisted core flax yarn. *Composites Part A: Applied Science and Manufacturing* 107: 260–270. doi.org/10.1016/j.compositesa.2018.01.003

Bar, M., A. Das, and R. Alagirusamy. 2019. Influence of flax/polypropylene distribution in twistless thermally bonded rovings on their composite properties. *Polymer Composites* 40(11): 4300–4310. doi.org/10.1002/pc.25291

Bar, M., R. Alagirusamy, and A. Das. 2020. Low velocity impact response of flax/polypropylene hybrid roving based woven fabric composites: Where does it stand with respect to GRPC?. *Polymer Testing* 89: 106565. doi.org/10.1016/j.polymertesting.2020.106565

Belgacem, M. N., P. Bataille, and S. Sapieha. 1994. Effect of corona modification on Cellulose/PP composites. *Journal of Applied Polymer Science* 53: 379–385. doi.org/10.1002/app.1994.070530401

Bledzki, A. K., A. A. Mamun, and A. Jaszkiewicz. 2010. Polypropylene composites with enzyme modified abaca fibre. *Composite Science Technology* 70: 854–860. doi.org/10.1016/j.compscitech.2010.02.003

Breuer, U., and M. Neitzel. 1996. High speed stamp forming of thermoplastic composite sheets. *Polymers and Polymer Composites* 4(2): 117–123.

Chang, I. Y., and J. K. Lees. 1988. Recent development in thermoplastic composites: A review of matrix systems and processing methods. *Journal of Thermoplastic Composite Materials* 1(3): 277–296. doi.org/10.1177/089270578800100305

Chawla, K. K. 1998. Composite materials science and engineering, Second Edition, Springer-Verlag, New York. doi.org/10.1007/978-1-4757-2966-5_14

Das, D., A. K. Pradhan, R. Chattopadhyay, and S. N. Singh. 2012. Composite nonwovens. *Textile Progress* 44(1): 1–84. doi.org/10.1080/00405167.2012.670014

Debnath, K., and I. Singh. (Eds). 2017. Primary and secondary manufacturing of polymer matrix composites. CRC Press, New York.

D'oria, F., P. Bourgin, and L. Coincenot. 1995. Progress in numerical modeling of the thermoforming process. *Advances in Polymer Technology: Journal of the Polymer Processing Institute*. 14(4): 291–301. doi.org/10.1002/adv.1995.060140403

Faruk, O, A. K. Bledzki, H. P. Fink, and M. Sain. 2012. Biocomposites reinforced with natural fibers: 2000–2010. *Progress in Polymer Science* 37: 1552–1596. doi.org/10.1016/j.progpolymsci.2012.04.003

Fernández, J. A., N. Le Moigne, A. S. Caro-Bretelle, R. El Hage, A. Le Duc, M. Lozachmeur, P. Bono, and A. Bergeret. 2016. Role of flax cell wall components on the microstructure and transverse mechanical behaviour of flax fabrics reinforced epoxy biocomposites. *Industrial Crops and Products* 85: 93–108. doi.org/10.1016/j.indcrop.2016.02.047

Fortea-Verdejo, M., E. Bumbaris, C. Burgstaller, A. Bismarck, and K. Y. Lee. 2017. Plant fibre-reinforced polymers: where do we stand in terms of tensile properties? *International Materials Reviews* 62(8): 1–24. doi.org/10.1080/09506608.2016.1271089

Gopanna, A., K. P. Rajan, S. P. Thomas, and M. Chavali. 2019. Polyethylene and polypropylene matrix composites for biomedical applications. In Materials for biomedical engineering. Elsevier, 175–216. doi.org/10.1016/B978-0-12-816874-5.00006-2

Hamad, K., M. Kaseem, M. Ayyoob, J. Joo, and F. Deri. 2018. Polylactic acid blends: The future of green, light and tough. *Progress in Polymer Science* 85: 83–127. 10.1016/j.progpolymsci.2018.07.001

Huber, T., U. Biedermann, and J. Muessig. 2010. Enhancing the fibre matrix adhesion of natural fibre reinforced polypropylene by electron radiation analyzed with the single fibre fragmentation test. *Compos Interfaces* 17: 371–381. doi.org/10.1163/092764410X495270

Hull, D., and T. W. Clyne. 1996. An introduction to composite materials, Second Edition, Cambridge University Press, New York.

John, M. J., and S. Thomas. 2008. Biofibres and biocomposites. *Carbohydrate Polymers* 71(3): 343–364. doi.org/10.1016/j.carbpol.2007.05.040

Johnson, T. 2018. History of composites. The evolution of lightweight composite materials. Available from: https://www.thoughtco.com/history-ofcomposites-820404

Kausar, A., and S. Anwar. 2018. Graphite filler-based nanocomposites with thermoplastic polymers: a review. *Polymer-Plastics Technology and Engineering* 57(6): 565–580. doi.org/10.1080/03602559.2017.1329438

Kim, D. W., Y. S. An, J. D. Nam, and S. W. Kim. 2003. Thermoplastic film infusion process for long-fiber reinforced composites using thermally expandable elastomer tools. *Composites Part A: Applied Science and Manufacturing* 34(7): 673–680. doi.org/10.1016/S1359-835X(03)00122-2

Krenchel, H. 1964. Fibre reinforcement; theoretical and practical investigations of the elasticity and strength of fibre-reinforced materials. Dissertation, Technical University of Denmark, Copenhagen.

Krzysik, A. M., J. A. Youngquist, G.E. Myers, I.S. Chahyadi, and P.C. Kolosick. 1990. Wood-polymer bonding in extruded and nonwoven web composite panels. In Proc. of Wood Adhesives Symposium, Madison, 183–189.

Laine, J. E., and D. A. I. Goring. 1977. Influence of ultrasonic irradiation on the properties of cellulosic fibers. *Cellulose Chemical Technology* 11: 561–567.

Lehtiniemi, P., K. Dufva, T. Berg, M. Skrifvars, and P. Jarvela. 2011. Natural fiber-based rein-
forcements in epoxy composites processed by filament winding. *Journal of Reinforced
Plastic and Composites* 30: 1947–1955. doi.org/10.1177/0731684411431019
Leonard, Y. M., and P.A. Martin. 2002. Chemical modification of hemp, sisal, jute and kapok
fibres by alkalisation. *Journal of Applied Polymer Science* 84(12): 2222–2234. doi.org/
10.1002/app.10460
Marshall, G. P., L. E. Culver, and J. G. Williams. 1973. Fracture phenomena in polystyrene.
International Journal of Fracture. 9(3): 295–309.
Matthews, F. L., and R. D. Rawlings. 1999. Composite materials: engineering and science.
Woodhead Publishing, Cambridge.
Michaeli, W., H. Kaufmann, H. Greif, and F. -J. Vosseburger. 1995. Training in plastics technology. Hanser Publishers, Ohio.
Mishra, S., A. K. Mohanty. L. T. Drzal, M. Misra, and G. Hinrichsen. 2004. A review on
pineapple leaf fibers, sisal fibers and their biocomposites. *Macromolecular Materials
and Engineering* 289(11): 955–974. doi.org/10.1002/mame.200400132
Misnon, M. I., M. M. Islam, J. A. Epaarachchi, and K. T. Lau. 2014. Potentiality of utilising
natural textile materials for engineering composites applications. *Materials and Design*
59: 359–368. doi.org/10.1016/j.matdes.2014.03.022
Mohanty, S., S. K. Nayak, S. K. Verma, and S. S. Tripathy. 2004. Effect of MAPP as a coupling agent on the performance of jute–PP composites. *Journal of Reinforced Plastic
and Composites* 23: 625–637. doi.org/10.1177/0731684404032868
Nisticò, R. 2020. Polyethylene terephthalate (PET) in the packaging industry. *Polymer
Testing* 90: 106707. doi.org/10.1016/j.polymertesting.2020.106707
Nofar, M., D. Sacligil, P. J. Carreau, M. R. Kamal, and M. C. Heuzey. 2019. Poly (lactic acid)
blends: Processing, properties and applications. *International Journal of Biological
Macromolecules* 125: 307–360. doi.org/10.1016/j.ijbiomac.2018.12.002
Piggott M. 2002. Load bearing fibre composites. Kluwer Academic Publishers, Boston.
Pleşa, I., P.V. Noţingher, C. Stancu, F. Wiesbrock, and S. Schlögl. 2019. Polyethylene nanocom-
posites for power cable insulations. *Polymers* 11(1): 24. doi.org/10.3390/polym11010024
Rana, S., and R. Fangueiro. (Eds.). 2015. Braided structures and composites: production,
properties, mechanics, and technical applications. CRC Press, New York.
Rowell, R. M. 2008. Natural fibers: types and properties. Properties and performance of
natural-fiber composites. Kim L. Pickering (Ed.), Woodhead Publishing Limited,
Cambridge, IBSN 978-1-85573-739-6.
Rudd, C. D., M. J. Owen, and V. Middleton. 1990. Mechanical properties of weft knit glass
fibre/polyester laminates. *Composites Science and Technology* 39(3): 261–277. doi.
org/10.1016/0266-3538(90)90045-7
Salazkin, S. N., and V. V. Shaposhnikova. 2020. Poly (arylene ether ketones): Thermostable,
heat resistant, and chemostable thermoplastics and prospects for designing various
materials on their basis. *Polymer Science, Series C* 62(2): 111–123. doi.org/10.1134/
S1811238220020125
Vaidya, U. K., and K. K. Chawla. 2008. Processing of fibre reinforced thermoplastic composites.
International Materials Reviews 53(4): 185–218. doi.org/10.1179/174328008X325223
Virk, A. S., W. Hall, and J. Summerscales. 2012. Modulus and strength prediction for natural
fibre composites. *Materials Science and Technology* 28(7): 864–871. doi.org/10.1179/1
743284712Y.0000000022
Wakeman, M. D., and C. D. Rudd. 2000. Compression moulding of thermoplastic compos-
ites, in Comprehensive composite materials. A. Kelly and C. Zweben (Eds.), Elsevier
Science Ltd, Oxford.
Wang, X. L., K. K. Yang, and Y. Z. Wang. 2003. Properties of starch blends with biodegradable polymers. *Journal of Macromolecular Science, Part C: Polymer Reviews*. 43(3):
385–409. doi.org/10.1081/MC-120023911

Wielage, B., T. Lampke, H. Utschick, and F. Soergel. 2003. Processing of natural-fibre rein-forced polymers and the resulting dynamic-mechanical properties. *Journal of Materials Processing Technology* 139(1–3): 140–146. doi.org/10.1016/S0924-0136(03)00195-X

Yuan, X., K. Jayaraman, and D. Bhattacharyya. 2004. Effect of plasma treatment in enhancing the performance of woodfibre-Polypropylene composite. *Composites Part A: Applied Science and Manufacturing*. 35: 1363–1374. doi.org/10.1016/j.compositesa.2004.06.023

Xie, Y., C. A. S. Hill, Z. Xiao, H. Militz, and C. Mai. 2010. Silane coupling agents used for natural fiber/polymer composites: a review. *Composites Part A: Applied Science and Manufacturing* 41: 806–819. doi.org/10.1016/j.compositesa.2010.03.005

2 Towpreg Manufacturing Techniques

Vijay Goud, R. Alagirusamy, and Apurba Das
Indian Institute of Technology Delhi
New Delhi, India

CONTENTS

DOI: 10.1201/9781003049715-2

2.1 INTRODUCTION

The reinforcement and the matrix are fundamental elements of a composite. Composites are often manufactured to yield the synergistic properties of the individual (reinforcement and matrix) constituents (Kaw 2005). The matrix is the primary phase of the composite and the reinforcement is the secondary phase. The reinforcement is dispersed in the matrix and is in a discontinuous phase in the composite while the matrix forms the continuous phase. The key function of the discontinuous phase is to deliver the mechanical strength to the composite whereas the continuous phase of the matrix is accountable for distributing the load among the fibers, shielding the fibers from the environmental damage and upholding position of the fibers within the composite. The resulting properties of the composite are governed by the properties of the reinforcement, the matrix and the effectiveness of the load transfer between the reinforcement and matrix (Bradley 1997).

On the basis of the matrices, the composites are classified as metal matrix composites, ceramic matrix composites and polymer matrix composites. Polymer matrix composites are further categorized as thermoset (epoxy, unsaturated polyester, vinyl ester etc.) and thermoplastic matrix (polyester, polypropylene, nylon, poly-ether-ether ketone, polyimide etc.). In spite of the advantages of the thermoplastic matrix such as ease of processing, ease of handling, unlimited shelf life at room temperature, recyclability and higher toughness, the majority of the composites being produced are thermoset polymer-based composites (Nunes, Silva, and Vieira 2008). The potential source for this hindered usage is the disproportionately high viscosity of the thermoplastic matrix that hampers the thorough infusion of the matrix within and between the fibers, at the juncture of yarns in woven fabrics or at the crossing points of loops in case of knitted fabrics (Alagirusamy and Ogale 2004). Elevated viscosity of the order of 10^2 to 10^4 Pa.s leads to resin-starved areas in the composite, a potential source of failure initiation for the material. Thus, achieving a meticulously consolidated composite with minimal void is the challenge for the processing of the advanced thermoplastic composite (Alagirusamy and Ogale 2005).

To achieve better impregnation, pre-deposition of the matrix on to the reinforcement is one of the methods explored by research groups (Muzzy and Kays 1984; Goud et al. 2018; Goud, Alagirusamy et al. 2019). The tow (i.e. an assortment of filaments) is converted to a towpreg when pre-deposited with matrix materials (Connor et al. 1995). Pre-deposition of the matrix on to the reinforcement ensures the presence of the matrix at the juncture points of woven and knitted fabrics. Production of towpreg significantly lowers the melt flow distance to the order of the sub-microns, warranting the ease of the coalescence of the matrix between the neighboring hybrid yarns (Goud et al. 2020). This diminished inter-diffusion distance for the matrices leads to well-consolidated composite structures with minimal void content and superior mechanical properties. Some commercial processes for accomplishing this reduced melt flow distance between neighboring matrices are classified into single- and multi-step impregnation processes (Vaidya and Chawla 2008). The processes of hot melt impregnation, solution coating and surface polymerization are clustered under the single-step impregnation process as reinforcement is combined with the matrix in one single step. The processes of ring spinning, rotor spinning, DREF spinning, commingling, air texturing, braiding, parallel winding, film stacking, wrap spinning and powder coating are categorized as multi-step impregnation process as the reinforcement is combined with the matrix in the first step in either dry or partially molten stage and final impregnation is carried out after the formation of the shape (Chang and Lees 1988; Ye, Klinkmuller, and Friedrich 1992; Alagirusamy et al. 2006; Svensson, Shishoo, and Gilchrist 1998; Goud, Ramasamy et al. 2019). The current review deals with the manufacturing, technological aspects, advantages and disadvantages of all of the above processes. Later in the chapter, different powder coating technologies of the conventional powder bed, conventional fluidized bed, recirculating powder bed, slurry coating, acoustic fluidized bed, electrostatic fluidized bed and electrostatic spray coating processes are deliberated and compared amongst each other.

2.2 SINGLE-STEP IMPREGNATION TECHNIQUES

2.2.1 MELT IMPREGNATION PROCESS

Melt impregnation process is also referred to as hot melt impregnation process. This process is suitable for production of long continuous fiber composites. The hot melt impregnation process involves, the impregnation of the reinforcement with the molten matrix under high pressure (Angell et al. 1989). The process involves unwinding of high-performance fibers like glass, carbon and aramid under appropriate tension. The unwound high-performance fiber is then impregnated with a molten thermoplastic matrix such as polypropylene, polyester, polycarbonate etc. Prior to impregnation with the matrix, the rovings of the high-performance fibers can be opened in order to ensure the penetration of the matrix in between the filaments. After the impregnation process is complete, the impregnated high-performance tow is made to pass through calendar rollers to squeeze out the excess matrix. The high-performance filament is finally cooled and taken up [Figure 2.1] (Tian et al. 2017). The polymer delivery rate, rate of take-up and calendaring pressure decide the volume fraction of

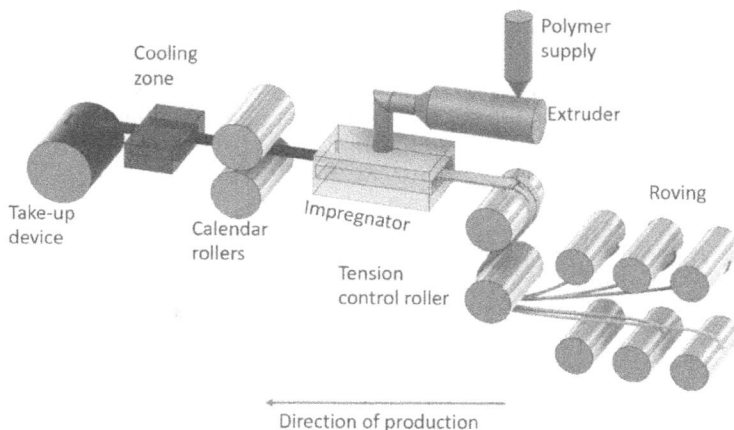

FIGURE 2.1 Hot melt impregnation process.

the delivered towpreg. The other simple way of hot melt impregnation is to dip the high-performance filament into the bath containing the molten matrix on its way to the take-up device. The high-performance filament is kept dipped in the immersion bath with the help of an immersion roller. The depth of the immersion roller decides the time for which the high-performance fiber remains in the bath, which in turn decides the amount of matrix picked up (Ahn and Seferis 1993).

Cogswell et al. formed fiber reinforced composites containing thermoplastic resins through melt impregnation. The authors pointed out the usefulness of the pultrusion process in which a fiber reinforced structure is produced by pulling a tow through a bath containing low viscosity thermoset resins. Heat curing of resin impregnated reinforcement is followed. The same principle is not viable for production of thermoplastic resin impregnated structures. A thorough impregnation with thermoplastic matrices is difficult, as it possesses substantially higher melt viscosities. The final product obtained has air entrapped in it and it leads to composites with poor mechanical properties. Also, the towpreg produced is stiff and it is not suitable for textile preforming operations of weaving, braiding and knitting. Hot-melt processing method is suitable for low viscosity resin. However, acceptable level of impregnation can be achieved with high viscosity resin, if high pressure and shear rates are imparted. Low processing speeds, possibility of matrix degradation and stiff towpreg are some of the limitations of the process. Use of high temperature and pressure to reduce viscosity as well as to achieve complete impregnation of reinforcement damages the reinforcing material in melt impregnation (Cogswell and Hezzell 1985).

2.2.2 Solution Coating

In solution coating, the resin material is dissolved in solvent to reduce its viscosity (Wu and Schultz 2000). After impregnation of fibers, solvent is evaporated. The schematic of the process is shown in Figure 2.2. However, these solvents are hazardous to the environment. Also, the presence of residual solvents leads to undesirable porosity in the final composite, which can reduce the service performance. The above-cited limitations preclude the use of solution coating for processing of thermoplastic matrices.

Iyer et al. reported that a solution processing technique will be highly undesirable environmentally and economically, as it is difficult to reclaim high boiling solvents

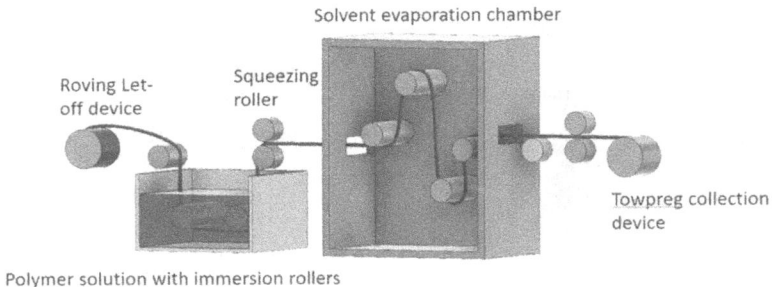

FIGURE 2.2   Solution impregnation.

that may contain volatile organic compounds and there are very few solvents for dissolving thermoplastic matrices to reduce their viscosity, limiting their use in high-volume applications (Iyer and Drzal 1990). Again, as with hot melt impregnation, the solution coating should produce a stiff towpreg. Thereby, production of textile preformed structures is rather difficult.

2.2.3 SURFACE POLYMERIZATION

In surface polymerization, the matrix polymerizes on the fiber, forming a coating of matrix over fiber. Solvent inhibitors and unwanted products of the polymerization reactions remain on the surface, which is undesirable. Electro-polymerization is a modified surface polymerization technique. However, large numbers of polymeric matrices of regular use are difficult to produce by electro-polymerization (Jogur et al. 2018). Due to the above-cited challenges in the processes of hot melt impregnation, solution coating and surface polymerization, the use of the matrix in dry form is explored.

2.3 MULTI-STEP IMPREGNATION PROCESS

2.3.1 RING SPINNING

Ring spinning dates back to 1828 with the introduction of traveler in 1830. Since then the basic concept of ring spinning has been same despite the modifications for enhancement of quality and productivity (Klein and Stalder 2014). One of the important modifications that the ring spinning has undergone is its potential for production of core sheath hybrid yarn. The basic sections of drafting, twisting and winding for production of conventional ring spun yarns with additional attachment for holding, tensioning and controlled feeding of the filament yarns form an essential element for production of hybrid yarns (Pourahmad and Johari 2011). The schematic of the ring spinning process for production of hybrid yarns is shown in Figure 2.3.

The filament yarn which forms the core in the hybrid yarn is fed at the controlled rate to the nip of the front drafting rollers, which are also fed with the staple fibers forming the sheath portion of the hybrid yarn. This filament yarn is wrapped with the staple fibers during the operation of twisting, which originates as a result of rotation of traveler. The filament-staple fiber core sheath hybrid yarn is finally wound onto the bobbin as a result of difference in surface speed of the spindle and traveler (Pourahmad and Johari 2009). Though the technique seems to be simple for production of hybrid yarn it suffers from a defect known as "barber pole," which is an incomplete coverage of the core by the sheath. This incomplete coverage is a result of slippage of the sheath fibers relative to the core, which leads to length of bare filament with a clump of fibers at one end. The possibility of end breakage in the subsequent processes is significantly pronounced in these incompletely covered regions. A high level of twist in order to increase the cohesion between the core and sheath can reduce the "barber pole" effect (Kim et al. 2009). The increase in twist leads to reduced production and thereby increased cost of production. The filament pretension has to be optimized in order to ensure that the core component adapts and remains in the axial position, thereby ensuring the complete coverage by the staple

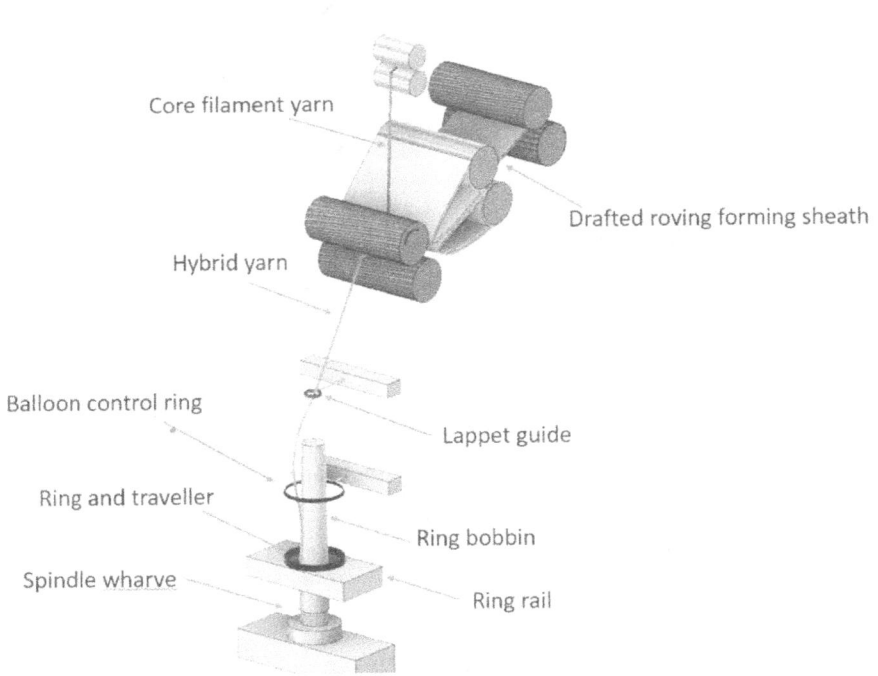

FIGURE 2.3   Ring spinning technology for production of hybrid yarns.

fibers in the sheath. The core stabilizer attachment with a special groove for filament core and a polished surface for wrap has been reported to produce superior quality of hybrid yarn with consistent core coverage. The filament charging principle is another way of production of hybrid yarn using ring spinning technology (Jou et al. 1996). The basic technology of ring spinning involves the twisting of the components of the yarn. This twist is also transferred to the core section. Any twist in the core is not suitable for composite applications as it reduces the mechanical properties. Moreover, the core is usually a high-performance fiber (carbon, glass, basalt), which performs poorly, leading to the breakage of individual filaments under the forces of flex and shear, the dominant forces during the process of twisting. These drawbacks have not led to popular use of ring spinning technology for production of hybrid yarns (Bar, Alagirusamy, and Das 2019).

2.3.2   ROTOR SPINNING

The rotor spinning process was the first commercial process to produce yarn without the use of traveler thereby overcoming the limitation associated with it. The rotor spinning process combines the process of spinning and winding in a single machine thereby eliminating the need to wind the small spinning cops. The yarn wound is waxed and cleared on the same machine as opposed to a completely different setup required for ring spun yarns. Ability to produce yarn out of sliver as opposed to roving is another advantage of rotor spinning. Thus, rotor spinning is a

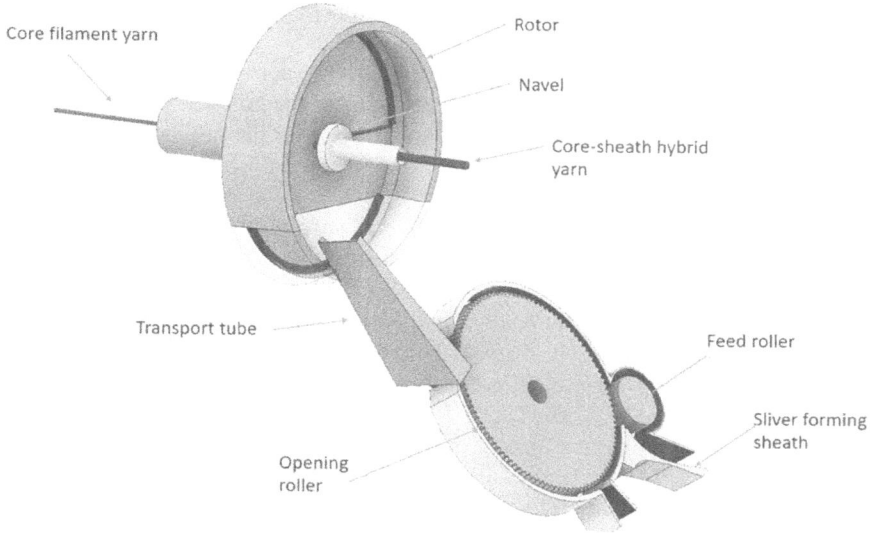

FIGURE 2.4 Rotor spinning process for production of hybrid yarn.

higher production, low cost alternative (elimination of roving frame and winding) to ring spinning though the quality is inferior in some of the quality parameters when compared to ring spun yarn. The same advantages can be translated to that of production of hybrid yarns (Ernst 2014). Similar to that of ring spinning, a slight modification in the existing setup of the rotor spinning for feeding up of the core filament yarn (through center of rotor) can assist in the production of core sheath hybrid yarn (Matsumoto et al. 2002). The schematic of the rotor spinning process for production of hybrid yarn is shown in Figure 2.4.

The hybrid yarn can be produced (as shown in Figure 2.4) by combining staple fibers with filament yarns under different levels of overfeed. The filament yarn is made to pass under appropriate tension by means of feed rollers (not shown in the figure) through the center of the rotor where it is combined with the staple fibers forming the sheath by means of the twist leading to the production of rotor spun hybrid yarn. The problems with the hybrid yarn produced from rotor spinning are similar to that of ring spinning. The misalignment of the core yarn and insertion of the twist in the core are the major drawbacks associated with the rotor spun hybrid yarn (Matsumoto et al. 2004).

2.3.3 DREF SPINNING

The DREF spinning system was developed by Dr. Ernst Fehrer in 1973. The initials of his title, first name and last name constituted DREF (Dr. Ernst Fehrer). DREF-I was the first of family of DREF machines and it worked on the principle of open-end spinning. It was followed by DREF-II for overcoming the problems associated with the slippage of fibers in the DREF-I spinning process. Later DREF-III was developed, which was used for the production of hybrid yarn. DREF-III spinning is not an open-end spinning process but instead produces a core-sheath structure which utilizes frictional forces in the yarn formation zone for the production of hybrid yarn. DREF-III spinning utilizes two drafting systems, one for the production of the core and other for the production of the sheath (Ishtiaque, Salhotra, and Gowda 2003). The process involves sliver feeding, sliver drafting, fiber opening, straightening and wrapping around the core. The produced hybrid yarn is wound on to the take-up package. The wrapping of yarn around the reinforcement is by means of two friction drums with perforations in them. These perforated rollers suck the fibers after their individualization and rotation of both the drums in the same direction wrap the fibers around the reinforcement. In case the reinforcement in the core is in the form of filament, it is fed through the bottom front drafting roller of draft unit II without exertion of any kind of draft. The DREF III spinning process is most suitable process for production of hybrid yarn among other spinning processes such as ring, rotor and wrap spinning (Jagatheesan et al. 2018). The schematic of DREF III spinning process for production of hybrid yarn is shown in Figure 2.5.

The main advantage of the DREF III spinning process is that the core is twisted less (false twist) thereby having higher contribution to the direction of applied load when compared to processes of ring and rotor where core undergoes significant twisting. Apart from twist less core the other advantages of the DREF III spinning process are very high production rate, less power consumption and easily achievable core to sheath ratio of up to 30:70. Though the core-sheath yarn can be produced at higher production rates and lower cost when compared to other processes, the problem of flaring off the sheath from the core due to abrasion during the process of weaving restricts the potential use of hybrid yarn from DREF spinning in technical applications (Bar, Alagirusamy, and Das 2018).

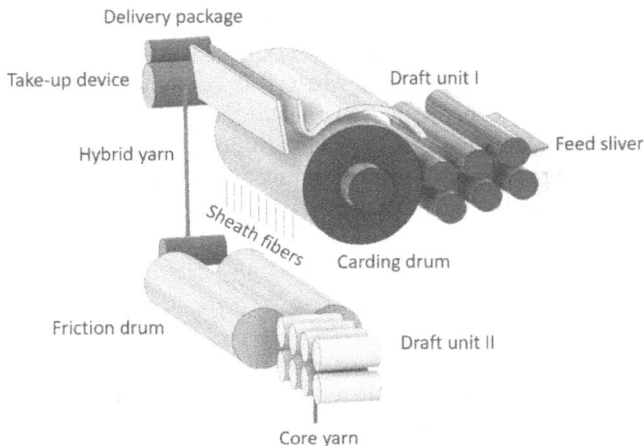

FIGURE 2.5   DREF III spinning process for production of hybrid yarn.

FIGURE 2.6 Film stacked laminate.

2.3.4 FILM STACKING

In film stacking, alternate layers of polymer films and reinforcement are stacked. Figure 2.6 shows reinforcement and matrix stacked (Weyenberg et al. 2006).

Application of high pressure on the stack causes the flow of resin in the transverse direction to reduce resin-starved areas. The major drawback is that complex shapes cannot be formed. A film stacking technique can be labor intensive, slow and thereby uneconomic.

2.3.5 WRAP SPINNING

The method of production of hybrid yarn using wrap spinning has both the core and the sheath components in filament form as shown in Figure 2.7.

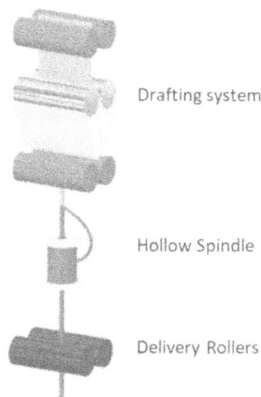

FIGURE 2.7 Wrap spinning for production of hybrid yarn.

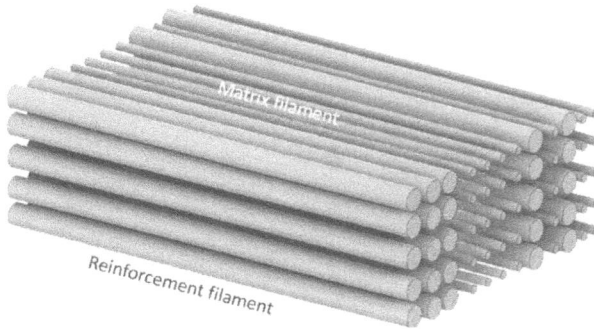

FIGURE 2.8   Parallel winding process for production of hybrid yarn.

The reinforcing filament forming the core is wrapped by the thermoplastic matrix filament as the sheath. These matrices forming sheath filament are melted during the process of consolidation. The process of wrap spinning suffers from the drawback of inhomogeneous distribution of the reinforcing and matrix yarns and requires higher temperature and pressure to bring about complete consolidation (Jean, Marc, and Guy 1999).

2.3.6   Parallel Winding

The schematic diagram for production of hybrid yarn using parallel winding is shown in Figure 2.8.

It is the simplest of the processes and involves placement of the reinforcement and matrix in the form of yarn side by side with each other. Achieving void free composites is a potential challenge in the process of parallel winding and requires severe consolidation conditions (Stolyarov et al. 2013).

2.3.7   Braiding

Braiding is another way for the production of the hybrid yarn. Braid is a complex structure, generally made by intertwining of three or more textile or flexible strands (Figure 2.9).

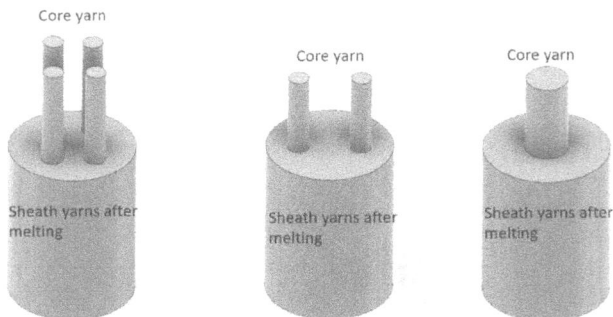

FIGURE 2.9   Core sheath braided structures.

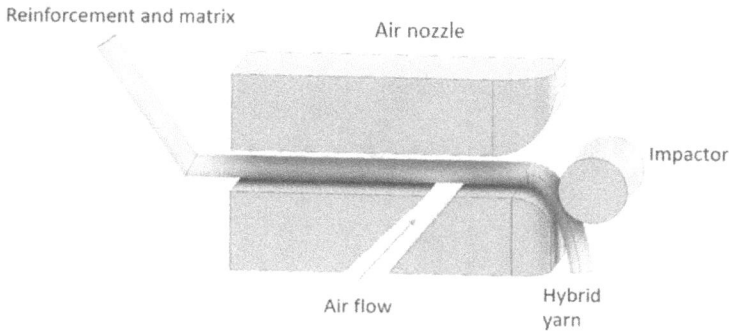

FIGURE 2.10 Air texturing process for production of hybrid yarn.

The hybrid yarn can be produced by placement of the reinforcement in the core as axial thread, which is wrapped by matrix filaments as braid. Braided composites are used extensively in friction bearings, aerospace applications and rocket launchers (Bradford and Bogdanovich 2008).

2.3.8 AIR JET TEXTURING

Air jet texturing is a mechanical process for combining reinforcement and matrix forming filaments for the production of hybrid yarn. The reinforcement and matrix are fed into the nozzle with an overfeed provided to the matrix as it has lower modulus and can be easily displaced to form the loops. The direction of impaction of compressed air is in the direction of moving of the filaments. This direction of impaction of compressed air distinguishes the process of commingling from that of air texturing (Bilgin, Versteeg, and Acar 1996). The schematic for the air texturing process is shown in Figure 2.10.

The direction of impaction of air causes the reinforcement and matrix forming filaments to mix in order to produce the hybrid yarn. The impaction of air produces the opened section of filaments followed by a mingling section. The nozzle is the heart of the air texturing machine. The nozzle produces supersonic, turbulent and non-uniform flow for entanglement of the reinforcement and matrix. The placement of the impact element at the exit of the nozzle can further improve the uniformity of mixing the reinforcement and matrix (Acar et al. 2006). The main problem associated with the process of air texturing is the formation of resin-rich and resin-starved regions as a result of segregation of reinforcement and matrix fibers during the textile preforming operation, such as weaving, braiding and knitting. This segregation is a result of differential load uptake between the reinforcement and matrix as a result of significantly different modulus exhibited by the reinforcement and matrix (Choi, Diestel, and Offermann 1999).

2.3.9 COMMINGLING

The mingling process involves impaction of rapidly moving air in a nozzle to generate entanglements among the reinforcement and matrix for production of the hybrid yarn.

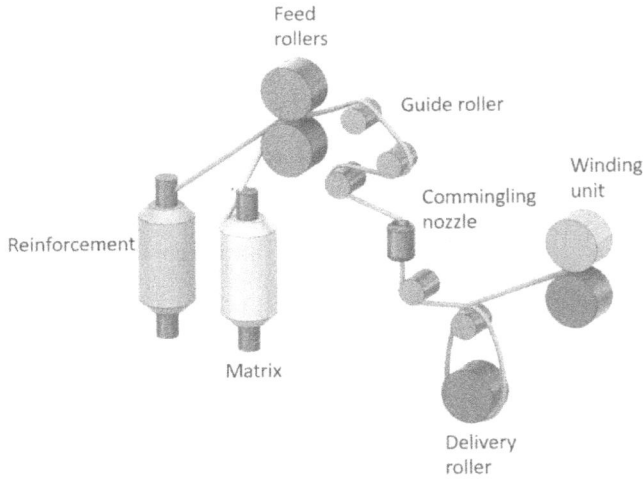

FIGURE 2.11   Commingling process for production of hybrid yarn.

There are different words used for the mingling process such as tangling, interlacing, intermingling, entangling and commingling. The term commingling refers to the entanglement of two or more yarns. The impaction of air is in a perpendicular direction in commingling as opposed to that in air texturing (Ogale and Alagirusamy 2007). The schematic of commingling process is shown in Figure 2.11.

The desired ratio of fiber to matrix can be obtained by varying process parameters such as feed roller speed, nozzle pressure, and take-up speed. Similar to air jet texturing, commingling makes use of supersonic, turbulent and non-uniform flow for entanglement of reinforcement with matrix (Ogale and Alagirusamy 2008). The flexibility of the commingled yarn makes it suitable for textile preforming operations of weaving, braiding etc. The problems associated with commingling are similar to that of air texturing; that is, the composites can easily produce resin-starved and resin-rich regions because of segregation of reinforcement and matrix in the preformed structure. To obtain the well consolidated composites, the stringent control over the process parameters is to be exercised. Also, the nip stability is an important parameter to be tested in order to ensure that segregation of reinforcement and matrix does not take place (Bernhardsson and Shishoo 2000).

In fiber commingling, the polymeric matrix and reinforcement are combined in fibrous form. Complete wetting of the fiber by the molten polymer is performed during consolidation. Thorough commingling of the matrix and fiber decides the effectiveness of impregnation. However, due to a difference in modulus of fiber and matrix, resin-rich and resin-starved areas are created during textile preforming operations. Higher external pressure is needed to consolidate the towpreg. High cost involved in producing thermoplastic matrix in filament form for commingling with high-performance filaments limits the scope of fiber commingling. Homogeneous commingled yarns are effectively produced under the trade name of Twintex® (Mäder, Rothe, and Gao 2007).

2.3.10 POWDER COATING TECHNOLOGIES

Powder coating of fibers is an environmentally friendly process as no solvent is required. The fibers are under less stress during impregnation. High utilization as well as recovery of powder, along with elimination of expensive solvents, is an economic advantage obtained through powder coating. In powder coating, the flow of the resin is along the fiber; thereby, complete wet out can be achieved at lower pressures.

There are number of powder processing technologies available, such as: conventional powder and fluidized bed, in which powder particles are mechanically deposited onto the reinforcement; recirculating fluidized bed, in which a curtain of powder is sprayed onto the reinforcement while the over-sprayed powder is recirculated; Liquid phase deposition pulls the reinforcement through the slurry of the powder where the powder particles mechanically adhere to the reinforcement; acoustically fluidized bed utilizing sound energy and nitrogen gas for deposition; electrostatic fluidized bed, in which electrostatically charged powder particles are deposited on the tow; electrostatic spray gun coating again utilizes electrostatic charge for charging powder particles for their deposition on the tow; (Muzzy and Colton 1993). The details of these powder coating lines are discussed below

2.3.10.1 Conventional Powder Bed

Price et al. passed the spread glass roving through the bed containing thermoplastic powder (without fluidization) (Figure 2.12), which upon getting coated was heated, to ensure flow and firm attachment of matrix around the fibers. Spreading of glass roving was assisted by the abrasive nature of powder particles. Importance of the abrasive nature of particles can be realized when compared to passage of roving through thermoplastic melt, emulsion or solution in which the emerging strand has resin deposited only on its periphery with little or no internal penetration (Roger 1973).

Such a roving with haphazard penetration is of limited use as starting material. Size and shape of powder particles were reported to have a significant effect on the amount of the spread of roving and deposition of powder within roving. A powder particle preferred for coating should have a particle size range, with certain quantity

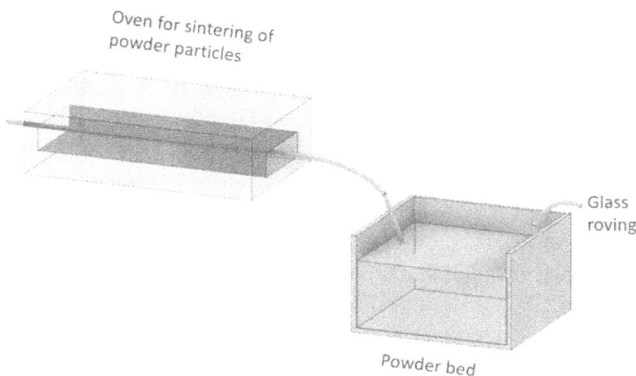

Oven for sintering of powder particles

Glass roving

Powder bed

FIGURE 2.12 Conventional powder bed for production of towpreg.

of powder falling below and above the mean particle size. Particles having size above the mean improves the abrasive nature of powder particles. Particles with lower size help in free flowing of powder thereby giving better penetration of roving. Particles with irregular and jagged shape improved the abrasive nature of powder particles when compared to spherical shaped powder particles. Passage of roving through the fluidized bed (figure not shown) led to lower haul off force required to pull the roving through the powdered bed. After powder coating, roving was passed through a heated tube, which had a forming die at the end through which roving emerges. The size of the die at the exit of the roving was used as a means to adjust the maximum polymer content of the roving. Inconsistent powder pick-up and non-permanent fixation of powder particles to the glass roving leading to fall off of the powder particles are some of the limitations of the conventional powder bed (Connor 1987).

2.3.10.2 Conventional Fluidized Bed

Ganga formed a towpreg, which has core of reinforcement impregnated with thermoplastic powder in a fluidized bed. The powder laden core is surrounded by continuous extruded or dip coated sheath. The thermoplastic powder can have the same or higher melting temperature as the continuous extruded sheath (Ganga 1986). The main advantage claimed by Ganga et al. is the pliability and flexibility of the towpreg formed. The flexibility is ensured because the powder particles deposited on the reinforcement are not melted and are in the raw form; thereby, the relative motion between filaments is not restricted. It is claimed in the invention that towpreg materials remain flexible, and can be bent and knotted without breaking. Thus, towpreg formed can be woven, braided and knitted with utmost ease. The reinforcement is coated with the matrix powder in the core as well as on the periphery and thorough mingling is achieved between them without the loss of the powder as a result of enclosure in a continuous extruded sheet. However, severe consolidation conditions are a must to achieve complete wet out. It requires fine thermoplastic powder similar to slurry coating, and production of fine thermoplastic powder can be extremely expensive (Ganga 1987).

2.3.10.3 Recirculating Fluidized Bed

Baucom et al. recognized the hurdle of embodying thermoplastics with continuous fiber tows for production of flexible towpregs. The authors reviewed the previous attempts made to combine polymer matrix and filamentary materials such as slurry coating, solvent coating and film coating. These methods could not ensure systematic, uniform deposition of polymer powder particles throughout the filamentary material, along with the possibility of failure to ensure complete removal of traces of carrier materials for the powder (as in slurry and solvent coating), which can lead to void in the composite. Baucom et al. patented recirculating powder bed to overcome aforementioned problems. The research team pneumatically spread the tow into a spreading unit and ensured uniform, continuous deposition of polymer powder particles and their subsequent sintering on the tow to produce towpregs, with minimal stress on filamentary material, without requiring extended, high temperature residence times for polymer. As in powder coating only a very small fraction of total powder is deposited on to the reinforcement, recirculation of oversprayed powder

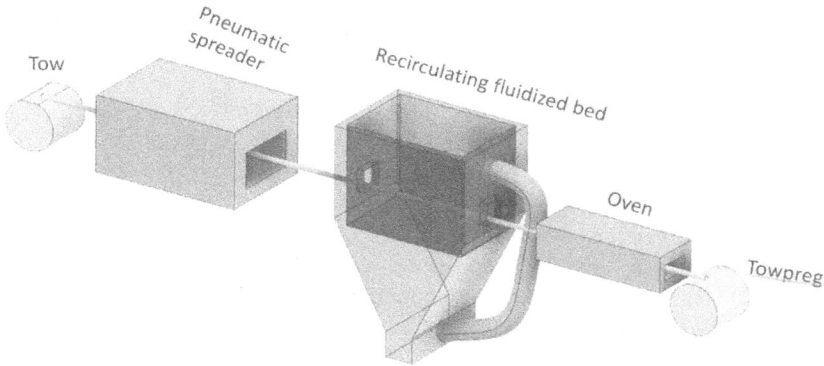

FIGURE 2.13   Recirculating fluidized bed for production of towpreg.

offers efficient utilization and thereby economic advantages. In order to have economic advantages and efficient utilization, the research team recirculated the powder as shown in Figure 2.13. The authors reported that sintering powder particles on the tow as a better method when compared to enclosement of the powder particles (film coating) into an extruded polymer sheath, because of the possibility of movement of powder deposited on the tow while it is being encased by a polymer sheath in the extruded tube (Robert, John, and Joseph 1991).

Nunes et al. carried out studies in three-point bend tests on unidirectional composites produced from glass polypropylene towpregs obtained from the recirculating powder bed. However, insufficient interlaminar adhesion was observed and the author suggested further research in consolidation techniques (Nunes et al. 2003).

2.3.10.4   Slurry Coating

Slurry coating requires higher consolidation parameters as the matrix in the form of powder does not wet out the reinforcement and is just merely deposited on it. Use of water for formation of emulsion leads to tacky towpregs, making it unsuitable for textile forming processes. Removal of traces of dispersants added to prevent agglomeration of powder particles can lead to entrapped air in the composite (Connor 1987). The slurry coating process for production of towpreg is shown in Figure 2.14.

For the emulsion process to be successful, powder particles have to be finer than the fiber diameter, which is a challenge as a result of the inherent crystalline structure of thermoplastics and may require severe cryogenic condition, shooting up the cost.

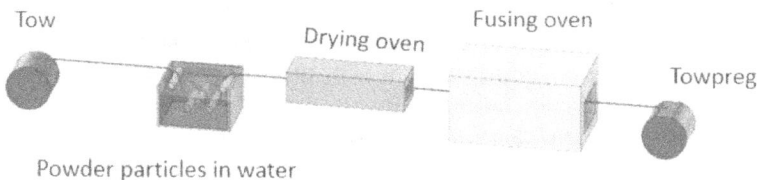

FIGURE 2.14   Slurry coating for production of towpreg.

Slurry processing usually produces composites with high void content, and producing a void free composite severe control over process parameters may be needed (Gordon 1981).

2.3.10.5 Acoustically Fluidized Bed

Iyer et al. used vibrating acoustic means in order to fluidize powder particles in a housing and to coat the powder particles on the fibers. The authors claimed the powder coating invention to have tremendous speed, cost, and technical and environmental advantages. These advantages are claimed by highlighting the limitation of previously available techniques for impregnation of reinforcement with the thermoplastic matrices. The research group claimed powder coating to be ideal process because of the following characteristics:

1. The powder coating process is independent of matrix viscosity. Powder coating coats the individual filament of the tow, thereby the distance the powder has to travel to coalesce after being heated for complete impregnation is of the order of submicrons as compared to other processes where distances are in centimeters from the outside to inside of fiber tows.
2. There is no use of binders or solvents to be evaporated, which can drastically reduce mechanical properties of the composite if complete removal is not ensured.
3. Powder particles of the same size as reinforcement can be obtained by severe cryogenic conditions, leading to uniform distribution of the matrix throughout the reinforcement.
4. Concentration of powder particles in the chamber can be controlled at all times.
5. The mechanism to adhere the powder particles to the fibers is controllable and independent of environmental conditions.
6. Powder coated towpregs have higher flexibility and drapability and thus complex preforms can be formed.
7. The powder coating process has lower energy consumption, is not labor intensive, can operate at higher speeds and is capable of scaling up to larger sizes.

This invention involves fluidizing the powder particles by using a vibrational energy, depositing the fluidized powder particles on the spread tow and then fusing it locally by passing it through the heater. The towpreg thus produced is flexible, drapable which can be later produced into a void free, fully densified composite. This invention differed from the previous inventions in the sense that it made use of acoustic energy, to overcome gravitational force exerted by powder particles for fluidization, as opposed to gas flow in conventional fluidization techniques. The most suitable means to have uniform deposition of powder particles on the reinforcement is to have powder particle size equivalent to fiber diameter. The usual filament diameter in a tow is less than 20 microns and the use of powder particles with size less than 20 microns will have significantly higher surface area, leading to agglomeration and channeling of cohesive powders making their fluidization extremely difficult by conventional gas fluidization techniques. Fluidization using acoustic energy can fluidize and entrain particles of any size at any level of concentration.

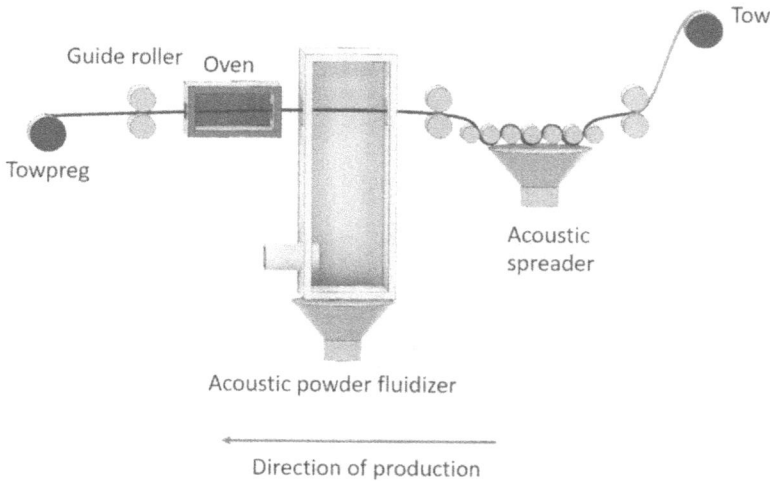

FIGURE 2.15   An acoustically fluidized bed for production of towpreg.

A schematic diagram of the machine is shown in Figure 2.15. Tow unwound from the spool is made to pass through a guide ring and nip roller, which ensures that the tow does not go out of the path of the spreader as it unwinds from the extreme end of the feed package. The spreader is an audio speaker having a high decibel sound level of about 80–130 db.

On entering the coating section, the spread tow is coated with powder particles, which are aerosolized by a combination of acoustic energy and an additional air input. The powder laden tow is made to pass through a heater and then through guides to the nip roller which controls the width of the towpreg. The towpreg formed is finally wound onto the drum (Shridhar, Drzal, and Krishnamurthy 1992).

2.3.10.6   Electrostatic Fluidized Bed

Muzzy et al. (John and Babu 1992) invented a flexible multiply towpreg that had suitability for weaving, braiding and knitting. The authors defined towpreg as consisting of continuous fibers/filaments combined with a matrix. A tow is a collection of filaments which can vary with respect to number of filaments per tow. Muzzy et al. emphasized on the use of thermoplastic matrices in aerospace applications, as these have better fracture toughness, environmental resistance and impact strength than their thermoset counterparts. Indefinite shelf life at room temperature, better quality, recyclability, regular storage and no refrigeration problems are advantages which make it cost effective and easier to handle. The major disadvantage with thermoplastic matrix hindering its use was attributed to high melt viscosity, which prevents uniform flow of the matrix around every filament in the tow. These when combined with extremely poor coating characteristics of carbon fibers can form a composite with a significant number of voids. If a thermoplastic towpreg could be produced by methods such as hot melt impregnation, it would be extremely rigid with higher bending curvatures and difficult to braid, knit and weave. The authors then compared the

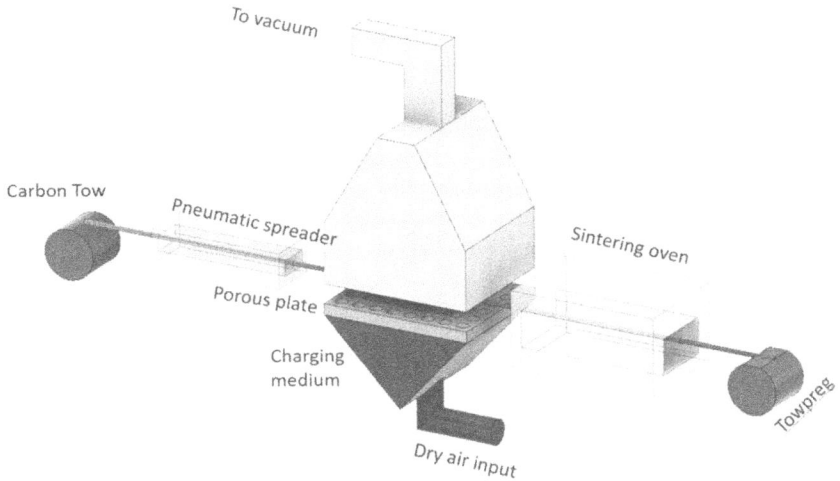

FIGURE 2.16   Electrostatic fluidized bed.

available techniques of forming thermoplastic towpregs of hot melt impregnation, solution coating, emulsion process, slurry coating, surface polymerization, commingling, film stacking and powder coating. The authors destined the powder coating technology to be more effective than other technologies for the aforementioned reasons. The schematic of the electrostatic fluidized bed is shown in Figure 2.16.

The tow was unwound in such a way that twisting of the filaments was avoided and then it passed through 6 rollers (not shown in the figure), 3 of which were placed before the pneumatic spreader and 3 after it. These rollers reduced the vibration of the tow during spreading and also provided tension to the tow, thereby imposing lateral constraint which prevented its collapsing. The tow passed over the first roller, under the second roller and over the third roller before and after spreader, imparting constant tension and reduced vibrations to the tow. These passage of tow over and under the rollers also ensured the spread to some extent prior to entering the spreader and the extent of spread further improved when the tow passed through pneumatic spreader. The preferred material for these rollers was Teflon as it gave considerable spreading and non-stickiness. One of the rollers prior to entering the coating section was made of steel so that the tow can be grounded. The tow then passed through an air comb spreader where forced air spread the tow to almost individual filament stage. The spread tow entered the fluidized bed whose dimensions were 6×6 inches. The electrostatic fluidized bed had the electrical output of 0–90 kV and air input of 0–12.5 cubic feet per minute. The electrostatic fluidized bed had an electrode made up of sharp points or small diameter wires. The air that passed through this charged electrode was ionized. The ionized air then passed through a porous plate above which powder particles were placed at a height of two to four inches.

The powder particles were then charged and fluidized simultaneously by incoming ionized air. The porous plate provided very high resistance to the passage of the air but permitted sufficient air needed for fluidization. The porous plate was made to vibrate through mechanical means so as to enhance fluidization and reduce air channeling and formation of powder lumps. The ionized air passing through the porous plate overcomes the frictional and gravitational force exerted by the powder particles, causing them to fluidize. During fluidization the ionized molecules of the air transfer their charge to the powder particles, which thereby repel each other and form a taut cloud. The cloud formation is assisted by vacuum created from the other end of the bed. The grounded or oppositely charged tow is made to pass through the cloud of the charged particles which thereby attract the powder particles. The powder particles are attached to the tow by firm electrostatic forces. The powder particles which are maintained to the depth of 50–100 mm expand to 50% to 150% of their unfluidized height after fluidization. The fluidized air used should be clean and dry. Contaminants in the air can block the porous portions of the plate, which may lead to uneven fluidization. The relative motion and hence fluidization of the powder will be affected in case the powder particles are wetted by the moisture in the air. Air dryer installation for removing excess moisture in the air along with attached filters for removing submicron impurities is a must. A negative polarity of 30 to 100 kV is usually employed depending on the powder being processed.

Polyester powders are reported to achieve better coating at 40 kV or less. Epoxy powders perform better in the range of 40 to 70 kV. Vinyl and PEEK powders gave better powder deposition at voltages of 60 to 75 kV and 70 to 90 kV respectively. As soon as the powder was laden on the tow it was made to pass through an oven to fuse the powder particles on the tow so that powder does not fall due to weakening of electrostatic force over duration of time. The temperature of the oven depends on powder being processed and it should be below the degradation temperature of the powder particles. The tubular oven used had a length of 1 m and inside diameter of two inches (John and Babu 1992).

2.3.10.7   Electrostatic Spray Coating Process

Throne et al. (James and Andrea 1994) pointed out that fiber reinforced thermoplastic composites have an increased market share in automobile and aircraft components. The authors also pointed out the inherent limitation of the thermoplastic matrix to bring about complete wet out of reinforcement, as a result of its high melt viscosity. The research group reported that other processes such as solution coating, slurry coating, melt impregnation and commingling were incapable of giving uniform distribution of reinforcement and matrix as well as void free composites. To overcome the problem of achieving thorough wet out of reinforcement, Throne et al. implemented electrostatic powder spray gun coating process for production of towpreg, which comprises a coating of long continuous fibers with a thermoplastic powder. Long continuous fibers are spread so as to expose every filament for powder coating. The fibers such as glass and carbon are spread and grounded, so that the electrostatically charged matrix in the form of powder deposits firmly on it and loss of powder as a result of tension fluctuation or abrasion is reduced. The resulting powder laden tow is made to pass through an oven, further ensuring firm attachment of powder particles to the tow and preventing its fall-off.

Ramani et al. carried out the consolidation experiments on towpregs produced by electrostatic spray coating of glass fibers with PEEK powder braids. About seven different unidirectional composites were produced and evaluated in the three-point bend test. The microstructural analysis of the produced composites showed resin-rich and resin-starved areas in the composites (Ramani, Woolard, and Duvall 1995).

2.4   COMPARISON OF MULTI-STEP IMPREGNATION PROCESSES (COMMINGLING & POWDER COATING)

Amongst all the methods of production of hybrid yarns commingling and powder coating have gained wide scale acceptance. This acceptance is attributed to significant reduction of melt flow distances enabling production of void free thermoplastic composites. Though numerous published literatures are available on the impregnation behavior of matrix in composites made of commingled and powder coated hybrid yarns, there are only a limited number of studies comparing them. The following description is among the earliest attempts to compare the properties of commingled yarns and powder coated towpregs along with the composites produced from them.

Alagirusamy et al. (Ramasamy 1998) produced carbon towpregs via the powder coating process. The authors used an electrostatic fluidized bed for powder coating. Powders used as matrix included epoxy, PA 6, PA 11 and PEEK. PA 6 was used to produce towpregs with different stiffness by simply varying the processing conditions. Uncoated tows and commingled tows were used for study, in order to compare their properties with powder coated towpregs. The authors characterized powder coated towpregs for friction coefficient, bending measurements and lateral compressive properties. The authors reported that friction coefficient and bending properties played a vital role for textile operations such as weaving, knitting and braiding. On the other hand, lateral compressive properties had a significant influence on the composite manufacturing process and played a vital role in deciding the complexity of preform formed and depth of mold that can be used as well as control of fiber orientation that can be achieved. Friction measurement tests carried on Instron revealed that wrap angle and test speeds did not affect coefficient of friction, whereas increased impeding force reduced the coefficient of friction. The research team attributed the reduction in coefficient of friction to the visco-elastic nature of textile materials.

All the powder coated tows showed higher coefficient of friction when compared to uncoated and commingled tows, thereby requiring greater care to be exercised during processing to avoid problems like breaking of filaments, entanglement of threads and powder fall off during textile operations. Bending characteristics of towpregs were analyzed using pure bending fixtures through the Kawabata Evaluation System and cantilever beam bending test. The authors reported superiority of pure bending tests, which exposes towpregs to higher bending curvatures which are similar to that in textile operations. Also, in pure bending tests towpregs are clamped at both of their ends and hence there is no splitting of towpregs as in case of cantilever beam tests. Powder coated tows were reported to have higher bending rigidity in comparison to uncoated and commingled tows. Among powder coated tows, PEEK powder nylon coated tows, as a result of larger particle size, showed higher bending rigidity. Wrapping of PA 6 filament to improve its process ability during textile operation led to increase in bending rigidity of towpregs whereas it did not affect the coefficient of friction. Lateral compressive tests carried out on the Instron testing machine with a special fixture indicated powder coated towpregs to have higher bulk, permanent set and lower resilience; when compared to uncoated and commingled tows. No change in compressibility was reported between powder coated, commingled and uncoated tows.

Alagirusamy et al. (Ramasamy, Wang, and Muzzy 1996b) conducted braiding studies to analyze the influence of friction coefficient and bending rigidity of powder coated towpregs, on behavior of towpregs during braiding as well as quality of braids produced. The authors carried out braiding experiments with powder coated and commingled tows. PA powder carbon coated tows involved Type I and Type II tows, where Type I had higher stiffness due to change in processing conditions. The research group reported frequent stoppages during braiding of commingled and powder coated tows which were not wrapped with a filament to improve their abrasion resistance. The research team reported absence of cohesion between filaments of the tows to be responsible for frequent interruptions during braiding operation. Wrapping of the tows with fine filament using hollow spindles significantly improved their braidability. Towpregs with lower friction coefficient and stiffness braided without significant stoppages. The authors designed a braiding simulator, which exposed towpregs to similar conditions as found in braiding, to evaluate the extent of damage the towpregs undergo during the operation. The results obtained from the simulator matched that of the actual braiding operation; thereby an offline system for braiding simulation tests can be used for determination of the extent of fiber damage during the actual process. The only limitation that comes forth for the braiding simulator is that the number of crossing towpregs at any given time are different and the possibility of filaments entangling the broken neighboring filaments is higher in the case of actual process than the braiding simulator.

Alagirusamy et al. (Ramasamy, Wang, and Muzzy 1996a) carried out the consolidation experiments on carbon fiber/nylon 6 powder coated and commingled tows braided to biaxial braids in compression molding machine. The authors used a 2^3 factorial design experiment which had three factors of temperature, time and pressure and two levels for each factor were chosen. Laminates formed were evaluated for laminate quality of density, void content and fiber volume fraction. Laminates were also tested for their mechanical properties of tensile, compression, flexural and interlaminar shear. The authors reported that for similar properties commingled tows required severe consolidation conditions, then powder coated tows. The research group ascertained this behavior to uniform distribution of powder particles in between carbon tows in powder coating in comparison to commingled tows where nylon fibers are not uniformly dispersed.

The authors carried out analysis of variance studies and reported that consolidation temperature and consolidation temperature-pressure interaction had significant effect on fiber volume fraction. The research group found that increase in fiber volume fraction with higher pressure was higher at lower temperature than at higher temperature due to excess resin squeeze out. The authors attributed this to significant reduction in melt viscosity at higher temperature. Consolidation temperature, consolidation pressure and their interaction have significant effect on void content in the composites. The authors reported that time did not have any significant effect on void as required resin flow for complete impregnation was achieved in less time than that chosen as the lower level for the factorial design experiment. Analysis of mechanical properties indicated powder coated tows to have better mechanical properties than commingled tows at similar processing conditions. This is attributed to uniform distribution of resin in powder coated towpreg when compared to commingled tow. This uniform distribution helps in even and firm attachment of the matrix to the fibers in powder coated towpreg, thereby reducing void content and better load transfer from the matrix to the matrix fiber interface to the fibers.

Commingled tows showed significant voids due to insufficient flow of resin and thereby reduction in strength and modulus thus obtained. Flexural and interlaminar properties are found to be better for powder coated towpregs than commingled tows. Temperature is the single factor showing significant difference for flexural and interlaminar properties, with properties improving from lower to higher temperature. Tape laminates produced with disposition of yarns in similar angles to the braided structure indicate that plane laminates have similar modulus but significantly reduced tensile strength. Fiber interlacement in braiding is responsible for higher strength in braided laminates when compared to plain laminates.

2.5 ELECTROSTATIC SPRAY COATING VS. OTHER POWDER COATING TECHNOLOGIES

Acoustically fluidized bed and electrostatic spray coating are advantageous for their medium cost and energy required in the process (Vaidya and Chawla 2008), for a better level of control over the fiber volume fraction in comparison to other processes. Also, no additional processing step is required. Acoustically fluidized bed, however, is more suitable for the use with extremely fine powder particles. Fine powder particles have a high surface area and thus agglomerate. To fluidize such powder particles, acoustic energy is more suitable than fluidization by means of air. Thus, depending on the particle size employed for coating a suitable powder coating technology should be chosen for production of towpregs.

2.6 SUMMARY

In this chapter, the advantages of thermoplastic matrices over thermoset matrices are discussed and the challenges of using thermoplastic resin as a matrix in composites are brought out. In order to overcome the impregnation problems due to high melt viscosity associated with thermoplastics, the idea of hybrid yarns or the flexible towpregs is introduced. Different techniques used for the production of hybrid yarns are discussed in length along with their merits and demerits. The properties of these hybrid yarns are analyzed with respect to their potential to be used for textile preforming.

REFERENCES

Ramasamy, A., Y. Wang, and J. Muzzy. 1998. "Braided Thermoplastic Composites from Powder-Coated Towpregs. Part I: Towpreg Characterization." *Polymer Composites* 17 (3): 497–504. doi:10.1002/pc.10639

Acar, M., S. Bilgin, H. K. Versteeg, N. Dani, and W. Oxenham. 2006. "The Mechanism of the Air-Jet Texturing: The Role of Wetting, Spin Finish and Friction in Forming and Fixing Loops." *Textile Research Journal* 76 (2): 116–25. doi:10.1177/0040517506062614

Ahn, K. J., and J. C. Seferis. 1993. "Prepreg Process Analysis." *Polymer Composites* 14 (4). John Wiley & Sons, Ltd: 349–60. doi:10.1002/pc.750140411

Alagirusamy, R., R. Fangueiro, V. Ogale, and N. Padaki. 2006. "Hybrid Yarns and Textile Preforming for Thermoplastic Composites." *Textile Progress* 38 (4). Woodhead Publishing Ltd.: 1–71. doi:10.1533/tepr.2006.0004

Alagirusamy, R., and V. Ogale. 2004. "Commingled and Air Jet-Textured Hybrid Yarns for Thermoplastic Composites." *Journal of Industrial Textiles* 33 (4): 223–43. doi:10.1177/1528083704044360

Alagirusamy, R., and V. Ogale. 2005. "Development and Characterization of GF/PET, GF/Nylon, and GF/PP Commingled Yarns for Thermoplastic Composites." *Journal of Thermoplastic Composite Materials* 18 (3): 269–85. doi:10.1177/0892705705049557

Angell, R., M. Michno, J. Konrad, and K. Hobbs. 1989. Hot-melt Prepreg Tow Process. US4,804,509A, issued February 14, 1989. https://patents.google.com/patent/US4804509

Mahadev, B., R. Alagirusamy, and A. Das. 2018. "Properties of Flax-Polypropylene Composites Made through Hybrid Yarn and Film Stacking Methods." *Composite Structures* 197 (8): 63–71. doi:10.1016/j.compstruct.2018.04.078

Mahadev, B., R. Alagirusamy, and A. Das. 2019. "Influence of Friction Spun Yarn and Thermally Bonded Roving Structures on the Mechanical Properties of Flax/Polypropylene Composites." *Industrial Crops and Products* 135 (9). Elsevier: 81–90. doi:10.1016/j.indcrop.2019.04.025

Bernhardsson, J., and R. Shishoo. 2000. "Effect of Processing Parameters on Consolidation Quality of GF/PP Commingled Yarn Based Composites." *Journal of Thermoplastic Composite Materials* 13 (4). Technomic Publishing Co., Inc.: 292–313. doi:10.1177/089270570001300403

Bilgin, S., H. K. Versteeg, and M. Acar. 1996. "Effect of Nozzle Geometry on Air-Jet Texturing Performance." *Textile Research Journal* 66 (2). Sage Publications, Thousand Oaks, CA: 83–90. doi:10.1177/004051759606600204

Bradford, P. D., and A. E. Bogdanovich. 2008. "Electrical Conductivity Study of Carbon Nanotube Yarns, 3-D Hybrid Braids and Their Composites." *Journal of Composite Materials* 42 (15). Sage Publications: London: 1533–45. doi:10.1177/0021998308092206

Bradley, J. 1997. "Consolidation of Thermoplastic Powder Coated Towpreg," PhD diss., Georgia Institute of Technology.

Chang, I. Y., and J. K. Lees. 1988. "Recent Development in Thermoplastic Composites: A Review of Matrix Systems and Processing Methods." *Journal of Thermoplastic Composite Materials* 1 (3): 277–96. doi:10.1177/089270578800100305

Choi, B. D., O. D. I. Diestel, and P. Offermann. 1999. "Commingled CF/PEEK Hybrid Yarns for Use in Textile Reinforced High Performance Rotors." In *International Conference on Composite Materials*: 796–806.

Cogswell, F., and D. Hezzell. 1985. Method for Impregnating Filaments with Thermoplastic. US4549920, issued 1985.

Connor, James. 1987. Reinforced Plastic. US4680224, issued 1987.

Connor, M., S. Toll, J. A. E. Månson, and A. G. Gibson. 1995. "A Model for the Consolidation of Aligned Thermoplastic Powder Impregnated Composites." *Journal of Thermoplastic Composite Materials* 8 (2): 138–62. doi:10.1177/089270579500800201

Ernst, H. 2014. *The Rieter Manual of Spinning*. Wintherthur: Rieter Machine Works Ltd. Copyright. www.rieter.com

Ganga, R. 1986. Flexible Composite Material and Process for Producing Same. US4614678, issued 1986.

Ganga, R. A. 1987. Apparatus for Producing Flexible Composite Material. US4713139, issued 1987.

Gordon, T. 1981. Method of Impregnating a Fibrous Textile Material with a Plastic Resin. US4292105, issued 1981.

Goud, V., R. Alagirusamy, A. Das, and D. Kalyanasundaram. 2018. "Dry Electrostatic Spray Coated Towpregs for Thermoplastic Composites." *Fibers and Polymers* 19 (2): 364–374. doi:10.1007/s12221-018-7470-7

Goud, V., R. Alagirusamy, A. Das, and D. Kalyanasundaram. 2019. "Influence of Various Forms of Polypropylene Matrix (Fiber, Powder and Film States) on the Flexural Strength of Carbon-Polypropylene Composites." *Composites Part B: Engineering* 166 (6): 56–64. doi:10.1016/j.compositesb.2018.11.135

Goud, V., A. Ramasamy, A. Das, and D. Kalyanasundaram. 2019. "Box-Behnken Technique Based Multi-Parametric Optimization of Electrostatic Spray Coating in the Manufacturing of Thermoplastic Composites." *Materials and Manufacturing Processes* 34 (14): 1638–1645. doi:10.1080/10426914.2019.1666991

Goud, V., D. Singh, A. Ramasamy, A. Das, and D. Kalyanasundaram. 2020. "Investigation of the Mechanical Performance of Carbon/Polypropylene 2D and 3D Woven Composites Manufactured through Multi-Step Impregnation Processes." *Composites Part A: Applied Science and Manufacturing* 130 (November 2019). Elsevier: 105733. doi:10.1016/j.compositesa.2019.105733

Ishtiaque, S. M., K. R. Salhotra, and R. V. M. Gowda. 2003. "Friction Spinning." *Textile Progress* 33 (2). Taylor & Francis Group: 1–68. doi:10.1080/00405160308688958

Iyer, S. R., and L. T. Drzal. 1990. "Manufacture of Powder-Impregnated Thermoplastic Composites." *Journal of Thermoplastic Composite Materials* 3 (4). Sage Publications: Thousand Oaks, CA: 325–55. doi:10.1177/089270579000300404

Jagatheesan, Krishnasamy, Alagirusamy Ramasamy, Apurba Das, and Ananjan Basu. 2018. "Electromagnetic Shielding Effectiveness of Carbon/Stainless Steel/Polypropylene Hybrid Yarn-Based Knitted Fabrics and Their Composites." *The Journal of the Textile Institute* 109 (11). Taylor & Francis: 1445–57. doi:10.1080/00405000.2018.1423883

James, T., and O. Andrea. 1994. Method of Depositing and Fusing Charged Polymer Particles on Continuous Filaments. US5370911, issued 1994.

Jean, G., F. Marc, and B. Guy. 1999. Hybrid Yarn for Composite Materials with Thermoplastic Matrix and Method for Obtaining Same. US5910361, issued 1999.

Jogur, G., A. N. Khan, A. Das, P. Mahajan, and R.Alagirusamy. 2018. "Impact Properties of Thermoplastic Composites." *Textile Progress* 50 (3): 109–83. doi:10.1080/00405167.2018.1563369

John, M., and V. Babu. 1992. Flexible Multiply Towpreg and Method of Production Therefor. US5094883, issued 1992.

Jou, G. T., G. C. East, C. A. Lawrence, and W. Oxenham. 1996. "The Physical Properties of Composite Yarns Produced by an Electrostatic Filament-Charging Method." *Journal of the Textile Institute* 87 (1). Taylor & Francis Group: 78–96. doi:10.1080/00405009608659058

Kaw, A. K. 2005. *Mechanics of Composite Materials*, Second Edition. New York: CRC Press.

Kim, H. J., J. S. Kim, J. H. Lim, and Y. Huh. 2009. "Detection of Wrapping Defects by a Machine Vision and Its Application to Evaluate the Wrapping Quality of the Ring Core Spun Yarn." *Textile Research Journal* 79 (17). SAGE Publications: London: 1616–24. doi:10.1177/0040517509103509

Klein, W., and H. Stalder. 2014. *The Rieter Manual of Spinning*. 4. Wintherthur: Rieter Machine Works Ltd. Copyright. www.rieter.com

Mäder, E., C. Rothe, and S. L. Gao. 2007. "Commingled Yarns of Surface Nanostructured Glass and Polypropylene Filaments for Effective Composite Properties." *Journal of Materials Science* 42 (19): 8062–70. doi:10.1007/s10853-006-1481-x

Matsumoto, Y. I., S. Fushimi, H. Saito, A. Sakaguchi, K. Toriumi, T. Nishimatsu, Y. Shimizu, H. Shirai, H. Morooka, and H. Gong. 2002. "Twisting Mechanisms of Open-End Rotor Spun Hybrid Yarns." *Textile Research Journal* 72 (8). Sage Publications: Thousand Oaks, CA: 735–40. doi:10.1177/004051750207200814

Matsumoto, Y.-I., H. Saito, A. Sakaouchi, K. Toriumi, T. Nishimatsu, Y. Shimizu, H. Shirai, H. Morooka, and H. Gong. 2004. "Combination Effects of Open-End Rotor Spun Hybrid Yarns." *Textile Research Journal* 74 (8). Sage Publications: Thousand Oaks, CA: 671–76. doi:10.1177/004051750407400803

Mitschang, P., M. Blinzler, and A. Wöginger. 2003. "Processing Technologies for Continuous Fibre Reinforced Thermoplastics with Novel Polymer Blends." *Composites Science and Technology* 63 (14): 2099–2110. doi:10.1016/S0266-3538(03)00107-6

Muzzy, J. D., and Colton Jonathan S. 1993. Non-woven Flexible Multiply Towpreg Fabric. US5198281A, issued 1993.

Muzzy, J. D., and A. O. Kays. 1984. "Thermoplastic vs. Thermosetting Structural Composites." *Polymer Composites* 5 (3): 169–72. doi:10.1002/pc.750050302

Nunes, J. P., J. F. Silva, and P. Vieira. 2008. "GF/PP Towpregs Production, Testing and Processing." *International Journal of Mechanics and Materials in Design 4 (6)*: 205–211. doi:10.1007/s10999-007-9050-2

Nunes, J. P., J. F. Silva, A. T. Marques, N. Crainic, and S. Cabral-Fonseca. 2003. "Production of Powder-Coated Towpregs and Composites." *Journal of Thermoplastic Composite Materials* 16 (3): 31–48. doi:10.1177/0892705703016003003

Ogale, V., and R. Alagirusamy. 2007. "Tensile Properties of GF-Polyester, GF-Nylon, and GF-Polypropylene Commingled Yarns." *Journal of the Textile Institute* 98 (1): 37–45. doi:10.1533/joti.2005.0181

Ogale, V., and R. Alagirusamy. 2008. "Properties of GF/PP Commingled Yarn Composites." *Journal of Thermoplastic Composite Materials* 21 (6). Sage Publications: London: 511–23. doi:10.1177/0892705708091281

Pourahmad, A., and M. S. Johari. 2009. "Production of Core-Spun Yarn by the Three-Strand Modified Method." *Journal of the Textile Institute* 100 (3). Taylor & Francis Group: 275–81. doi:10.1080/00405000701763865

Pourahmad, A., and M. S. Johari. 2011. "Comparison of the Properties of Ring, Solo, and Siro Core - Spun Yarns" 5000. doi:10.1080/00405000.2010.498170

Roger, P. 1973. Production of Impregnated Rovings. US3742106, issued 1973.

Ramani, K., D. E. Woolard, and M. S. Duvall. 1995. "An Electrostatic Powder Spray Process for Manufacturing Thermoplastic Composites." *Polymer Composites* 16 (6). John Wiley & Sons, Ltd: 459–69. doi:10.1002/pc.750160604

Ramasamy, A., Y. Wang, and J. Muzzy. 1996a. "Braided Thermoplastic Composites from Powder-Coated Towpregs. Part III: Consolidation and Mechanical Properties." *Polymer Composites* 17 (3): 515–22. doi:10.1002/pc.10641

Ramasamy, A., Y. Wang, and J. Muzzy. 1996b. "Braided Thermoplastic Composites from Powder-Coated Towpregs. Part II: Braiding Characteristics of Towpregs." *Polymer Composites* 17 (3). John Wiley & Sons, Ltd: 505–14. doi:10.1002/pc.10640

Robert, B., S. John, and M. Joseph. 1991. Process for Application of Powder Particles to Filamentary Materials. US5057338, issued 1991.

Shridhar, I., L. Drzal, and J. Krishnamurthy. 1992. Method Coating Fibers with Particles by Fluidization in a Gas. US5102690, issued 1992.

Stolyarov, O. N., I. N. Stolyarov, T. A. Kryachkova, and P. G. Kravaev. 2013. "Hybrid Textile Yarns and Thermoplastic Composites Based on Them." *Fibre Chemistry* 45 (4): 217–20. doi:10.1007/s10692-013-9515-z

Svensson, N., R. Shishoo, and M. Gilchrist. 1998. "Manufacturing of Thermoplastic Composites from Commingled Yarns – A Review." *Journal of Thermoplastic Composite Materials* 11 (1): 22–56. doi:10.1177/089270579801100102

Tian, X., T. Liu, Q. Wang, A. Dilmurat, D. Li, and G. Ziegmann. 2017. "Recycling and Remanufacturing of 3D Printed Continuous Carbon Fiber Reinforced PLA Composites." *Journal of Cleaner Production* 142 (January). Elsevier: 1609–18. doi:10.1016/J.JCLEPRO.2016.11.139

Vaidya, U. K., and K. K. Chawla. 2008. "Processing of Fibre Reinforced Thermoplastic Composites." *International Material Reviews* 53 (April). Elsevier: 185–218. doi:10.1179/174328008X325223

Weyenberg, I. V. d., T. Chi Truong, B. Vangrimde, and I. Verpoest. 2006. "Improving the Properties of UD Flax Fibre Reinforced Composites by Applying an Alkaline Fibre Treatment." *Composites Part A: Applied Science and Manufacturing* 37 (9). Elsevier: 1368–76. doi:10.1016/J.COMPOSITESA.2005.08.016

Wu, G. M., and J. M. Schultz. 2000. "Processing and Properties of Solution Impregnated Carbon Fiber Reinforced Polyethersulfone Composites." *Polymer Composites* 21 (2). John Wiley & Sons, Ltd: 223–30. doi:10.1002/pc.10179

Ye, L., V. Klinkmuller, and K. Friedrich. 1992. "Impregnation and Consolidation in Composites Made of GF/PP Powder Impregnated Bundles." *Journal of Thermoplastic Composite Materials* 5 (1): 32–48. doi:10.1177/089270579200500103

3 Natural Fibres-Based Hybrid Towpregs

Pierre Ouagne and Mahadev Bar
Université de Toulouse, Tarbes, France

CONTENTS

3.1 INTRODUCTION

Over the last few decades, considerable interest has been observed in the use of natural fibres as polymer composite reinforcements (Faruk et al. 2012; Pickering, Efendy and Le 2016). It is mainly due to the advantages posed by the natural fibres over the synthetic fibres, such as lower density compared to glass fibre, abundant availability at a reasonable price, less abrasiveness, environmental friendliness, zero CO_2 fingerprint, being non-abrasive to the processing equipment etc. (John and Thomas, 2008). In addition, issues like global warming, greenhouse gas emission and government's norms related to solid waste disposal have bolstered the interest in natural fibre as

DOI: 10.1201/9781003049715-3

polymer composite reinforcement (Fortea-Verdejo et al. 2017). As a result, companies such as Libeco Lagae (Belgium), Lineo (Belgium) and NPSP (Netherlands) have shown their interest in natural fibre composites (NFC) and the market size of natural fibre composites is growing at a rate of 20-25% every year (Bar, Alagirusamy, and Das 2017).

However, the natural fibre has some drawbacks which reduce its potential as composite reinforcement. For instance, natural fibre and polymer matrices are not compatible, which leads to the development of a composite with poor fibre-matrix interface. Secondly, the natural fibres have short and discrete fibre architecture, which makes the control of fibre orientation in the composite structure difficult. Further, the natural fibre-based textile structures (for instance spun yarn, twisted yarn-based textile fabrics) as composite reinforcements disturb the uniform resin distribution in the composite structures, especially in the case of the thermoplastic matrix system. All these factors lower the mechanical properties of the resultant composites. Surface modification of the natural fibres through suitable chemical treatment or by physical means improves the mechanical properties of the resultant composite followed by improving the fibre-matrix interaction (de Farias et al. 2017; Xie et al. 2010). However, the problem of controlling fibre orientation and uniform fibre-resin distribution, especially in cases of thermoplastic composite, is still a challenging task for the composite researchers and manufacturers. Use of hybrid towpregs during composite manufacturing is often considered to overcome the above problems. This present chapter seeks to provide an overview about various hybridization approaches involved in natural fibre composite manufacturing.

3.2 NATURAL FIBRES

Fibres are a class of hair-like material, generally available in continuous filament form or discrete short staple form. Depending on their origin, fibrous materials are of two types, namely the synthetic fibres and the natural fibres. In general, the synthetic fibres are manufactured by extruding the fibre forming polymer through the spinneret where the polymer may originate from a petroleum by product, or it may be synthesized using natural resources. On the other hand, natural fibres are extracted from different natural sources. Based on their origin, natural fibres are classified in some sub-groups. The classification of textile fibres is shown in Figure 3.1. Among these natural fibres, the ligno-cellulosic vegetable fibres are mostly used as a composite reinforcing material (Jawaid and Abdul 2011).

3.2.1 STRUCTURE AND PROPERTIES OF LIGNO-CELLULOSIC NATURAL FIBRES

Natural fibres are considered as a naturally occurring composite in which cellulosic fibrils are embedded in the lignin matrix. Fibrils are aligned along the direction of fibre length and determine the mechanical and other properties of the natural fibres. Besides cellulose and lignin, hemicellulose, pectin and wax are the other significant components of ligno-cellulosic natural fibres. Cellulose is an abundantly available natural polymer, mainly comprising D-anhydro glucose ($C_6H_{11}O_5$) repeating units joined by 1, 4-β-D-glycosidic linkages at the C1 and C4 position, whereas

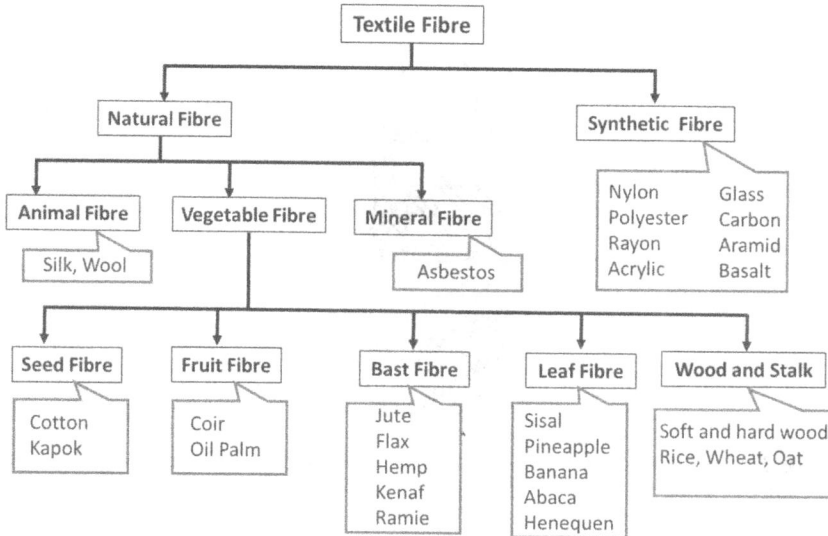

FIGURE 3.1   Classification of textile fibres.

hemicellulose is a low molecular weight copolymer of glucose, glucuronic acid, xylose, etc. but it is not a form of cellulose. Hemi-cellulose is highly hydrophilic and undergoes easy dissolution under the alkaline condition. Lignin is considered as a three-dimensional copolymer of aliphatic and aromatic constituents. The main problem with lignin chemistry is that it cannot be isolated to its native state. It is entirely amorphous and hydrophobic in nature and becomes soluble under hot alkaline conditions. Pectins and waxes are two other major constituents of natural fibres. Pectins are a collective name for hetero poly-saccharides which determine fibre flexibility, whereas waxes consist of different types of alcohols (Jayaraman 2003).

Most of the ligno-cellulosic fibres are multicellular, i.e. a single fibre is made up of multiple cells. The cell-wall of these cells are made up of fibrils which are composed of cellulose and these fibrils in the cell-wall are held together by means of lignin, hemicellulose and pectin (John and Thomas 2008). The wall of each cell has a complex, double-layered structure made of a thin primary wall which encircles the secondary wall. The secondary wall is further composed of three layers named S_1, S_2 and S_3. Among these three, the middle layer, i.e. S_2, is thicker than the other two layers and determines the mechanical properties of that fibre. All layers of a cell-wall are composed of cellular micro-fibrils which are made of long-chain cellulose molecules. In a primary cell-wall, the fibrils are arranged randomly while in the secondary wall they are arranged at an angle to the fibre axis, known as the micro-fibrillar angle. The micro-fibrillar angle determines stiffness and other mechanical properties of a natural fibre. Figure 3.2 shows the structural constitution of a fibril (Rong et al. 2001). Fibres are ductile when the microfibrils are spirally orientated to the fibre axis, and they become rigid and inflexible when micro-fibrils are parallel to the fibre axis (Rong et al. 2001; Satyanarayana et al. 1986). Mechanical properties of

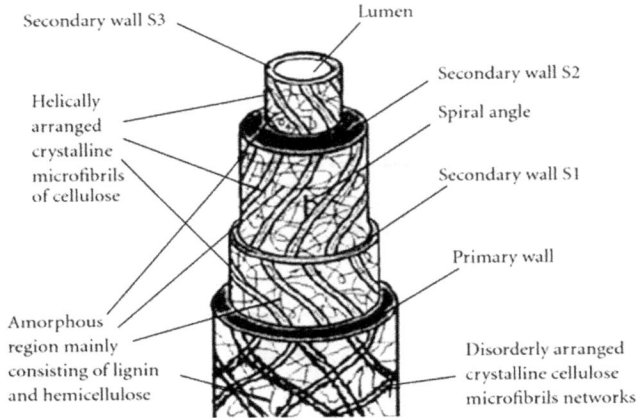

FIGURE 3.2   Structural constitution of a fibril. (Reproduced with permission from Rong et al. 2001.)

TABLE 3.1
Properties of Selected Natural and Manmade Fibres

Fibre	Density (g/cm³)	Breaking Elongation (%)	Tensile Strength (MPa)	Tensile Modulus (GPa)
Flax	1.5	2.7–3.2	500–1500	30–50
Jute	1.39	1.5–1.8	390–800	13–54
Hemp	1.47	2–4	600–900	40–60
Kenaf	1.45	1.6	700–1000	40–55
Bamboo	1.5	1–4	450–800	11–30
Sisal	1.5	2–2.5	500–635	9.4–22
Coir	1.25	30	165–222	4–6
E-glass	2.5	0.5	2000–3500	70
Aramid	1.4	3.3–3.7	3000–3150	63–67
Carbon	1.4	1.4–1.8	4000	230–240

some natural fibres and some synthetic fibres which are used as composite reinforcing materials are summarized in Table 3.1.

3.3   HYBRID COMPOSITES

The word "hybrid," extensively used in different scientific fields, originated from the Latin word "hybrida," which means something of mixed origin or composition. Fibre reinforced polymer hybrid composites are fourth-generation composites, which evolved in the early 1990s. At present, hybrid composites are gaining the attention of many researchers as a way of enhancing the performance of a composite.

In the context of fibre reinforced polymer composites, hybridization can be done in the following ways.

Matrix hybridization: A system in which one kind of reinforcing material is incorporated in a mixture of different matrices (blends) (Yu, Dean, and Li 2006).

Reinforcement hybridization: A system in which two or more reinforcing and filling materials are present in a single matrix (Fu, Xu, and Mai 2002).

Structural hybridization: A composite system which is made up of hybrid structures such as hybrid yarns (Alagirusamy et al. 2006).

The structural hybridization is mainly carried out to enhance the morphological behaviour of a composite system such as enhancement of fibre/matrix distribution, fibre orientation in the composite structure, decrement of void content etc. Hybrid yarn structures made of natural fibres and their effect on different composite properties are discussed later. On the other hand, the matrix or the reinforcing material hybridization is carried out to improve the various properties of the existing composite materials such as mechanical behaviour, bio-degradability, flame retardant performance, weight reduction, cost reduction etc. Among these two approaches, the fibre/reinforcing component hybridization is more popular than the matrix hybridization.

3.3.1 MATRIX HYBRIDIZATION

The matrix is one of the main components of any polymer composite system. The characteristics of a polymer matrix determine the shape, surface appearance, environmental tolerance, stress transfer between fibres and the durability of a polymer composite system. Based on their thermal behaviour, polymer matrices are of two types, namely thermoset and thermoplastic. Further, depending on their bio-degradation characteristics polymers are classified as biodegradable polymers and non-biodegradable polymers. In general, a single polymer is used as a matrix for composite manufacturing. However, with growing high-end and specific application of polymer composite and to overcome the drawbacks of a single polymer as a matrix, such as poor mechanical performance of the polymers from renewable resources or to offset the high price of synthetic biodegradable polymers etc., various polymer blends as a composite matrix have been developed over the last few decades.

Epoxy is one of the most popular resins used for composite manufacturing. Fibre reinforced epoxy composites have found wide application in the areas of automotive, aerospace, construction etc. Epoxy has excellent mechanical and thermal properties but due to its high density of cross-linkages, the ductility and fracture toughness of epoxy is very poor (Sultan and McGarry 1973). Hayes et al. (2007) have dissolved a thermoplastic polymer (polybisphenol-A-co-epichlorohydrin) in an epoxy resin to create a miscible blend and have used that blend as a matrix for glass fibre-reinforced laminates. They have observed that the blended polymer reinforced glass composite can heal the internal damage if the damaged sample is heated at 130°C. There is much similar literature available where an amorphous or semi crystalline thermoplastic polymer such as cyclic olefin copolymer (Mahmood, Dorigato, and Pegoretti 2020), poly(ε-caprolactone) (Luo et al. 2009), poly (ethylene-co-(methacrylic acid)) (Meure, Wu, and Furman 2009) etc. are added to epoxy resin and the polymer blend is used as a self-healing composite matrix. Intrinsic self-healing systems are based on specific molecular structures that may induce crack healing with the need of external stimuli such as heat, light, UV etc. (Garcia 2014; Wu, Meure, and Solomon 2008; Van der Zwaag et al. 2014; Zhong and Post 2015). Reddy et al. (2020) have developed a hybrid composite by reinforcing an epoxy/polyester blend with alkali-treated cordia dichotoma fibre and granite powder. The above composite exhibits very good tensile, flexural and impact properties and is thermally stable up to a temperature of 430°C.

Over last few decades, polymers from renewable resources are drawing the attention of the composite researchers and the manufacturers. It is predominately due to two reasons: firstly, the growing environmental awareness and secondly the realization that our petroleum resources are limited. However, the natural polymers have some drawbacks. For instance, most of the natural polymers are moisture sensitive; moist environment raises their degradability and the speed of degradation. The natural polymers have poor mechanical properties and they generally have a high cost. The advanced manufacturing technology and the bulk manufacturing have lowered down the cost of natural polymers. The blending of natural polymers with other polymers can develop a new low-cost products with better mechanical properties. PLA polymer, derived from renewable resources, is an interesting candidate for composite matrix. PLA is compostable and shows attractive physico-mechanical properties in terms of tensile strength and stiffness. Nevertheless, PLA shows poor impact strength and low heat distortion temperature. To overcome the drawbacks of PLA, many attempts have been reported in the literature. Recently, poly (methyl methacrylate) (PMMA) has received great interest as a blend candidate for PLA due to its higher glass transition temperature (Tg) and complementary mechanical properties (Anakabe et al. 2015; Arrieta et al. 2015). Orue, Eceiza, and Arbelaiz (2018) have prepared a PLA and PMMA blend (blend ratio 80:20) by the melt mixing method and have impregnated alkali treated sisal fibre to develop a hybrid matrix composite. It is observed that the heat deflection temperature (HDT), which is the maximum temperature at which a polymer system can be used as a rigid material, barely increased 4°C with respect to the unreinforced system while an annealing process can raise HDT of the composite system up to 38°C. Larguech et al. (2020) have develop a hybrid matrix composite using a PLA and poly (butylene succinate) (PBS) blend reinforced with jute fibres. It is observed that the incorporation of jute fibres in the PLA/PBS blend matrix gave rise to other dielectric relaxations associated with the water dipoles and interfacial (jute fibres/blend matrix) polarization effects.

3.3.2   Fibre Hybridization

Fibre hybridization involves the use of more than one type of fibre as reinforcement in the same composite structure. At present, the popularity of the hybrid composites is increasing constantly due to their capability of providing freedom when tailoring

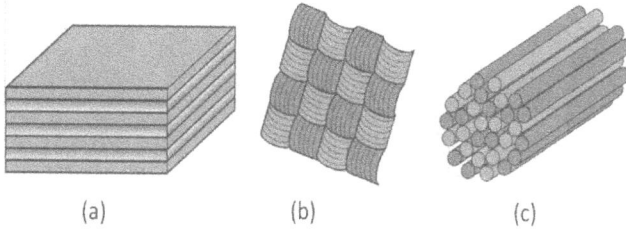

FIGURE 3.3   Main hybrid configurations (a) interlayer or layer-by-layer, (b) intralayer or yarn-by-yarn, (c) intra-yarn or fibre-by-fibre. (Reproduced with permission from Swolfs et al. 2014.)

composites and achieving properties that cannot be attained in composite systems containing only one type of reinforcement. The typical aim of fibre hybridization is to maintain the advantages of a particular fibres and ease some disadvantages. The other aim of fibre hybridization is to reduce the cost of the final composite product. Although the price of natural fibres is cheaper as compared to its synthetic counterpart, the cost varies from one fibre to another. For instance, the price of wood fibre is as much as fourteen times lesser than that of flax fibres (Väisänen, Das, and Tomppo 2017). Thus, a partial substitution of flax fibre with wood fibre could provide a significant advantage in terms of production cost and performance of the composite. The selection of components for a hybrid composite system and their compositions are mainly determined by the purpose of hybridization or by the ultimate uses of the resultant composite. Among these two types of fibre of a hybrid composite system, the one fibre generally has low elongation (LE) while the other has high elongation (HE). Among these two the LE fibre normally fails first but its failure strength is always higher than the HE-fibre. The LE and HE fibres can be placed in a composite structure through different configurations. The three most important configurations, which are interlayer or layer-by-layer, intra-layer or yarn-by-yarn and intrayarn or fibre-by-fibre, are visualized in Figure 3.3 (Swolfs, Gorbatikh, and Verpoest 2014). Among these three methods the layer-by-layer method is the easiest and cheapest method of hybrid composite manufacturing. However, a more complex configuration can be obtained by combining two of these three configurations. For instance, a yarn made of two fibres can be woven together with a homogeneous yarn. The ultimate properties of a hybrid composite are determined by the dispersion of the two fibre types. This is a measure for how well the two fibre types are mixed and is defined as the reciprocal of the smallest repeat length (Kretsis 1987). Among these structures the best fibre dispersion in the composite structure is achieved by intrayarn or fibre-by-fibre hybridization technique (Swolfs, Gorbatikh, and Verpoest 2014).

3.3.2.1   Fibre Hybridization Influencing Factors

The most basic definition of the hybrid effect is the apparent failure strain enhancement of the LE fibre in a hybrid composite compared to the failure strain of an LE fibre-reinforced non-hybrid composite. This definition is schematically illustrated

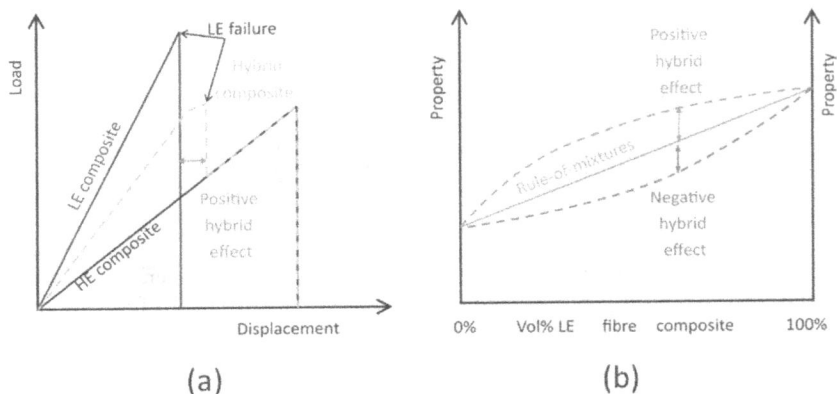

FIGURE 3.4 Illustration of the definitions of the hybrid effect: (a) the apparent failure strain enhancement of the LE fibres, under the assumption that relative volume fraction is 50/50 and that the hybrid composite is twice as thick as the reference composites and (b) a deviation from the rule of mixtures. (Reproduced with permission from Swolfs et al. 2014.)

in Figure 3.4 (Swolfs, Gorbatikh, and Verpoest 2014). Many factors influence the failure strain of a hybrid fibre composite having LE and HE fibres. These factors are mentioned below.

a. *Relative amount of fibres:* The relative volume of the constituent fibres is a crucial parameter for the hybrid effect. Kretsis (1987) has reported that a prominent hybrid effect in the failure strain of the composite is observed when the relative volume of the LE fibre is lower than all other fibres. This conclusion is confirmed by the model of Jones and Dibenedetto (1994), who showed an increase in the apparent breaking strength of carbon fibre by 92% if the carbon fibres are isolated from each other by the addition of glass fibres.

b. *Elastic properties of the fibres:* The elastic properties of both constituent fibres are important, as they influence (1) the static stress concentrations (Zweben, 1977), (2) the ineffective length, and (3) the dynamic stress concentrations (Xing, Hsiao, and Chou 1981).

c. *Failure strain ratio:* The ratio of the average failure strains of both fibre types was indicated by Zweben's (1977) model to play a crucial role in the hybrid effect. His definition of the hybrid effect was based on fracture of an HE fibre near a broken LE fibre. In contrast, Fukuda (1984) defined the hybrid effect based on fracture of an LE fibre near a broken LE fibre. In that case, the ratio of the failure strains has no effect on the hybrid effect. Both Zweben's (1977) and Fukuda's (1984) models are simplified and require several assumptions, which probably means the reality lies somewhere in the middle. If the failure of the HE fibre is close to that of the LE fibre, then some portion of HE fibre will fail prior to complete failure of the LE fibres. This will result in poor hybrid effect. On the other hand, if the failure strain

of the HE fibres is much higher than that of LE fibres, then both fibres will
act independently and a better hybrid effect can be expected.

d. *Fibre strength distribution:* Textile fibres are inhomogeneous; their property varies within the same set of samples. This variation is more prominent
in case of natural fibres. The fibre strength distribution or the variation in
fibre strength plays an important role in the hybrid effect. Fukunaga, Tsu-
Wei, and Fukuda (1984) have revealed that the hybrid effect is insignificant
when there is no variation in the strength of the LE fibre. Conversely, the
hybrid effect is expected to be larger when the LE fibre has a larger strength
deviation.

e. *Degree of dispersion:* The degree of LE and HE fibre dispersion in the
composite structure plays an important role in deterring the hybrid effect.
Fukunaga, Tsu-Wei, and Fukuda (1984) have proved that at a constant fibre
volume fraction, the hybrid effect increases with decreasing bundle size.
In a similar study, Mishnaevsky and Dai (2014) have showed that a finer
dispersion leads to slower development of internal damage.

f. *Matrix properties:* The properties of the matrix also impact the hybrid
effect, through their influence on the stress concentrations and ineffective
length. The influence of the matrix on the ineffective length is determined
by its shear modulus while the stress concentration is affected by the fibre
cross-sections and modulus (Zweben 1977). This is due to the shear lag
theory, which assumes that the matrix does not carry axial loads. Pan and
Postle (1996) have shown in their model that an increased matrix shear
yield strength can increase the hybrid effect, but only at a high fraction of
LE fibres. The matrix properties are hence expected to have only a secondary effect.

g. *Other parameters:* Along with the above factors, there are several other
parameters such as fibre–matrix interface strength, inter-laminar strength
and inter-laminar fracture toughness etc. that also influence the hybrid
effect (Swolfs, Gorbatikh, and Verpoest 2014). So far, none of the models
consider these properties and it is therefore difficult to judge their importance. More advanced models are required to establish the importance of
these parameters.

3.3.2.2 Mechanical Properties of Hybrid Fibre Composite

The prediction of mechanical behaviour of any hybrid composite depends on its
constituent parameters, i.e. on mechanical properties of reinforcements and on the
matrix mechanical properties. Along with this, the distribution and depression of
reinforcements, reinforcement volume fraction, interaction between the reinforcement and matrix and the test conditions also influence the prediction. Rule of mixture is often used to predict the mechanical behaviour of the hybrid composite
materials. In literature, several models based on the rule of mixture are reported
for the prediction of the mechanical properties of the hybrid composite materials.
These theoretical models include Voigt, Reuss, Hirsch and Tsai–Pagano (Essabir
et al. 2016). Yusoff, Takagi, and Nakagaito (2016) have reported that the modulus

of a single polymer reinforced with one type of reinforcement can be determined by using Equation (3.1).

$$E_C = E_f V_f + E_m V_m \tag{3.1}$$

$$\text{Where, } V_f + V_m = 1 \text{ or } V_m = (1 - V_f) \tag{3.2}$$

$$\text{Or, } E_C = E_f V_f + E_m (1 - V_f) \tag{3.3}$$

Where, E_C, E_f, E_m are the modulus of the composite, fibre and matrix respectively. V_f and V_m are the volume fraction of fibre and matrix in the composite system. In a fibre hybridized composite, more than one type of reinforcing fibres is present in the composite system. Hence, the young modulus of a hybrid composite is the sum of two composite systems, i.e. the polymer composite reinforced with fibre 1 and the polymer composite reinforced with fibre 2. The modulus of a hybrid composite can be predicted using Equation (3.4).

$$E_{hc} = E_{c1} V_{c1} + E_{c2} V_{c2} \tag{3.4}$$

$$V_{c1} = \frac{V_{f1}}{V_t} \tag{3.5}$$

$$V_{c2} = \frac{V_{f2}}{V_t} \tag{3.6}$$

$$V_t = V_{c1} + V_{c2} \tag{3.7}$$

Where, E_{hc}, E_{c1}, E_{c2} are the modulus of the hybrid composite, composite 1 and composite 2 respectively, while V_{c1} and V_{c2} are the fibre volume fraction of the fibre 1 and fibre 2 respectively. The modulus of the hybrid composite can also be expressed using Equation (3.8).

$$E_{hc} = E_{f1} V_{f1} + E_{f2} V_{f2} + E_m (1 - V_{f1} - V_{f2}) \tag{3.8}$$

The above equations work based on the following assumptions (Venkateshwaran, Elayaperumal, and Sathiya 2012):

   i. There is no interaction between the fibres of the hybrid composite.
   ii. Fibres are well aligned, dispersed and distributed throughout the polymer matrix.
   iii. Strong adhesion between the fibre and matrix.
   iv. Fibres are aligned to composite's loading direction.

Based on fibre origins, hybridization can be performed in three different ways, i.e. by combining different synthetic fibres, or by combing different natural fibres, or by combining natural and synthetic fibres (Safri et al. 2017). Synthetic fibre-synthetic fibre hybridization is out of the scope of this present chapter; thus, two other type of fibre hybridization are discussed below.

3.3.2.3   Synthetic Fibre-Natural Fibre Hybridization

A combination of synthetic and natural fibres as composite reinforcement is mainly used to enhance the performance of the resultant material mainly in terms of reduced moisture absorption, costs of production, lower carbon footprint and environmental impact (John, and Thomas 2008). Natural fibres are hydrophilic and they swell in contact with moisture, which results in unsatisfactory composite properties. Surface modification of natural fibres can mitigate the problem but the process of natural fibre surface modification involves harsh chemicals and is energy intensive. Hybridization offers an alternative to surface treatment by combining natural fibres with synthetic hydrophobic fibres (Safri et al. 2017). This approach improves the properties of the resultant composite followed by the simultaneous reduction of water absorption. Synthetic-natural fibre hybridization is also carried out to enhance the mechanical and thermal performance of the resultant composite. Lots of work has been reported in the literature on synthetic-natural fibre hybrid composite and in most of the cases either glass or carbon or an aramid fibre is mixed with various natural fibres. Akil et al. (2014) have developed a hybrid composite by reinforcing a polyester matrix with glass and jute fibres respectively. They have observed that the addition of glass fibre in the composite system reduces the water absorption and significantly improves the tensile and flexural strength and modulus of the resultant composite.

A similar result is observed when the properties of a hybrid composite developed by reinforcing epoxy resin with flax and carbon unidirectional fabric was studied (Kureemun et al 2018; Ramana and Ramprasad 2017). Yahaya et al. (2014) have developed a hybrid composite using epoxy as matrix and Kevlar and kenaf as reinforcement. It is observed that the mechanical properties of the resultant hybrid composite decrease with increasing kenaf percentage, which is mainly due to the lower strength of the kenaf as compared to Kevlar. Some researchers have studied the synthetic-natural fibre-based hybrid composite made with the thermoplastic matrix system. For instance, Samal, Mohanty, and Nayak (2009) have produced a hybrid composite using MAgPP treated glass and banana fibres as reinforcement and PP as a matrix. They have observed that the addition of glass fibre to a banana-PP composite improved the flexural and impact strength of the resultant composite.

Haneefa et al. (2008) have studied the tensile and impact properties of the glass/banana impregnated polystyrene composite. The authors concluded that, as the compatibility between glass and polystyrene resin is better than that of the banana fibre and polystyrene resin, a boost in mechanical properties of the hybrid composite is observed with increasing volume percent of glass fibres in the composite system. However, the increase in glass fibre content decreases the elongation at break due to the brittle nature of the glass fibres, which in turn diminishes the impact properties of the resultant composite.

3.3.2.4   Natural Fibre-Natural Fibre Hybridization

In the context of ecological concern, the incorporation of a synthetic fibre in a bio-composite structure is not preferable as it affects the biodegradability of the composite. Some researchers have tried to enhance the mechanical performance of the bio-composites by using two or more types of natural fibres as reinforcement. This approach can balance the cost of natural fibre composites and can reduce the inconsistency of the natural fibre composite properties in addition to the benefit of protecting the environment. Edhirej et al. (2017) reported that the optimum tensile strength of a natural fibre-natural fibre reinforced hybrid composite can be obtained when high strain is achieved. Das (2017) has developed a hybrid composite after impregnating woven jute fabric and waste paper in polyester resin. It is observed that the mechanical properties of the hybrid composite are inferior to that of the jute-polyester composite but are superior to that of waste paper-polyester composite. The waste paper has shorter fibre length than that of the jute fibres. Because of this shorter fibre length of the waste paper the hybrid composite exhibits inferior mechanical properties than that of jute-polyester composite. Maslinda et al. (2017) have developed varieties of hybrid composites after reinforcing epoxy resin with kenaf-jute fibres and kenaf-hemp fibres. They observed that all hybrid composites exhibit higher flexural and tensile properties than that of the jute fibre reinforced epoxy composite, kenaf fibre reinforced epoxy composite and the hemp fibre reinforced epoxy composites.

At present, the development of green hybrid composites and their life-cycle study have been a research hotspot. A green hybrid composite should consist of a biodegradable polymer matrix and natural fibres. Sarasini et al. (2017) have developed a green composite after reinforcing biodegradable polycaprolactone (PCL) with ramie and borassus fibres. It is observed that the incorporation of ramie fibre in the PCL matrix improves its tensile properties. However, the addition of borassus fibres in that system does not improve its tensile properties and has insignificant influence on the hardness of the hybrid composite. Asaithambi, Ganesan, and Kumar (2014) have studied the mechanical properties of the banana/sisal fibre reinforced PLA hybrid composite and compared with those of virgin PLA laminates. It is observed that hybrid fibre reinforced PLA composites show better mechanical properties than those of virgin PLA and they further improve with the alkali and peroxide treatment of the natural fibres. Jumaidin et al. (2017) have developed a hybrid composite after reinforcing seaweed and sugar palm fibres (SPF) in starch matrix. It is observed that the hybrid composite exhibit higher tensile and flexural strength than that of individual fibre composites. In addition, the hybridization of seaweed and SPF led to slower biodegradation activity.

3.4 STRUCTURAL HYBRIDIZATION

The structural hybridization is a third way of producing hybrid composite. Unlike previous two approaches, in the present approach the composites are made of generally but not limited to one type of fibre and one type of matrix. Structural hybridization is mainly carried out to enhance the morphological behaviour of a composite system such as enhancement of fibre/matrix distribution, fibre orientation in the composite structure, decrement of void content etc. Structural hybridization can be carried out in two stages, i.e. in fabric stage and in yarn stage. The present chapter mainly focuses on yarn hybridization, i.e. on the natural fibre-based hybrid yarns used for composite manufacturing. Hybrid yarns are yarns having two or more fibre components arranged uniquely in the yarn structure. These yarns fall into the category of blended yarns, but there is a dominance of one type of fibre in one region of the yarn structure. Hence, it can be said that all hybrid yarns are blended yarns but all blended yarns are not hybrid yarns. In the context of thermoplastic polymer composites, a hybrid yarn is a yarn in which both the matrix as well as the reinforcing components are present. The main aim of a hybrid yarn is to provide uniform distribution of matrix and reinforcing fibres within the yarn structure along with a reduction in fibre damage to reinforcing fibres. Based on their structures, hybrid yarns can be broadly classified into two groups, namely, core-sheath structured hybrid yarn and mingled structured hybrid yarn (Sawhney et al. 1992; Ye et al. 1995). Natural fibre-based hybrid yarns can be manufactured by different methods, including co-wrapping, core spinning, commingling and also by the conventional spinning methods such as ring spinning, rotor spinning, wrap spinning, and friction spinning etc. Different ways of hybrid yarn manufacturing and the pros and cons of these techniques are already discussed in Chapter 2 of this book.

3.5 INFLUENCE OF TWIST ON YARN AND COMPOSITE PROPERTIES

Natural fibres (except silk) are available in a short staple form which makes their handling very difficult. Hence, these fibrous materials are often twisted together in a suitable spinning system to produce a continuous yarn. Fibres in a twisted yarn structure are held together by means of fibre-to-fibre cohesion which is derived from the helical fibrous path (twist) developed during spinning. Twist enhances the processability of the fibrous strand and has a profound influence on the yarn strength (Gegauff 1907). In a low twisted yarn, fibres are oriented more towards the yarn axis and the cohesion between fibres generated by twist is low. As a result, a low-twisted yarn fails due to fibre slippage. On the other hand, fibres in a high twisted yarn are inclined to the yarn axis and therefore fibre slippage is largely prevented by high fibre-to-fibre friction. Conversely, high twist diminishes the fibre strength contribution to the yarn strength due to fibre obliquity (Rao 1966). The influence of yarn twist on tensile properties of short fibre yarn is represented schematically in Figure 3.5.

Fibres in a composite structure are held together by means of resin which prevents the fibre slippage. Hence, the yarn strength generated from fibre friction is only needed during composite fabrication. Once the fibres are reinforced in a composite structure, the helical fibre arrangement contributes negatively to the composite's mechanical properties, which are determined by the Krenchel fibre orientation factor (η_o) (Krenchel 1964). The modified rule of mixture equation and the Krenchel fibre orientation factor (η_o) are mentioned in the Equation 3.9 and 3.10, respectively.

$$E_c = \eta_o \eta_l V_f E_f + (1 - V_f) E_m \tag{3.9}$$

$$\eta_o = \sum a_n \cos^4 \theta \tag{3.10}$$

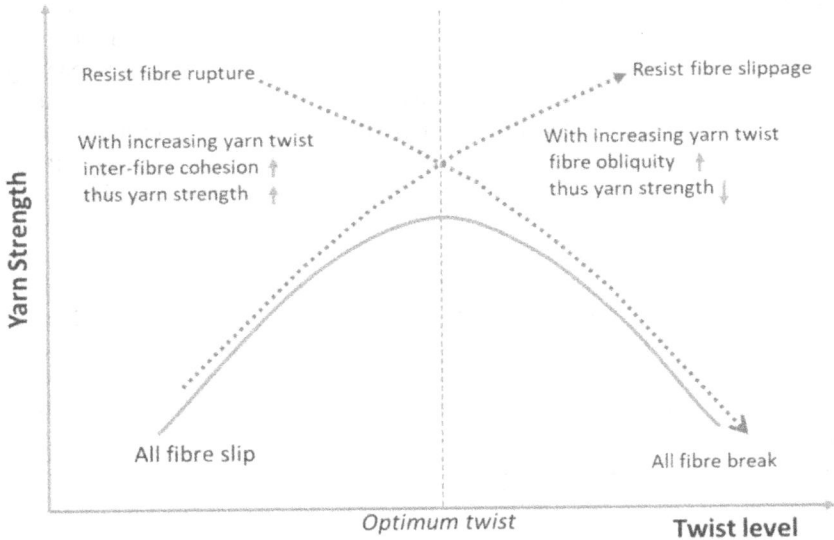

FIGURE 3.5　Influence of yarn twist on the tensile properties of short fibre yarn.

Where, E_c is the composite mechanical response, E_m and E_f are the corresponding response of the fibre and matrix respectively. V_f is the fibre volume fraction. η_l and η_o are the factors related to fibre length and orientation respectively; and a_n is the fraction of fibre oriented at an angle θ with respect to the composite loading direction. Secondly, twist reduces the yarn production rate and the twisting of short fibres is an energy consuming process which ultimately increases the cost of yarn reinforced composites. The twist also tightens the yarn structure, which hinders the matrix penetration in the reinforced yarn structure. The uneven resin penetration results in a composite with a high amount of voids resulting in a composite with inferior mechanical properties. Thus, it can be concluded that use of twist-less yarn for composite reinforcement is very useful and economical.

Some studies have already been carried out where an effort has been made to utilize full properties of the fibre to the final composite by reducing twist in the yarn. Ma, Li, and Wang (2014) have produced sisal reinforced phenolic composite using sisal yarn of different twist levels. It is observed that high twist level of the reinforcing sisal yarn impaired the mechanical properties of the resultant composites due to changes in orientation angles between the reinforcing fibres and applied load.

Baets et al. (2014) have studied the tensile behaviour of the composites made from flax fibres taken from different steps of the fibre extraction and yarn preparation process. They have observed that as the number of processes increases the dry fibre, bundle strength increases due to the high level of twist, which has a negative impact on composite properties. Gu and Miao (2014) have observed that the optimum twist for a ply yarn is 28% lower than that of a single yarn. Hence, the use of a ply yarn as reinforcing component offers better fibre orientation in the composite structure. Moothoo et al. (2014) have developed a flat flax tow as an alternate to flax spun yarn in composite processing. Fibres in the flax tows are highly oriented towards tow axis and held together by means of a natural binder. Further, flax reinforced epoxy composites are manufactured using twist-less flax tows and their mechanical properties have been investigated. Studies on natural fibre composites to understand the effect of reinforcing yarn twist on composite properties are mainly carried out using twisted yarn of different twist-levels. However, there are some technologies available for twist-less yarn manufacturing which are discussed below.

3.6 TWIST-LESS YARN

As mentioned earlier, in a conventional yarn, twist holds the constituent fibres together and imparts strength to the yarn. In case of a twist-less yarn, the binding material or adhesive works similarly to twist for a twisted yarn. Several technologies have been developed in order to produce a good quality twist-less yarn and most of these are patented. Different techniques of twist-less yarn manufacturing are discussed below.

3.6.1 ADHESIVE BONDING METHOD

Adhesive bonding is one of the important techniques for producing twist-less yarn. In this process, staple fibres are held together by means of an adhesive to produce the adhesive bonded yarn structure. A method of adhesive bonded yarn manufacturing is disclosed in British Pat. No. 1,380,004 (Werf 1975) for producing twist-less yarns of glued fibres, wherein the fibres are fitted together by adding a solution of glue to staple fibrous material before or during its drawing when wet. The drawn fibrous materials are false twisted followed by drying and winding. US Pat. No. 4038813 (Heap and Naarding 1977) explains similar methods of twist-less yarn manufacturing. However, in all these methods, the glue component cannot be uniformly distributed along the length of the fibrous material, and when it touches the rollers of the drawing apparatus and the devices for false twisting, it sticks to them, thus creating a condition wherein the guiding of fibres is uncontrolled. In the US Pat. No. 4107914 (Dimitrov 1985) a method of twist-less yarn manufacturing is disclosed where potential adhesive staple fibres are used as binding material. In Dutch Pat. No. 144679 (Dort 1975), an inactive binding element either in the form of staple fibres or filaments is added to a staple fibre strand. Later in the process, these inactive binding materials are activated and made to glue with the adjacent fibres after the application of heat or suitable solvent or both. The use of adhesive fibre or inactive binding material during twist-less yarn manufacturing can solve the processing problems such as fibre accumulation, fibre sticking and improve the adhesive bonded yarn quality. There are several techniques available for twist-less yarn manufacturing; among these the Tek-ja, Twilo, Pavena and Bobtex methods are mentionable. In the Tek-ja process, normal starch is used as a binder while in the Twilo and Pavena methods, an adhesive is used for binding fibres and a polymer is used in the Bobtex method (Lord and Mohamed 1973). However, the presence of adhesive in a yarn structure makes it inflexible and restricts the resin penetration during composite fabrication. Moreover, all the adhesive bonded yarn manufacturing processes evaporate some solvents or hazardous gases, which are not environmentally friendly, all processes are limited with regard to yarn count and they are too expensive for commercial use.

3.6.2 THERMAL BONDING METHOD

Thermal bonding is another way of manufacturing twist-less yarn. In this process, at least one component has to be a heat activated binding material. US Pat. No. 7189295 (Bowers 1986) describes a method where non-binding fibres and heat activated binding materials are twisted together either in a ring spinning or wrap spinning system. The twisted assembly is then subjected to a thermal treatment where the heat activated material is melted, followed by a cooling process to solidify the binder material. The produced yarn enhances the wear and appearance properties of a carpet when it is tufted into a carpet structure.

In the US Pat. No. 3877214 (Werf 1972), a process is disclosed where a twist-free yarn is manufactured using a sliver comprised of at least two components out of which one is a polyamide fibre, melts under a relatively low temperature and acts as a bonding component. The sliver is initially drafted to form a thinner fibrous strand and subsequently subjected to a thermal treatment to develop a thermally bonded twist-free yarn. However, this thermally bonded yarn is very rigid in nature due to the presence of a polyamide binder throughout the yarn cross-section. Hence, it is not usable for composite preforming purposes.

3.7 EVALUATION OF TOWPREG

Towpregs or hybrid yarns are a type of blended yarn. Like other blended yarns, the hybrid yarns also have two or more fibre components in their structure. The characteristics of a blended yarn depend on several factors; among those, the blending ratio and the blend homogeneity are the most important. In the context of apparel textiles, blending is done to improve the functional properties of fabric such as tensile strength, uniformity, better appearance, increased wear life, wrinkle resistance, dimensional stability, elasticity, comfort and aesthetic values etc. Besides this, blending also improves the processing performance of carding, drawing, spinning etc. For example, when a longer and finer synthetic fibre is blended with cotton it enhances its spinning limit and makes it possible to spin finer count. In a conventional short fibre spinning line (ring spinning), blending can be done in different stages, i.e. in blow room, carding, drawing, roving or during final yarn spinning. The degree of bending increases with the increasing number of processing stages.

The evenness of fibre blending in a hybrid yarn structure is assessed in two directions, i.e. in longitudinal direction and transverse direction. When there is unevenness in the longitudinal direction, successive yarn portions exhibit different percentage distributions of the individual fibre components as shown in Figure 3.6.

Where there is unevenness in the transverse direction, the fibres are poorly distributed in the yarn section as shown in Figure 3.7. This irregularity leads to an uneven appearance of the finished product. The determination of the evenness of a blend of synthetic and natural fibres is a troublesome task. Hence, in the mean process, one component is usually dissolved out or coloured differently.

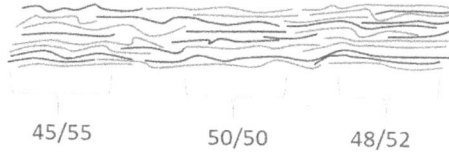

45/55 50/50 48/52

FIGURE 3.6 Unevenness of the hybrid yarn in the longitudinal direction.

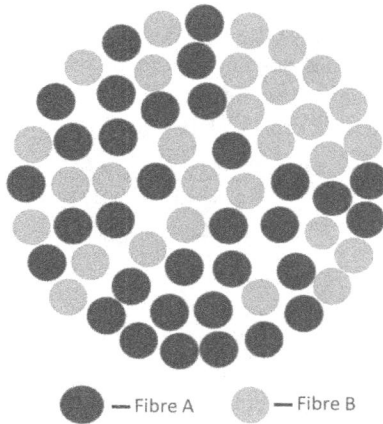

● —Fibre A ● —Fibre B

FIGURE 3.7 Unevenness of the blend in the transverse direction.

3.7.1 Indices of Blending

The indices of blending are generally expressed in terms of degree of blending and Blend irregularity, which are discussed below.

3.7.2 Degree of Mixing

This is a statistical parameter, which estimates the inherent intimacy of a blend. DeBarr and Walker (1957) made a series of yarns from a blend of black and white fibres and examined their cross-sections for the number of groups of fibres. They assumed that the fibre distribution in yarn sections would become random as the number of doublings approached ∞. If g_∞ is the number of groups of white fibres, one can write,

$$g_\infty = np(1-p)$$ (3.11)

Where n is the number of fibres in the average cross-section of the yarn and p is the average proportion of white fibres in the blend. In practice, the number of groups of white fibres is less than g_∞ since the number of doublings in use never achieves perfect randomization. If g represents the actual number of groups of white fibres, the degree of mixing (γ) is given as,

$$\gamma = g/g_\infty$$ (3.12)

A value of 1 for γ shows a thorough random distribution of fibres. A value of less than 1 means less than random mixing and a value of more than 1 means that the mixing is better than random. As this value increases, the blend approaches a perfect blend. In practice, the value of γ is less than 1. Finer yarns therefore require fewer doublings to achieve random distribution (Townsend and Cox 1951).

3.7.3   BLEND IRREGULARITY

This is another statistical measure used to assess the degree of randomness of fibre distribution. This index was developed by Coplan and Bloch (1955) for variation in the blend proportion against the theoretical value for random mixing. Blend irregularity is generally expressed in terms of IBI value.

$$\text{IBI} = \sqrt[2]{\frac{1}{m} \sum \frac{(T_i p - W_i)}{T_i pq}} \qquad (3.13)$$

Where T_i = total no. of fibre in a given cross-section.

W_i = no. of fibre of component W at that section
P = average fraction of component w for all sections
q = $(1 - p)$
M = number of sections examined

An IBI value of unity represents complete randomness and a value more than 1 means less homogeneity than complete randomness while an IBI value of zero represents a perfect blending. This index can also be used to calculate the fibre cluster size. An index of 1.3 has the physical meaning that the blend suffers from causative factors that increase the homogeneity by 30% over what can be expected from a purely random process operating on single fibre elements. By considering fibres to be in clusters, replacing Ti by Ti/C where C is cluster size (no. of fibres in the cluster) and substituting 1 for IBI, an estimate of cluster size can be obtained. Coplan and Bloch (1955) expanded on the longitudinal variation method and developed two other methods of evaluation viz., radial distribution and rotational distribution of fibres.

    a. *Radial distribution*
       This describes the fibre motion across the cross-sections of the yarn. To estimate this, the yarn cross-section is divided into usually four concentric circles of either equal area or thickness. Figure 3.8 shows circles of equal thickness along with the blend ratio plotted in the form of a bar diagram. Ideally, the blend ratio should be the same in all four zones from inside to outside.
    b. *Rotational distribution*
       For estimating the variation in the rotational distribution, the yarn cross-section is divided into four to six segments. The blend ratio is calculated and plotted as bar diagram as shown in Figure 3.9. Ideally the ratio should be the same for all the segments.

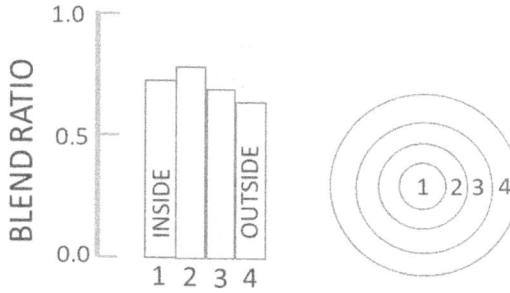

FIGURE 3.8   Radial distribution of fibres.

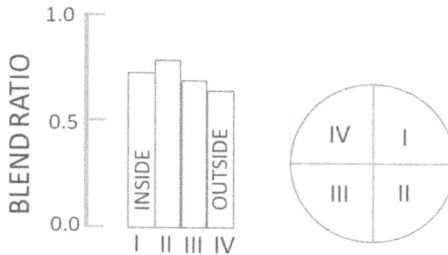

FIGURE 3.9   Rotational distribution of fibres.

3.8   NATURAL FIBRE-BASED HYBRID YARN COMPOSITES

Among all yarn hybridization techniques, commingling is one of the simplest and most cost-effective methods. Natural fibre has shorter fibre length, hence the mingling of the reinforcing and matrix forming components can be done either in the fibre state or the yarn state. George et al. (2012) have developed a bio-commingled composite using PP filaments and twisted jute yarn, where both components are mixed intimately to avoid the damage of the natural fibres. Bio-commingling improves the wetting of reinforcing jute fibres, which ultimately results in better stiffness and load-bearing capacity of the resultant composites than their pure matrix counterpart. Aslan (2013) has manufactured a similar composite structure by commingling flax and polyethylene terephthalate (PET) yarns in a winding machine. The tensile failure and the fracture behaviour of these composites are characterized by acoustic emission techniques. It is observed that the flax/PET composites show high strength, which is suitable for low-load-bearing structural applications, but with increasing flax content the tensile strength and stiffness decreases due to high void content. In another study, Angelov et al. (2007) have developed pultruded composite using a flax/PP commingled yarn. The discontinuous flax and PP fibres are mixed at the fibre level and twisted together to manufacture the commingled yarn. Further, they have investigated the effect of different pultrusion parameters on various composite properties.

Although the use of commingled yarn during composite manufacturing improves the mechanical properties of the resultant composites, this method is suitable for producing lab scaled simple structures only. Researchers have developed natural fibre/thermoplastic resin-based micro-braided yarns to overcome the issues of using commingled hybrid yarns as composite reinforcing material.

The hybrid yarn made through micro-braiding techniques has a stable, core-sheath type yarn structure and it is drawing more attention from the research community as it is suitable for producing any complex structured textile preform. Khondker et al. (2006) have developed a micro-braided yarn using twisted jute yarn as a core and PP filament as a sheath and produced a unidirectional composite sample after consolidating the same. It is observed that braided yarn improves the wettability of the reinforcing jute yarn, which ultimately enhances the tensile and flexural strength of the resultant composites. They have also found that along with fibre wetting, the other factors such as fibre orientation and fibre/matrix distribution also play a major role in determining the tensile and flexural properties of the present composite sample. Kobayashi and Takada (2013) have also observed similar kinds of improvement in case of hemp/PLA braided yarn unidirectional composites made using compression moulding. Memon and Nakai (2013) have pultruded jute-PLA braided yarn to produce tubular composites and compared the same with jute-PLA commingled yarn pultruded tubular composites. The four-point bending behaviour and the fibre-resin distribution in both composites are evaluated. It is observed that braided yarn pultrusion developed a more uniform composite structure with fewer efforts.

Kobayashi, Takada, and Nakamura (2014) have developed hemp/PLA micro braided yarn-based plain woven fabric compressed composites and compared it with cross-ply laminates and weft knitted fabric compressed composites. It is observed that cross-ply laminates exhibit better mechanical properties than that of textile reinforced composites due to undulated structures with textile composites. Although the use of hybrid braided yarn as reinforcing component significantly improves the mechanical properties of the resultant composites through enhancing resin penetration and fibre/resin distribution in the composite structure, these methods are not commercially viable as the rate of production of micro-braided yarn is very low as compared to other hybrid yarn manufacturing methods. Moreover, in most of these above-mentioned studies, the researchers have used natural fibres in twisted yarn form, which is not preferable for a high strength composite. The effect of yarn twist on composite properties will be discussed in the later sections of this book.

Zhang, and Miao (2010) have designed and fabricated a composite using flax/PP commingled fibre-based wrap spun yarn in which the flax fibres remain in a twist-less state. It is observed that the wrap yarn significantly improves fibre/resin distribution in the composite structure, which enhances the flexural performance of the resulting composite. Baghaei, Skrifvars, and Berglin (2013) have developed a similar kind of wrap spun yarn structure using hemp and PLA fibres. They have arranged the wrap spun yarns in a 0/90 bidirectional way and consolidated them in a compression molding machine under heat and pressure to manufacture the resultant composites. It is observed that the tensile, flexural and Izod impact properties of the resultant composites improve significantly as compared to neat PLA. However, in the process of fabric formation, wrap spun yarns are not suitable as warp and even as weft in high-speed weaving due to their poor mechanical behaviour. The poor weaving performance of this hybrid wrap spun yarn limits its potential as a composite reinforcing component.

3.9 SUMMARY

At present, the natural fibre composites are gaining the interest of the composite researchers and the manufacturers, but they have some drawbacks. Non-uniform fibre resin distribution in the composite structure and inadequate resin penetration in the reinforced fibre bundles are some such drawbacks. Use of hybrid towpregs during composite manufacturing and other hybridization approaches can overcome the above problem. During bio-composite manufacturing hybridization can be performed in three ways, which are matrix hybridization, reinforcement hybridization and structural hybridization. Hybrid towpreg can be produced through different methods. Among those methods, the hybrid towpregs produced through the DREF spinning method and through the wrap spinning methods are the most suitable structures for composite reinforcement. The natural fibre in the existing hybrid yarn structure is present in a twisted form. Twist reduces the reinforcing efficiency of the natural fibres; thus twist-less yarns are preferred for the composite reinforcement. Approaches for hybrid towpreg evaluation and the influence of different natural fibre-based hybrid towpregs on composite properties studied by various researchers are discussed in this present chapter.

REFERENCES

Akil, H. M., C. Santulli, F. Sarasini, J. Tirillò, and T. Valente. 2014. Environmental effects on the mechanical behaviour of pultruded jute/glass fibre-reinforced polyester hybrid composites. *Composite Science and Technology* 94: 62–70. doi.org/10.1016/j.compscitech.2014.01.017

Alagirusamy, R., R. Fangueiro, V. Ogale, and N. Padaki. 2006. Hybrid yarns and textile preforming for thermoplastic composites. *Textile Progress* 38(4): 1–71. doi.org/10.1533/tepr.2006.0004

Anakabe, J., A. M. Zaldua Huici, A. Eceiza, and A. Arbelaiz. 2015. Melt blending of polylactide and poly (methyl methacrylate): thermal and mechanical properties and phase morphology characterization. *Journal of Applied Polymer Science* 132: 42677. doi.org/10.1002/app.42677

Angelov, I., S. Wiedmer, M. Evstatiev, K. Friedrich, and G. Mennig. 2007. Pultrusion of a flax/polypropylene yarn. *Composites Part A: Applied Science and Manufacturing* 38: 1431–38. doi.org/10.1016/j.compositesa.2006.01.024

Arrieta, M. P., E. Fortunati, F. Dominici, J. López, and J. M. Kenny. 2015. Bionanocomposite films based on plasticized PLA-PHB/cellulose nanocrystal blends. *Carbohydrate Polymers* 121: 265–75. doi.org/10.1016/j.carbpol.2014.12.056

Asaithambi, B., G. Ganesan, and S. A. Kumar. 2014. Bio-composites: Development and mechanical characterization of banana/sisal fibre reinforced poly lactic acid (PLA) hybrid composites. *Fibers and Polymers* 15(4): 847–54. doi.org/10.1007/s12221-014-0847-y

Aslan, M. 2013. Investigation of damage mechanism of flax fibre LPET commingled composites by acoustic emission. *Composite Part B: Engineering* 54: 289–97. doi.org/10.1016/j.compositesb.2013.05.042

Baets, J., D. Plastria, J. Ivens, and I. Verpoest. 2014. Determination of the optimal flax fibre preparation for use in unidirectional flax–epoxy composites. *Journal of Reinforced Plastics and Composites* 33: 493–502. doi.org/10.1177/0731684413518620

Baghaei, B., M. Skrifvars, and L. Berglin. 2013. Manufacture and characterisation of thermoplastic composites made from PLA/hemp co-wrapped hybrid yarn prepregs. *Composites Part A: Applied Science and Manufacturing* 50: 93–101. doi.org/10.1016/j.compositesa.2013.03.012

Bar, M., R. Alagirusamy, and A. Das. 2017. Studies on low-twist flax-pp based hybrid yarns for thermoplastic composite reinforcement. *Journal of Reinforced Plastics and Composites* 36(11): 818–31. doi.org/10.1177/0731684417693428

Bowers, C. E., 1986. United States Patent No. 7189295, U.S. Patent.

Coplan, M. J., and M. G. Bloch. 1955. A study of blended woolen structures: Part II: Blend distribution in some wool-nylon and wool-viscose yarns. *Textile Research Journal* 25(11): 902–22.

Das, S. 2017. Mechanical properties of waste paper/jute fabric reinforced polyester resin matrix hybrid composites. *Carbohydrate Polymers* 172: 60–7. doi.org/10.1016/j.carbpol.2017.05.036

de Farias, J. G. G., R. C. Cavalcante, B. R. Canabarro, H. M. Viana, S. Scholz, and R.A. Simão. 2017. Surface lignin removal on coir fibers by plasma treatment for improved adhesion in thermoplastic starch composites. *Carbohydrate Polymers* 165. 429–36. doi.org/10.1016/j.carbpol.2017.02.042

Debarr, A. E., and P. G. Walker. 1957. A measure of fiber distribution in blend yarns and its application to the determination of the degree of mixing achieved in different processes. *Journal of Textile Institute* 58. 405. doi.org/10.1080/19447025708660101

Dimitrov, M. D. 1985. United States Patent No. 4499717, U.S. Patent.

Dort, J. M. V. 1975. Dutch Patent No. 144679, Netherland.

Edhirej, A., S. M. Sapuan, M. Jawaid, and N. I. Zahari. 2017. Cassava/sugar palm fiber reinforced cassava starch hybrid composites: Physical, thermal and structural properties. *International Journal of Biological Macromolecules* 101: 75–83. doi.org/10.1016/j.ijbiomac.2017.03.045

Essabir, H., M. O. Bensalah, D. Rodrigue, R. Bouhfid, and A. Qaiss. 2016. Structural, mechanical and thermal properties of bio-based hybrid composites from waste coir residues: Fibers and shell particles. *Mechanics of Materials* 93: 134–44. doi.org/10.1016/j.mechmat.2015.10.018

Faruk, O., A. K. Bledzki, H. P. Fink, and M. Sain. 2012. Biocomposites reinforced with natural fibers: 2000–2010. *Progress in Polymer Science* 37: 1552–96. doi.org/10.1016/j.progpolymsci.2012.04.003

Fortea-Verdejo, M., E. Bumbaris, C. Burgstaller, A. Bismarck, and K. Y. Lee. 2017. Plant fibre-reinforced polymers: where do we stand in terms of tensile properties?. *International Materials Reviews* 62(8): 1–24. doi.org/10.1080/09506608.2016.1271089

Fu, S. Y., G. Xu, and Y. W. Mai. 2002. On the elastic modulus of hybrid particle/short-fiber/polymer composites. *Composites Part B: Engineering* 33(4): 291–99. doi.org/10.1016/S1359-8368(02)00013-6

Fukuda, H. 1984. An advanced theory of the strength of hybrid composites. *Journal of Materials Science* 19(3): 974–82. doi.org/10.1007/BF00540468

Fukunaga, H., C. Tsu-Wei, and H. Fukuda. 1984. Strength of intermingled hybrid composites. *Journal of Reinforced Plastics and Composites* 3(2): 145–60. doi.org/10.1177/073168448400300204

Garcia, S. J. 2014. Effect of polymer architecture on the intrinsic self-healing character of polymers. *European Polymer Journal* 53: 118–25. doi.org/10.1016/j.eurpolymj.2014.01.026

Gegauff, G. 1907. Force et elasticite des files en cotton. *Bulletin de la Société Industrielle de Mulhouse* 77: 153. doi.org/10.1016/j.eurpolymj.2014.01.026

George, G., E. T. Jose, K. Jayanarayanan, E. R. Nagarajan, M. Skrifvars, and K. Joseph. 2012. Novel bio-commingled composites based on jute/polypropylene yarns: Effect of chemical treatments on the mechanical properties. *Composites Part A: Applied Science and Manufacturing* 43: 219–30. doi.org/10.1016/j.compositesa.2011.10.011

Gu, H., and M. Miao. 2014. Optimising fibre alignment in twisted yarns for natural fibre composites. *Journal of Composite Materials* 48: 2993–3002. doi.org/10.1177/0021998313504322

Haneefa, A., P. Bindu, I. Aravind, and S. Thomas. 2008. Studies on tensile and flexural properties of short banana/glass hybrid fiber reinforced polystyrene composites. *Journal of Composite Materials* 42(15): 1471–89. doi.org/10.1016/j.eurpolymj.2014.01.026

Hayes, S. A., W. Zhang, M. Branthwaite, and F. R. Jones. 2007. Self-healing of damage in fibre-reinforced polymer-matrix composites. *Journal of the Royal Society Interface* 4: 381–87. doi.org/10.1098/rsif.2006.0209

Heap, S. A., and W. J. Naarding. 1977. United States Patent No. 4038813, U.S. Patent.

Jawaid, M., and H. K. Abdul. 2011. Cellulosic/synthetic fibre reinforced polymer hybrid composites: A review. *Carbohydrate Polymers* 86: 1–18. doi.org/10.1016/j.carbpol.2011.04.043

Jayaraman, K. 2003. Manufacturing sisal-polypropylene composites with minimum fibre degradation. *Composite Science and Technology* 63: 367–74. doi.org/10.1016/S0266-3538(02)00217-8

John, M. J., and S. Thomas. 2008. Biofibres and biocomposites. *Carbohydrate Polymers* 71: 343–64. doi.org/10.1016/j.carbpol.2007.05.040

Jones, K. D., and A. T. Dibenedetto. 1994. Fiber fracture in hybrid composite systems. *Composite Science and Technology* 51(1): 53–62. doi.org/10.1016/0266-3538(94)90156-2

Jumaidin, R., S. M. Sapuan, M. Jawaid, M. R. Ishak, and J. Sahari. 2017. Thermal, mechanical, and physical properties of seaweed/sugar palm fibre reinforced thermoplastic sugar palm starch/agar hybrid composites. *International Journal of Biological Macromolecules* 97: 606–15. doi.org/10.1016/j.ijbiomac.2017.01.079

Khondker, O. A., U. S. Ishiaku, A. Nakai, and H. A. Hamada. 2006. Novel processing technique for thermoplastic manufacturing of unidirectional composites reinforced with jute yarns. *Composites Part A: Applied Science and Manufacturing* 37, 2274–84. doi.org/10.1016/j.compositesa.2005.12.030

Kobayashi, S., and K. Takada. 2013. Processing of unidirectional hemp fiber reinforced composites with micro-braiding technique. *Composites Part A: Applied Science and Manufacturing* 46: 173–9. doi.org/10.1016/j.compositesa.2012.11.012

Kobayashi, S., K. Takada, and R. Nakamura. 2014. Processing and characterization of hemp fiber textile composites with micro-braiding technique. *Composites Part A: Applied Science and Manufacturing* 59: 1–8. doi.org/10.1016/j.compositesa.2013.12.009

Krenchel, H. 1964. Fibre reinforcement, theoretical and practical investigations of the elasticity and strength of fibre-reinforced materials. Copenhagen: Akademisk Forlag.

Kretsis, G. 1987. A review of the tensile, compressive, flexural and shear properties of hybrid fibre-reinforced plastics. *Composites* 18: 13–23. doi.org/10.1016/0010-4361(87)90003-6

Kureemun, U., M. Ravandi, L. Q. N. Tran, W. S. Teo, T. E. Tay, and H. P. Lee. 2018. Effects of hybridization and hybrid fibre dispersion on the mechanical properties of woven flax-carbon epoxy at low carbon fibre volume fractions. *Composite Part B: Engineering* 134: 28–38. doi.org/10.1016/j.compositesb.2017.09.035

Larguech, S., A. Triki, M. Ramachandran, and A. Kallel. 2020. Dielectric properties of jute fibers reinforced poly (lactic acid)/Poly (butylene succinate) blend matrix. *Journal of Polymers and the Environment* 29: 1–7.

Lord, P. R., and M. H. Mohamed. 1973. Twist-less yarns and woven fabrics. *Textile Research Journal* 43, 96–102. doi.org/10.1177/004051757304300206

Luo, X., R. Ou, D. E. Eberly, A. Singhal, W. Viratyaporn, and P. T. Mather. 2009. A thermoplastic/thermoset blend exhibiting thermal mending and reversible adhesion. *ACS Applied Materials and Interfaces* 1(3): 612–20. doi.org/10.1021/am8001605

Ma, H., Y. Li, and D. Wang. 2014. Investigations of fiber twist on the mechanical properties of sisal fiber yarns and their composites. *Journal of Reinforced Plastics and Composites* 33: 687–96. doi.org/10.1177/0731684413520187

Mahmood, H., A. Dorigato, and A. Pegoretti. 2020. Thermal mending in novel epoxy/cyclic olefin copolymer blends. *Express Polymer Letters* 14: 368–83. doi.org/10.3144/expresspolymlett.2020.31

Maslinda, A. B., M. A. Majid, M. J. Ridzuan, M. Afendi, and A. G. Gibson. 2017. Effect of water absorption on the mechanical properties of hybrid interwoven cellulosic-cellulosic fibre reinforced epoxy composites. *Composite Structures* 167: 227–37. doi.org/10.1016/j.compstruct.2017.02.023

Memon, A., and A. Nakai. 2013. The processing design of jute spun yarn/PLA braided composite by pultrusion molding. *Advances in Mechanical Engineering* 5: 816513. doi.org/10.1155/2013/816513

Meure, S., D. Y. Wu, and S. Furman. 2009. Polyethylene-co-methacrylic acid healing agents for mendable epoxy resins. *Acta Materialia* 57(14): 4312–20. doi.org/10.1016/j.actamat.2009.05.032

Mishnaevsky Jr., L., and G. Dai. 2014. Hybrid carbon/glass fiber composites: Micromechanical analysis of structure–damage resistance relationships. *Computational Materials Science* 81: 630–40. doi.org/10.1016/j.commatsci.2013.08.024

Moothoo, J., S. Allaoui, P. Ouagne, and D. Soulat. 2014. A study of the tensile behaviour of flax tows and their potential for composite processing. *Materials and Design* 55: 764–72. doi.org/10.1016/j.matdes.2013.10.048

Orue, A., A. Eceiza, and A. Arbelaiz. 2018. Preparation and characterization of poly (lactic acid) plasticized with vegetable oils and reinforced with sisal fibers. *Industrial Crops and Products* 112: 170–80. doi.org/10.1016/j.indcrop.2017.11.011

Pan, N., and R. Postle. 1996. The tensile strength of hybrid fibre composites: a probabilistic analysis of the hybrid effects. *Philosophical Transactions of the Royal Society of London. Series A: Mathematical, Physical and Engineering Sciences* 354(1714): 1875–97. doi.org/10.1098/rsta.1996.0082

Pickering, K. L., M. G. A. Efendy and T.M. Le. 2016. A review of recent developments in natural fibre composites and their mechanical performance. *Composites Part A: Applied Science and Manufacturing* 83: 98–112. doi.org/10.1016/j.compositesa.2015.08.038

Ramana, M. V., and S. Ramprasad. 2017. Experimental investigation on jute/carbon fibre reinforced epoxy based hybrid composites. *Materials Today Proceedings* 4: 8654–64. doi.org/10.1016/j.matpr.2017.07.214

Rao, R. N. 1966. Estimation of single yarn twist and the relation between the estimated and mechanical twists. *Textile Research Journal* 36: 65–9. doi.org/10.1177/004051756603600108

Reddy, B. M., R. M. Reddy, D. M. Reddy, N. Ananthakrishna, and P. V. Reddy. 2020. Development and characterisation of Cordia Dichotoma Fibre/Granite filler reinforced polymer blended (Epoxy/Polyester) hybrid composites. *Advances in Materials and Processing Technologies* 1–17. doi.org/10.1080/2374068X.2020.1815138

Rong, M. Z., M. Q. Zhang, Y. Liu, G. C. Yang, and H. M. Zeng. 2001.The effect of fiber treatment on the mechanical properties of unidirectional sisal-reinforced epoxy composites. *Composites Science and Technology* 61: 1437–47. doi.org/10.1016/S0266-3538(01)00046-X

Safri, S. N. A., M. T. H. Sultan, M. Jawaid, and J. Kandasamy. 2017. Impact behaviour of hybrid composites for structural applications: a review. *Composite Part B: Engineering* 133: 112–21. doi.org/10.1016/j.compositesb.2017.09.008

Samal, S. K., S. Mohanty, and S. K. Nayak. 2009. Banana/glass fiber-reinforced polypropylene hybrid composites: Fabrication and performance evaluation. *Polymer-Plastics Technology and Engineering* 48(4): 397–414. doi.org/10.1080/03602550902725407

Sarasini, F., J. Tirillò, D. Puglia, F. Dominici, C. Santulli, K. Boimau, T. Valente, and L. Torre. 2017. Biodegradable polycaprolactone-based composites reinforced with ramie and borassus fibres. *Composite Structures* 167: 20–9. doi.org/10.1016/j.compstruct.2017.01.071

Satyanarayana, K. G., K. K. Ravikumar, K. Sukumaran, P. S. Mukherjee, S. G. K. Pillai, and A. K. Kulkarni. 1986. Structure and properties of some vegetable fibres. Part 3. Talipot and palmyrah fibres. *Journal of Materials Science* 21: 57–63. doi.org/10.1016/0026-0800(86)90073-X

Sawhney, A. P. S., G. F. Ruppenicker, L. B. Kimmel, and K. Q. Robert. 1992. Comparison of filament-core spun yarns produced by new and conventional methods. *Textile Research Journal* 62: 67–73. doi.org/10.1177/004051759206200202

Sultan, J. N., and F. J. McGarry. 1973. Effect of rubber particle size on deformation mechanisms in glassy epoxy. *Polymer Engineering and Science*. 13: 29–34. doi.org/10.1002/pen.760130105

Swolfs, Y., L. Gorbatikh, and I. Verpoest. 2014. Fibre hybridisation in polymer composites: A review. *Composites Part A: Applied Science and Manufacturing* 67: 181–200. doi.org/10.1016/j.compositesa.2014.08.027

Townsend, M. W. H., and D. R. Cox. 1951. The analysis of yarn irregularities. *Journal of Textile Institute* 42: 107–13. doi.org/10.1080/19447015108663777

Väisänen, T., O. Das, and L. Tomppo. 2017. A review on new bio-based constituents for natural fiber-polymer composites. *Journal of Cleaner Production* 149: 582–96. doi.org/10.1016/j.jclepro.2017.02.132

Van der Zwaag, S., A. M. Grande, W. Post, S. J. Garcia, and T. C. Bor. 2014. Review of current strategies to induce self-healing behaviour in fibre reinforced polymer based composites. *Materials science and technology* 30(13): 1633–41. doi.org/10.1179/1743284714Y.0000000624

Venkateshwaran, N., A. Elayaperumal, and G. K. Sathiya. 2012. Prediction of tensile properties of hybrid-natural fiber composites. *Composite Part B: Engineering* 43: 793–96. doi.org/10.1016/j.compositesb.2011.08.023

Werf, H. A. V. 1972. United States Patent No 3877214, U.S. Patent.

Werf, H. A. V. 1975. Great Britain Patent No. 1380004A, London: G. B. Patent.

Wu, D. Y., S. Meure, and D. Solomon. 2008. Self-healing polymeric materials: A review of recent developments. *Progress in Polymer Science* 33: 479–522. doi.org/10.1016/j.progpolymsci.2008.02.001

Xie, Y., C. A. S. Hill, Z. Xiao, H. Militz, and C. Mai. 2010. Silane coupling agents used for natural fiber/polymer composites: a review. *Composites Part A: Applied Science and Manufacturing* 41: 806–19. doi.org/10.1016/j.compositesa.2010.03.005

Xing, J., G. C. Hsiao, and T. W. Chou. 1981. A dynamic explanation of the hybrid effect. *Journal of Composite Materials* 15: 443–61. doi.org/10.1177/002199838101500504

Yahaya, R., S. M. Sapuan, M. Jawaid, Z. Leman, and E. S. Zainudin. 2014. Quasi-static penetration and ballistic properties of kenaf–aramid hybrid composites. *Materials & Design* 63: 775–82. doi.org/10.1016/j.matdes.2014.07.010

Ye, L., K. Friedrich, J. Kastel, Y. M. Mai. 1995. Consolidation of unidirectional CF/PEEK composites from commingled yarn prepreg. *Composite Science and Technology* 54: 349–58. doi.org/10.1016/0266-3538(95)00061-5

Yu, L., K. Dean, and L. Li. 2006. Polymer blends and composites from renewable resources. *Progress in Polymer Science* 31: 576–602. doi.org/10.1016/j.progpolymsci.2006.03.002

Yusoff, R. B., H. Takagi, and A. N. Nakagaito. 2016. Tensile and flexural properties of poly-lactic acid-based hybrid green composites reinforced by kenaf, bamboo and coir fibers. *Industrial Crops and Products* 94: 562–73. doi.org/10.1016/j.indcrop.2016.09.017

Zhang, L., and M. Miao. 2010. Commingled natural fibre/polypropylene wrap spun yarns for structured thermoplastic composites. *Composite Science and Technology* 70: 130–135. doi.org/10.1016/j.compscitech.2009.09.016

Zhong, N., and W. Post. 2015. Self-repair of structural and functional composites with intrin-sically self-healing polymer matrices: A review. *Composites Part A: Applied Science and Manufacturing* 69: 226–239. doi.org/10.1016/j.compositesa.2014.11.028

Zweben, C. 1977. Tensile strength of hybrid composites. *Journal of Material Science* 12(7): 1325–37. doi.org/10.1007/BF00540846

4 Properties of Bio-Composites Made with Thermally Bonded Roving Towpregs

Mahadev Bar[a], R. Alagirusamy[b], and Apurba Das[b]
[a]Université de Toulouse
Tarbes, France
[b]Indian Institute of Technology Delhi
New Delhi, India

CONTENTS

4.1 INTRODUCTION

Over the last few decades, in the arena of fibre reinforced polymer composite, an increasing interest has been observed in the use of natural fibres over glass and other synthetic fibres as composite reinforcements (Faruk et al. 2012). It is mainly due to

DOI: 10.1201/9781003049715-4

the increasing environmental awareness and the matchless attributes of the natural fibres over the synthetic fibres such as low cost, abundant availability, high specific properties etc. (Pickering, Efendy, and Le 2016). These factors resulted in huge consumption of natural fibre composites (NFCs) in last several years. Although NFCs have very good potential in many sectors ranging from consumer goods to construction, but their present consumption in all sectors except automobile sector is minimal (Wambua, Ivens, and Verpoest 2003). It is mainly due to the inadequate mechanical performances of the NFCs. Mechanical properties of the fibre reinforced polymer composites are dictated by the nature of the fibre-matrix interface, properties of the constituents and by the fibre architecture, which includes fibre volume fraction, fibre orientation, fibre geometry and their packing arrangement in the composite structure (Netravali and Chabba 2003; John and Thomas 2008). In case of thermoplastic NFCs, resin penetration in the reinforcing fibre bundle plays an important role. Often, the manufacturer uses heavy mould or high processing temperature to improve the fibre resin distribution in the composite structure. However, the high processing temperature degrades the reinforcing natural fibre thermally and the use of heavy mould disturbs the alignment of the reinforcing fibres in the composite structures. In both cases, it ultimately resulted a composite with poor mechanical properties. To overcome this problem, researchers suggest using hybrid yarn or hybrid yarn-based textile preform for fibre reinforced thermoplastic composite fabrication. Different types of hybrid yarn and their influence on various composite properties have been described in the other chapters of this book. However, the present chapter discusses about the development of thermally bonded roving (TBR), which is a novel hybrid yarn, and its influence on bio-composite properties. The above-mentioned TBR is made up of flax and PP fibres which are highly aligned to the roving axis.

4.2 SHORTCOMINGS IN TWISTED HYBRID YARNS AS COMPOSITE REINFORCEMENT

Hybrid yarn is a yarn which contains both the matrix forming and reinforcing components in its structure. During composite fabrication, hybrid yarn reduces the effective polymer flow distance and ultimately improves the thermoplastic polymer distribution in the composite structure (Alagirusamy et al. 2006; Bar, Alagirusamy, and Das 2017). Some studies on natural fibre-based hybrid yarn reinforced thermoplastic composites have already been carried out where the hybrid yarns are manufactured through DREF spinning (Bar, Alagirusamy, and Das 2018a, 2018b), wrap spinning (Zhang and Miao 2010; Baghaei, Skrifvars, and Berglin 2013), microbraiding (Khondker et al. 2006) and bio-commingling (George et al. 2012) methods respectively. The use of hybrid yarns during composite manufacturing improves the thermoplastic resin distribution in the composite structure and eventually enhances the mechanical performances of the resultant composite.

The natural fibres in the above hybrid yarn structures are present in a twisted state, which is not an ideal reinforcement for the fibre reinforced composites (Khondker et al. 2006). Zhang and Miao (2010) have designed and fabricated a composite using flax-PP-based comingled wrap spun yarn where both the flax and PP fibres are in a twist-free state and a PP filament is spirally wrapped around the assembly of fibres, binding them together to provide the yarn integrity. A similar study is carried out by Miao and Shan (2011), where the flax-PP blended sliver is subjected to a thermal treatment at 140°C before wrapping. Since the filament binding replaces twist in a wrap spun yarn, better utilization of fibre strength to the final composite is possible (Zhang and Miao 2010). However, a disadvantage of the wrapped yarn structure is that the assembly of the fibres become crimped or wavy owing to the tension of the wrapping thread. Therefore, full utilisation of the natural fibre mechanical properties is still not achieved (Akonda, Shah, and Gonga 2020).

Further, in the process of woven fabric formation, wrap spun yarns are not suitable as warp and even as weft in high-speed weaving due to their poor abrasion resistance (Baghaei, Skrifvars, and Berglin 2015). Thermally bonded yarns have very good potential as composite reinforcing components. McGregor et al. (2017) have manufactured a pre-consolidated tape using flax yarn (twisted) and PLA binder. This tape yarn has good potential as a reinforcing component for bio-composite but it cannot produce a composite with optimum mechanical properties as the starting flax yarn has a twisted structure. In some recent studies a semi-consolidated flax-PP tape is developed in which the flax fibres are highly aligned to the tape axis. This tape yarn has good potential as composite reinforcing component but is not good for textile preforming as it is rigid by nature due to the presence of thermoplastic resin throughout the tape yarn cross-section (Akonda, Shah, and Gonga 2020).

4.3   SURFACE MODIFICATION OF FLAX FIBRES

Natural fibres are hydrophilic and they do not adhere strongly with the hydrophobic thermoplastic matrices. This results in poor fibre-matrix interfacial bonding leading to a composite with poor mechanical properties (Torres and Cubillas 2005; de Farias et al. 2017). Hence, prior to composite fabrication, the natural fibre surfaces are often modified with suitable chemical, physical or enzymatic treatment. In this study, the flax fibres are treated with 5 wt. % MAgPP to improve the flax-PP interaction in composites. The MAgPP treatment of flax slivers are carried out in hank form in a boiling xylene medium at 140°C for 5–7 min. After the treatment, flax hanks are washed a number of times with hot xylene at a temperature just above 90°C to remove the unfixed MAgPP from the flax surface. Finally, the washed flax samples are dried in an air oven at 110°C overnight. The SEM images of the untreated and MAgPP treated flax fibres are shown in Figure 4.1.

The SEM images show that untreated flax fibres have a clean surface while after MAgPP treatment, a continuous layer of MAgPP as well as some micro particles of MAgPP are observed over the flax fibre surface. Figure 4.2 illustrates the IR-spectra of MAgPP treated and untreated flax fibres. The IR-spectra of the untreated flax fibre displays strong peaks located at 3376 cm^{-1}, 2896 cm^{-1} and 1642 cm^{-1}, which are assigned to the hydroxyl stretching vibration, C-O stretching vibration and C=O stretching vibration of the alpha-keto carbonyl groups respectively (Ma, Li, and Luo 2011).

These functional groups are mainly present in the cellulose and lignin (although the flax cell-walls have very low lignin content) of the flax fibre. The IR spectra of MAgPP treated flax fibres indicates that the peak assign for hydroxyl stretching

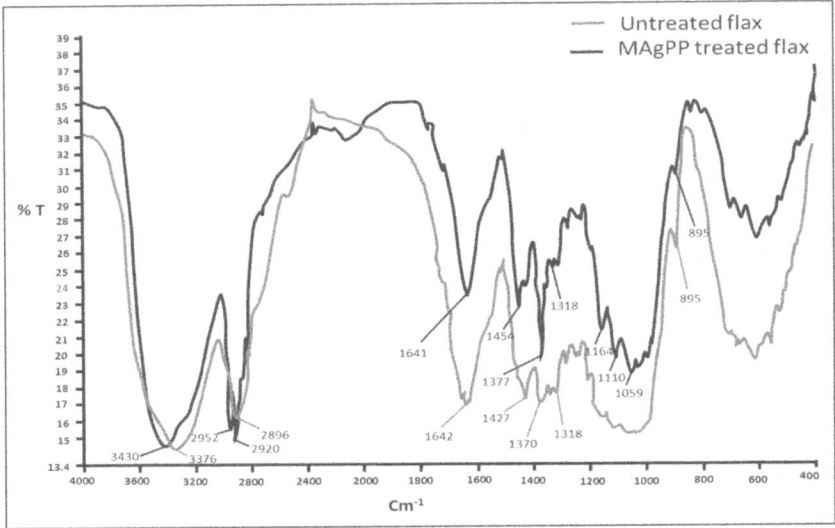

FIGURE 4.1 SEM images of (a) untreated flax fibre and (b) MAgPP treated flax fibre. (Reproduced with permission from Bar, Alagirusamy, and Das. 2018a.)

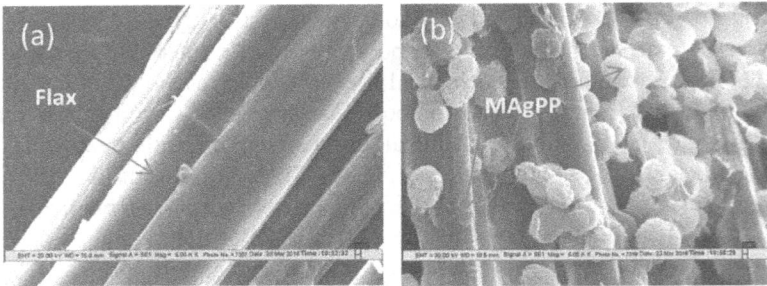

FIGURE 4.2 IR-spectra of MAgPP treated and untreated flax fibres. [Reproduced with permission from Bar, Alagirusamy, and Das. (2018a).]

vibration, C-O stretching vibration and C=O stretching vibration of the alpha-keto carbonyl groups are shifted to 3430 cm^{-1}, 2952 cm^{-1} and 1778 cm^{-1} respectively. Beside this, some new peaks are also observed between 1200 cm^{-1} and 1000 cm^{-1}, i.e. at 1278 cm^{-1}, 1164 cm^{-1}, 1110 cm^{-1} and 1059 cm^{-1}, respectively. The literature suggests that the above-mentioned peaks arise due to the ether linkage formation (Kobayashi and Takada 2013; Omrani et al. 2017). Hence, it can be concluded from the changes observed in the SEM images and in the IR-spectra that the MAgPP reacts with the various functional groups of flax fibre through ether linkage formation. This MAgPP treated flax fibres are used further in this present study.

4.4 PREPARATION OF THERMALLY BONDED ROVING (TBR)

Flax-PP-based TBRs are produced in a specially designed machine which mainly consists of one gill drawing unit and one thermal treatment unit. A schematic diagram stating the path of materials in the machine and the cross-section of the TBR is shown in Figure 4.3. The photographs of the gill drawing unit and the thermal treatment unit are illustrated in Figure 4.4(a and b), respectively.

The drawing unit of the present machine is basically a conventional drawing system used for bast fibre drawing. During TBR formation, the drawing unit improves the degree of mixing of flax and PP fibres and also improves their orientation towards the roving axis. The thermal treatment unit of the present machine has three main parts, namely condenser, heater and temperature controller with a temperature sensor. The condenser is heated from outside through a ring heater, tightly fitted around

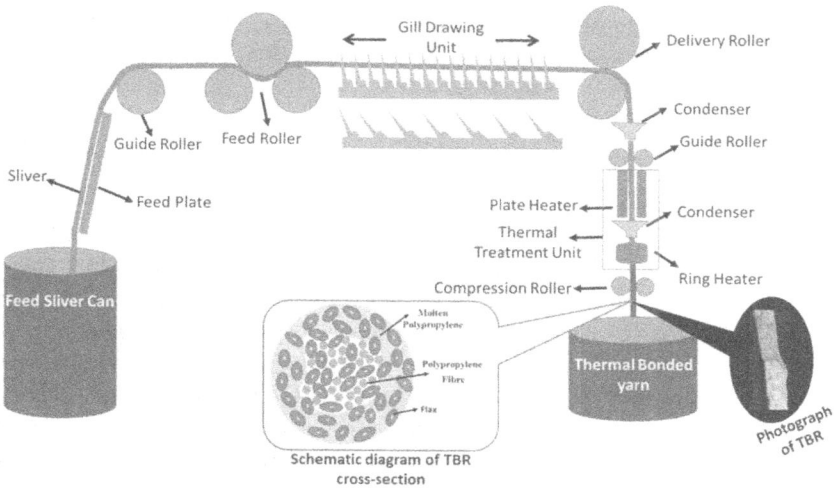

FIGURE 4.3 Schematic diagram of the passage of material in the thermally bonded roving manufacturing machine.

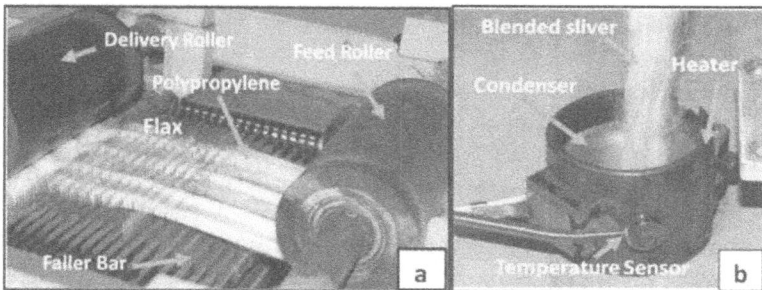

FIGURE 4.4 Photographs of the (a) gill drawing unit and (b) thermal treatment unit. (Reproduced with permission from Bar, Alagirusamy, and Das. 2019.)

the condenser. The temperature sensor is fixed at the outer surface of the condenser and maintains the set temperature in the main heating zone. When a flax-PP blended roving passes through the hot condenser, the PP fibres present on the roving surface touch the hot surface and are melted. As a result, a novel flax-PP-based hybrid roving structure is developed in which the flax fibres are held together by means of PP resin at the roving surface while the core remains unboned (as shown in the schematic diagram of the TBR cross-section in Figure 4.3). Unlike existing thermally bonded rovings/yarns, the newly developed roving is flexible enough for structural prepreg formation as its core remains unboned.

4.5   PROPERTIES OF THERMALLY BONDED ROVING (TBR)

TBR samples are produced using a specially designed machine as shown in Figure 4.3. The main purpose of these TBRs is to produce woven textile preforms for composite fabrication purpose. Tensile and weavabilty performance of the TBR samples are studied by Bar, Alagirusamy, and Das (2019a) using a Box and Behnken experimental design and the results are discussed.

4.5.1   TENACITY AND MODULUS OF THERMALLY BONDED ROVINGS

Tensile properties of the flax-PP-based thermally bonded, flexible roving are measured using the Instron Universal Testing Machine, Model no. 4042 with 10 kN load cell at 300 mm/min testing speed. All the tests are carried out at 27 ± 2°C room temperature and 65% RH. However, there is no specific standard available for the tensile characterization of such roving. Moothoo et al. (2014) have studied the tensile behaviour of epoxy impregnated flax tows by developing a laboratory experiment-based standard protocol which is quite similar to ASTM D2256 standard. The study concludes that the gauge length for the tensile test specimen should be beyond the maximum length of the fibre present in the sliver. The maximum fibre length of the flax fibres used in this study is around 42 cm. Considering the maximum flax fibre length, the gauze length for the tensile test of flax-PP-based thermally bonded roving specimens is kept at 50 cm. Each roving sample is tested twenty-five times and the average values of roving tenacity and Young's modulus are reported.

The response surface plots in Figure 4.5(a–c) show the effect of natural fibre content, number of drawing passages and thermal treatment temperature on the tenacity of thermally bonded roving.

3D surface plots demonstrate that at constant temperature and number of drawing passages, roving tenacity increases with increasing natural fibre content. However, the rate of increment in roving tenacity with respect to increasing natural fibre content is less at low numbers of drawing passages, but an exponential rate of increment in roving tenacity is observed at high numbers of drawing passages. At constant temperature and low levels of natural fibre content, the roving tenacity decreases with increasing drawing passages, but at high level of natural fibre content and constant temperature, i.e. at 195°C, the roving tenacity increases with increasing number of drawing passages. However, the roving tenacity increases with increasing thermal treatment temperature, keeping fibre content and the number of drawing passages constant at all levels.

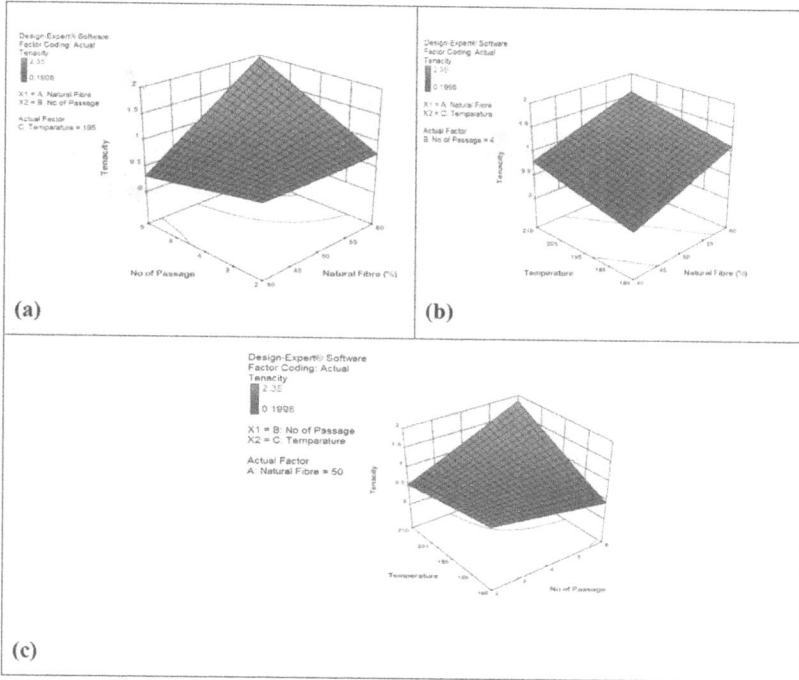

FIGURE 4.5 Effect of process parameters on thermally bonded roving's tenacity. (a) No. of Drawing Passage – Natural Fibre Content; (b) Temperature – Natural Fibre Content; (c) Temperature – No. of Drawing Passage. (Reproduced with permission from Bar, Alagirusamy, and Das. 2019b.)

The response surface plots in Figure 4.6(a–c) show the effect of natural fibre content, number of drawing passages and thermal treatment temperature on the modulus of thermally bonded roving.

The 3D surface plots demonstrate that at constant temperature (at 195°C) and number of drawing passages (at both high as well as low levels), roving modulus initially decreases with increasing natural fibre content and then starts increasing with further increase in natural fibre content. However, at constant temperature and low level of natural fibre content, the roving modulus decreases with increasing number of drawing passages, but at high level of natural fibre content, the roving modulus increases with increasing drawing passages. At constant natural fibre content (50%) and number of drawing passages (at both high as well as low levels), the roving modulus increases with increasing thermal treatment temperature.

The tenacity and modulus of the present thermally bonded roving mainly depends on the continuity of the flax-PP bonding surface and on the total area of flax-PP bonding, i.e. the total surface area of flax fibre covered by the PP resin. In the present study, both the flax and PP fibres in sliver form are taken as starting materials and mixed together in a gill drawing machine. The schematic diagram in Figure 4.7 shows the effect of increasing drawing passages on flax and PP fibre distribution into the blended sliver.

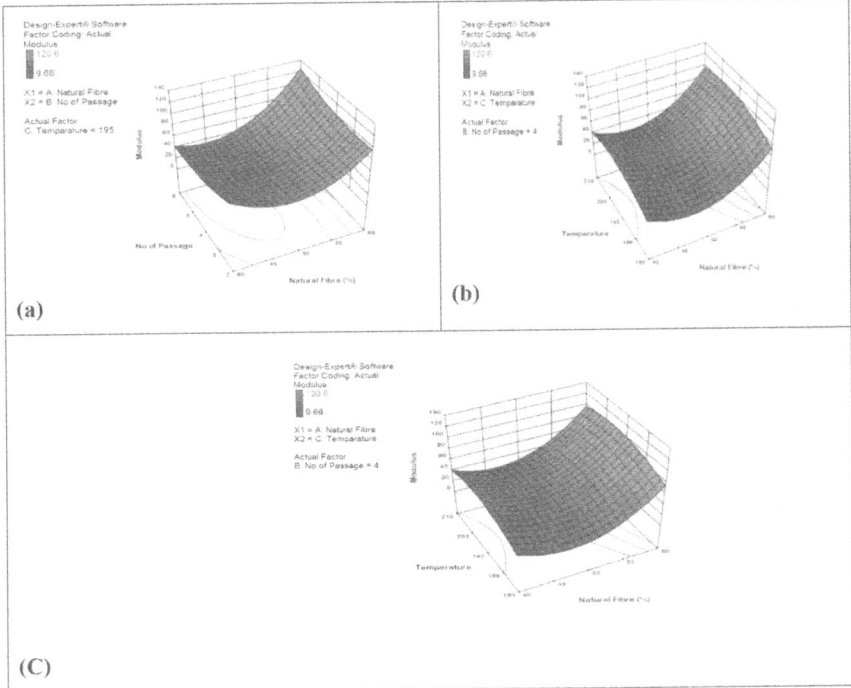

FIGURE 4.6 Effect of process parameters on thermally bonded roving's modulus. (a) No. of Drawing Passage – Natural Fibre Content; (b) Temperature – Natural Fibre Content; (c) Temperature – No. of Drawing Passage. (Reproduced with permission from Bar, Alagirusamy, and Das. 2019b.)

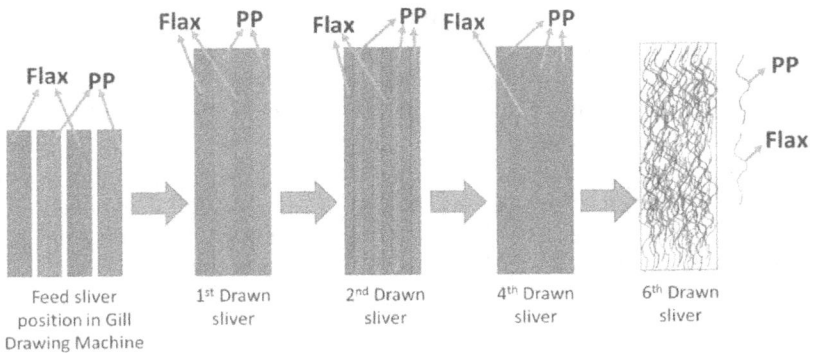

FIGURE 4.7 Effect of increasing number of drawing passages on flax and PP fibre distribution into the blended sliver structure. (Reproduced with permission from Bar, Alagirusamy, and Das. 2019b.)

Referring to Figure 4.7, after one drawing passage the flax and PP slivers come closer where both the fibre strands still have the continuity and two flax fibre strands are separated by a PP fibre strand. On further increase in drawing passages, initially the PP fibre strand starts losing its continuity and penetrates into the flax and then the dimensions of PP fibre bundles become smaller and eventually a well-mixed flax-PP sliver is obtained. Hence, the continuity of flax-PP fibre interphase initially decreases with increasing drawing passages (i.e. from 2 to 4) and then starts increasing with further increase in drawing passages due to improved fibre mixing. As a result, the modulus of the thermally bonded roving also decreases initially and then starts increasing with further increase in drawing passages. However, the tenacity of thermally bonded roving increases with increasing drawing passages since the area of the flax PP bonding surface increases with increasing drawing passages. But at low flax content (when the volume of PP is high), the PP fibres tend to agglomerate due to lack of availability of flax fibres. This results in the formation of voluminous PP clusters and ultimately reduces the area of flax-PP contact. Hence, at low flax content, the tenacity of the thermally bonded roving decreases with increasing drawing passages. In the present study, flax to PP ratios are varied at three different levels, i.e.60:40; 50:50 and 40:60. At higher flax content, more flax surface area is available to form flax-PP bondage which further increases on increasing drawing passages due to better mixing of flax-PP fibre. Hence, the tenacity of the thermally bonded roving increases with increasing flax fibre content and it shows an exponential rate of increment when the number of drawing passages is high. However, the modulus of thermally bonded roving initially decreases and then starts increasing with increasing flax fibre content while the number of drawing passages and thermal treatment temperatures is kept constant at 4 and 195°C respectively. After 4 drawing passages the blended sliver has a discontinuous flax-PP interphase. The volume of PP fibres with respect to constant flax is more in case of the sliver having 40% flax than in the sliver having 50% flax. Hence, the area of flax-PP interphase is more in case of the sliver having 40% flax content. As a result, with increasing flax content, the modulus of the thermally bonded roving initially decreases with increasing flax content.

During drawing, the movement of PP fibres is controlled by the long flax fibres which are gripped by the front delivery roller. Hence, after a certain number of drawing passages, the degree of mixing of flax and PP fibre will be more in case of the sliver having 60% flax than the sliver having 50% flax fibre due to the presence of higher numbers of long flax fibres. As a result, the area of flax-PP interphase should be more in case of the sliver having 60% flax fibre than the sliver having 50% flax fibre. Eventually, the roving modulus starts increasing when the natural fibre content goes beyond 50%. The degree of bonding between the adjacent flax fibres increases with increasing thermal treatment temperature due to higher degree of melting of PP fibres at high temperature. As a result, the roving tenacity and modulus increase with increasing thermal treatment temperature.

4.5.2 Weavability of Thermally Bonded Rovings

The weaving potential of flax-PP-based thermally bonded, flexible roving is tested using Reutligen Webtester, made by Sulzer Ruti, Germany. This instrument simulates

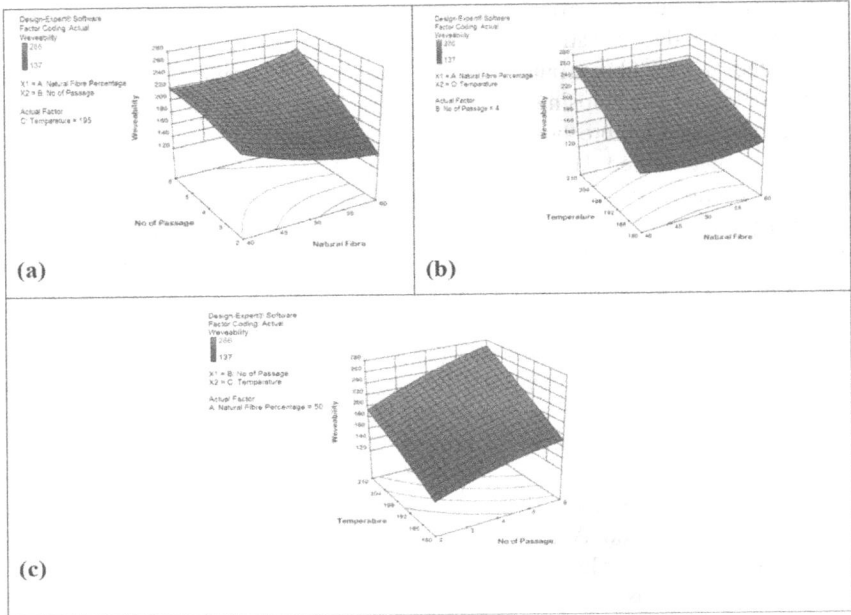

FIGURE 4.8 Effect of process parameters on thermally bonded roving's weavability performance. (a) No. of Drawing Passage – Natural Fibre Content; (b) Temperature – Natural Fibre Content; (c) Temperature – No. of Drawing Passage. (Reproduced with permission from Bar, Alagirusamy, and Das. 2019b.)

all major stresses occurring during weaving such as cyclic extension, axial abrasion, flexing and bending, excluding beat up and yarn/roving entanglement. This instrument reports weavability potential of a yarn in terms of number of cycles required to break or distort the same. The response surface plots in Figure 4.8(a–c) show the effect of natural fibre content, number of drawing passages and thermal treatment temperature on the weavability behaviour of thermally bonded roving.

The 3D surface plots demonstrate that at constant temperature (195°C) and low number of drawing passages, roving weavability decreases with increasing natural fibre content. Conversely, at constant temperature (195°C) and high number of drawing passages, roving weavability increases with increasing natural fibre content. Weavability is a surface phenomenon. At a low number of drawing passages, the degree of mixing of flax and PP is lower and the area of flax-PP interphase mainly depends on the volume of PP fibres. At high numbers of drawing passages, the degree of mixing of flax and PP is higher. Hence, the area of flax-PP interphase further increases with increasing natural fibre content due to the improved degree of mixing of flax and PP. As a result, a converse effect on the weavability of thermally bonded roving with increasing natural fibre content is observed at high and low numbers of drawing passages. At constant temperature (at all levels) and constant natural fibre content (50%), the roving weavability increases with increasing drawing passages due to better mixing of flax and PP fibres. However, the roving weavability increases with increasing thermal treatment temperature due to a high degree of PP melting at high temperature, keeping fibre content and number of drawing passages constant at any levels.

In general, it is observed that both the tensile and weavability performance of the TBR specimens increases with increasing thermal treatment temperature, with increasing flax content and with increasing the degree of flax-PP mixing. Thus, the TBR made after passing 6 times through the gill-drawing zone, having 60% flax content and processing at 210°C temperature is the most suitable TBR for woven textile preform formation.

4.6 TBR-BASED UNIDIRECTIONAL COMPOSITES

The influence of TBR structure on unidirectional composite properties was studied based on five sets of different TBR-based unidirectional composite. The thermal treatment temperature during TBR formation does not have a significant impact on composite properties (Bar, Alagirusamy, and Das. 2019b). Thus, only the flax content and the number of drawing passages are altered during the TBR formation. The flax content in the TBR varies at 40 wt.%, 50 wt.% and 60 wt.% levels while the number of drawing passages varies for two times, four times and six times respectively. The details of all TBR making parameters are reported in Table 4.1.

4.6.1 COMPOSITE PREPARATION

Flax-PP-based TBR composites are produced through a hot compression method which involves the following steps. Initially, the flax-PP-based TBR samples are wound in a parallel configuration over a spring-loaded metallic frame in a layer-by-layer manner. This spring mechanism adjusts the roving tension variation caused by thermal shrinkage during consolidation. The hybrid yarn bundle is then wrapped with a thin PP film and subsequently tied with a PP filament as shown in Figure 4.9. After winding, the roving bundle is placed in the grove of the female mould part and subjected to hot compression after closing the male mould half for 5 minutes at 190°C temperature, 8 bar consolidation as well as vacuum pressure respectively. During hot compression, some polymer comes out from the mould through both open ends. Keeping that squeeze out matrix in view, the corresponding volume of

TABLE 4.1
Details of Thermally Bonded Roving Making Parameters

Sample ID	NF (flax) Content (wt. %)	Polypropylene Content (wt. %)	No. of Drawing Passages	Thermal Treatment Temperature (°C)
P-2, NF-60%	60	40	2	
P-4, NF-60%	60	40	4	
P-6, NF-60%	60	40	6	210
P-6, NF-50%	50	50	6	
P-6, NF-40%	40	60	6	

FIGURE 4.9   Photographs of thermally bonded roving-based UD-composite fabrications. (Reproduced with permission from Bar, Alagirusamy, and Das. 2019.)

PP in film form is pre-added, which ultimately compensates the matrix loss. After curing, the whole set-up is cooled down to below 100°C without releasing the consolidation pressure. During cooling, the molten PP matrix solidifies and holds the fibres in definite positions.

4.6.2   Tensile Properties

The tensile properties of the composite samples are tested according to the ASTM D638 test method using the Zwick-Roell universal testing machine, Model no. Z052 with 50 KN load cell attachment. Tensile testing is carried out at 2 mm/min test speed after clamping the composite sample over an area of 40 mm ×20 mm at both ends leaving 100 mm gauge length. The strain of all composite specimens along the loading direction is measured using an extensometer, Model no. 3560-BIA-025M-010-ST (Epsilon Technology Corp, Jackson, WY, USA) of 25 mm gauze length. The photographs of the tensile failure surfaces of the thermally bonded roving reinforced composite specimens are shown in Figure 4.10. Referring to Figure 4.10, some PP rich regions at the tensile fracture end of the composite specimens made of 60% flax fibre are observed when the number of drawing passages for the constituent roving is 2 and 4 respectively. However, no such matrix rich region at the tensile fracture end of the composite specimen having 60% flax is observed when the number of drawing passages for the constituent roving is 6. Then again in case of six passage drawn sliver composite, some PP rich regions in the composite fracture end are observed when the flax content in the constituent rovings is 40% and 50% respectively. The degree of flax-PP mixing is poor at low numbers of drawing passages (i.e. at 2 and 4 passages) and with a constant number of drawing passages the flax-PP fibre mixing decreases with increasing PP content (as the PP fibres tend to agglomerate during drawing due to their short fibre length). Hence, the PP rich regions in the composite fracture ends are observed when the number of drawing passages for the constituent roving is low and the PP content in the constituent roving is high.

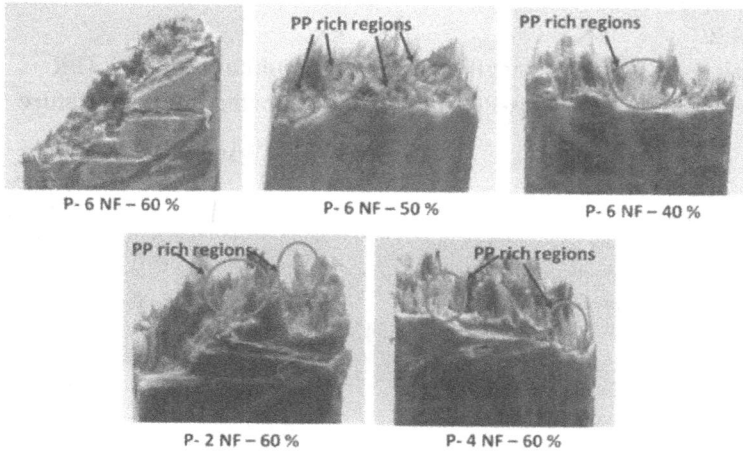

FIGURE 4.10   Photographs of the tensile failure surfaces of the TBR compressed composite specimens.

The ultimate tensile strength and initial modulus of the TBR compressed unidirectional composite specimens are reported in Figure 4.11. With increasing flax content, the TBR compressed composite specimens exhibit a common trend as observed in cases of existing ligno-cellulosic thermoplastic composites (George et al. 2012).

Around 57% improvement in tensile strength and around 60% improvement in tensile modulus of the TBR-compressed composite specimens are observed when the flax content in the respective composite specimens increases from 40% to 60% level. At constant flax content (60%), around 10% improvement in tensile strength and around 5% improvement in tensile modulus of the unidirectional composites are

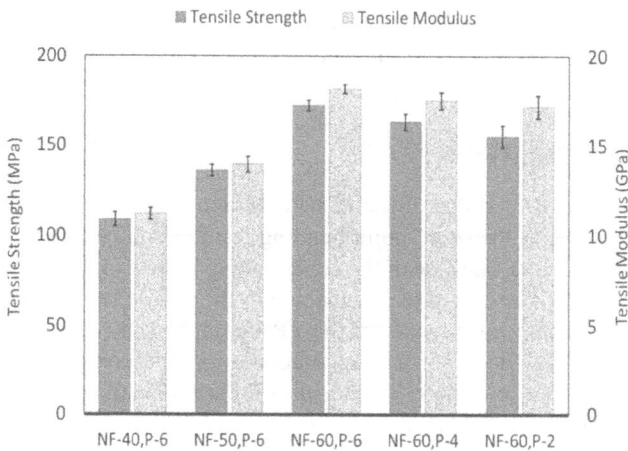

FIGURE 4.11   Tensile strength and modulus of the TBR compressed unidirectional composites.

TABLE 4.2

Comparison of Tensile Strength and Tensile Modulus between TBR Composites and Other Flax-PP Composites Reported in the Literature

S. No	Composite Description	Strength (MPa)	Modulus (GPa)	Reference
1.	Low twisted flax yarn reinforced PP composite prepared through film stacking method.	84	9.4	Bar, Alagirusamy, and Das. 2018b
2.	PP melt impregnated unidirectional fibre mat compressed composites	77.2	5.1	Oksman, 2000
3.	Flax-PP-based inter-woven UD-fabric compressed composites.	71	5.8	Kannan et al. 2012
4.	Flax-PP needle punched nonwoven composites made in compression moulding method.	78	5.3	John and Anandjiwala 2009
5.	DREF yarn reinforced composite (low twisted yarn)	93	8.5	Bar, Alagirusamy, and Das 2018a
6.	Flax-PP TBR compressed composites (present study)	108	11.5	

N.B.: All composite specimens which are mentioned in the table have 40 wt. % fibre content.

observed when the number of drawing passages for the constituent TBR increases from 2 passages to 6 passages respectively. However, the present composite samples exhibit better tensile properties compared to flax-PP composites reported in literature. It is mainly due to better fibre orientation, fibre-matrix distribution and fibre-matrix bonding of the present composite samples. Table 4.2 compares the tensile behaviour of the present flax-PP composite samples with the flax-PP composites having 40% flax reported in literature.

4.6.3 FLEXURAL PROPERTIES

Flexural strength and modulus of the flax-PP TBR compressed composite samples are tested according to the three-point bending test method (ASTM D790). During three-point bending test, one side of the composite specimen is subjected to the compressive force while the other side is subjected to the tensile force. The photographs of the tensile and compression sides of the TBR compressed composite samples are shown in Figure 4.12. Some matrix accumulation in the compression side of the flexural tested composite sample (as shown in Figure 4.12(a)) is observed when the flax content in the composite specimens are 40% and 50% respectively. However, no such matrix accumulation is observed in case of 60% flax reinforced composites. A prominent line of fracture is observed on the tensile side of all flexural tested TBR compressed composites.

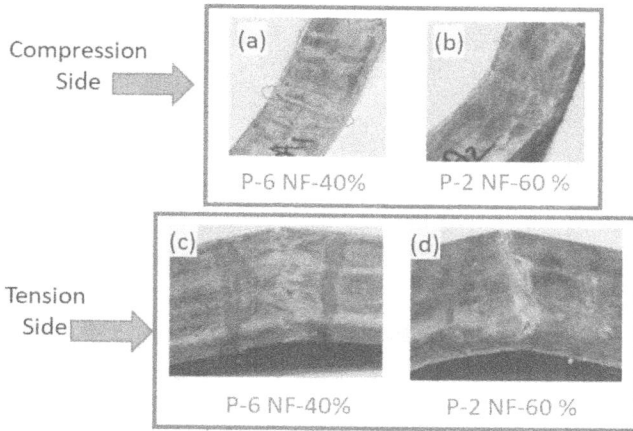

FIGURE 4.12 Photographs of the compression and tensile side of the three-point bending test unidirectional composite specimens.

Flexural stress-deflection curves of the composite samples made from flax-PP-based TBR having different flax content and the composite samples made with TBR having different numbers of drawing passages are exhibited in Figure 4.13(a and b), respectively. The flexural strength and modulus of the TBR compressed composite specimens are reported in Figure 4.14.

Referring to Figure 4.13, a catastrophic failure (i.e the sample breaks suddenly once it reaches to the maximum stress level) in the flexural stress-deflection curves of the TBR compressed composite samples is observed when the flax content in the composite sample is 60%. However, no such sudden fall in the flexural stress-deflection curves is observed in case of the composites having 40% and 50% flax content respectively. In these two cases, the flexural stress reaches a maximum level and then remains almost at the same stress level with further deflection. The differences in the flexural stress-deflection curves of the composite samples having

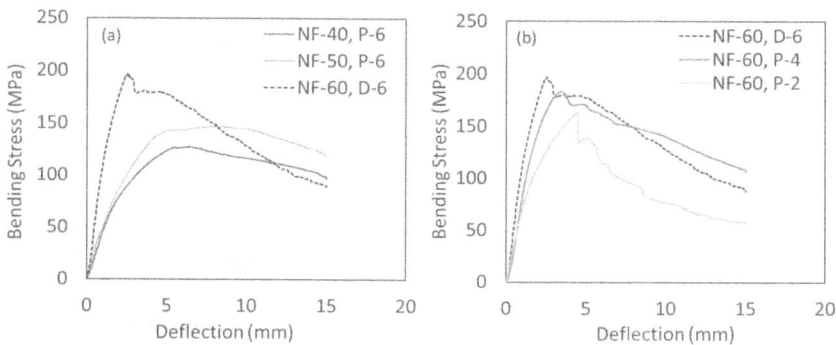

FIGURE 4.13 Flexural stress-deflection curves of the unidirectional composites made of TBR having (a) different flax contents (b) different numbers of drawing passages.

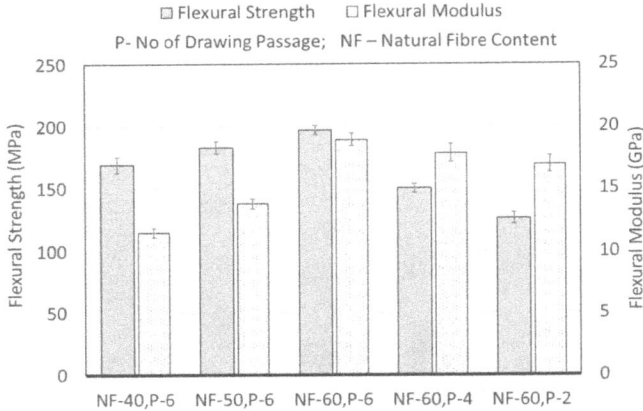

FIGURE 4.14 Flexural strength and modulus of the TBR compressed unidirectional composites.

different flax content are mainly attributed to the variances in flax-PP distribution in the respective composite structure. Referring to Figure 4.14, the flexural strength and modulus of the flax-PP TBR compressed composite samples increase with increasing flax content. Around 56% improvement in flexural strength and around 60% improvement in flexural modulus are observed when the flax content in the composite samples increases from 40% to 60%. However, at constant flax content, around 15% improvement in flexural strength and around 10% improvement in flexural modulus are observed when the number of drawing passages for the constituent roving increases from 2 passages to 6 passages.

It is observed that the flexural and tensile properties (i.e. strength and modulus) of the flax-PP-based TBR compressed composite specimens increase with increasing flax content in the composite structure. Further, it improves with increasing degree of flax-PP mixing, i.e. the number of drawing passages in order to obtain better stress transfer from the matrix to the reinforcing component. Thus, the TBR made after passing 6 times through the gill-drawing zone, having 60% flax content and processing at 210°C temperature, is the most suitable TBR for composite reinforcement. In consequence, it is the most suitable TBR for woven textile preform formation. Hence, the above-mentioned TBR is considered for the further studies on TBR-based woven fabric composites.

4.7 TBR-BASED WOVEN FABRIC COMPOSITES

Composite materials are anisotropic and show the maximum mechanical properties in the leading fibre orientation direction. Hence the properties of a unidirectional composite should be different from a woven textile preform reinforced composites. Thus, it is very much necessary to study the effect of these novel hybrid yarn-based woven fabric architectures on their composite properties and to compare it with glass reinforced polypropylene composites (GRPC) (as at present, glass reinforced polymer composites are mostly used in various applications). Two types of fabrics with

Properties of Bio-Composites

101

TABLE 4.3
Physical Parameters of Different Fabrics

Property	Glass-PW	Roving-PW	Roving-UD
Warp/Weft Constituents	Glass/Glass	TBR/TBR	PP/TBR
EPI/PPI	7/7	3/5	2/6
Thickness (mm)	0.63	3.65	2.4
Areal density (g/m²)	612	840	630

(EPI – ends per inch, PPI – Picks per inch, Glass-PW: Plain Woven Glass fabric, Roving-PW: Plain Woven Roving fabric, Roving-UD: Unidirectional Roving fabric.)

different fabric architectures, i.e. plain woven and unidirectional, are fabricated on a handloom fitted with dobby shedding mechanisms. In the case of plain-woven fabric, TBRs are used both in warp and weft directions while in the case of unidirectional fabric PP filaments are used as warp and the TBR as weft. All details of the TBR-based woven fabrics and the other fabrics used in this study are summarized in Table 4.3 and their photographic images are presented in Figure 4.15. Composite laminates are fabricated using the above-mentioned fabric specimens and the tensile, flexural and impact properties of the resultant composites are evaluated and compared.

4.7.1 COMPOSITE LAMINATE FABRICATION

TBR-based unidirectional fabric laminates (roving-UD), TBR-based plain woven fabrics laminates (roving-PW) and the plain-woven glass fabric reinforced PP laminates (Glass-PW) are fabricated in a vacuum-assisted compression moulding machine after heating the stacked fabrics and PP films for 10 minutes under 10 bar pressure at 200°C. Although the TBR-based woven fabrics contain both the reinforcement (flax) and matrix (PP) forming components, the laminates made after compressing only TBR-based woven fabrics have a lot of macro void (as shown in Figure 4.16) due to lack of PP resin in the yarn crossover points of the woven fabrics. To overcome the above-mentioned problem, PP films and TBR-based fabric layers are stacked alternatively during laminate formation. This brings down the fibre to matrix ratios of the resultant composites to 50:50 level from 60:40 level (in mass).

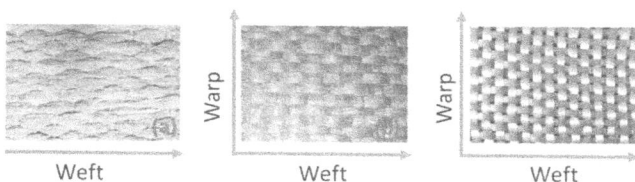

FIGURE 4.15 Photographic images of (a) TBR-based unidirectional fabric, (b) TBR-based plain woven fabric, and (c) Plain woven glass fabric.

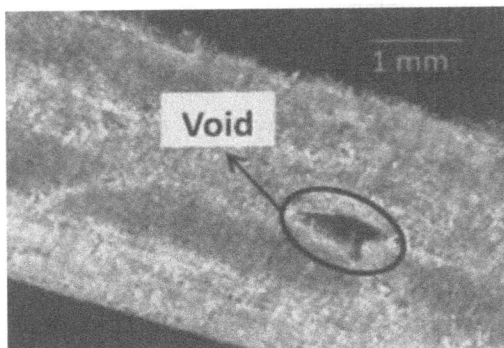

FIGURE 4.16   Flax-PP-based TBR compressed composite with macro void. (Reproduced with permission from Bar et al. 2020.)

All the laminates produced in this study have the same fibre to matrix ratio, and same thickness. Hence, the laminates with different fabric configurations are produced with different numbers of fabric layers. Three layers of fabric in the case of roving-PW composites and four layers of unidirectional fabrics (arranged in 0/90 configuration) in the case of roving-UD composites are consolidated together to produce a laminate having 3.5±0.1 mm thickness. Six plain woven glass fabric layers are used to produce glass-PW laminates of same thickness and same fibre to matrix ratios (i.e. 50:50 in mass).

4.7.2   Tensile and Flexural Properties

The ultimate tensile and flexural properties of the flax-PP TBR-based woven fabric laminates and glass-PP laminates are reported in Table 4.4. It is observed that the glass PW composite has higher tensile strength and modulus than both the TBR-based woven fabric composites. This is mainly attributed to the better tensile properties of glass fibre than that of flax. The roving UD composite reaches up to 70% tensile strength and 75% tensile modulus of glass PW composites, whereas the roving-PW composite reaches up to 58% tensile strength and 48% tensile modulus

TABLE 4.4
Comparison of the Mechanical Properties of the TBR-Based Woven Fabric Composites with Glass-PP Composites

Sample ID	Fibre vol. Fraction (%)	Tensile Strength (MPa)	Tensile Modulus (GPa)	Flexural Strength (MPa)	Flexural Modulus (GPa)
Glass PW	26.0 ± 0.5	82.41 ± 2.76	11.41 ± 0.55	129 ± 2.36	5.32 ± 0.29
Roving PW	37.8 ± 0.5	48.21 ± 3.17	5.41 ± 0.49	103 ± 3.33	3.33 ± 0.35
Roving UD	38.2 ± 0.5	57.30 ± 3.21	8.63 ± 0.68	121 ± 4.32	2.36 ± 0.48

of glass PW composites. The flax fibres in the roving UD composite structure have better orientation (due to crimp-less structure of unidirectional fabric) than the roving PW composites. This is attributed to better tensile behaviour of the roving UD composite than those of roving PW composites.

Like tensile properties, the flexural strength and modulus of the Glass-PW composites are higher than those of TBR-based woven fabric composites. It is mainly attributed to the high modulus and the brittle nature of glass fibre. Interestingly, the roving-UD composite reaches to approximately 90% flexural strength and 45% flexural modulus of glass-PW composites, whereas the roving-PW composite reaches approximately 80% flexural strength and 60% flexural modulus of glass-PW composites. Although the roving-UD composite has higher bending stress, it fails catastrophically once it reaches the pick load. On the other hand, the roving-PW composite can sustain a high degree of deformation at pick load level, which is an added advantage for load bearing applications. In the case of roving-PW composites, withstanding deformation at pick loading could be attributed to its interlaced woven fabric structure.

4.7.3 Low Velocity Impact (LVI) Properties

The present study investigates the LVI performances of the roving-UD and roving-PW laminates and compares it to the LVI performances of glass-PP composite laminates. Initially the laminates are tested for impact responses and the impact damages are analysed to understand the impact failure mechanism. Finally, the damage tolerance of the composite laminates is evaluated by comparing their flexural strength before and after impact testing.

4.7.3.1 Impact Response Parameters

The parameters measured during impact testing are peak force, peak energy, peak deformation and total energy. The damage degree, which is the ratio between the absorbed energy and the impact energy, is then calculated. The impact energy is the kinetic energy of the impactor right before the contact with sample takes place. The absorbed energy is the energy dissipated by the system through several mechanisms occurring after the impactor's contact, such as elastic deformation, friction, plastic deformation and most importantly, those specific to the material such as matrix cracking, de-bonding, fibre pull-out and fibre breakage (Sarasini et al. 2013). The key parameters related to the impact responses of the glass-PP and flax-PP composite laminates are summarized in Table 4.5.

Figure 4.17 shows the force-displacement curves of all flax-PP and glass-PP composite laminates. The force-displacement curves exhibit a closed pattern and the area under the closed loop refers to the energy absorbed during the impact. The absorbed energy-displacement curves are shown in Figure 4.18. As expected the Glass-PW composite absorbs more energy and reaches the peak energy level at a lower deformation than all flax-PP composites. This above observation is mainly attributed to higher modulus, tenacity and work of rupture of the glass fibres than that of the flax fibres..

TABLE 4.5
Impact Response Parameters of Glass-PP and Flax-PP Composites

Laminate	Peak Force (KN)	Peak Energy (J)	Peak Deformation (mm)	Ductility Index	Damage Degree	Total Absorbed Energy (J)
Glass-PW	2.7 ± 0.02	8.7 ± 0.31	6.0 ± 0.33	55	0.11	23 ± 1.11
Roving-PW	1.9 ± 0.05	8.2 ± 0.47	6.4 ± 0.31	57	0.09	19 ± 1.29
Roving-UD	1.6 ± 0.07	6.1 ± 0.35	7.6 ± 0.43	62	0.06	14 ± 1.53

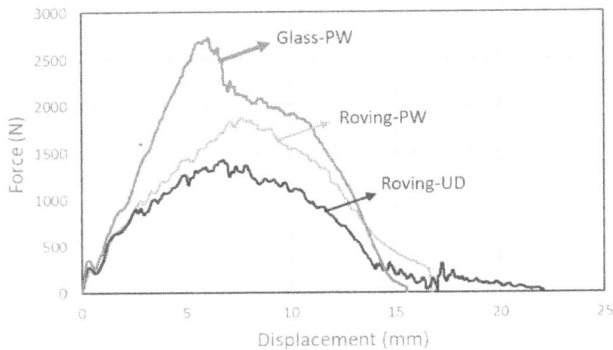

FIGURE 4.17 Force-displacement curve of flax-PP and glass-PP composites.

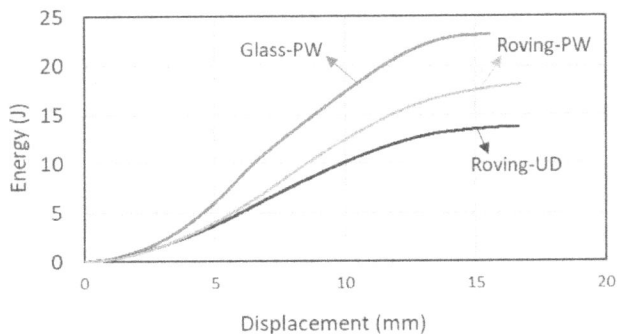

FIGURE 4.18 Energy-displacement curve of flax-PP and glass-PP composites.

Figures 4.17 and 4.18 also reveal that the roving-PW composites absorb more energy and reach the peak load at a higher deformation level than roving-UD composites. This may be due to the interlaced structure of plain-woven fabric, which provides more resistance during an impact failure than a unidirectional fabric.

Figure 4.19 shows the variation in load during impact with respect to time. It is observed that the roving-UD laminate breaks at lower loads as compared to

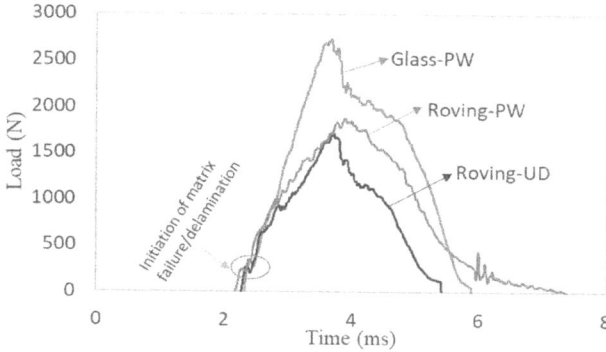

FIGURE 4.19   Impact force-time response of the flax-PP and glass-PP laminates.

roving-PW and glass-PW laminates. At the initial portion of all curves (Figure 4.19) a sudden drop in load is observed. This represents the initiation of matrix cracking or delamination (Sutherland and Soares 2005). It also indicates the transition of the specimen from an intact state to a damaged state (Sevkat et al. 2010; Bandaru et al. 2016). The loading and unloading cycle are almost identical in the case of roving-UD laminates while it is not identical in case of roving-PW and glass-PW laminates. It is because of the significant change in the striker velocity during the impact testing of roving-PW and glass-PW laminates (Reyes and Sharma 2010). Around 0.4 m/s change in impactor velocity is observed in the above-mentioned cases while the change is around 0.1 m/s in cases of roving-UD laminates. The peak load under the impactor is an indication of the load buffering capacity of the composite laminates, directly related to the stiffness of the materials. From Table 4.5 it is observed that the energy absorption of flax-PP roving-UD and flax-PP roving-PW composites is around 60% and 81% and peak force is around 59% and 70% respectively of the corresponding values exhibited by the glass-PW composite.

For a better understanding, the specific loads and specific energies of all composite laminates are presented in Figure 4.20. It is observed that the specific load of the

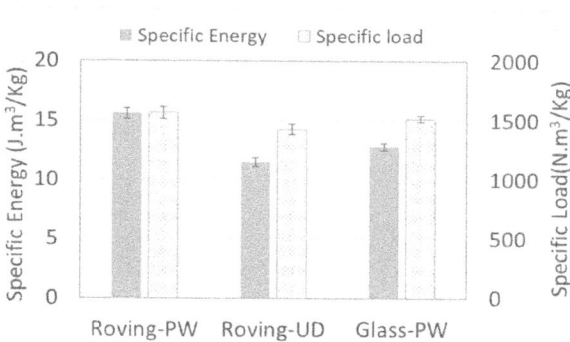

FIGURE 4.20   Specific energy and specific force of the flax-PP and glass-PP laminates.

glass-PP composites is 6% higher than the flax-PP TBR-based UD-fabric composites while it is 3% lower than that of flax-PP TBR-based plain woven fabric composites. On the other hand, the absorbed specific energy of the TBR-based plain woven fabric composites exceeded by 21% that of glass-PP composites while the absorbed specific energy of the TBR-based UD-fabric composites is around 11% less than that of glass-PP composites. The results given here therefore show that for an equivalent mass of fibre, the resistance to impact is improved when using flax-PP fabrics in comparison to glass-based composites. One can therefore think that the flax fibres present advantageous impact resistance properties for low velocity impact.

4.7.3.2 Failure Mechanisms

When a projectile hits the target, various failure modes including delamination, matrix cracking, fibre failure and edge delamination can occur. These failure modes depend on the mass and profile of the projectile, sample architecture, sample constituents and on other testing parameters (Safri et al. 2017). Figures 4.21 and 4.22

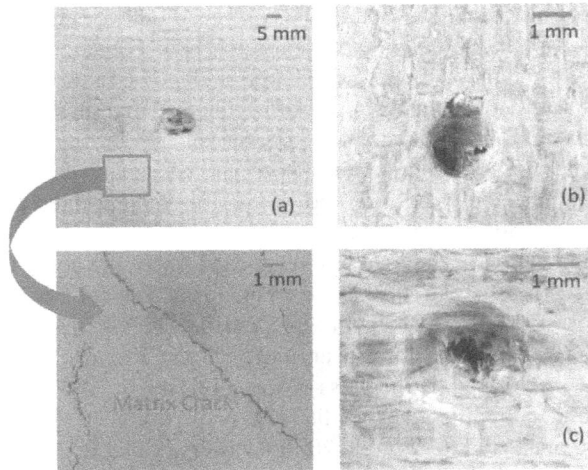

FIGURE 4.21 Front side damage fragmentation of composite laminates. (a) Glass-PW (b) Roving-PW (c) Roving-UD. (Reproduced with permission from Bar et al. 2020.)

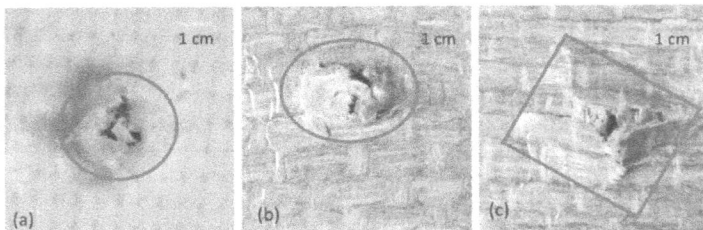

FIGURE 4.22 Back side damage fragmentation of the composite laminates. (a) Glass-PW (b) Roving-PW (c) Roving-UD. (Reproduced with permission from Bar et al. 2020.)

demonstrate the typical post impact failure modes observed in flax-PP TBR-based unidirectional and plain-woven fabric laminates and in GRPC laminates. The failure zone in the front face of the laminate is due to direct contact of the projectile whereas the failure at the back side of the laminate is due to energy absorption followed by damage propagation.

Matrix cracking takes place in the front side of the GRPC laminate (shown in Figure 4.21). The cracks emanate radially from the impact zone. However, no such matrix cracking is observed in the surface of flax-PP TBR-based fabric composites. It is mainly due to better fibre/matrix interaction owing to better mingling of flax and PP fibre in the TBR structure and due to MAgPP treatment of the flax fibre before TBR formation. The damage pattern at the back side of the roving-UD composite is in the shape of a diamond. It is mainly due to inter-laminar deformation (layer delamination). On other hand, no such inter-laminar deformation is observed in the case of glass-PW and roving-PW laminates. The average damaged area at the back side of the glass-PW, roving-PW and roving-PW laminates are 452.16 mm², 512.11 mm², 960.73 mm² respectively. During perforation impact test of a fibre reinforced polymer composite, the fibres have to break in order to achieve perforation. Hence, the stronger the fibre is, more predominant is the fibre fracture energy to the total absorbed energy. Glass has higher modulus and strength than flax. Hence, the fracture area at the back side of the impacted Glass-PW composite is lower than all flax-PP composites. In the back surface of the impacted glass-PP composite specimen, significant fibre pull out is observed but no such fibre pull out is observed in case of flax-PP TBR-based plain woven or UD-fabric compressed composite. Hence, it can be concluded that during the impact testing of Glass-PW laminates, fibre breakage, matrix cracking and fibre pull out are the major mechanisms of energy absorption. In the case of flax-PP roving-PW laminates, fibre breakage is the only mechanism of energy absorption while in case of roving-UD laminates, fibre breakage and laminar deformation are the major mechanisms of energy absorption.

4.7.3.3   Damage Tolerance Investigation

The composite's property after withstanding an impact thrust is very critical in order to predict its impact thrust sustaining ability. Post-impact properties of a composite laminate also give an idea about how the load bearing capacity of the laminate is affected, whether it needs to replace or repair etc. The damages occurring as a result of impact thrust are more sensitive to compression than to tensile loading because during compression the bulking outwards of the delaminated parts takes place. However, a global buckling due to low stiffness of flax fibre composite occurs during compression loading, which is not acceptable according to the compression after impact test standard (ASTM 7136). As an alternative, flexural properties before and after impact test are compared to evaluate the impact thrust sustaining ability of the present composite materials. Flexural strength and modulus of the glass-PP and TBR-based PW and UD composites, before and after impact thrust, are reported in Figure 4.23. After the impact thrust, around 7% and 10% loss in flexural strength are observed in case of roving-PW and roving-UD composites respectively while around 14% deterioration in flexural strength of glass/PP composite is observed after the impact thrust. After impact thrust, the proportionate deterioration in flexural strength is lower in case of

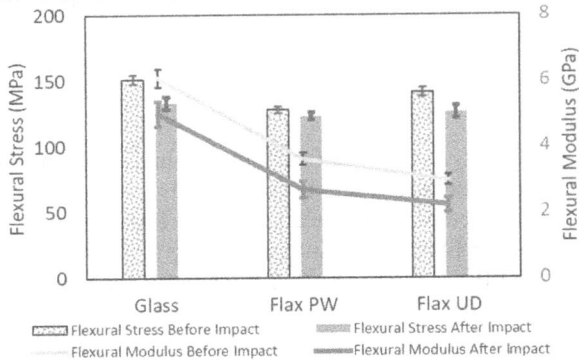

FIGURE 4.23   Flexural stress and modulus of the impacted and non-impacted composite laminates.

TBR-based fabric compressed composites as compared to glass/PP composites. It is mainly due to better fibre-matrix adhesion (as a result of MAgPP treatment of flax fibre) and better fibre-matrix distribution in the case of TBR compressed composites as compared to glass/PP composites. However, the drop in flexural modulus after impact thrust is comparable for all composites, which is around 15% in case of all TBR compressed composites and 17% in case of glass/PP composites.

4.8   SUMMARY

In this present study, a novel, flax-PP-based twist-less, thermally bonded, flexible roving has been developed using a specially designed machine. Flax and PP fibres in this roving structure remain in a mingled state and are highly oriented towards the roving axis. Fibres in this twist-less roving structure are held together by means of thermoplastic resin at the roving surface while the roving core remain unaffected. During roving formation, the process parameters such as natural fibre content, number of drawing passages and thermal treatment temperature are varied at different levels. Finally, the effect of these process parameters on TBR properties relevant to fabric formation such as tenacity, modulus and weavability have been studied using the Box and Behnken design. It is observed that the preforming ability of TBR samples improves with increasing thermal treatment temperature, flax fibre content and with increasing the degree of flax-PP mixing. Further, the influence of TBR structure on their unidirectional composite properties are evaluated after varying the flax content and the degree of flax-PP fibre mixing at three different levels. Produced TBRs are consolidated in a compression moulding machine and the resultant composites are tested for tensile and flexural properties. The composite produced using TBR, having 60% flax, 6 drawing passages and processed at 210°C temperature exhibits the best results over all other TBR-based composite specimens. Using this optimized TBR, two different fabric architectures, namely plain woven and unidirectional (UD) are produced. These woven prepregs are consolidated in a compression moulding machine to produce composite laminates. Finally, different properties of the produced composite laminates such as LVI, flexural and tensile properties have been tested and compared with glass/PP laminates at same thickness and same fibre to matrix ratio (on mass basis). It is observed that the TBR-based woven composites can reach up to 70% tensile strength, 75% tensile modulus, 90% flexural strength, 60% flexural modulus and 81% energy absorption (during LVI testing) of the Glass-PW composite. Therefore, one can replace the plain-woven glass-PP film stacked composites with TBR-based flax-PP laminates in cases of low and medium load bearing applications.

REFERENCES

Akonda, M. H., D. U. Shah, and R. H. Gonga. 2020. Natural fibre thermoplastic tapes to enhance reinforcing effects in composite structures. *Composites Part A: Applied Science and Manufacturing* 131: 105822. doi.org/10.1016/j.compositesa.2020.105822

Alagirusamy, R., R. Fangueiro, V. Ogale, and N. Padaki. 2006. Hybrid yarns and textile preforming for thermoplastic composites. *Textile Progress* 38(4): 1–71. doi.org/10.1533/tepr.2006.0004

Baghaei, B., M., Skrifvars, and L. Berglin. 2015. Characterization of thermoplastic natural fibre composites made from woven hybrid yarn prepregs with different weave pattern. *Composites Part A: Applied Science and Manufacturing* 76: 154–161. doi.org/10.1016/j.compositesa.2015.05.029

Baghaei, B., M. Skrifvars, and L. Berglin. 2013. Manufacture and characterisation of thermoplastic composites made from PLA/hemp co-wrapped hybrid yarn prepregs. *Composites Part A: Applied Science and Manufacturing* 50: 93–101. doi.org/10.1016/j.compositesa.2013.03.012

Bandaru, A. K., V. V. Chavan, S. Ahmad, R. Alagirusamy, and N. Bhatnagar. 2016. Low velocity impact response of 2D and 3D Kevlar/Polypropylene composites. *International Journal of Impact Engineering* 93: 136–143. doi.org/10.1016/j.ijimpeng.2016.02.016

Bar, M., R. Alagirusamy, A. Das, and P. Ouagne. 2020. Low velocity impact response of flax/polypropylene hybrid roving based woven fabric composites: Where does it stand with respect to GRPC?. *Polymer Testing* 89: 106565. doi.org/10.1016/j.polymertesting.2020.106565

Bar, M., R. Alagirusamy, and A. Das. 2019a. Influence of friction spun yarn and thermally bonded roving structures on the mechanical properties of Flax/Polypropylene composites. *Industrial Crops and Products* 135: 81–90. doi.org/10.1016/j.indcrop.2019.04.025

Bar, M., R. Alagirusamy, and A. Das. 2019b. Development of flax-PP based twist-less thermally bonded roving for thermoplastic composite reinforcement. *Journal of Textile Institute*. doi.org/10.1080/00405000.2019.1610997

Bar, M., R. Alagirusamy, and A. Das. 2018a. Effect of interface on composites made from DREF spun hybrid yarn with low twisted core flax yarn. *Composites Part A: Applied Science and Manufacturing* 107: 260–270. doi.org/10.1016/j.compositesa.2018.01.003

Bar, M., R. Alagirusamy, and A. Das. 2018b. Properties of flax-polypropylene composites made through hybrid yarn and film stacking methods. *Composite Structure* 197: 63–71. doi.org/10.1016/j.compstruct.2018.04.078

Bar, M., R. Alagirusamy, and A. Das. 2017. Studies on low-twist flax-pp based hybrid yarns for thermoplastic composite reinforcement. *Journal of Reinforced Plastics and Composites* 36(11): 818–831. doi.org/10.1177/0731684417693428 de Farias, J. G. G., R. C. Cavalcante, B. R. Canabarro, H. M. Viana, S. Scholz, and R. A. Simão. 2017. Surface lignin removal on coir fibers by plasma treatment for improved adhesion in thermoplastic starch composites. *Carbohydrate Polymers* 165: 429–436. doi.org/10.1016/j.carbpol.2017.02.042

Faruk, O., A. K. Bledzki, H. P. Fink, and M. Sain. 2012. Biocomposites reinforced with natural fibers: 2000–2010. *Progress in Polymer Science* 37: 1552–1596. doi.org/10.1016/j.progpolymsci.2012.04.003

George, G., E. T. Jose, K. Jayanarayanan, E. R. Nagarajan, M. Skrifvars, and K. Joseph. 2012. Novel bio-commingled composites based on jute/polypropylene yarns: Effect of chemical treatments on the mechanical properties. *Composites Part A: Applied Science and Manufacturing* 43: 219–230. doi.org/10.1016/j.compositesa.2011.10.011

John, M. J., and R. D. Anandjiwala. 2009. Chemical modification of flax reinforced polypropylene composites. *Composites Part A: Applied Science and Manufacturing* 40(4): 442–448. doi.org/10.1016/j.compositesa.2009.01.007

John, M. J., and S. Thomas. 2008. Biofibres and biocomposites. *Carbohydrate Polymers* 71: 343–364. doi.org/10.1016/j.carbpol.2007.05.040

Kannan, T. G., C. M. Wu, B. K. Cheng, and C. Y. Wang. 2012. Effect of reinforcement on the mechanical and thermal properties of flax/polypropylene interwoven fabric composite. *Journal of Industrial Textiles* 42(4): 417–433. doi.org/10.1177/1528083712442695

Khondker, O. A., U. S. Ishiaku, A. Nakai, and H. A. Hamada. 2006. Novel processing technique for thermoplastic manufacturing of unidirectional composites reinforced with jute yarns. *Composites Part A: Applied Science and Manufacturing* 37, 2274–2284. doi.org/10.1016/j.compositesa.2005.12.030

Kobayashi, S., and K. Takada. 2013. Processing of unidirectional hemp fiber reinforced composites with micro-braiding technique. *Composites Part A: Applied Science and Manufacturing* 46: 173–179. doi.org/10.1016/j.compositesa.2012.11.012

Ma, H., Y. Li, and Y. Luo. 2011. *The effect of fiber twist on the mechanical properties natural fiber reinforced composites*. 18th International Conference on Composite Materials, Jeju, Korea, 116.

McGregor, O. P. L., M. Duhovic, A. A. Somashekar, and D. Bhattacharyya. 2017. Preimpregnated natural fibre-thermoplastic composite tape manufacture using a novel process. *Composites Part A: Applied Science and Manufacturing* 101: 59–71. doi.org/10.1016/j.compositesa.2017.05.025

Miao, M. and M. Shan. 2011. Highly aligned flax/polypropylene nonwoven preforms for thermoplastic composites. *Composite Science and Technology* 71: 1713–1718. doi.org/10.1016/j.compscitech.2011.08.001

Moothoo, J., S. Allaoui, P. Ouagne, and D. Soulat. 2014. A study of the tensile behaviour of flax tows and their potential for composite processing. *Materials and Design* 55: 764–772. doi.org/10.1016/j.matdes.2013.10.048

Netravali, A. N., and S. Chabba. 2003. Composites get greener. *Materials Today* 6(4): 22–29. 10.1016/S1369-7021(03)00427-9

Oksman, K. 2000. Mechanical properties of natural fibre mat reinforced thermoplastic, *Applied Composite Materials* 7: 403–414. doi.org/10.1023/A:1026546426764

Omrani, F., P. Wang, D. Soulat, and M. Ferreira. 2017. Mechanical properties of flaxfibre-reinforced preforms and composites: Influence of the type of yarns on multi-scale characterization. *Composites Part A: Applied Science and Manufacturing* 93: 72–81. doi.org/10.1016/j.compositesa.2016.11.013

Pickering, K. L., M. G. A. Efendy and T. M. Le. 2016. A review of recent developments in natural fibre composites and their mechanical performance. *Composites Part A: Applied Science and Manufacturing* 83: 98–112. doi.org/10.1016/j.compositesa.2015.08.038

Reyes, G., and U. Sharma. 2010. Modeling and damage repair of woven thermoplastic composites subjected to low velocity impact. *Composite Structure* 92(2):523–531. doi.org/10.1016/j.compstruct.2009.08.038

Safri, S. N. A., M. T. H. Sultan, M. Jawaid, and J. Kandasamy. 2017. Impact behaviour of hybrid composites for structural applications: a review. *Composite Part B: Engineering* 133: 112–121. doi.org/10.1016/j.compositesb.2017.09.008

Sarasini, F., J. Tirillo, M. Valente, L. Ferrante, S. Cioffi, and S. Iannace. 2013. Hybrid composites based on aramid and basalt woven fabrics: impact damage modes and residual flexural properties. *Materials and Design* 49: 290–302. doi.org/10.1016/j.matdes.2013.01.010

Sevkat, E., B. Liaw, F. Delale, and B. Raju. 2010. Effect of repeated impacts on the response of plain-woven hybrid composites. *Compos Part B Engineering* 41(5):403–413. doi.org/10.1016/j.compositesb.2010.01.001

Sutherland, L. S., and C. G. Soares. 2005. Impact characterization of low fibre-volume glass reinforced polyester circular laminated plates. *International Journal of Impact Engineering* 31(1):1–23. doi.org/10.1016/j.compositesb.2010.01.001

Torres, F. G., and M. L. Cubillas. 2005. Study of the interfacial properties of natural fibre reinforced polyethylene. *Polymer Testing* 24: 694–698. doi.org/10.1016/j.polymertesting.2005.05.004

Wambua, P., J. Ivens, and I. Verpoest. 2003. Natural fibres: can they replace glass in fibre reinforced plastics?. *Composite Science and Technology* 63: 1259–1264. doi.org/10.1016/S0266-3538(03)00096-4

Zhang, L., and M. Miao. 2010. Commingled natural fibre/polypropylene wrap spun yarns for structured thermoplastic composites. *Composite Science and Technology* 70: 130–135. doi.org/10.1016/j.compscitech.2009.09.016

5 Commingled Towpregs and Their Composites

Ganesh Jogur[a], Vinayak Ogale[b],
and R. Alagirusamy[a]
[a]Indian Institute of Technology
New Delhi, India
[b]IIT Madras Research Park
Chennai, India

CONTENTS

5.1   INTRODUCTION

Thermoplastic hybrid yarns (TPH) are thermoplastic prepegs formed by the homogeneous hybridisation of reinforcement and matrix-forming material, having sufficient flexibility to go through textile preforming without getting damaged and reducing the problems associated with a high melt viscosity of polymer resin (Alagirusamy et al. 2006). These hybrid yarns are the necessary semi-finished materials used for the consolidation of continuous fibre-reinforced thermoplastic composites. Since TPH yarns consist of both reinforcing fibre and matrix-forming polymers in the same

DOI: 10.1201/9781003049715-5

yarn structure before textile performing, they play a critical role in the consolidation and fabrication of thermoplastic composites. The primary purpose of structural yarn hybridisation is to achieve a homogeneous distribution of fibre and matrix.

Hybrid yarns with good blend quality reduce the mass transfer distance of matrix polymers during composite processing and lead to fast and complete impregnation of the reinforcement fibres. Various techniques are available in the textile industries to manufacture different types of hybrid yarns. The most commonly used methods are commingling, powder coating, co-wrapping, core/sheath spinning, and braiding. Among these techniques, the commingling method of hybrid yarn formation is more versatile, economical, and produces soft, flexible, and drapable hybrid yarn. These unique features make the commingling technology more suitable in textile preforming to be used in high-performance composites (Jogur et al. 2018). The commingling technique also offers the best solution to overcome the high polymer resin viscosity associated problems by minimising the melt flow distance of liquid resin during composite consolidation.

5.2   COMMINGLING TECHNOLOGY

The mingling process involves the manufacturing of hybrid yarn by mixing two or more fibre strands. Interlacing, tangling, entangling, intermingling, and commingling are the commonly used terms in industries to replace the word mingling. However, the term commingling is more popular. The commingling process consists of reinforcing fibres and matrix polymers in the continuous multi-filament form (Alagirusamy and Ogale 2004). The commingled hybrid yarns can be obtained in many ways. In the first method, both fibre bundle and matrix polymers can be spread out on top of each other and subsequently blended by water. This method offers a hybrid yarn having reinforcement fibres not subjected to any significant stresses (Svensson, Shishoo, and Gilchrist 1998). The commingling techniques well-known in the composite industries are air-jet associated commingling and online commingling.

5.2.1   Air-Jet Associated Commingling

Several researchers have developed commingling machines and optimised air-jet nozzles to produce commingled hybrid yarns in the past. Alagirusamy and Ogale have developed a laboratory model of a commingling machine and an air-jet nozzle (Ogale and Alagirusamy 2007). Selver et al. (2016) used an air-jet commingling nozzle (having two 90° inlets and one 45° inlet) in a Gemmill & Dunsmore twisting fancy yarn machine to produce hybrid yarn. Mankodi and Patel (2009) have also developed a commingling machine that provides flexibility to deliver both mingled and co-wrapped towpregs separately. The device runs at 800 m/min, and has simple and positive feeding with overfeed achievable up to 4%. Figures 5.1 and 5.2 show schematics of the commingling operation and nozzle used in the hybrid yarn production.

In this method of commingling process, reinforcing and matrix-forming multifilaments are simultaneously subjected via pair of feed rollers and guide rollers to the air turbulence created inside the nozzle. After mingling action, the commingled

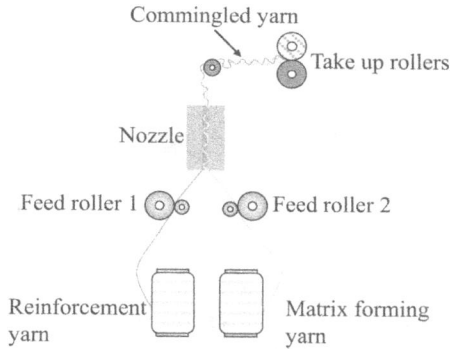

FIGURE 5.1 Schematic of the commingling process.

hybrid yarn is wound onto a final package. The air-jet nozzle is the heart of commingling machine as it facilitates fibre mingling. It consists of an oblique air inlet at the entry of filaments followed by two perpendicular air inlets in the opposite direction. Yarn bundles are made to travel through the hollow yarn channel of the nozzle under high air pressure. During the commingling process, the yarns in the form of multi-filaments are directed to the nozzle under defined tension. They get mingled under the action of perpendicular or nearly perpendicular high-pressure air stream. The turbulence created by the high-pressure air helps to open and individualise the filament bundle by rapidly vibrating them between upper and lower vortices. An oblique angle air inlet situated at the nozzle entrance offers a non-uniform division of vortices. It thus takes the responsibility of pre-opening as well as filament transportation inside the yarn channel. The other two right-angle air inlets ensure the intermingling of transported filaments. The fibre reinforcement and matrix forming filaments are separated into small bundles and form balloon shapes (open section). Inside the nozzle, small bundles of matrix-forming filaments will change their position and get trapped inside the opened bundles of fibre-forming filaments. In such a way, both filaments will undergo intertwining and mingling action to produce

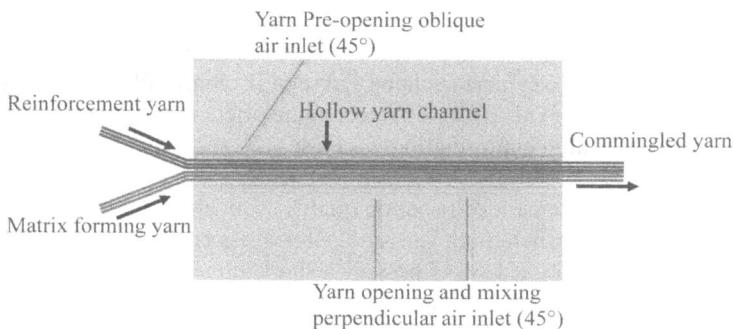

FIGURE 5.2 Air-jet nozzle configuration of the commingling machine. (Adapted from Selver et al. 2016.)

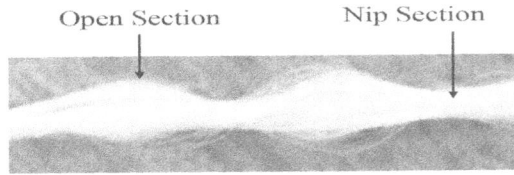

Open Section   Nip Section

FIGURE 5.3   Typical commingled hybrid yarn structure with open and nip sections.

compact sections in the hybrid yarn known as nips. The typical hybrid yarn structure will have alternate open segments and nips across the length, as shown in Figure 5.3.

The commingled yarns formed by using air-jets offer various advantages such as i) thorough distribution of reinforcing and matrix-forming yarn, ii) production of soft, flexible, and drapable yarns, iii) possiblity of processing almost any weavable reinforcing fibre and most spinnable polymers, iv) flexibility in choosing any desired fibre to matrix ratio just by changing the feed roller speeds and filament numbers, and v) cost-effective process. Other than merits, these commingled yarns do possess some limitations such as i) improper mingling results in resin-rich and resin-weak areas due to the aggregation of fibres and matrix, ii) process suitable only for the yarns which are in continuous multi-filament form, and iii) hybrid yarn structure with poor nip stability leads to de-mingling under the various stresses during textile preforming (Long, Wilks, and Rudd 2001; Richter et al. 2014; Selver et al. 2016). Commingling and co-wrapping processes can be combined to overcome the de-mingling associated problems. Here, commingled yarns are further wrapped by a continuous matrix-forming yarn to produce a hybrid yarn with excellent matrix to reinforcement distribution. Such yarn also helps in protecting the reinforcing fibres (Jogur et al. 2018).

5.2.2   ONLINE COMMINGLING TECHNIQUE

Online commingling, also known as in-situ commingling, is the most promising technique, which significantly differs from the air-jet textured commingling process, is shown in Figure 5.4. The process involves commingling both matrix and reinforcement filaments by simultaneous extrusion and size application during the melt spinning technique. An efficient technology called SpinCom was developed at the Leibniz-Institut für Polymerforschung Dresden (Leibniz-IPF) by employing the same principle (Richter et al. 2014). This technique offers many advantages over the traditional air-jet textured commingling, such as:

   i. Fibre to matrix filament distribution quality is much better since mingling is done at a stage where both yarns do not possess pronounced integrity.
   ii. The online commingled yarns possess a short impregnation path, low void content, and improved consolidation quality over air-jet textured hybrid yarns.
  iii. Excellent integrity of online commingled hybrid yarn as the size paste is applied simultaneously after extrusion. Thus, these hybrid towpregs offer superior protection when subjected to external weather conditions.

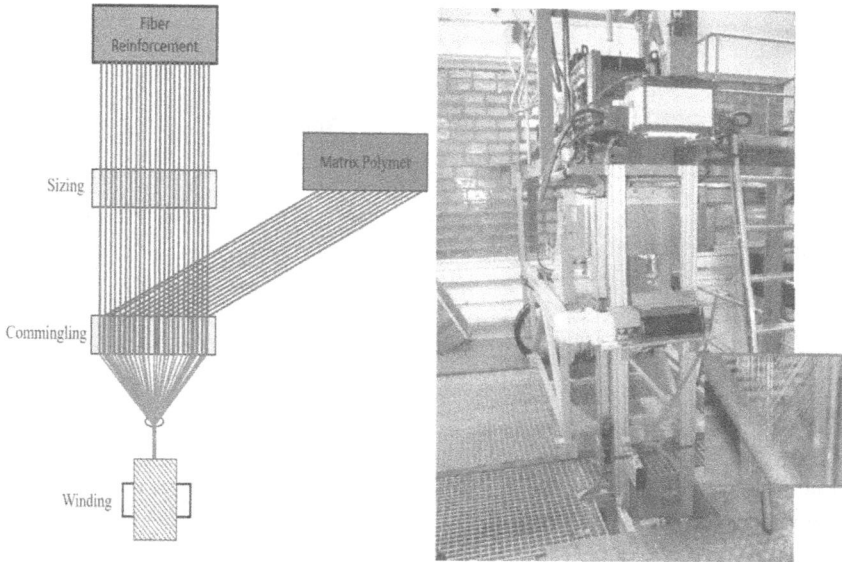

FIGURE 5.4 The principle of the online commingle process. (Adapted Mäder, Rausch, and Schmidt 2008.)

iv. Unlike air-jet textured commingling, the mechanical stress on the yarns during online commingling is very negligible.
v. The process provides hybrid yarns of no fibre breakage/damage and non-stretched polymer filaments.
vi. The process yields hybrid yarns of higher tensile strength and no thermal shrinkage during consolidation (Mäder et al. 2007; Mäder, Rothe, and Gao 2007; Mäder, Rausch, and Schmidt 2008; Wiegand and Mäder 2017).

Homogeneously blended commingled hybrid yarns are expertly produced from different manufacturers under different trade names. The most common commercial hybrid yarns are from Twintex®, COMFIL®, and Coats Synergex. Glass manufacturing company Vetrotex in France developed its first GF/PP commingled yarns in a one step process and patented it under the trade name Twintex®. COMFIL® developed and manufactured tailor-made thermoplastic composite yarns/roving and fabrics. The COMFIL® commingling technology uses compressed air to blend the high-performance fibres such as carbon, glass, aramid with thermoplastic polymers LPET, PET, PA, and PPS. Custom-made comingled hybrid yarns are produced based on the customer demand for matrix type, reinforcement type, fibre to matrix volume ratio or weight ratio, and tex/denier value for yarn/roving. Similarly, Synergex composite technology from the company Coats intimately commingles carbon fibres with different thermoplastic yarns such as nylon, PET, PEEK, PP, PPS and UHMWPE. A typical illustration of glass and PET fibre distribution across the hybrid yarn cross-section is shown in Figure 5.5.

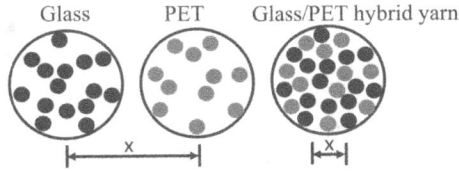

FIGURE 5.5 Cross-section of glass, PET and glass/PET commingled hybrid yarn. (COMFIL® thermoplastic composites from commingled yarn.)

5.3   FACTORS AFFECTING THE STRUCTURE AND PROPERTIES OF AIR-JET TEXTURED COMMINGLED HYBRID YARNS

The parameters influencing structure and properties of air-jet textured or commingled yarns are classified as i) raw material parameters such as filament denier, number of filaments, cross-sectional shape of fibres, filament rigidity, and frictional characteristics of filaments, ii) process parameters such as air pressure, feed ratio, take-up speed, and winding tension, iii) nozzle design and settings such as nozzle type, venturi, and baffle settings, yarn entry angle, and nozzle guide setting.

5.3.1   EFFECT OF RAW MATERIALS

5.3.1.1   Filament Fineness

Yarns with finer filaments (low diameter) are more suitable for air-jet texturing or commingling due to lower bending and torsional stiffness. As a result, smaller air drag forces are required to change the position of the filaments inside the nozzle, and they bend more easily to undergo mixing. In contrast, a higher air drag force is needed for coarser filaments to overcome their inertial resistance.

5.3.1.2   Number of Filaments

With an increase in the number of filaments, the potential for mutual filament entanglements also increases, causing yarns with a higher number of filaments to mingle better than those with fewer filaments. However, after a certain threshold limit, a further increase in the number of filaments might not give satisfactory mixing and stable yarn structure.

5.3.1.3   Filament Cross-Section Shape

Elliptical and hollow circular cross-sectional filaments mingle better than those with a solid circular cross-section. Therefore, filament yarns with non-circular cross-sections are more suitable for improved air-jet texturing/commingling than the yarns with circular cross sections (Acar, Turton, and Wray 1986).

5.3.1.4   Fibre Linear Density

With an increase in linear filament density, instability in yarns first increased and then decreased with a further rise in filament yarn linear density (Demir, Acar, and Wray 1988). Rengasamy, Kothari, and Sengupta (1990) observed that coarser filament textured yarns had lower loops, less bulk, and better retention of parent yarn tensile strength. The authors also commented that the fine denier filament yarns had better mass uniformity than coarser filament yarns. However, finer filaments below certain deniers may not be suitable for some nozzles and process parameters. These filaments produce yarns with higher bulk, lower instability, and higher loop frequency than coarser filaments. Fine denier filaments also produce yarns with higher nip frequency compared to coarser filaments (Sengupta, Kothari, and Sensarma 1996). Fibre distribution and blend quality also enhanced when commingled with the higher number of filaments having a low linear density.

5.3.1.5 Filament Modulus

With an increase in the initial filament modulus, flexural rigidity increases, resulting in reduced loop frequency, poor bulk and increased nip frequency, instability, and CV% of loop frequency. Such yarns would have an overall deterioration in textured yarn properties (Sengupta, Kothari, and Sensarma 1996). High modulus yarns may also need higher air pressure and more overfeed and low take-up speed to mingle with each other. In addition, more drag force is required for shifting the position of filament bundles as these yarns have more bending rigidity.

5.3.2 Effect of Process Parameters

Air pressure, overfeed, and delivery speed are the critical process parameters of the commingling technique, influencing the structure and properties of the produced hybrid yarns. Several authors have conducted studies to understand the effect of these process parameters (Kravaev et al. 2013; Stolyarov et al. 2016). Among them Choi, Diestel, and Offermann (1999) studied the effect of nozzle type and air pressure on the produced hybrid yarn's tensile strength and linear density. It is reported that the use of lower air pressure resulted in reduced carbon fibre damage and thus increased hybrid yarn's relative strength and yarn count. When all other process parameters were kept constant, the air texturing nozzle showed lower filament damage (due to gentle fibre treatment principle of air-jet texturing) than its inter-mingling nozzle counterpart. However, filament distribution across the yarn cross-section lacked in the air texturing nozzle compared to the intermingling nozzle. Therefore, blend quality was found to be good with the intermingling nozzle. Kravaev et al. (2014) reported that air pressure and degree of overfeeding are the main parameters influencing the degree of freedom of each filament in the continuous yarn. Higher overfeed presents loose filament bundle into the air-jet. Higher air pressure opens and spreads these free filaments and forms nip and open sections in the hybrid yarn. Therefore, blend quality was found to be better at high pressure, high overfeed conditions. They also concluded that the influence of delivery speed increases with the increase in the overfeed. The low delivery speed offered proper fibre distribution and blended quality in the yarn cross-section. In contrast, Iemoto, Chono, and Sawazaki (1986) concluded that use of high overfeed reduces the yarn tension within the nozzle. As a result filaments might be blown out of the potential air-fibre interaction zone, leading to poor nip frequnecy. Author found maximunm nip frequency 1% overfeed.

Mankodi and Patel (2009) concluded that individual processing parameters and their interaction behaviour decide the quality of commingled hybrid yarn. The hybrid yarn structural properties such as nip frequency, nip stability and nip regularity mainly depend on the interaction effect of air pressure, overfeed, and delivery speed. They reported that hybrid yarn count and blend quality increases with the increased overfeed and air pressure. In contrast, high air pressure with low overfeed gives low strength and poor quality yarn due to apparent fibre damage. Therefore, it is always recommended to maintain a balance between the blend quality and filament damage.

During commingling, the yarn must experience low tension to vibrate freely and move over each other to mingle inside the nozzle. Low yarn tension can be achieved either by increasing the overfeed or by lowering the delivery speed. Hence at constant delivery speed, with increased air pressure and overfeed, the hybrid yarn tenacity increases due to the increased number of nips and nip stability. Similar results can be obtained when the yarn is processed at low delivery speed and high air pressure. However, when both overfeed and delivery speed are low at the constant air pressure, the process gives better yarn tenacity but with poor nip characteristics (due to too low tension in yarn), which improves with higher air pressure. The increase in the delivery speed leads to a decrease in nip frequency and yields yarn with poor tenacity and non-regularity. Nip frequency in the hybrid yarn structure increases with the increase in air pressure at constant delivery speed.

Similarly, high nip frequency can also be achieved with higher air pressure with low delivery speed. The air pressure and delivery speed are the critical parameters. The interaction effect of these two plays a significant role in deciding the nip frequency. In contrast, the interaction effect between overfeeding and delivery speed has no impact on nip frequency. Nip stability is another structural property of hybrid yarn. Higher the nip stability, easier the textile preforming and better the impregnation and consolidation quality. This property is governed by the interaction effect of all three processing parameters. Higher take-up speed with low air pressure leads to better nip stability. A similar trend can be obtained at higher air pressure with a lower take-up rate. Poor nip stability results when the yarn is processed at low overfeed and low air pressure, a similar trend for low take-up speed with low overfeed. Nip regularity is enhanced when the hybrid yarn is processed at higher air pressure; the same result can be accomplished at 1% overfeed with less take-up speed (Mankodi and Patel 2009).

Ogale and Alagirusamy (2007) studied the effects of commingling on the axial and non-axial tensile behaviour of hybrid yarns. The study reported that the modulus of commingled hybrid yarn was dominated by the glass fibre, while tenacity, by the thermoplastic polymer yarns. Air pressure did not affect the tenacity of the hybrid yarns. However, in the case of GF-PET and GF-nylon hybrid yarn, an increased air pressure led to a significant drop in modulus. In GF-PP, air pressure does not affect the modulus significantly. The tenacity of hybrid yarn would reduce as the volume fraction of thermoplastic polymer increases in the hybrid yarn.

5.3.3 Effect of Nozzle Design and Settings

Many researchers in the past have conducted systematic experiments to understand the airflow profile behaviour inside the air-jet nozzle and tried to optimise the ideal nozzle configuration. Many special purpose nozzle configurations have been developed by conducting proper experiments rather than by informed theoretical considerations. They optimised the different parameters associated with the nozzle design. The parameters include the length and diameter of the hollow yarn channel, the cross-section of the air inlet orifice, the number of air entry inlets, the shape and cross-section of the yarn channel, the angle between the air inlet orifice and yarn channel, and the compressed air pressure. The performance of nozzles can also be assessed by measuring the linear density, instability, elongation at break, and tenacity of the hybrid yarn samples (Schubert 1980; Weinsdorfer 1981; Lazauskas, Lukosaitis, and Matukonis 1987; Lunenschloss and Zilg 1992; Weichun, Iemoto, and Chono 1995; Acar and Versteeg 2000). Most researchers find the circular yarn channel more effective than semi-circular or rectangular (Lazauskas, Lukosaitis, and Matukonis 1987; Weichun, Iemoto, and Chono 1995). Contradictory conclusions have been drawn on the effect of channel length on the commingling performance. Lazauskas, Lukosaitis, and Matukonis (1987) used different cross-section nozzles and reported poor mingling when a short yarn channel length is used. Iemoto, Chono, and Tanida (1986) found that with the increased nozzle length, intensity of the air-jet force acting on the filament bundle decreases, thus reduces the formation of open sections. In contrast, Lunenschloss and Zilg (1992) concluded that intermingling density improves with increased yarn channel length.

Iemoto and Chono (1987) observed that the pressure in the yarn duct does not always increase with supplied air pressure, but it predominantly depends on the diameter ratio between the yarn channel and the air entry orifice. Chono and Iemoto (1992) conducted a study on the yarn motion behaviour in an intermingling nozzle and found almost symmetric yarn motion in the axial direction of the yarn duct. Weichun, Iemoto, and Chono (1995) studied the effect of the diameter of the yarn channel and air inlet orifice on the nips' frequency. They concluded that nips per unit length do not monotonously change with the diameter of yarn channels and inlet orifices. Schubert (Schubert 1980) found that a single air inlet provides maximum nip frequency over small narrow inlets. The author also suggested that using an air inlet with slight deviation at the yarn entry side of the yarn channel would increase the nozzle effectiveness.

Most of the above studies used texturing jets that produce more energetic vortices at the exit end than the entry end of the jet. Thus, this nozzle design offers higher forwarding capacity with gentle and less commingling action. However, such forwarding action is not suitable in the production of compact commingled yarn that has a homogeneous distribution of reinforcing fibre and matrix filaments. Therefore, it is crucial to design and develop a suitable jet design, which will produce compact and stable commingled hybrid yarns with better fibre to matrix blend quality (Peckinpaugh, Stables, and Biron 1974). Numerical simulation studies can be conducted to understand the airflow pattern inside the nozzle and its effect on the high-performance fibres (carbon, glass, aramid). Some researchers have used computational fluid dynamics (CFD) to understand the airflow profile generated inside the jets. Different type of nozzles was developed using CFD as a numerical tool. Air-jet spinning jets (Zeng and Yu 2003), intermingling nozzles (Rewi and Pai 2001), and air-jet texturing jets (Rewi and Pai 2002) have been designed using CFD.

Alagirusamy et al. (2005) conducted a numerical simulation study on three different nozzle configurations to obtain an optimised nozzle design for the production of hybrid yarns from high-performance fibres. The CFD technique is used to study

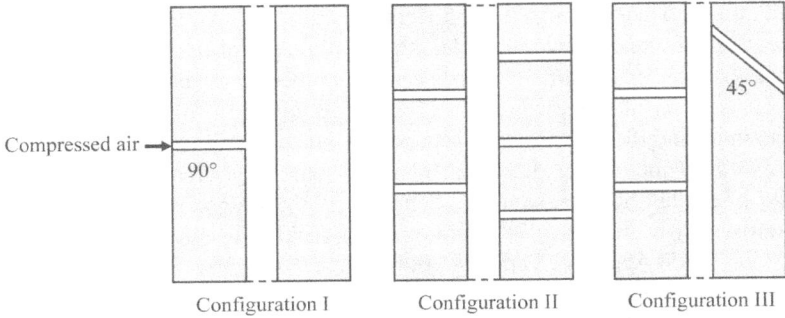

FIGURE 5.6 Three different air-jet nozzle configurations. (Adapted from Alagirusamy et al. 2005.)

the airflow profile inside the nozzle and its effect on the commingled yarn properties. The authors concluded that the airflow pattern inside the yarn channel has the most significant influence on the mixing behaviour of reinforcing fibres and thermoplastic yarns. The CFD results showed that nip frequency and degree of the interlacing of GF/nylon commingled hybrid yarns depend on the nozzle configurations. The study suggested an optimised nozzle design, i.e., configuration III, having a total of three air inlets. Among these inlets, an air inlet is kept inclined at 45° angle away from the yarn entry side. The other two air nozzles are positioned at a perpendicular direction (90°) to the axis of the yarn channel in the opposite direction. Three different nozzle configurations are illustrated in Figure 5.6.

The study concluded that Configuration III with the combination of 45° and 90° air inlets gives the non-uniform distribution of air vortices. The angle of air inlet to the yarn channel mainly governs the division of vortices into two branches. Figure 5.7

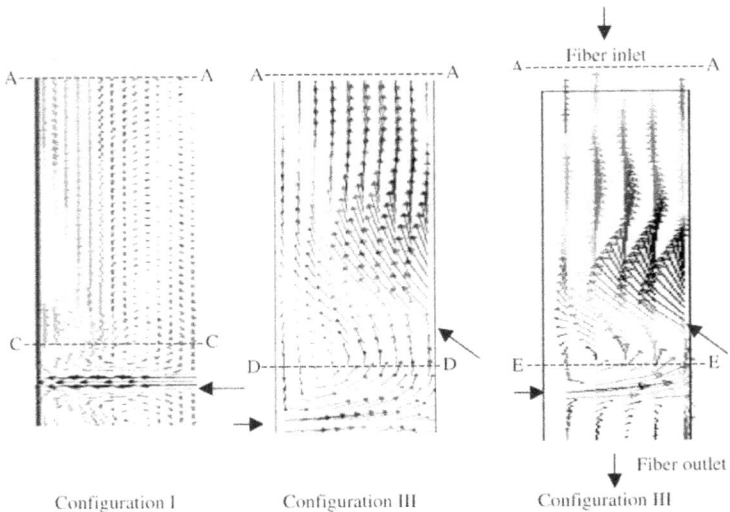

FIGURE 5.7 Airflow profile of nozzle Configuration III. (Adapted from Alagirusamy et al. 2005.)

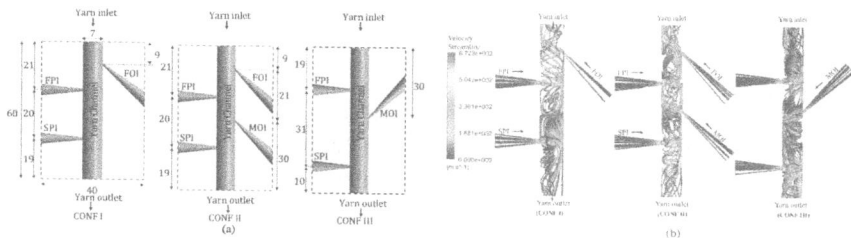

FIGURE 5.8 (a) Different nozzle designs considered for the optimization study and (b) The contours of 3D streamlines showing airflow profiles of nozzle designs. (CONF-configuration, FPI-first perpendicular inlet, SPI-second perpendicular inlet, FOI-first oblique inlet and MOI-middle oblique inlet).

depicts the vector field airflow of Configuration III. A non-uniform division of vortices and recirculation region exists at section AA-DD. Furthermore, both radial flow as well as counter-rotating vortices are present in the AA-EE area and thus become the most favourable region for the yarn bundle mixing. This section has air in semi-spiral motions and is stronger at the centre of the yarn channel. Therefore, a feed yarn bundle in the yarn channel took movement in a semi-rotatory pattern and is subjected to complex air action. The 45° oblique air inlet helps to pre-open the yarn into small bundles, while the first 90° inlet (next to the 45° inlet) helps to open the yarn bundles to form the open section. The second 90° inlet allows further binding of the filaments. The combined effect of all these inlets helps to achieve favourable mixing conditions for the commingling of high-performance fibres and polymers Jogur et al. (2021) used 3D CFD simulation and experimental approaches and developed new air-jet nozzle (CONF III) which functions with forward airflow principle. The nozzle CONF III showed high vorticity magnitude (1/s), turbulence kinetic energy (m^2/s^2) and turbulence intensity (%) and radial velocity (m/s) as compared to other nozzle designs. As a result, the optimized nozzle is capable of producing stable yarn structure with maximum nip frequency and blend quality. This nozzle design can be adopted in commercial and academically developed commingling machines to produce compact, stable, and uniformly blended commingled hybrid yarns. The nozzle designs considered for the optimization study and their airflow pattern are illustrated in Figure 5.8(a and b).

5.4 CHARACTERISATION OF COMMINGLED HYBRID YARNS

Thermoplastic commingled hybrid yarns are considered as the basic unit (semi-finished product) for textile preforming structures. The properties of the textile structure are mainly influenced by the fibre type (reinforcement and polymer), configuration, and arrangement of the constituent fibres in the architecture. Commingled yarn formed by blending the high-performance fibre with matrix polymer translates individual fibre material properties into the fabric properties. Furthermore, fabric properties such as drape, wrinkling, flexibility, permeability, and thickness are mainly affected by how the hybrid yarns are arranged in the preforming structures. Depending on the end-use applications, the geometry and design of textile architectures can be tailored by modifying the structure and properties of fibres and polymer yarns. The composite properties are also influenced by the arrangement of reinforcing fibres, uniform distribution of fibre and matrix, and impregnation of fibre and matrix. Even though commingled yarns have commercial importance and possess many valuable features, they tend to show a high amount of resin-rich and dry fibre areas when processed into a composite. These yarns tend to de-mingle under the stresses imparted during textile preforming operations due to the difference in the modulus of constituent materials. Therefore, it is necessary to produce compact and highly stable commingled hybrid yarns. Several researchers have characterised commingled yarns for their tensile, melt flow distance, nip stability, nip frequency, degree of interlacing, blend quality, etc.

Lauke, Bunzel, and Schneeider (1998) reported that the impregnation quality of the reinforcing fibre with a polymer resin mainly depends on the average flow distance of the polymer, which again depends on the homogeneous mixing of fibre and resin. Therefore, the authors developed different hybrid yarns and ranked them according to the increasing flow distance as schappe yarn (SCH), commingled yarn (COM), kemafil yarn (KEM), side-by-side yarn (SBS), and friction spun yarn (FS). SCH and COM hybrid yarns showed less melt flow distance and better mechanical properties, followed by SBS and KEM yarn. At the same time, FS hybrid yarn performed poorly. The authors also concluded that COM hybrid yarn structure with slight disturbance in fibre orientation significantly impacts inter-laminar and intralaminar delamination of uni-directional (UD) thermoplastic composites. In addition, the effect of fibre bridging in the UD composites offered the highest fracture toughness for crack initiation and the highest crack resistance during crack propagation. Mankodi and Patel (2009) also reported better mixing quality in the commingled hybrid yarns compared to the FS and hollow spindle wrapped yarns. Choi, Diestel, and Offermann (1999) reported that the mechanical performance of the Carbon/PEEK composites made from commingled yarn was superior over SBS yarns.

Two main parameters that define the quality of fibre disposition in the commingled yarn cross-section are fibre distribution and blend quality. Homogeneity of the component distribution in the yarn cross-section defines the fibre distribution, while the level of mixing between the components defines the blend quality. It should be noted that a homogenous distribution of fibres in the cross-section does not necessarily lead to a uniform blending of the elements. Fibre distribution and blend quality can be quantified by the image processing of cross-section of the hybrid yarns. The processed images will be analysed for packing density, radial distribution index (RDI), and lateral distribution index (LDI). Only a few researchers have exclusively investigated the blend quality and fibre distribution in the COM hybrid yarns. The cross-sectional analysis of COM hybrid yarn using the existing method is challenging and complicated as the yarn consists of high modulus fibres; some are highly cut resistant. Kravaev et al. (2014) analysed the cross-section of the COM hybrid yarns produced under optimum processing conditions to quantify the fibre distribution and blend quality. First, COM hybrid yarns were examined by an optical microscope. The micrographs were then captured using a digital camera. Later, the image processing technique was utilised for reliable recognition of the filament position. Various filtering processes were employed to eliminate the noise and to distinguish the fibre from the background. The original grey images were changed into binary images. After the image processing, the different filament components were identified using a coordinate system.

Both RDI and LDI of GF and PP filaments were calculated. Both indexes showed a superior disposition of the components when the applied air pressure was high. Further, Kravaev et al. (2013) introduced a new method for investigating the blend quality along the length of the COM yarn structures. The "blending index" was introduced as a new measure of the quality of the yarns. This index combines the existing radial and lateral distribution indices of fibres. The author implemented the new blending index in the range of 0 to 1 for GF/PA COM hybrid yarn. A yarn with index value of 0 indicates the perfect distribution of yarn components in the cross-section. Similarly, a yarn with index value > 0.2 gives SBS alignment of yarn components with poor blending quality. This method can be used as part of the quality management system to manufacture COM thermoplastic composites and compare different yarn structures.

Schäfer et al. (2016) investigated the effect of commingling process parameters on the filament distribution and blend uniformity in the cross-section of the produced hybrid yarns. The results revealed that there exists a correlation between process parameters and blend uniformity and fibre distribution. Both radial and lateral distribution indices showed that the distribution of carbon filaments remains constant and almost independent of the production speed. However, the increase in air pressure with lower production speed shifts the PPS (polyphenylene sulfide) filaments towards the core of the yarn cross-section. It thus leads to a better blending of the components.

Commingled hybrid yarns undergo different stresses during textile preforming. Since high-performance fibres are the constituents of these yarns and have very high tensile strength, they do not pose any problem in such conditions. However, the high-performance fibres are fragile in compression and flexural properties. Due to this, COM yarns have to negotiate acute curvatures and complex yarn paths during textile preforming, particularly during the formation of multi-axial weaves, knits and braids. Therefore, the non-axial behaviour of commingled yarns will play a critical role in deciding preforming performance. The axial and non-axial tests give a good idea about the characteristics of the structure of the COM yarns. Ogale and Alagirusamy (Ogale and Alagirusamy 2007) therefore evaluated the non-axial performance of COM yarns in terms of the ratio of loop strength or knot strength to the axial tensile strength. The authors concluded that the COM yarns in the looped or knotted form retained almost 55%–60% of the axial strength. Furthermore, COM yarns with the nips in the entangled, braid entangled, and braided form have better glass filament dispersion and thus showed higher knot strength than the SBS and core types of nips, in which glass filaments are not at all dispersed. Therefore, these non-axial tests also give an idea about the mixing of filaments in the yarn cross-section.

Brünig et al. (2003) reported the production of fine PEEK filaments with reduced filament diameter to obtain optimised hybrid yarns. The authors, in the study, explain the difficulties associated with the production of ideal hybrid yarns. They used a commercial PEEK/CF hybrid yarn of 40/60 volume %. The diameter of the carbon and PEEK filaments used was 7 μm and 39 μm, respectively. The number of individual carbon and PEEK filaments was f300 and f6. The study reveals that the mixture of PEEK of 39 μm diameter and CF of 7 μm diameter would not produce flawless thermoplastic hybrid yarns. The reason is the significant difference in the diameters of hybrid yarn components (fibre and matrix). Furthermore, when such hybrid yarns are hot-pressed, the molten PEEK polymer filaments cannot reach each reinforcing CF filament. Therefore, the thermoplastic composites made from such yarns lead to more or less unimpregnated reinforcing filament clusters.

Beyreuther, Brünig, and Vogel (2002) reported that the mechanical performance of COM hybrid yarns is influenced by the various parameters of the fibre component and the technological conditions adopted during the commingling and hot press moulding. Among different fibre parameters, the diameter ratio of matrix and reinforcing yarn determines the blend quality and performance of thermoplastic composites produced from the same hybrid components. To better understand the effect of diameter ratio of hybrid yarn, the author developed two different COM hybrid yarns, CF/PEEK and GF/PP. Furthermore, the author considered a theoretical consideration where each fibre was effectively coated with sufficient thermoplastic material during consolidation. The parameters calculated were the filament ratios and the achieved volume ratios of CF/PEEK and GF/PP hybrid yarns. The study demonstrated that a small range of difference between matrix diameter and reinforcing fibre diameter seems to be the best choice for the production of faultless COM yarns. In other words, the hybrid yarn must have the same number of matrix and fibre filaments and the same filament surfaces (volume ratio) of the components of hybrid yarn. Golzar et al. (2007) also concluded that the diameter of matrix polymer yarn must be equal to or near the reinforcement fibre's diameter for effective impregnation during hot press moulding.

Ramasamy, Wang, and Muzzy (1996a) developed the nylon powder coated towpregs and nylon/carbon commingled yarns. The yarns were characterised for flexibility and friction coefficients, which are critical for textile processability and lateral compressibility, which is essential for mouldability. Powder coated towpregs showed significantly higher friction coefficient, bending stiffness, and yarn bulk factor as compared to the uncoated and commingled yarns. In the later studies (Ramasamy, Wang, and Muzzy 1996b), they revealed that the friction and flexural properties of the towpregs affect the braidability. It is found that comingled yarns were easy to braid as compared to the powder-coated towpregs. The lower coefficient of friction and bending rigidity of the commingled yarns demonstrated higher braidability and less fibre damage. In addition, wrapping of these commingled yarns with fine nylon filaments further enhanced the braidability significantly. The authors compared the composite laminates made from nylon powder coated and commingled hybrid yarns in further studies (Ramasamy, Wang, and Muzzy 1996c). The study concluded that the commingled composites require more severe consolidation conditions than the powder coated towpregs to obtain comparable properties. It is because, in powder coated towpregs, the resin particles are well dispersed between the carbon fibres. In contrast, in commingled towpregs, nylon fibres are not well distributed in the cross-section. The mechanical properties along the braid axis of the laminates reveal that commingled yarns performed poorly against the powder coated tows even at similar processing conditions.

The processing of COM hybrid yarns in the textile preforming is quite challenging since the yarns may de-mingle, and individual yarn components may get separated. These yarns should have enough stability to withstand the various acting forces imparted during preforming. The yarns having good nip stability offers some resistance to such actions. However, the open structure of the yarn and poorly stabled yarn causes frequent production stoppages. The COM hybrid yarns can be twisted to a specific limit to avoid such problems to ensure no filament separation and production stoppages. Torun, Hoffmann, and Mountasir (2012) performed an experiment where the GF/PP COM hybrid yarns were twisted in different levels. The effect of twist on the yarn's mechanical properties and the UD composites produced were recorded. It should be noted that the structure and surface of the commingled yarns are different from the multi-filament yarns. The COM yarns have high unevenness caused by air turbulence, and the twist insertion increases the yarn's surface evenness.

Furthermore, the application of twist reduced the average yarn diameter and created a more compact yarn structure. Such twisted COM yarns increase the productivity in dense woven preforming. The authors concluded that the twisted COM yarns showed a decreasing tendency of the yarn coefficient friction. The decrease was almost linear for yarn/yarn friction. The COM yarn twisted up to 40 tpm (twist per meter) had nearly the same modulus of elongation and breaking force values as the reference yarn sample (no twist). The yarns E-modulus began to reduce after 60 tpm. A twist level of up to 20 tpm can be inserted, where a UD tensile strength reduction of 10% can be tolerated without any loss of E-modulus reduction. All these results concluded that the twist insertion significantly enhanced the intermingling of filaments and hence the mechanical properties.

5.5 CONSOLIDATION OF COMMINGLED YARN COMPOSITES

Moulding of thermoplastic composites involves primarily three factors: applied pressure, consolidation time, and temperature, which help to obtain fully impregnated fibre bundles. During the moulding process, an external load is used in the transverse direction of the composite laminates. This applied pressure removes the entrapped air and squeezes out the resin to suppress voids, making the fibre volume fraction uniform. The applied pressure in conjunction with the mould tool and its surface establishes the dimensions and surface finish of the composite parts. This consolidation process plays a significant role in governing the mechanical properties of the composites. As the pressure applied increases, the reinforcement fibres start to move towards each other. However, the further displacement of these fibres is prevented through the molten polymer matrix. After melting, the polymer droplets bear the hydrostatic pressure and thus oppose the fibre movement. The molten polymer then drives into the gaps between the reinforcing fibres due to the applied force. Therefore, compaction and the vacuum reduce the voids by filling molten polymer around the fibres and by squeezing out the air entrapped.

Consolidation time is defined as the total amount of time taken by the polymer to impregnate the dry reinforcing fibres fully. The consolidation time must be minimised to have maximum production rate and process repeatability rate. It is essential to understand the material properties and consolidation process parameters since the consolidation time mainly depends on them. Material properties include properties of fibre, yarn and fabric structures, and degree of fibre/matrix mixing in the hybrid yarn. The process parameters, including the applied pressure, rate of pressure, temperature, and time, affect the rate at which the matrix impregnation and fibre wetting take place.

To better understand the consolidation phenomenon of COM hybrid yarns, Ye, Friedrich, and Kästel (1994) explored the impregnation and consolidation behaviour CF/PP commingled yarn. The authors proposed an impregnation model and studied the effect of processing variables on the consolidation process of commingled yarn. Because of the difference in the distribution of matrix and reinforcing fibres in the preforms, the consolidation process of CF/PP commingled yarn is quite different compared to the prepreg sheets or powder/sheath fibre bundles.

The typical consolidation mechanism involves four steps, as mentioned by the authors. The first step applies pressure to the textile preforms so that the individual filaments in the hybrid yarns get flattened and try to move towards each other. As the processing temperature is raised above the melt temperature (T_m) of the PP polymer, it begins to melt and separates out of the bundles and covers the space between the fibre bundles. At this stage, one can clearly identify the border of the dry reinforcing fibre tow. In the second step, the molten polymer impregnates the dry fibre tow with the help of applied pressure and increased processing time. The process removes the entrapped air between the fibre bundles. The process continues until the complete wetting of the fibre tow (meaning full consolidation). It is assumed that the consolidation procedure of COM composites is mainly dominated by the sequence mentioned above. However, not all the molten polymer indeed leaves the fibre bundle to form a matrix pool around the fibre tow at the first step. The steps can describe the impregnation of the composites that occur in a single representative fibre bundle, as the processing steps will be the same in all other bundles. The impregnation quality is strongly influenced by the degree of commingling i.e., D_{com}. The higher the D_{com}, the faster the consolidation, and the lower the void content. Therefore, producing a stable commingled yarn preform and maintaining its status during transportation helps to achieve high impregnation and consolidation quality during subsequent manufacturing of the composites.

5.6   MECHANICAL PROPERTIES OF THE COMMINGLED YARN COMPOSITES

Ye, Friedrich, and Kästel (1994) concluded that their impregnation model could predict the void content as a function of bundle geometry and processing variables. Since the consolidation quality largely depends on the void content, the study reports an optimum processing cycle depending on the desired minimum level of void content. The authors suggested that the best mechanical properties can be achieved with only 2% void content in GF/PP COM composites. The processing temperature was about 185° C with a holding time of 18 minutes at an applied pressure of 1 MPa. A similar study was conducted to establish an optimum consolidation cycle for the Carbon/PEEK COM composites (Ye et al. 1995). Highly consolidated thermoplastic

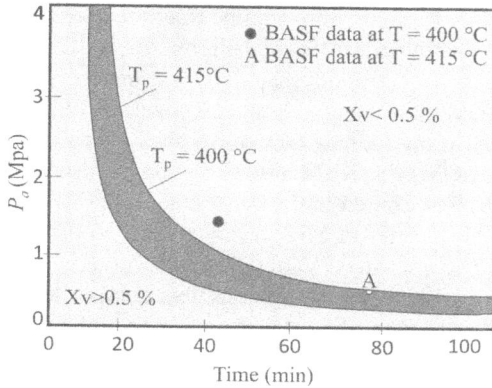

FIGURE 5.9   Processing window of Carbon/PEEK yarn. (Adapted from Ye et al. 1995.)

composites with only 0.5% void content were obtained with the CF/PEEK COM yarns processed at about 400° C temperature with over 60 min. holding time at 0.7 MPa pressure. The processing window for the Carbon/PEEK is shown in Figure 5.9.

Ogale and Alagirusamy (2008) compared the GF/PP composite laminates made from the COM, FS, and SBS techniques. Figure 5.10 depicts the tensile test results of the same composites. The results revealed that the degree of interlacing has a substantial effect on void content, impregnation quality, and tensile properties of the laminates produced from the COM yarns. On the other hand, the laminates made from SBS and FS showed more voids, dry fibres, and low tensile strength against COM composites. The poor performance of SBS and FS composites was linked to the more considerable melt flow distance between GF and PP.

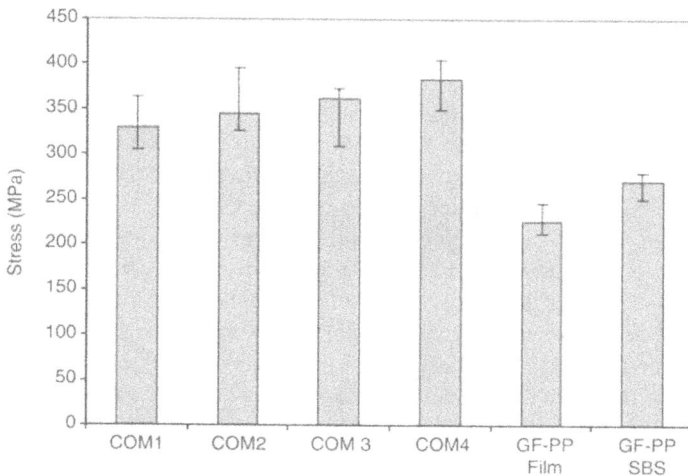

FIGURE 5.10   Tensile strength values of different COM composites, FS composites and SBS composites. (Adapted from Ogale and Alagirusamy 2008.)

Parameters like fibre strength, fibre volume fraction, fibre orientation, and discontinuous fibre reinforcement mainly affect the properties of UD composites. These, in turn, cause local stress concentrations that may cause matrix failure, interface, or fibre fracture. To understand these factors, Lauke, Bunzel, and Schneider (1998) investigated the mechanical properties of UD composites prepared from various hybrid yarns such as SCH, COM, FS, KEM and SBS. The longitudinal tensile strength of these UD composites was graded in the decreasing order as SBS, FS, KEM, COM, and SCH. The UD composites of SBS yarn with low V_f 0.46 showed much higher strength than the SCH composites (V_f of 0.56). This result is verified again using the rule of mixture relationship for the continuous fibre composites. Considering the rule of mixture theory, the strength of SCH composites should have been higher by 20% than the SBS composites. As discussed earlier, the fibre arrangement in the composite had played a critical role here. The possibility of flow of discontinuous fibre arrangement in the SCH yarn can lead to different fibre orientations—similar results in the COM yarn bundle that possess waviness. Therefore, the UD composites made of SCH and COM yarns performed poorly in the directional fibre properties. In contrast, fibres in the other variants (SBS, FS, and KEM) had more or less unidirectional orientation, hence showing good tensile strength. A similar trend was observed for the longitudinal tensile modulus.

The microstructure of the UD composites significantly influences the transverse strength and modulus. In other words, these transverse directional properties mainly depend on the reinforcement fibres orientation and their distribution in the yarn cross-section. Additionally, they get influenced by the impregnation quality that decides how good the fibre tows are wetted and firmly bonded with the surrounding matrix. The composites of SCH and COM yarn structures showed the highest transverse strength and modulus as these yarn structures consist of discontinuous fibres and misoriented fibres, respectively. In contrast, other hybrid yarns, having fibres oriented more or less in the fibre axis direction showed poor performance in transverse directional properties. The composites made from FS yarn recorded the worst transverse features due to the parallel arrangement of fibres and matrix, which resulted in insufficient impregnation because of the highest melt flow distance.

The parameters that affect the transverse properties also influence the shear strength values of the composite. The author ranked the hybrid yarns based on the decrease in interlaminar and intralaminar shear strength: SCH > COM > SBS > KEM > FS. The SCH composites offered the highest shear strength, followed by COM composites. The shear strength increments in SCH and COM hybrid yarns is due to the discontinuous and waved fibres present in the yarn, respectively. It should be noted that the shear strength value decreases either with improved fibre alignment/orientation or with the increased melt flow distance.

The authors observed the highest mode I (critical energy release rates) and mode II values to crack initiation for the laminate made with COM yarns as compared to other hybrid yarns, as shown in Figure 5.11(a and b). It is understood from Figure 5.12 that the COM composites offered superior crack propagation resistance. The primary mechanism contributing to the energy dissipation process here is fibre bridging, followed by fibre fracture. The crack propagation in the UD COM composites leads to de-bonding between the disoriented fibres and matrix. However, the same fibres

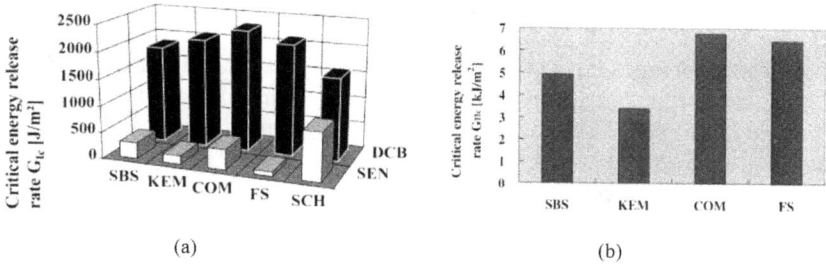

(a)   (b)

FIGURE 5.11 (a) Critical energy release rate G_{Ic} and 10 (b) G_{IIc} of different hybrid yarn composites. (Adapted from Lauke, Bunzel, and Schneider 1998.)

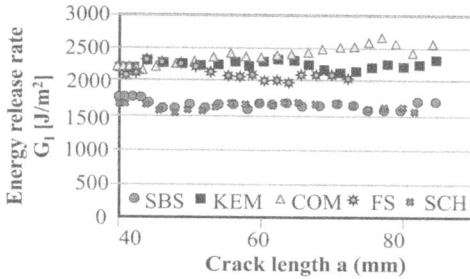

FIGURE 5.12 Crack resistance curve (R-curve) of different hybrid yarn composites. (Adapted from Lauke, Bunzel, and Schneider 1998.)

maintain connectivity on both sides of the crack and thus form fibre bridging. In other words, fibre bridging makes sure to connect the misoriented fibres that lose adhesion with the matrix even under crack propagation. Therefore, this mechanism offers the highest crack propagation resistance compared to other forms of energy dissipation mechanisms. The composites that fail with the fibre bridging mechanism thus show the highest crack resistance. The SCH composites also show this mechanism; however, peel off lengths is limited due to fibre length (discontinuous fibres). In KEM and FS composites, the delamination crack propagation occurs through fibre bundles and matrix rich regions.

Hasan et al. (2015) investigated the mechanical properties of the UD composites made of commingled CF/PA 6,6. The authors concluded that the preheating of carbon fibre at 430° C before commingling could improve fibre mixing and interface adhesion properties between PA 6,6 and CF. The pre-heating removes the thin coated size layer from the fibre surface. An important observation of this study was the significant enhancement in the tensile strength at 90° direction and apparent improvement in interlaminar shear strength due to the online preheating of fibres. However, the heat treatment resulted in lower tensile, bending, and impact strength. A systematic modification of IR (infrared) setup can improve the properties mentioned above.

5.7 SUMMARY AND FUTURE PROSPECTS

Thermoplastic composites (TPCs) are gaining importance in many industries, including aerospace, defence, medical, transportation, marine, etc. Currently, TPCs are commercially successful owing to their excellent mechanical and impact performance. However, the production of TPCs possesses some challenges that include higher viscosity of the polymeric resin and the need for higher pressure, time, and temperature for consolidation. Hybrid yarns consisting of both reinforcing fibres and matrix polymers in the same yarn structure provide the solution to mitigate the viscosity related problems. Hybrid yarns could be flexible, drapable, and soft; the preforms made from these yarns can be effortlessly processed into composite laminates. The melt flow distance of the polymeric resin plays a crucial role in determining the consolidation quality. In the cross-section of the hybrid yarn, both reinforcing fibres and matrix fibres are distributed close to each other homogeneously. This distribution reduces the melt flow distance, as the polymeric fibre is just placed next to or near the reinforcing fibres. With the application of heat and pressure, the polymer melts and immediately impregnates the surrounding fibre.

Thermoplastic COM yarns are the one variety of hybrid yarns. These yarns can be produced either using an air-jet texturing machine with some modifications or by online in-situ commingling techniques. In the cases of in-situ commingling, both fibres and matrix fibres are co-extruded and then intermingled simultaneously. The online commingling technique offers hybrid yarns with uniform fibre distribution, excellent fibre-matrix mixing, no fibre breakage, or polymer stretch. In addition, these hybrid yarns possess no thermal shrinkage and offer short melt flow distance, hence low void content and good consolidation quality. Thus, the composites made from these yarns provide better mechanical properties over air-jet textured commingled yarns. Even though online commingled yarns are superior to air-jet commingled yarns, very few studies have reported them. The challenge associated with online commingling is the co-extrusion of two different fibres.

Many researchers have conducted several studies on the air-jet textured commingled yarns. These yarns are formed by subjecting both fibre and polymer filaments to the larger air turbulence inside the nozzle. The yarn structure is identified by the formation of nips and open segments. The commingled yarns are characterised by their tensile strength, flexural strength, non-axial strength, blend quality, melt flow distance, nip frequency, nip stability, and degree of interlacing. However, these yarn properties are mainly influenced by various factors, such as process parameters, raw materials, and nozzle design. These factors are discussed in detail in Section 5.3. Therefore, it is necessary to understand the effect of these factors before starting commingling production.

In general, COM yarns offer proper fibre/matrix distribution across yarn cross-sections; however, improper selection of the parameters (mentioned earlier) leads to poor blend quality and causes resin-rich and resin weak areas after consolidation. The commingled yarns offer short melt flow distance due to better fibre and matrix mixing. The nip frequency, nip stability, and degree of the interlacing of COM yarn must be high, as they offer excellent stability to the yarn structure during preforming. The nip stability of these yarn structures can be enhanced either by heat setting the yarns above the Tg (glass transition temperature) of the polymer or by imparting a certain amount of twist or size coating to the hybrid yarn. A COM yarn with excellent nip stability, flexibility and low co-efficient of friction makes the textile performing operation much more effortless. However, the tensile properties of COM yarns are always low as compared to the parent yarns. The dis-orientation of the fibre and matrix reduces the tensile strength. In brittle fibres like glass and carbon, the tensile strength reduces due to fibre damage.

Section 5.5 details the effect of moulding parameters and the consolidation mechanism of COM yarns. The effect of different hybrid yarn structures on the composite properties such as tensile, transverse strength and modulus, shear strength, crack propagation resistance, and melt flow behaviour during consolidation are critically discussed in Section 5.6. The commingled yarns outperform other hybrid yarn structures concerning intra-laminar and inter-laminar shear strength and impact performance due to the effect of the fibre bridging mechanism. The COM yarns show a unique phenomenon called fibre bridging that helps composites resist crack propagation and hence absorb more impact energy.

Section 5.7 highlights the summary and future scopes associated with commingling technology. For future research, a new state of the art commingling machines can be fabricated with a new optimised nozzle configuration for the production of optimum hybrid yarns. The following modifications can be introduced into the machine, such as i) addition of post-yarn heating chamber and yarn tension regulator, ii) option to save process parameters, iii) sensors to determine the length/weight of the yarn wound on the final package, etc. Additionally, numerical simulation (CFD) and experimental studies can be conducted to develop new nozzle design, which can combine different fibres with dissimilar properties and yield hybrid yarns with excellent structure stability, weavability, nip frequency and mixing quality. Limited research has been reported on the enhancement of hybrid yarn stability by twisting and heat setting. Therefore, in future research, the effect of twist and heat setting on the different fibre-polymer hybrid yarn structures and properties can be explored.

Similarly, minimal articles are available on the effect of COM yarn structures on the tribological properties. Therefore, the benefit of different hybrid yarns with multiple fibres and matrix combinations can be explored in this area. Furthermore, a detailed systematic investigation can be done by comparing COM yarns with other hybrid yarns such as powder-coated towpregs, friction spun yarns, side-by-side yarns, film stacking, schappe yarns, and kemafil yarns. Characterisation of all these hybrid yarns for the ease of processability, melt behaviour, ease of textile preforming, blend quality (in yarn cross-section and along the yarn length), and consolidation quality can be studied. Hybrid yarn composites performance for mechanical, impact, fatigue and tribological properties can be studied.

Globally, comingling technology has gained some commercial success; however, it is not up to the mark. A few companies have the commercialised hybrid yarns in the international market, namely Twintex®, Comfil®, and Coats Synergex. These companies produce tailor-made COM yarns according to consumer requirements. Therefore, there is a need for the introduction of commingling technology in the Indian and international market. The technology is very versatile and can be adapted easily at low capital investment. Yet there are no Indian companies that commercially produce and market commingled yarns. Therefore, Indian composite industries need to think and act immediately.

REFERENCES

Acar, M., and H. K. Versteeg. 2000. "Air Jet Texturing and Intermingling Nozzle Designs: Easy to Manufacture and Maintain." *Chemical Fibers International* 50 (5): 502–5.

Acar, M., R. K. Turton, and G. R. Wray. 1986. "18—an Analysis of the Air-Jet Yarn-Texturing Process Part IV: Fluid Forces Acting on the Filaments and the Effects of Filament Cross-Sectional Area and Shape." *The Journal of The Textile Institute* 77 (4): 247–54. https://doi.org/10.1080/00405008608658417

Alagirusamy, R., R. Fangueiro, V. Ogale, and N. Padaki. 2006. "Hybrid Yarns and Textile Preforming for Thermoplastic Composites." *Textile Progress* 38 (4): 1–68. https://doi.org/10.1533/tepr.2006.0004

Alagirusamy, R., Vinayak Ogale, Abhijeet Vaidya, and P. M. V. Subbarao. 2005. "Effect of Jet Design on Commingling of Glass/Nylon Filaments." *Journal of Thermoplastic Composite Materials* 18 (3): 255–68. https://doi.org/10.1177/0892705705049555

Alagirusamy, Ramasamy, and Vinayak Ogale. 2004. "Commingled and Air Jet-Textured Hybrid Yarns for Thermoplastic Composites." *Journal of Industrial Textiles* 33 (4): 223–43. https://doi.org/10.1177/1528083704044360

Beyreuther, R., H. Brünig, and R. Vogel. 2002. "Preferable Filament Diameter Ratios of Hybrid Yarn Components for Optimized Longfiber Reinforced Thermoplastics." *International Polymer Processing* 17 (2): 153–7. https://doi.org/10.3139/217.1681

Brünig, H., R. Beyreuther, R. Vogel, and B. Tändler. 2003. "Melt Spinning of Fine and Ultra-Fine PEEK-Filaments." *Journal of Materials Science* 38 (10): 2149–53. https://doi.org/10.1023/A:1023719912726

Choi, B.-D., Olaf Dr Ing Diestel, and Peter Offermann. 1999. "Commingled CF/PEEK Hybrid Yarns for Use in Textile Reinforced High Performance Rotors." In *International Conference on Composite Materials.*

Chono, S., and Y. Iemoto. 1992. "Study on Interlaced Yarn. VII – Yarn Motion and Production Mechanism of Tangling Part." *Journal of the Textile Machinery Society of Japan* 45 (2): T19–25.

Demir, A., Acar, M., and Wray, G. R. 1988. Air-jet Textured Yarns: The Effects of Process and Supply Yarn Parameters on the Properties of Textured Yarns. *Textile Research Journal* 58(6): 318–28.

Golzar, M., H. Brünig, and E. Mäder. 2007. "Commingled Hybrid Yarn Diameter Ratio in Continuous Fiber-Reinforced Thermoplastic Composites." *Journal of Thermoplastic Composite Materials* 20 (1): 17–26. https://doi.org/10.1177/0892705707068069

Hasan, M. M. B., E. Staiger, M. Ashir, and C. Cherif. 2015. "Development of Carbon Fibre/Polyamide 6, 6 Commingled Hybrid Yarn for Textile-Reinforced Thermoplastic Composites." *Journal of Thermoplastic Composite Materials* 28 (12): 1708–24. https://doi.org/10.1177/0892705715604677

Iemoto, Y., S. Chono, and M. Tanida. 1986. "Model Experiment on Interlaced Yarn. I. Single Blowing Method." *Journal of the Textile Machinery Society of Japan* 39 (8): T123–130.

Iemoto, Y., Chono, S., and Sawazaki, K. 1986. Study on Interlaced Yarn Part 2: Variation of Yarn Tension, and Relation between Tangling Number and Yarn Tension. *Sen'i Kikai Gakkaishi (Journal of the Textile Machinery Society of Japan)* 39(7): T115–120.

Iemoto, Y. and S. Chono. 1987. "Study on Interlaced Yarn. III. Air Flow in Yarn Path." *Journal of the Textile Machinery Society of Japan* 40 (5): T47–56.

Jogur, Ganesh, Ashraf Nawaz Khan, Apurba Das, Puneet Mahajan, and R. Alagirusamy. 2018. "Impact Properties of Thermoplastic Composites." *Textile Progress* 50 (3): 109–83. https://doi.org/10.1080/00405167.2018.1563369

Jogur, G., Mourya, D., Alagirusamy, R., and Das, A. 2021. Numerical Simulation of Airflow Behaviour and Nozzle Geometry on Commingling for Thermoplastic Composites. *Composites Part B: Engineering* 223: 109118.

Kravaev, Plamen, Oleg Stolyarov, Gunnar Seide, and Thomas Gries. 2013. "A Method for Investigating Blending Quality of Commingled Yarns." *Textile Research Journal* 83 (2): 122–29. https://doi.org/10.1177/0040517512456760

Kravaev, Plamen, Oleg Stolyarov, Gunnar Seide, and Thomas Gries. 2014. "Influence of Process Parameters on Filament Distribution and Blending Quality in Commingled Yarns Used for Thermoplastic Composites." *Journal of Thermoplastic Composite Materials* 27 (3): 350–63. https://doi.org/10.1177/0892705712446167

Lauke, B., U. Bunzel, and K. Schneider. 1998. "Effect of Hybrid Yarn Structure on the Delamination Behaviour of Thermoplastic Composites." *Composites Part A: Applied Science and Manufacturing* 29 (11): 1397–1409. https://doi.org/10.1016/S1359-835X(98)00059-1

Lazauskas, V., A. Lukosaitis, and A. Matukonis. 1987. "Effect of the Design Parameters of the Aerodynamic Device on the Tangle Lacing Intensity of Filament Yarns." *Tekhnologiya-Tekstil'noi-Promyshlennosti* 177 (3): 27–30.

Long, A. C., C. E. Wilks, and C. D. Rudd. 2001. "Experimental Characterisation of the Consolidation of a Commingled Glass/Polypropylene Composite." *Composites Science and Technology* 61 (11): 1591–1603. https://doi.org/10.1016/S0266-3538(01)00059-8

Lunenschloss, J., and J. P. Zilg. 1992. "Intermingling of Synthetic Filament Yarns in Air Flow." *Textile Praxis International* (9): 934–36 & VII–IX.

Mäder, E., J. Rausch, and N. Schmidt. 2008. "Commingled Yarns - Processing Aspects and Tailored Surfaces of Polypropylene/Glass Composites." *Composites Part A: Applied Science and Manufacturing* 39 (4): 612–23. https://doi.org/10.1016/j.compositesa.2007.07.011

Mäder, Edith, Christina Rothe, Harald Brünig, and Thomas Leopold. 2007. "Online Spinning of Commingled Yarns-Equipment and Yarn Modification by Tailored Fibre Surfaces." *Key Engineering Materials* 334: 229–32. https://doi.org/10.4028/www.scientific.net/kem.334-335.229

Mäder, Edith, Christina Rothe, and Shang-Lin Gao. 2007. "Commingled Yarns of Surface Nanostructured Glass and Polypropylene Filaments for Effective Composite Properties." *Journal of Materials Science* 42 (19): 8062–70. https://doi.org/10.1007/s10853-006-1481-x

Mankodi, Hireni, and Pravin Patel. 2009. "Study the Effect of Commingling Parameters on Glass/Polypropylene Hybrid Yarns Properties." *Autex Research Journal* 9 (3): 70–73.

Ogale, V., and R. Alagirusamy. 2007. "Tensile Properties of GF-Polyester, GF-Nylon, and GF-Polypropylene Commingled Yarns." *Journal of the Textile Institute* 98 (1): 37–45. https://doi.org/10.1533/joti.2005.0181

Ogale, Vinayak, and R. Alagirusamy. 2008. "Properties of GF/PP Commingled Yarn Composites." *Journal of Thermoplastic Composite Materials* 21 (6): 511–23. https://doi.org/10.1177/0892705708091281

Peckinpaugh, F., Stables, W., Biron, R. 1974. Commingling Jet For Multifilament Yarn. US3828404. *United States Patent*, issued 1974.

Ramasamy, A., Youjiang Wang, and John Muzzy. 1996. "Braided Thermoplastic Composites from Powder-Coated Towpregs. Part I: Towpreg Characterization." *Polymer Composites* 17 (3): 497–504.

Ramasamy, A., Youjiang Wang, and John Muzzy. 1996. "Braided Thermoplastic Composites from Powder-Coated Towpregs. Part II: Braiding Characteristics of Towpregs." *Polymer Composites* 17 (3): 505–14.

Ramasamy, A., Youjiang Wang, and John Muzzy. 1996. "Braided Thermoplastic Composites from Powder-Coated Towpregs. Part III: Consolidation and mechanical properties." *Polymer Composites* 17 (3): 505–14.

Rengasamy, R. S., V. K. Kothari, and A. K. Sengupta. 1990. "Effect of Specific Charecteristics of the Supply Yarn on Structure and Properties of Airjet Textured Yarns. I. Effect of Filament Fineness." *Melliand Textilberichte* 71 (9): 655–659+E301-302.

Rewi, S. P., and H. Pai. 2001. "Fluid Simulation of the Air Flow in Interlacing Jets." *Textile Research Journal* 71 (7): 630–34.

Rewi, S. P., and H. Pai. 2002. "Fluid Simulation of the Airflow in Texturing Jets." *Textile Research Journal* 72 (6): 520–25.

Richter, E., K. Uhlig, A. Spickenheuer, L. Bittrich, E. Mäder, and G. Heinrich. 2014. "Thermoplastic Composite Parts Based on Online Spun Commingled Hybrid Yarns with Continuous Curvilinear Fibre Patterns." *16th European Conference on Composite Materials, ECCM 2014*, no. June: 22–26.

Schäfer, Jens, Oleg Stolyarov, Rana Ali, Christoph Greb, Gunnar Seide, and Thomas Gries. 2016. "Process–Structure Relationship of Carbon/Polyphenylene Sulfide Commingled Hybrid Yarns Used for Thermoplastic Composites." *Journal of Industrial Textiles* 45 (6): 1661–73. https://doi.org/10.1177/1528083715569372

Schubert, G. 1980. "Vortexing of Flat Yarns." *Chemiefasern/Textilindustrie* 30/82 (10): 820–23 & E92-93.

Selver, Erdem, Prasad Potluri, Paul Hogg, and Costas Soutis. 2016. "Impact Damage Tolerance of Thermoset Composites Reinforced with Hybrid Commingled Yarns." *Composites Part B: Engineering* 91: 522–38. https://doi.org/10.1016/j.compositesb.2015.12.035

Sengupta, A. K., V. K. Kothari, and J. K. Sensarma. 1996. "Effects of Filament Modulus and Linear Density on the Properties of Air-Jet Textured Yarns." *Textile Research Journal* 66 (7): 452–55. https://doi.org/10.1177/004051759606600706

Stolyarov, Oleg, Rana Ali, Christoph Greb, Gunnar Seide, and Thomas Gries. 2016. "Process–Structure Relationship of Carbon/Polyphenylene Sulfide Commingled Hybrid Yarns Used for Thermoplastic Composites." https://doi.org/10.1177/1528083715569372

Svensson, N., R. Shishoo, and M. Gilchrist. 1998. "Manufacturing of Thermoplastic Composites from Commingled Yarns - A Review." *Journal of Thermoplastic Composite Materials* 11 (1): 22–56. https://doi.org/10.1177/089270579801100102

Torun, A. R., G. Hoffmann, A. Mountasir, and C. Cherif. 2012. "Effect of Twisting on Mechanical Properties of GF/PP Commingled Hybrid Yarns and UD-Composites." *Journal of Applied Polymer Science* 123: 246–56. https://doi.org/10.1002/app.29812

Weichun, L., Y. Iemoto, and S. Chono. 1995. "Size Effect of Interlacer – III. Formulation of Number of Tangles." *Journal of the Textile Machinery Society of Japan* 48 (2): 33–42.

Weinsdorfer, H. 1981. "Mechanisms of the Intermingling of Filament Yarns." *Chemiefasern/Textilindustrie* 31/83 (3): 198–202 & E21-22.

Wiegand, Niclas, and Edith Mäder. 2017. "Commingled Yarn Spinning for Thermoplastic/Glass Fiber Composites." *Fibers* 5 (3). https://doi.org/10.3390/fib5030026

Ye, Lin, Klaus Friedrich, and Joachim Kästel. 1994. "Consolidation of GF/PP Commingled Yarn Composites." *Applied Composite Materials* 1 (6): 415–29. https://doi.org/10.1007/BF00706502

Ye, Lin, Klaus Friedrich, Joachim Kästel, and Yiu Wing Mai. 1995. "Consolidation of Unidirectional CF/PEEK Composites from Commingled Yarn Prepreg." *Composites Science and Technology* 54 (4): 349–58. https://doi.org/10.1016/0266-3538(95)00061-5

Zeng, Y. C., and C. W. Yu. 2003. "Numerical Simulation of Air Flow in the Nozzle of an Air Jet Spinning Machine." *Textile Research Journal* 73 (4): 350–56.

6 Powder Coated Towpregs and Their Composites

Vijay Goud[a], Dinesh Kalyanasundaram[bc], and R. Alagirusamy[b]

[a]3B-the Fibreglass Company
Goa, India
[b]Indian Institute of Technology Delhi
New Delhi, India
[c]All India Institute of Medical Sciences
New Delhi, India

CONTENTS

6.1 INTRODUCTION

Thermoset resins are extensively utilized as a primary phase in the polymer matrix composites. The chief thermoset polymers employed as resins in composites are unsaturated polyesters, phenolic resins, vinyl esters, and epoxy resins. Nevertheless, thermoplastic polymers of polypropylene, polyamide, and PEEK present a formidable alternative to the traditional matrix materials due to straightforward processing,

DOI: 10.1201/9781003049715-6

effortless handling, indefinite shelf life at room temperature, briefer consolidation cycle, and superior energy absorption characteristics (Liu et al. 2015; Shin et al. 2017). In spite of the virtues offered, processing of thermoplastic composites remained a hurdle to obtain void free high quality composites with similar production methods employed for thermoset matrices (Sorrentino et al. 2017). Due to lower viscosity (flow nature similar to that of water), thorough impregnation of the thermoset matrix in long fiber composites can be achieved.

On the other hand, viscosity of the thermoplastic matrix is thousands of times higher than the thermoset matrix, and use of analogous processing technologies to that of the thermoset matrix will lead to fabrication of highly porous composites (Bar, Das, and Alagirusamy 2017). The overwhelmingly high viscosity bars the impregnation of the resin into woven, knitted, and braided textile preforms to displace the air as well as produce the thoroughly impregnated, well consolidated, void free composite structures (Silva et al. 2008). This necessitates production of towpregs or prepregs prior to the production of ultimate composites (Novo et al. 2016). The methods of hot-melt impregnation, solution impregnation, and surface polymerization impregnate the viscous thermoplastic matrices with the reinforcing fibers prior to production of preformed textile structure (Steggall-Murphy et al. 2010; Vaidya and Chawla 2008).

Hot-melt technique is apt with low viscosity matrices, although a satisfactory impregnation exercising severe consolidation (high temperature, shear) can be accomplished with high viscosity resin. Probability of degradation of reinforcement as well as matrix, reduced line speeds, and inflexible towpregs limits the use of the hot melt technique with thermoplastic matrices (Kim, Weaver, and Potter 2014). In solution coating, viscosity of a resin is lowered by dissolving it in a fitting solvent. Nevertheless, these solvents are perilous to the surroundings. Moreover, the traces of solvents left in the ultimate composites despite evaporation reduces the mechanical performance of the composite. Thus, solution coating finds limited acceptance with processing of thermoplastic matrices (Liu, Xu, and Bao 2015). In surface polymerization the fiber and matrix are embodied on the fiber itself. Undesirable byproducts as a result of polymerization reaction impact performance of the composites. Electro-polymerization is an amended surface polymerization technique, although it is not compatible with most of the commonly used thermoplastic matrices (Shabani Shayeh et al. 2015). Thus, use of the matrix in dry form gained attention.

The techniques that combine the thermoplastic matrix in dry form are ring spinning, rotor spinning, DREF spinning, fiber commingling, film stacking, powder coating etc. Ring spinning and rotor spinning combine reinforcement and matrix in fibrous form. These processes have a tendency to insert the twist in the core of the hybrid yarn. The core being a high performance fiber, any torsional or flexural stresses on this high strength and modulus fibers are reported to deteriorate the mechanical properties of the ultimate composites produced. Similarly, in fiber commingling, the polymeric matrix and reinforcement are combined in fibrous form. Complete wetting of the fiber by the molten polymer is performed during consolidation. Thorough commingling of the matrix and fiber decides the effectiveness of impregnation. However, due to a difference in modulus of fiber and matrix as well as due to textile preforming techniques, resin-rich and resin-starved areas are created.

Higher external pressure is needed to consolidate the towpregs (Schäfer et al. 2016). In film stacking, alternate layers of polymer films and reinforcement are stacked. Application of high pressure on the stack causes the flow of resin in the transverse direction to reduce resin-starved areas. The major drawback is that complex shapes cannot be formed (Han, Oh, and Kim 2014). Powder coating of fibers is an effective process due to better impregnation of the matrix into fiber tows even at lower pressures (Mohanty, Drzal, and Misra 2002). High utilization, recovery of powder and elimination of pricely solvents are some of the quintessential gains of powder coating.

There are several articles available on the process efficiency of matrix deposition and impregnation by powder coating, commingling and film stacking as aforementioned. However, there are only a limited number of studies reported for comparison of matrix infiltration efficiency among processes such as powder coating, DREF spinning and film stacking (Alagirusamy 2010; Kravaev et al. 2014; Krucińska et al. 2009). It is known that the DREF spinning technique has advantages of very high rates of production with zero twist in the core yarn. DREF spinning can produce hybrid yarns with a wide range of core to sheath ratio. As no twist is imparted in the core, DREF spinning imparts minimal damage to the high performance fiber in the core. Thus, the DREF spinning process is chosen for production of core-sheath structured hybrid yarn. Similarly, powder coating nonetheless a drastically slow production process in comparison to DREF spinning deposits the powder particles in layers homogeneously over the core without changing the packing of the filaments. Powder coating has even higher flexibility of towpreg as well as a wide range of core-sheath ratio that can be accomplished in comparison to that of the DREF spinning process.

Thus, this chapter describes production of towpregs (carbon-polypropylene) using electrostatic spray coating (powder coating) and DREF spinning. Furthermore, the subsequent sections of the chapter report the production of unidirectional composites from film stacking methods as well as from towpregs obtained through the electrostatic spray coating process and DREF spinning process. Also, investigation into produced composites laminate properties in order to study the efficiency of matrix impregnation is reported. Additionally, the chapter deals with the preparation of textile preforms through 2D and 3D weaving for thermoplastic composites using powder coated towpregs and DREF spun hybrid yarn, as well as consolidation of the produced fabrics in the compression molding process to produce 2D and 3D composite laminates and investigation of their properties.

6.2 MANUFACTURING OF TOWPREGS USING THE ELECTROSTATIC SPRAY COATING SYSTEM

To achieve a flexible towpreg, carbon tow (7 μm, 350 tex, 6K, Toray Japan) is unwound with steady tension and imperiled with polypropylene powder (Newgen Specialty Plastics Limited, Noida, India, with 90% of particles of a size less than 520 μm in diameter) in the powder coating booth. The electrostatic spray coating process involves fetching the powder particles from a fluidizing hopper. The venturi sucked and conveyed polypropylene powder particles are charged to a high negative voltage by ion pounding and transferred onto the carbon tow through the amalgamation of electrostatic, aerodynamic and gravitational force. Aerodynamic forces convey the powder particles to the carbon tow and in the vicinity of the tow electrostatic forces dominate leading to the bondage of the powder particles to the tow. The tow deposited with powder particles is made to pass through a hot air convective oven where long term embodiment as a result of fusion of powder particles is attained. The process flow chart and schematic diagram for the electrostatic dry powder coating process is shown in Figures 6.1 and 6.2.

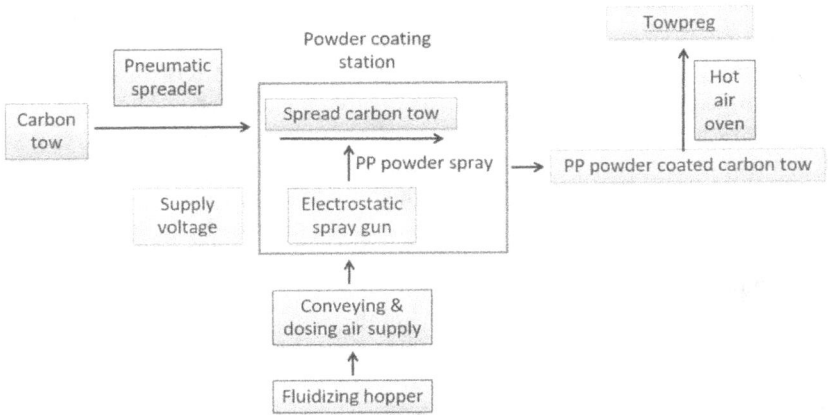

FIGURE 6.1 Flowchart for electrostatic powder coating process.

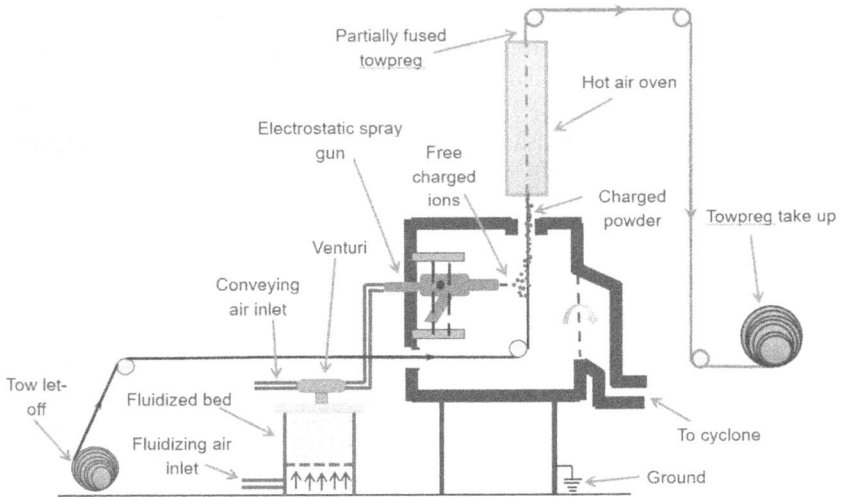

FIGURE 6.2 Illustration of electrostatic dry powder coating process.

6.3 MANUFACTURING OF HYBRID YARN BY DREF SPINNING

Three polypropylene slivers (procured from Zenith Fibres Private Limited; Gujarat, India) each of hank of 0.12 were drafted through a draft unit I, as shown in Chapter 2, Figure 2.5 of DR Ernst Fehrer friction spinning (DREF III FT, Fehrer AG, Austria) system. The slivers on drafting were individualized by the action of the carding drum. Carbon tow was made to pass through the bottom front drafting roller of draft unit II. The speed of the carbon tow is controlled by the surface speed of the take-up roller. The fibers opened to the single fiber stage by the action of the carding drum were wrapped as sheaths onto the carbon tow by means of two friction drums rotating in the same direction. The friction drums have perforations with suctions incorporated over certain portion of their surface area to help in collection and wrapping of the PP fibers over the carbon tow (Merati 2010).

6.4 PRODUCTION OF UNIDIRECTIONAL COMPOSITES (UDC)

Unidirectional composites (UDC) with carbon fiber as reinforcement and polypropylene (PP) as matrix were produced by means of three diverse approaches. In the first approach, PP powders were electrostatically charged and uniformly sprayed over the carbon tow and subsequently consolidated to form UDC-P. In the second method, a PP matrix was wrapped over the carbon tow to produce a DREF core-sheath hybrid yarn. A pile of yarns was consolidated to yield UDC-D composites. In the third approach, PP films were placed between layers of carbon tows and subjected to heat and pressure to form UDC-F. The net fiber volume fraction was ~50% in all the three UDCs. The detailed production is described below.

DREF spun hybrid yarns and powder coated towpregs were looped in parallel arrangement about a steel plate (25.5 cm (length) × 19.5 cm (width)) with 112 turns in 8 layers and subjected to heat and pressure in compression molding to yield UDC-D and UDC-P composites respectively. For the case of UDC-F, 12 PP films of 50 μm were placed between the 8 layers of the parallel wound carbon fiber tows to achieve a fiber volume fraction of 50%. A single film was placed between each layer of carbon fiber. At the top, bottom and middle layer 2 PP films were placed. The film stacking arrangement from top to bottom is of the order: 2, 1, 1, 1, 2, 1, 1, 1, 2. Compression molding was used to consolidate the composites. Consolidation was performed to eliminate voids by application of heat, pressure, and vacuum. In the curing cycle, the wound towpregs were placed in a mold between two platens of the compression molding machine and pre-heated to a temperature of 190°C. The towpregs were compressed at 10 bar pressure for 10 minutes (Bradley.J 1997). The composites were then cooled down to 100°C without the removal of the pressure. The dimensions of the finally produced composites were 20 mm (width) × 160 mm (length) × 3.2 ± 0.2 mm (thickness).

The mold comprises of a female and male part, and a cavity between them. The pressure was applied on the specimen placed in the cavity, by top and bottom platen of the compression molding machine. The sample was made to undergo three breathing cycles of 10 seconds prior to curing cycle.

6.5 FLEXURAL STRENGTH BY THREE-POINT BEND TESTS AND INTERLAMINAR SHEAR STRENGTH BY SHORT BEAM TESTS

The three-point bend results for UDC-D, UDC-P, and UDC-F are shown in Figure 6.3(a). UDC-F revealed the lowest load bearing ability whereas UDC-P showed the highest. Similarly, the modulus for UDC-P was noted to be maximum amongst the produced composites. Numerical values for flexural strength and modulus are given in Table 6.1. This superior flexural strength and flexural modulus of the composite was ascribed to better impregnation of the PP powder in UDC-P.

The Interlaminar Shear Strength (ILSS) plot for UDC-D, UDC-P, and UDC-F is illustrated in Figure 6.3(b). The trend was similar to that reported for the three-point bend test. From the illustrated plot, it can be inferred that the maximum load bearing ability for UDC-P is approximately four times that of UDC-F. The strength values are reported in Table 6.1. It was hypothesized that superior impregnation of the matrix inside the unidirectional composites (UDC-P) led to prevention of interlaminar delamination, therefore revealing superior short beam strength.

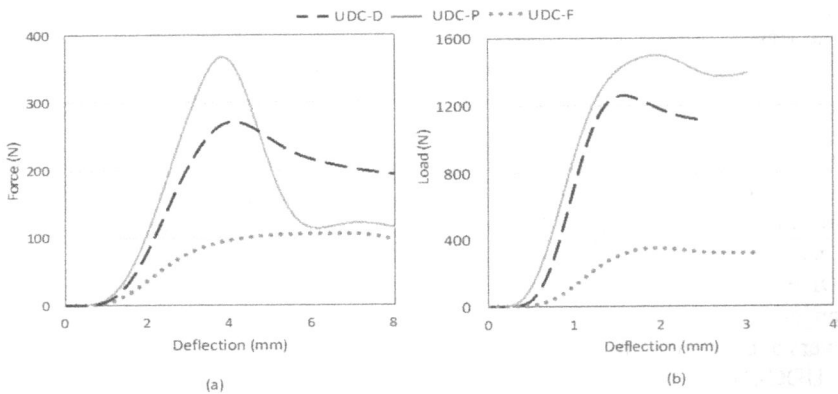

FIGURE 6.3 Load-deflection plot in (a) three-point bend test and (b) short beam test for UDC-D, UDC-P and UDC-F.

TABLE 6.1
Three-Point Bend Test Results of UDC-D, UDC-P and UDC-F
($n = 5$)

Sample	Three-Point Bend Test		Short Beam Test
Unidirectional Composite	Average Flexural Strength (MPa)	Average Flexural Modulus (GPa)	Short Beam Strength (MPa)
UDC-D	188	33	12
UDC-P	232	43	14
UDC-F	73	8	3

6.5.1 MICROSTRUCTURAL ANALYSIS

Scanning electron microscope images of UDC-D, UDC-P, and UDC-F are illustrated in Figure 6.4. A clear distinction between, the carbon fibers (denoted as F) and the matrix region (denoted as M) can be spotted for UDC-D (Figure 6.4(a1 and a2)) and UDC-F (Figure 6.4(c1 and c2)). On the other hand, for UDC-P, the fibers and matrix cannot be distinctly identified (Figure 6.4(b1 and b2)). This reveals superior penetration of polypropylene matrix into the carbon fiber in UDC-P than the other two UDCs. In UDC-D and UDC-F, the load exerted on the compression side of a three-point bend sample was not effectually conveyed between the polypropylene resin and the carbon reinforcement and hence the load bearing ability was inferior. The superior load withstanding ability of UDC-P can be attributed to its lower void of 2.84% content in comparison to 4.54% for UDC-D and 10.34% for UDC-F. The presence of air pockets (voids) in the composites drastically reduce the mechanical properties of the composites as a result of load concentration in the resin deprived region. The powder coating process can be concluded to be superior to the DREF spinning and film stacking process. Thus, to address the issue of impregnation of high viscosity thermoplastics into the continuous fiber reinforcements, the powder coating process should be preferred.

The micro-CT pictures for tested samples of UDC-D, UDC-P, and UDC-F are depicted in Figure 6.5. The matrix is denoted in green color while the fibers are represented in black. In the case of UDC-D, fiber breakage was observed in the top layer whereas bottom layer remained intact. This indicates poor transfer of applied load

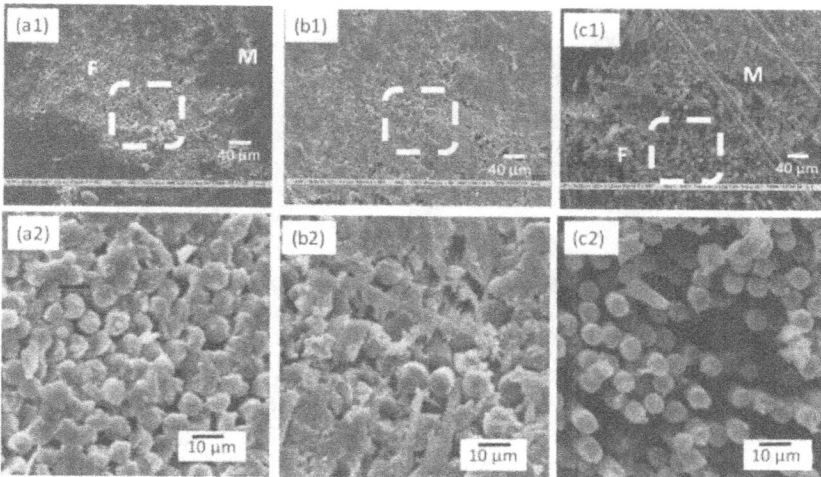

FIGURE 6.4 (a1) SEM image of a cross-section of a UDC-D (a2) SEM image at higher magnification of the encircled/highlighted area of (a1). (b1) SEM image of a cross-section of a UDC-P (b2) SEM image at higher magnification of the encircled/highlighted area of (b1). (c1) SEM image of a cross-section of a UDC-F (c2) SEM image at higher magnification of the encircled/highlighted area of (c1).

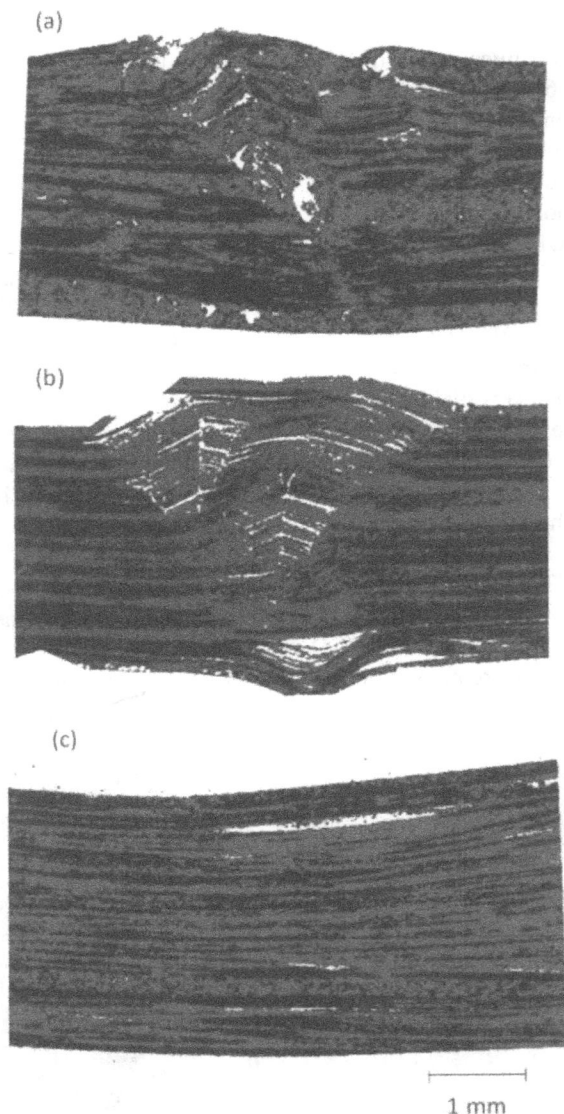

(a)

(b)

(c)

1 mm

FIGURE 6.5   Micro-CT image of the bending point (a) UDC-D, (b) UDC-P and (c) UDC-F.

from the compression side to the extension side of the specimen. This poor transfer of load can be attributed to the incomplete penetration of matrix into the compact carbon core.

On the other hand, the micro-CT image of UDC-P illustrates fiber breakage on the compression as well as the extension side. Thus, considerable amounts of load were transferred from the compression side to the extension side of the specimen. Therefore, it can be concluded that UDC-P was competent to sustain higher flexural load as well as energy due to improved fiber-matrix adhesion. This specifies significant penetration of PP powder into the individual carbon tow during compression molding. In the micro-CT image of UDC-F, no fiber breakage is noted in the compression or extension side. The UDC-F specimen failed at lesser loads (nearly $1/3^{rd}$ compared to UDC-P) in interlaminar delamination. Due to inferior adhesion between fibers and matrix, stress concentrated in the matrix deprived areas thereby leads to failure of the specimen with minimal load transfer from the compression to the extension side of the specimen.

The core-sheath DREF spun hybrid yarns used in UDC-D involved tight wrapping of a sheath of PP fibers around the core of carbon tow as shown in Figure 6.6(a1). Thus, the carbon tow in the core was subjected to significant compaction thereby restricting the impregnation of the PP matrix. During consolidation, the matrix was not able to soak into the deep layers of the carbon tow due to above cited factors of compacted carbon fibers as well as the wrapped sheath of the PP fibers. In the case of UDC-P, PP powder was uniformly sprayed on the carbon tow. (Figure 6.6(b1–b3)). Spraying of PP powder does not form a wrapped sheath around the core of the carbon as with PP fibers in DREF spinning. Thus, on melting, PP powder particles impregnate into the spaces between the carbon fibers thereby lowering the voids and increasing the adhesion between the resin and the fibers (Figure 6.6(b2)).

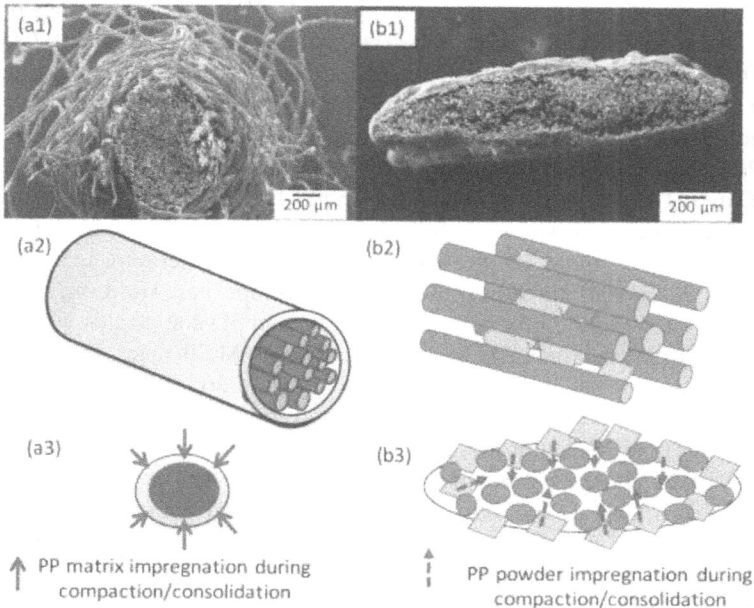

FIGURE 6.6 (a1) SEM image of the DREF hybrid yarn before consolidation to form UDC-D, (a2) schematic of PP fibers around carbon tows in the form of sheath, (a3) schematic of the impregnation of sheath during consolidation, (b1) SEM image of the PP powder coated towpreg before consolidation to form UDC-P, (b2) schematic of powder coated carbon tows, (b3) schematic of the impregnation of powder during consolidation.

Due to densification of carbon fibers in DREF spinning, the area for the impregnation of the carbon tows in UDC-D was nearly $1/3^{rd}$ the carbon tows of UDC-P (Figure 6.6(a1 versus b1)). The PP sheath around the squashed carbon tow cannot impregnate into the core (Figure 6.6(a2 and a3)) regardless of the fact that PP fibers were having lower viscosity than PP powders. The viscosity of PP powders was higher by 42% at 100 rad/s and 600% at 0.1 rad/s than PP fibers. Hence, the diminished strength of UDC-D is ascribed to the reduced interface between PP fibers and carbon tows. However, in the case of UDC-P, the enhanced impregnation of the carbon tows with PP powder forms a sounder interface and thus offers a superior load transfer. For film stacked composites (UDC-F), consisting of alternate layers of the carbon tows and PP films, SEM images reveal the matrix opulent and matrix deprived regions due to the stacking sequence of each constituent layer by layer. Profound viscosity of PP films as well as minimal MFI do not support impregnation of PP into the various layers. Therefore, it can be concluded that the UDC-P reveals high flexural strength upon consolidation.

6.6 INVESTIGATION OF MECHANICAL PROPERTIES OF 2D AND 3D WOVEN COMPOSITES FROM DREF SPUN HYBRID YARN AND POWDER COATED TOWPREGS

3D woven composites are reported to be superior to 2D woven composites in terms of resistance to impact and resistance to delamination (Castaneda et al. 2017; Gerlach et al. 2012; Liu et al. 2017; Seltzer et al. 2013). Recently, 3D woven composites have been finding increased applications in the industries such as aerospace, submarines, automobile, sporting goods, civil infrastructure etc. (Hufenbach et al. 2011; Pei et al. 2015; Pendhari, Kant, and Desai 2008; Warren, Lopez-anido, and Goering 2015). Both 2D and 3D weaving have been widely researched. Plain, twill and satin/sateen are the basic 2D weaves whereas solid structures of orthogonal, angle interlock and warp interlock are simple 3D weaves. Presently, limited literature is available on comparison of 2D and 3D woven thermoplastic composites. However, substantial studies have been reported for comparison of 2D and 3D composites with thermoset matrices (Cox, Dadkhah, and Morris 1996; Quinn, McIlhagger, and McIlhagger 2008). However, the reported results are ambiguous. Certain findings indicate 2D composites to be superior to 3D composites; however others conclude 3D composites to be worthier than 2D composites.

Hence, thorough investigation has been carried out in this section to resolve the ambiguity. Thus, two 2D and four 3D composites were woven using powder coated towpregs as well as DREF spun hybrid yarns. The static and impact mechanical properties of the six variations of the composites at a fiber volume fraction of 0.54 ± 0.03 were evaluated.

6.6.1 PRODUCTION OF FABRICS AND COMPOSITES

The cross-sectional view of the 2D and 3D woven fabrics are shown in Figure 6.7. The design plan for the production of 2D and 3D woven fabrics are given in Figure 6.8.

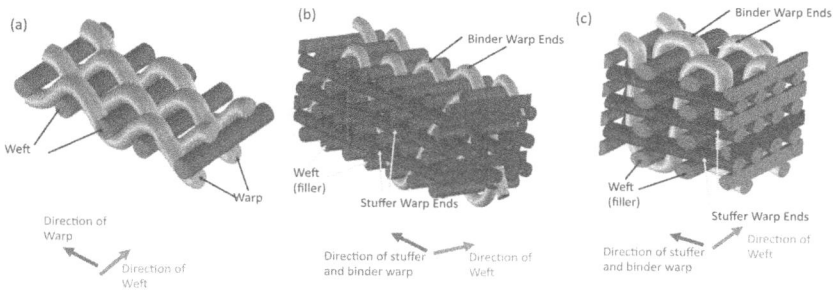

FIGURE 6.7 Three-dimensional view of (a) 2D weave pattern for 2DWF, (b) 3D angle interlock weave pattern for 3DWAF, (c) 3D orthogonal weave pattern for 3DWOF.

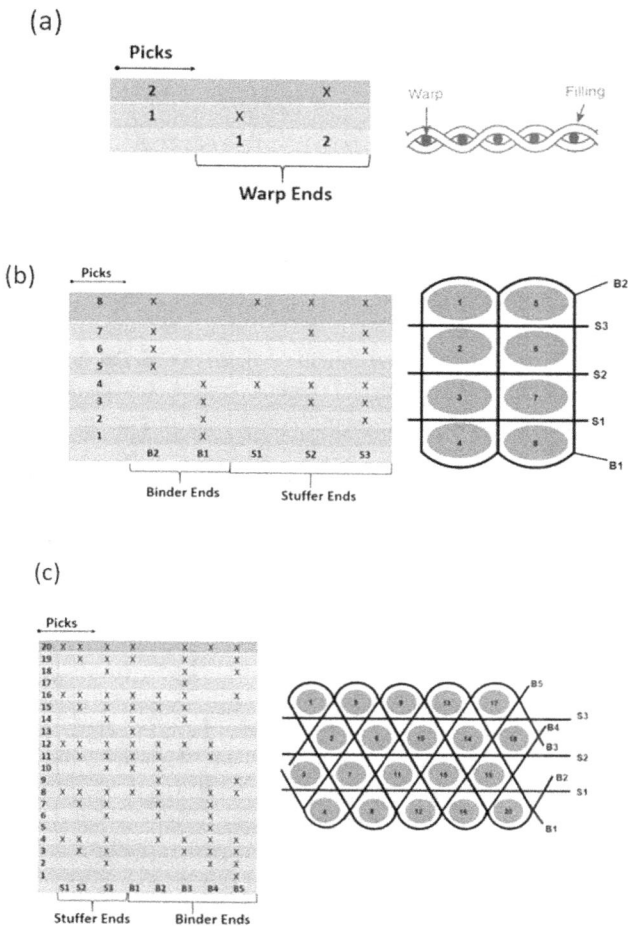

FIGURE 6.8 Peg plan for production of (a) plain woven fabric, (b) 3D woven orthogonal structure, (c) 3D woven angle interlock structure.

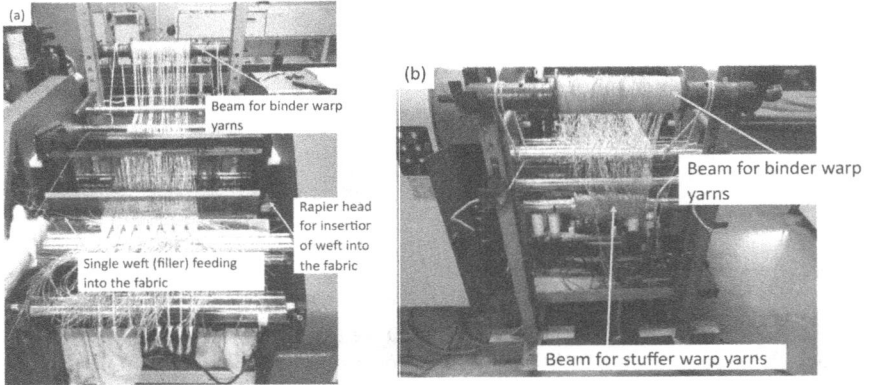

FIGURE 6.9 (a) Loom for the production of 2D and 3D woven fabrics, (b) beam arrangement for stuffer and binder warp ends.

Both DREF III spun heat set hybrid yarns and powder coated towpregs were woven into 2D and 3D fabrics using a weaving machine (Model SL8900S, CCI Tech Inc., Taiwan) as shown in Figure 6.9. During the process of weaving, the sheaths were found flaring off the surface, mainly due to abrasion at the healds and reed (image not shown). To prevent the sheath from dislodging off the carbon fiber surface and to improve the weavability of the hybrid yarns into 2D and 3D structures, the DREF III spun hybrid yarns were heat treated. The schematic of the process is shown in Figure 6.10(a). The heat treatment equipment consists of two flat heater plates with the guide rollers between them. The DREF III spun hybrid yarn was made to pass through the pre-heated surfaces set at the temperature of 180°C. Figure 6.10(c) in comparison to that of Figure 6.10(b) shows a lower number of fibers projecting away from the body thereby ensuring better weavability for the heat set DREF spun hybrid yarn.

The 2D woven fabric consists of yarns in two mutually perpendicular direction i.e. warp (x direction) and weft (y direction). A single beam is sufficient for weaving of 2D fabrics. 3D woven fabrics consist of stuffer warp yarns (x direction), filler or weft yarns (y direction) and binder warp yarns (z direction). The bottom beam was positively driven and supplied stuffer warp (x direction) yarns. The negatively driven top beam supplied binder warp yarns (y direction). Thus, two warped beams were used in the production of the 3D fabrics as shown in Figure 6.9(b). The binder warp yarns (z direction) are oriented in different angles in angle interlock and orthogonal structure. During the weaving of the angle interlock, the z binder warp yarns weave at an angle to the stuffer warp threads. However, in the case of orthogonal structure, the z binder warp yarns weave orthogonally to the stuffer warp threads. The specifications for the production of fabric were chosen such that the grams per meter square (GSM) obtained for 3D woven fabrics and 3 stacked layers of 2D woven fabrics are the same.

For 2D and 3D woven composites, initially both powder coated towpregs and DREF spun hybrid yarn were woven into fabrics: 2DWF (3 stacked layers), 3DWAF and 3DWOF with the same grams per square metre (GSM) of ~1400 and later

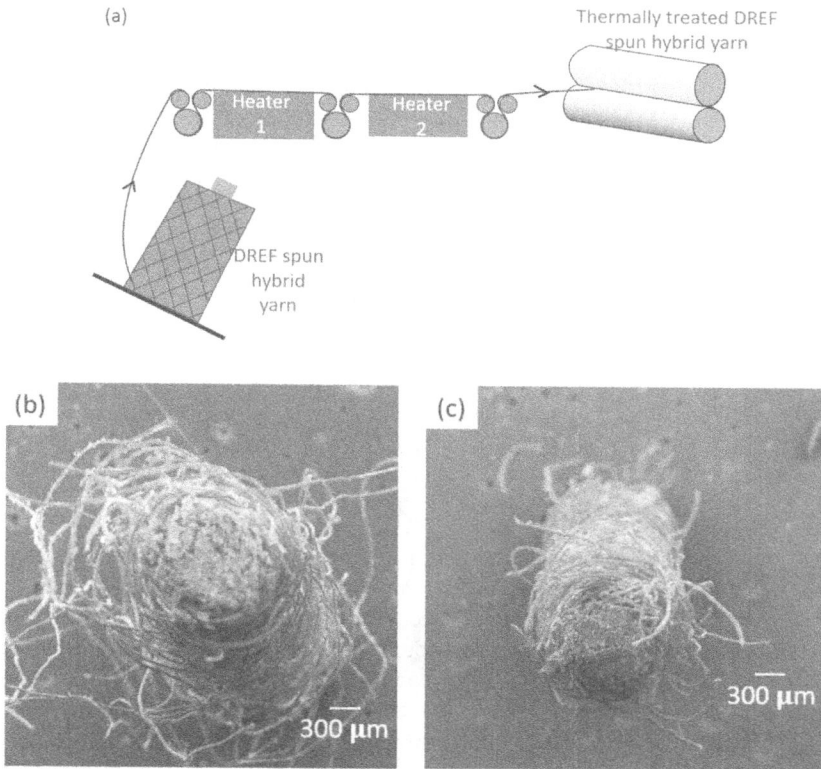

FIGURE 6.10 (a) Thermal treatment of DREF III spun hybrid yarn, (b) SEM image of DREF III spun hybrid yarn, (c) SEM image of DREF III spun heat set hybrid yarn.

consolidated using a compression molding machine to yield two variations of 2D and four variations of 3D composites. The fabric samples were cut to a size of 300 mm × 300 mm and placed between the two flat platens. Antistatic spray (Afra heavy duty silicone spray, Uttar Pradesh, India) was applied on the platens and the cut samples were placed in the aluminium foil in order to ensure easy, non-sticky removal of consolidated samples. The samples were consolidated using the breathing pressure of 10 bar, consolidation pressure of 15 bar for a duration of 15 minutes. For each of the six variations of composites, five samples were tested via a universal testing machine.

6.6.2 Tensile Properties

The tensile characteristics of the composites are specified in Table 6.2. Stress-strain curves for each type of the composites are illustrated in Figure 6.11(a). The average tensile strength and the modulus attained are represented alongside the error bar in Figure 6.11(b and c). The specimen for all the six variations of composites were loaded in the warp direction. 2DWC-P resisted maximum stress among the

TABLE 6.2

Tensile Properties for 2DWC, 3DWAC and 3DWOC ($n = 5$)

Sample	Tensile Properties	
2D and 3D Woven Composites	Tensile Strength (MPa) Mean ± (Std. dev.)	Tensile Modulus (GPa) Mean ± (Std. dev.)
2DWC-D	75 ± (3)	11 ± (1)
3DWAC-D	43 ± (4)	5 ± (1)
3DWOC-D	42 ± (4)	4 ± (1)
2DWC-P	84 ± (3)	13 ± (1)
3DWAC-P	54 ± (4)	7 ± (1)
3DWOC-P	47 ± (4)	5 ± (1)

composites produced from powder coated towpregs. The strength of 2DWC-P was found to be higher than that of 3DWAC-P and 3DWOC-P.

Likewise for the results obtained from composites of the powder coated towpregs, the maximum stress for 2DWC-D was found to be greater than that of 3DWAC-D and 3DWOC-D. As aforementioned, 3 piled layers of 2D fabrics (2DWF) were consolidated in the production of 2D composites (2DWC). On concentration of stress on such composites in which all the warp threads of the 3 layers were oriented in the same direction of the stress, the resistance to endured load is significantly higher. Whereas, for 3D composites, merely the stuffer warps resist the load as the binder warp had undertaken severe deflection/crimp and thus is incapable of opposing the applied load. Therefore, the strength of the 2D composites was superior to that of the 3D structures.

The crimp in the binder threads is expressed using Equation 6.8.

$$Crimp\ (\%) = \frac{Extended\ length\ of\ yarn - length\ of\ yarn\ in\ fabric}{length\ of\ yarn\ in\ fabric} \times 100 \qquad (6.8)$$

Binder warps for 3DWAF-P and 3DWOF-P possessed 7% and 30% of crimp respectively. The maximum strain that carbon tow resists is ~ 1% (Jones 1994). Thus as soon as the stuffer warps break the load is transferred to the binder warps. The binder warps in an attempt to straighten under the transferred load immediately fail. Amongst the 3D composites, 3DWAC has an lower crimp in comparison to that of 3DWOC.

Though the average strength of the 3DWAC was greater than 3DWOC, the values were statistically insignificant. Analogous results were observed for 2D versus 3D composites produced from DREF spun hybrid yarns.

Along with crimp, the arrangement of the successive binder warps also contribute to the strength of the composite. The optical images of the failed tensile specimens

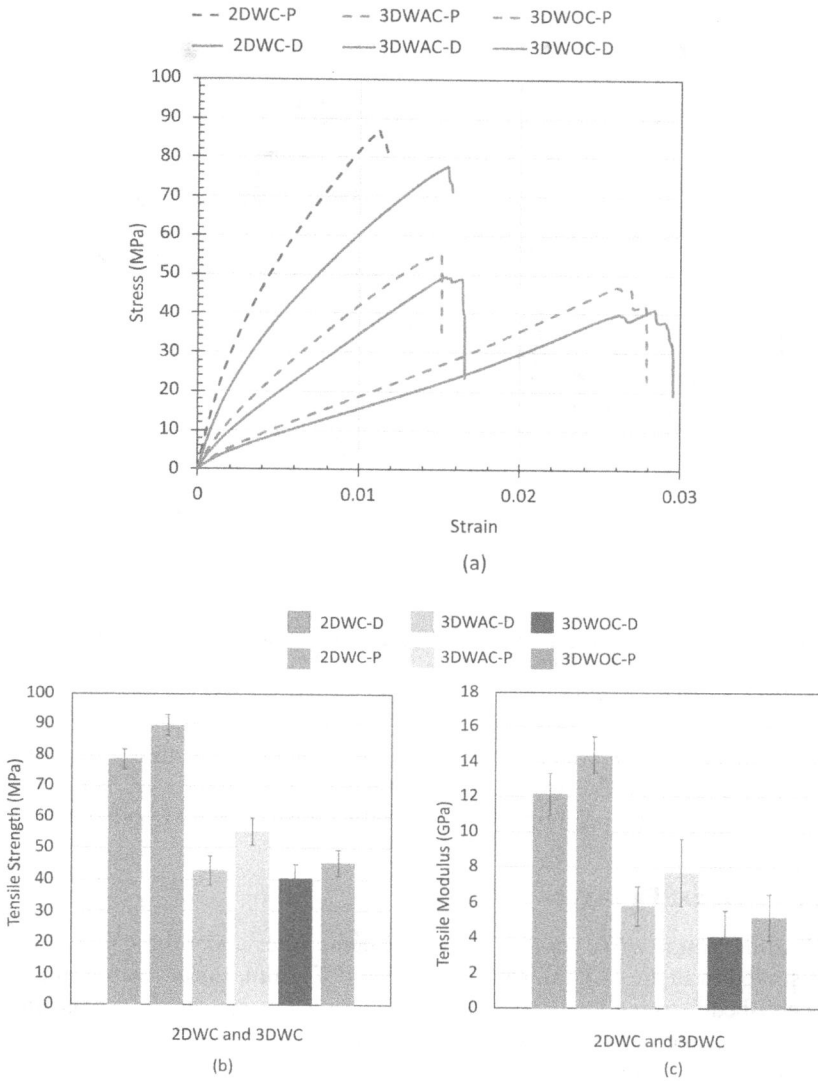

FIGURE 6.11   (a) Stress-strain plot, (b) tensile strength, and (c) tensile modulus for 2DWC, 3DWAC and 3DWOC.

of 2DWC-P, 3DWAC-P and 3DWOC-P are shown in Figure 6.12(a1–a3) and pictures of merely the fabrics 2DWF, 3DWAF and 3DWOF (i.e. without the matrix) are shown in Figure 6.12(b1–b3). The path for the specimen split-up is lengthier in the case of 2DWC trailed by 3DWAC. Thus, the tensile strength also pursues the identical pattern, i.e. $\sigma_{2DWC} > \sigma_{3DWAC} > \sigma_{3DWOC}$. Composites produced from DREF hybrid yarns (2DWC-D, 3DWAC-D and 3DWOC-D) also revealed similar paths of material separation.

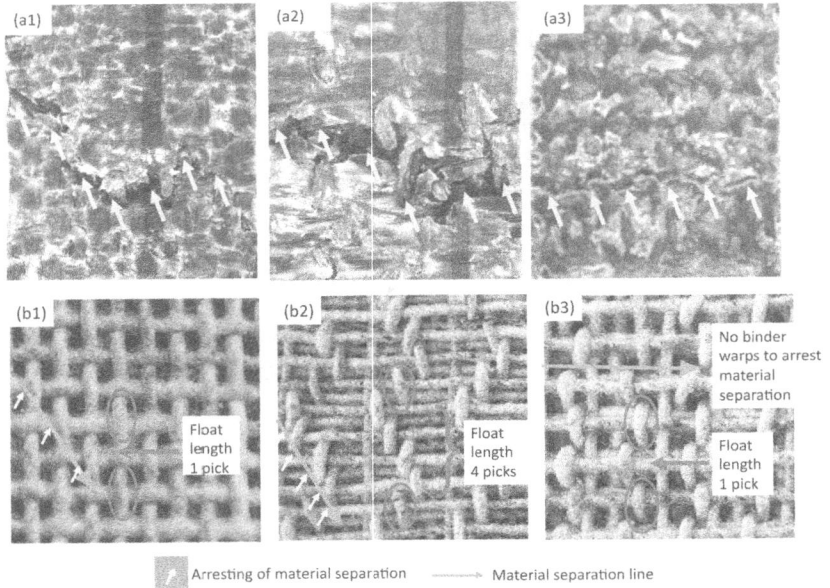

FIGURE 6.12   (a1) Warp breakage in 2DWC, (a2) stuffer warp pull out and binder warp breakage in 3DWAC, (a3) stuffer and binder warp breakage in 3DWOC; float length and material separation line in (b1) 2DWC, (b2) 3DWAC, and (b3) 3DWOC.

Also, the Young's modulus of 2DWC-P was found to be higher than that of 3DWAC-P and 3DWOC-P for the same rationale suggested above. Similarly, the modulus for 2DWC-D was found to be higher than that 3DWAC-D and 3DWOC-D.

6.6.3   FLEXURAL PROPERTIES

The load-deflection curve for the 2D and 3D composites under three point bending is illustrated in Figure 6.13(a). Upto a deflection of 2 mm, linear relationship could be established between load and deflection. Beyond 2 mm, a small increase in load showed drastic increase in deflection of the specimen. The curve is ductile in nature, a typical behavior for theromplastic composites loaded in a 3 point bend. The mean values of flexural strength and modulus are depicted in Figure 6.13(b and c) respectively. The average and standard deviation of the flexural properties of five samples tested for each of the six variations of composites are given in the Table 6.3. The flexural strength of 2DWC-P was higher than 3DWAC-P and 3DWOC-P.

Likewise, flexural strength of 2DWC-D was more than 3DWAC-D and 3DWOC-D. Also, 2DWC-D depicted higher flexural modulus than 3DWAC-D and 3DWOC-D. In the case of composites produced from powder coated towpregs, the modulus of 2DWC-P was greater than 3DWAC-P and 3DWOC-P. Overall, 2D composites presented better flexural strength and modulus in contrast to the 3D composites. This is accredited to the large number of load bearing towpregs in-line to the direction of the applied flexural load for 2D composites in comparison to 3D composites.

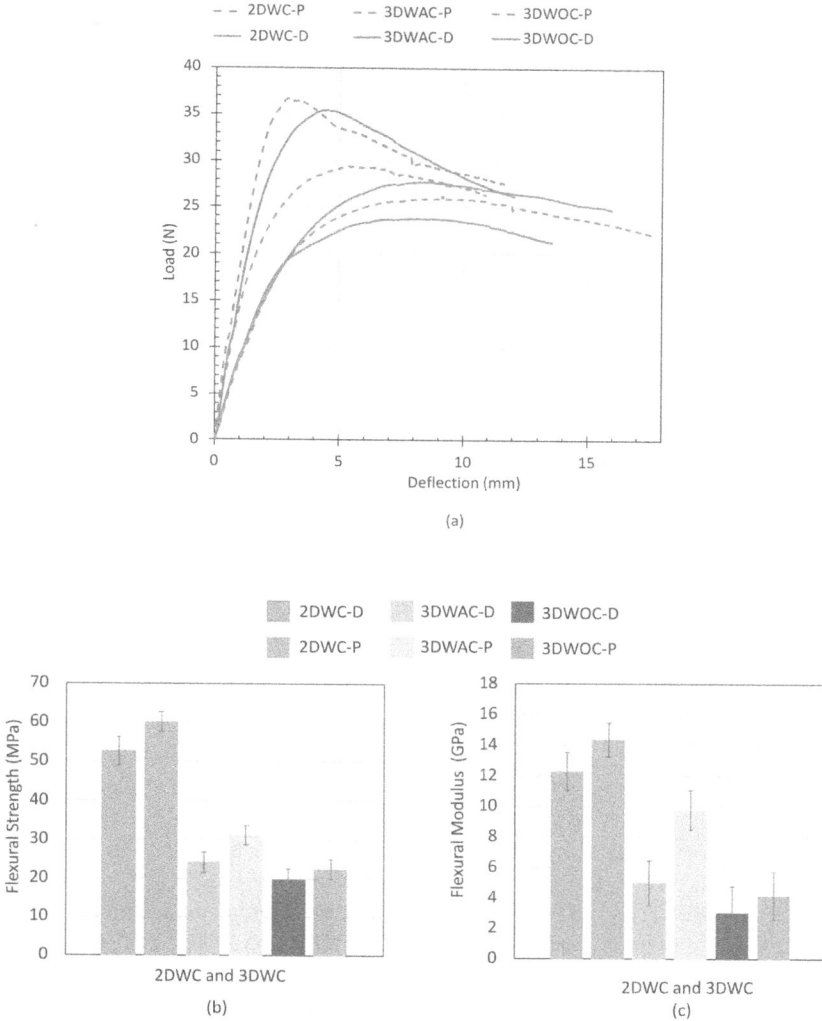

FIGURE 6.13 (a) Load-deflection plot, (b) flexural strength, and (c) flexural modulus of 2DWC, 3DWOC, and 3DWAC in 3-point bend test.

2DWC showed higher flexural strength and modulus than 3DWC. In Figure 6.13(a), a drastic drop in load was observed upon reaching peak load for 2DWC whereas the drop in load for 3DWAC and 3DWOC was not that significant. This indicates that 3DWAC and 3DWOC exhibited slow material separation on failure. This can be ascribed to the presence of binder warp threads which resists sudden failure for 3DWAC and 3DWOC, unlike 2DWC. Comparing the 3D weave structures, 3DWAC-P revealed higher flexural strength than 3DWOC-P. Also, the 3DWAC-D showed greater flexural strength than 3DWOC-D. On similar lines, the modulus of 3DWAC-P was better than 3DWOC-P.

TABLE 6.3

Flexural Properties for 2DWC, 3DWOC and 3DWAC

Sample	Flexural Properties	
	Flexural Strength (MPa)	Flexural Modulus (GPa)
2D and 3D Woven Composites	$S = \dfrac{3PL}{2bd^2}$	$E_B = \dfrac{l^3 m}{4bd^3}$
	Mean ± (Std. dev.)	Mean ± (Std. dev.)
2DWC-D	50 ± (3)	11 ± (1)
3DWAC-D	24 ± (2)	5 ± (1)
3DWOC-D	21 ± (2)	3 ± (1)
2DWC-P	57 ± (2)	13 ± (1)
3DWAC-P	31 ± (2)	9 ± (1)
3DWOC-P	23 ± (2)	4 ± (1)

6.6.4 NOTCH IMPACT PROPERTIES

The mean of impact energy for all the different composites produced is illustrated in Figure 6.14(a). The numerical values are reported in Table 6.4. In contrast to the tensile and the flexural properties, the average notch impact energy of 3DWOC was greater than that of the 2DWC. 3DWOC-P soaked more energy than 2DWC-P while 3DWOC-D absorbed more energy than 2DWC-D. The micro CT images are shown in Figure 6.14(b–f). The images indicate the failure of the 2DWC specimen because of warp breakage whereas the breakage of straight stuffer warp ends led to failure in 3DWAC and 3DWOC.

In both DREF and powder coated composites, the average values of notch impact energy of 3D orthogonal weave structure 3DWOC-P and 3DWOC-D were slightly higher than the corresponding angle interlock structures 3DWAC-P and 3DWAC-D respectively but not statistically significant. The difference in the path followed by binder warp threads in orthogonal and angle interlock structure leads to difference in energy absorption. Tighter and closer wrapping in an orthogonal structure enables it to absorb higher energy. This is noticed in Figure 6.14(c versus d). Figure 6.14(c) of angle interlock structure 3DWAC-D exhibits drastic shifting of the binder warp threads that are not perceived in Figure 6.14(d) corresponding to orthogonal weave structure 3DWOC-D structures.

This *displacement* of binder warp threads in angle interlock structure points to the looseness in the wrapping of the structure. The influence of wrapping can be appreciated via the parameter *float length* (see Figure 6.12) that refers to the number of fillers between the entry and exit point of binder warp. The float length is *1* in 3DWOF which results in a tight structure whereas 3DWAF has a float length of *4* resulting in a relatively loose structure. Thus, angle interlock structures (3DWAC) had impact absorption similar to 2DWC.

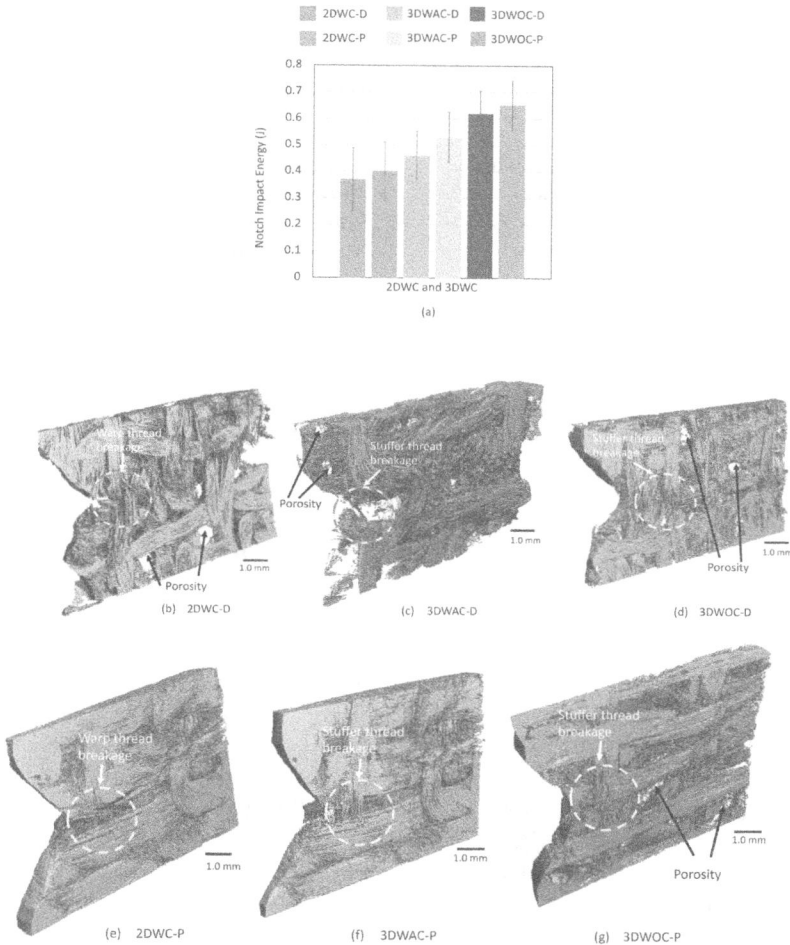

FIGURE 6.14   (a) Notch impact energy plot for all the composites. Micro-CT images of post notch-test samples of 2D and 3D woven composites from DREF spun hybrid yarn; 2DWC-D (b) sectioned slice showing fiber breakage and porosity; 3DWAC-D (c) sectioned slice showing fiber breakage and porosity; 3DWOC-D (d) sectioned slice showing fiber breakage and porosity; 2DWC-P (e) sectioned slice showing fiber breakage and porosity; 3DWAC-P (f) sectioned slice showing fiber breakage and porosity; 3DWOC-P (g) sectioned slice showing fiber breakage and porosity.

6.6.5   DREF-BASED COMPOSITES VERSUS POWDER
COATED TOWPREG-BASED COMPOSITES

Composites 2DWC-P and 3DWAC-P displayed higher tensile strength as compared to 2DWC-D and 3DWAC-D as reported in Table 6.2. The better tensile properties of 2DWC-P and 3DWAC-P are attributed to better impregnation of the PP matrix with the reinforcements as compared to the DREF counterparts. The impregnation

TABLE 6.4
Notch Impact Properties for 2DWC,
3DWOC and 3DWAC

2D and 3D Woven Composites	Notch Impact Energy (J) Mean ± (Std. dev.)
2DWC-D	0.35 ± (0.12)
3DWAC-D	0.46 ± (0.091)
3DWOC-D	0.66 ± (0.085)
2DWC-P	0.37 ± (0.11)
3DWAC-P	0.52 ± (0.095)
3DWOC-P	0.67 ± (0.092)

quality can be observed via micro-CT images (see Figure 6.14). 3DWOC-P in comparison to 3DWAC-P and 2DWC-P showed higher porosity and thereby did not show significant difference in tensile strength when compared to 3DWOC-D.

The cross-sectional SEM images of carbon tow, DREF spun hybrid yarn and powder coated towpreg are shown in Figure 6.15. The *melt-flow-distance* refers to the distance the matrix has to travel to coalesce from one reinforcement to the neighboring reinforcement. The *melt-flow-distances* are quite critical and are dependent on the manufacturing process. In the DREF spinning process, the carbon fibers in the core were wrapped with PP fibers forming a cylindrical sheath as shown in Figure 6.15(b). Hence, in the DREF spinning process, a significant compaction of carbon tow occurred as a result of wrapping of PP fibers. On the other hand, the electrostatic spray coating has not changed the elliptical cross-section of the carbon tow as shown in Figure 6.15(c). Given the circularity of the DREF spun hybrid yarn, the matrix has to flow longer distances to coalesce with the neighboring hybrid yarn. Another important aspect of discussion is the viscosity of PP fiber versus PP powder. The viscosity of PP fiber was lower by 42% at 100 rad/s and 600% at 0.1 rad/s than PP powder, as mentioned earlier. However, powder coated towpreg-based 2D and 3D composites including 2DWC-P and 3DWAC-P showed higher flexural strength than the corresponding DREF-based composites of 2DWC-D and 3DWAC-D as reported in Table 6.3. Despite higher viscosity of the PP powder matrix than PP fiber, better tensile properties and flexural properties were obtained in composites made from

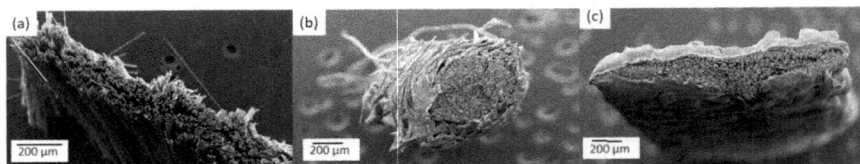

FIGURE 6.15 (a) SEM image of the carbon tow; (b) SEM image of the carbon tow wrapped with PP fibers by DREF III spinning; (c) SEM image of the carbon tow coated with PP powder by electrostatic spray coating process.

PP powder-based composites compared to PP fiber-based composites. This proves that the manufacturing process of powder coating is highly advantageous.

The micro CT images for the six variations of the composites are given in Figure 6.14. These images indicate significant porosity and less integrity in the 2DWC-D sample as compared to 2DWC-P. Higher porosity was observed in the samples from 3DWAC-D (Figure 6.14(c)) than 3DWAC-P (Figure 6.14(f)). Figure 6.14(d) shows similar results of higher porosity in 3DWOC-D in comparison to that of 3DWOC-P, shown in Figure 6.14(g). Hence, porosity is a common feature in DREF-based composites and leads to lower energy absorption.

Comparing Figure 6.14(c and d), it can be noticed that angle interlock was more prone to porosity and shock loading (3DWAC-D) than orthogonal structure (3DWOC-D). However, in the composites produced from powder coated towpregs, the differences in the angle interlock and orthogonal structure, i.e. 3DWAC-P (Figure 6.14(f)) versus 3DWOC-P (Figure 6.14(g)) are minimal and largely look quite similar. In other words, the powder coating technique helps in better integration and helps in overcoming the disadvantages of the type of weave.

6.7 CONCLUSIONS

- The effects of impregnation on the carbon reinforcement with the matrix in three different forms of fiber, powder and film were studied. The three-point bend test indicates higher flexural strength for UDC-P as compared to UDC-D and UDC-F. The superiority in transverse mechanical properties for UDC-P was attributed to higher and more uniform matrix impregnation compared to UDC-D and UDC-F. The electrostatic spray process is thereby able to consolidate the composites with superior mechanical properties at low pressures of 1 MPa. These results clearly indicate that the powder coating can be the preferred manufacturing technique for manufacturing high temperature thermoplastic composites with matrices such as PEEK, PEI and PES having high melt viscosity as compared to conventional film stacking method.

- Among the 2D versus 3D composites, 2DWC outperformed 3DWOC and 3DWAC in tensile and flexural tests whereas higher notch impact energy was observed for 3DWOC and 3DWAC. Better tensile properties for the 2DWC were attributed to the higher number of threads resisting in the direction of the applied load in addition to its lower undulation in the warp threads. Furthermore, higher flexural strength and flexural modulus along with a brittle failure were observed in 2DWC whereas lower flexural strength and flexural modulus were noticed in 3DWOC and 3DWAC along with a ductile behavior. The superior notch impact properties for 3DWOC and 3DWAC are attributed to presence of the binder warp ends that encircle the filler and stuffer warp ends.

- Among DREF yarn-based composites and powder coated towpreg-based composites, the latter yielded better results in all the mechanical properties due to better impregnation of the matrix within the structure as confirmed from the micro-computed tomography. Furthermore, porosity was quite a common feature of DREF yarn-based composites that displayed breakage of stuffer warp ends in 3DWOC and 3DWAC composites. Among the different methods of hybrid yarn production, composites manufactured from powder coated towpregs performed better than the DREF counterparts due to better matrix impregnation.

ACKNOWLEDGMENT

Part of this article was published in *Composites Part B: Engineering*, 166, Vijay Goud, Ramasamy Alagirusamy, Apurba Das, Dinesh Kalyanasundaram, Influence of various forms of polypropylene matrix (fiber, powder and film states) on the flexural strength of carbon-polypropylene composites, 56-64, Copyright Elsevier, 2019, and the other half of the article was published in *Composites Part A: Applied Science and Manufacturing*, 130, Vijay Goud, Ramasamy Alagirusamy, Apurba Das, Dinesh Kalyanasundaram, Investigation of the mechanical performance of carbon/polypropylene 2D and 3D woven composites manufactured through multi-step impregnation processes, 105733, Copyright Elsevier, 2020.

REFERENCES

Alagirusamy, R. 2010. *Hybrid Yarns for Thermoplastic Composites. Technical Textile Yarns*. Woodhead Publishing Limited, Cambridge, England. doi:10.1533/9781845699475.2.387

Bar, Mahadev, Apurba Das, and R. Alagirusamy. 2017. "Studies on Flax-Polypropylene Based Low-Twist Hybrid Yarns for Thermoplastic Composite Reinforcement." *Journal of Reinforced Plastics and Composites* 36 (11). SAGE Publications: London, England: 818–31. doi:10.1177/0731684417693428

Bradley, J. 1997. "Consolidation of Thermoplastic Powder Coated Towpreg," PhD diss., Georgia Institute of Technology.

Castaneda, Nestor, Brian Wisner, Jefferson Cuadra, Shahram Amini, and Antonios Kontsos. 2017. "Composites : Part A Investigation of the Z-Binder Role in Progressive Damage of 3D Woven Composites." *Composites Part A: Applied Science and Manufacturing* 98. Elsevier Ltd: 76–89. doi:10.1016/j.compositesa.2016.11.022

Cox, Brian N., Mahyar S. Dadkhah, and W. L. Morris. 1996. "On the Tensile Failure of 3D Woven Composites." *Composites Part A: Applied Science and Manufacturing* 27 (6). Elsevier: 447–58. doi:10.1016/1359-835X(95)00053-5

Gerlach, Robert, Clive R. Siviour, Jens Wiegand, and Nik Petrinic. 2012. "In-Plane and Through-Thickness Properties, Failure Modes, Damage and Delamination in 3D Woven Carbon Fibre Composites Subjected to Impact Loading." *Composites Science and Technology* 72 (3). Elsevier Ltd: 397–411. doi:10.1016/j.compscitech.2011.11.032

Han, Song Hee, Hyun Ju Oh, and Seong Su Kim. 2014. "Evaluation of Fiber Surface Treatment on the Interfacial Behavior of Carbon Fiber-Reinforced Polypropylene Composites." *Composites Part B: Engineering* 60: 98–105. doi:10.1016/j.compositesb.2013.12.069

Hufenbach, W., R. Bohm, M. Thieme, A. Winkler, E. Mäder, J. Rausch, and M. Schade. 2011. "Polypropylene/Glass Fibre 3D-Textile Reinforced Composites for Automotive Applications." *Materials and Design* 32 (3). Elsevier Ltd: 1468–76. doi:10.1016/j.matdes.2010.08.049

Jones, Frank R. 1994. *Handbook of Polymer-Fibre Composites*. Longman Scientific & Technical, England.

Kim, Byung Chul, Paul M. Weaver, and Kevin Potter. 2014. "Manufacturing Characteristics of the Continuous Tow Shearing Method for Manufacturing of Variable Angle Tow Composites." *Composites Part A: Applied Science and Manufacturing* 61 (June). Elsevier: 141–51. doi:10.1016/J.COMPOSITESA.2014.02.019

Kravaev, Plamen, Oleg Stolyarov, Gunnar Seide, and Thomas Gries. 2014. "Influence of Process Parameters on Filament Distribution and Blending Quality in Commingled Yarns Used for Thermoplastic Composites." *Journal of Thermoplastic Composite Materials* 27 (3): 350–63. doi:10.1177/0892705712446167

Krucińska, Izabella, Eulalia Gliścińska, E. Mäder, and R. Haler. 2009. "Evaluation of the Influence of Glass Fibre Distribution in Polyamide Matrix During the Consolidation Process on the Mechanical Properties of GF/PA6 Composites." *Fibres and Textiles in Eastern Europe* 72 (1): 81–86.

Liu, B., A. Xu, and L. Bao. 2015. "Preparation of Carbon Fiber-Reinforced Thermoplastics with High Fiber Volume Fraction and High Heat-Resistant Properties." *Journal of Thermoplastic Composite Materials* 30: 3–15. doi:10.1177/0892705715610408

Liu, Dong, Yingdan Zhu, Jiangping Ding, Xinyao Lin, and Xinyu Fan. 2015. "Experimental Investigation of Carbon Fiber Reinforced Poly(Phenylene Sulfide) Composites Prepared Using A Double-Belt Press." *Composites Part B: Engineering* 77 (August). Elsevier: 363–70. doi:10.1016/j.compositesb.2015.03.062

Liu, Gang, Li Zhang, Licheng Guo, Qimei Wang, and Feng Liao. 2017. "A Modified V-Notched Beam Test Method for Interlaminar Shear Behavior of 3D Woven Composites." *Composite Structures* 181 (December). Elsevier: 46–57. doi:10.1016/J.COMPSTRUCT.2017.08.056

Merati, A. A. 2010. "Friction Spinning." In *Advances in Yarn Spinning Technology*, 274–314. Elsevier. doi:10.1533/9780857090218.2.274

Mohanty, A. K., L. T. Drzal, and M. Misra. 2002. "Engineered Natural Fiber Reinforced Polypropylene Composites: Influence of Surface Modifications and Novel Powder Impregnation Processing." *Journal of Adhesion Science and Technology* 16 (8). Taylor & Francis Group: 999–1015. doi:10.1163/156856102760146129

Novo, P. J., J. F. Silva, J. P. Nunes, and A. T. Marques. 2016. "Pultrusion of Fibre Reinforced Thermoplastic Pre-Impregnated Materials." *Composites Part B: Engineering* 89 (March). Elsevier: 328–39. doi:10.1016/J.COMPOSITESB.2015.12.026

Pei, Xiaoyuan, Jialu Li, Kuifu Chen, and Gang Ding. 2015. "Composites : Part B Vibration Modal Analysis of Three-Dimensional and Four Directional Braided Composites." *Composites Part B* 69. Elsevier Ltd: 212–21. doi:10.1016/j.compositesb.2014.10.001

Pendhari, Sandeep S., Tarun Kant, and Yogesh M. Desai. 2008. "Application of Polymer Composites in Civil Construction: A General Review." *Composite Structures* 84 (2). Elsevier: 114–24. doi:10.1016/J.COMPSTRUCT.2007.06.007

Quinn, J. P., A. T. McIlhagger, and R. McIlhagger. 2008. "Examination of the Failure of 3D Woven Composites." *Composites Part A: Applied Science and Manufacturing* 39 (2). Elsevier: 273–83. doi:10.1016/J.COMPOSITESA.2007.10.012

Schäfer, Jens, Oleg Stolyarov, Rana Ali, Christoph Greb, Gunnar Seide, and Thomas Gries. 2016. "Process–Structure Relationship of Carbon/Polyphenylene Sulfide Commingled Hybrid Yarns Used for Thermoplastic Composites." *Journal of Industrial Textiles* 45 (6): 1661–73. doi:10.1177/1528083715569372

Seltzer, R., C. González, R. Muñoz, J. Llorca, and T. Blanco-varela. 2013. "Composites : Part A X-Ray Microtomography Analysis of the Damage Micromechanisms in 3D Woven Composites under Low-Velocity Impact." *Composites Part A: Applied Science and Manufacturing* 45. Elsevier Ltd: 49–60. doi:10.1016/j.compositesa.2012.09.017

Shabani Shayeh, J., A. Ehsani, M. R. Ganjali, P. Norouzi, and B. Jaleh. 2015. "Conductive Polymer/Reduced Graphene Oxide/Au Nano Particles as Efficient Composite Materials in Electrochemical Supercapacitors." *Applied Surface Science* 353 (October). North-Holland: 594–99. doi:10.1016/J.APSUSC.2015.06.066

Shin, Hyunseong, Byungjo Kim, Jin-Gyu Han, Man Young Lee, Jong Kyoo Park, and Maenghyo Cho. 2017. "Fracture Toughness Enhancement of Thermoplastic/Epoxy Blends by the Plastic Yield of Toughening Agents: A Multiscale Analysis." *Composites Science and Technology* 145 (June). Elsevier: 173–80. doi:10.1016/J.COMPSCITECH.2017.03.028

Silva, R. F., João F. Silva, João Pedro Nunes, Carlos A. Bernardo, and António Torres Marques. 2008. "New Powder Coating Equipment to Produce Continuous Fibre Thermoplastic Matrix Towpregs." *Materials Science Forum* 587–588 (June). Trans Tech Publications Ltd: 246–50. doi:10.4028/www.scientific.net/MSF.587-588.246

Sorrentino, Luigi, Fabrizio Sarasini, Jacopo Tirillò, Fabienne Touchard, Laurence Chocinski-Arnault, David Mellier, and Pietro Russo. 2017. "Damage Tolerance Assessment of the Interface Strength Gradation in Thermoplastic Composites." *Composites Part B: Engineering* 113: 111–22. doi:10.1016/j.compositesb.2017.01.014

Steggall-Murphy, Claire, Pavel Simacek, Suresh G. Advani, Shridhar Yarlagadda, and Shawn Walsh. 2010. "A Model for Thermoplastic Melt Impregnation of Fiber Bundles during Consolidation of Powder-Impregnated Continuous Fiber Composites." *Composites Part A: Applied Science and Manufacturing* 41 (1). Elsevier: 93–100. doi:10.1016/J.COMPOSITESA.2009.09.026

Vaidya, U. K., and K. K. Chawla. 2008. "Processing of Fibre Reinforced Thermoplastic Composites." *International Materials Reviews* 53 (4). Taylor & Francis: 185–218. doi:10.1179/174328008X325223

Warren, Kyle C., Roberto A. Lopez-anido, and Jonathan Goering. 2015. "Behavior of Three-Dimensional Woven Carbon Composites in Single-Bolt Bearing." *Composite Structure* 127. Elsevier Ltd: 175–84. doi:10.1016/j.compstruct.2015.03.022

7 Characterization of Flexible Towpregs

Ganesh Jogur, Ashraf Nawaz Khan,
and Apurba Das
Indian Institute of Technology
New Delhi, India

CONTENTS

7.1   INTRODUCTION

The structural properties of thermoplastic hybrid yarns used in the fabrication of textile preforms have considerable influence on the resultant fabric properties and preforming operation. The fabric structural properties such as flexibility, drapability, permeability, formability, thickness etc. are dependent on the structure of hybrid yarns. These properties will play a significant influence during the consolidation of composites made from the preforms. The advantage of hybrid yarns lies in their ability to tailor the required geometry and structural properties in preforms just by modifying fibre or yarn structural properties. Generally, these yarns are characterized by tensile strength, flexural stiffness, friction, compression resilience, linear density, yarn stability, regularity, and dimensional stability. From a processing point of view, they are characterized for their ease of processability in textile preforming operations such as weaving, braiding, knitting and non-woven. Other properties, namely melt flow distance, fibre to matrix volume ratio, and matrix deposition/blend quality play an important role during the composite consolidation process. The fabrics prepared from hybrid yarns having poor formability, flexibility and drapability create difficulties in fabricating complex-shaped composites. Similarly, poor permeability causes uneven matrix impregnation during compression moulding, eventually leading to poor mechanical properties. Hybrid yarns with good quality parameters will ease the textile preforming difficulties and make the process effortless.

DOI: 10.1201/9781003049715-7

Additionally, composites with good structural properties can be obtained with high quality hybrid yarns (Alagirusamy and Ogale 2004; Alagirusamy et al. 2006; Jogur et al. 2018).

The hybrid yarns can be prepared from various techniques, to name a few, powder coating, commingling, friction spinning, side-by-side, Kemafil, solution coating etc. Each of these methods produces hybrid yarns with unique structural characteristics. It is therefore essential to qualitatively analyse and understand the properties of each hybrid yarns. In this regard, the current chapter highlights the influence of different variables on the structure and properties of hybrid yarns.

7.2 FACTORS AFFECTING STRUCTURE AND PROPERTIES OF FLEXIBLE TOWPREGS

The influencing parameters that govern the quality and structure of the different hybrid yarns can be divided into two categories: namely

  i. Raw material parameters – filament linear density, number of filaments, cross sectional shape of fibres, flexural rigidity and frictional properties of filaments.
 ii. Processing parameters – type of technique and processing conditions used during hybrid yarn manufacturing (Alagirusamy at al. 2006).

The influence of these variables on hybrid yarns properties is critically discussed in Alagirusamy and Ogale (2004) and Alagirusamy et al. (2006).

7.3 FIBRE DISTRIBUTION IN HYBRID YARNS

The composite properties differ with the type of hybrid yarns used for the preforming and consolidation. The difference in properties occurs depending on how well the reinforcing and matrix-forming materials are arranged/mixed across the hybrid yarns. The homogeneous distribution of reinforcement in the composite also depends on the structure of hybrid yarns (Alagirusamy et al. 2006). The ideal hybrid yarns consist of matrix material arranged next to each fibre reinforcement. The matrix can be deposited over the yarn surface and between the individual fibres of yarn as in powder coating (Goud et al. 2018, 2019, 2020), commingling (Merter, Baüer, and Tanoğlu 2016; Ogale and Alagirusamy 2008), core-sheath deposition (Bar, Das, and Alagirusamy 2017), solution coating (Wu 2000), melt coating (Angell et al. 1986), etc. Several studies have been conducted to understand the fibre-matrix distribution across the hybrid yarn structure and composites.

Among those studies, Lauke, Bunzel, and Schneider (1998) investigated the fibre distribution quality of different yarn structures by determining their melt flow distance, which is a critical parameter in deciding composite consolidation quality. The authors ranked yarns according to the increment in average flow distance – Schappe yarn, commingled yarn, Kemafil yarn, side-by-side yarn and lastly friction spun yarn. As per their findings, Schappe and commingled hybrid yarns show the best degree of mixing of composite components. Figure 7.1 shows the different hybrid yarn structures used in composite applications with the ascending order of resin melt flow during consolidation.

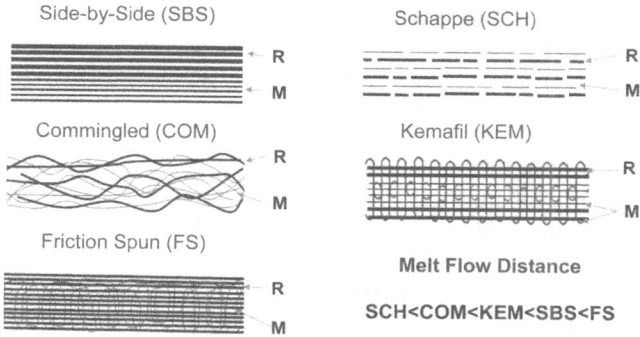

FIGURE 7.1 Longitudinal views of different hybrid yarn structures with their resin melt flow rate in ascending order (Adapted from Lauke, Bunzel, and Schneider 1998). [Note: R denotes reinforcement fibre and M denotes matrix material.

Goud et al. (2019) compared the mechanical properties of laminates prepared from electrostatic powder coating, DREF spinning, and film stacking techniques. The micro-computed tomography and scanning electron microscopy studies revealed better impregnation of PP in powder coated UD carbon composite (UDC-P) as compared to DREF (UDC-D) and film stacked (UDC-F) composites, respectively. Therefore, UD composites made from powder coating techniques showed better performance in 3-point bending and short beam tests. The poor load bearing capacity and non-uniform load transfer were observed for the composites made from UDC-D and UDC-F due to the higher void percentage and improper resin impregnation as compared to UDC-P. The void content observed was 2.84%, 4.54% and 10.34% for UDC-P, UDC-D and UDC-F composites, respectively. The lower void content of powder coated composites showed the superiority of the powder coating process over other two process.

The micro-city scan studies of these composites also reveal the superiority of UDC-P. The powder coated composites were able to absorb higher flexural load and energy due to the substantial impregnation of PP into the individual carbon fibres. The high flexural load and energy were associated with the fibre as well as matrix breakage in the top layer due to compression. The fibre slack was observed in the bottom layer indicating stretch under tensile forces. On the other hand, UDC-D showed substantial fibre breakage in top layers under compressional force and breakage was observed under tensile forces in bottom layers. This signifies the partial transfer of load between the fibre and matrix, which was due to the incomplete impregnation of resin into the compact carbon core. In contrast, UDC-F showed no fibre breakage either on top layers or on bottom layers, and simply failed at lower flexural load with inter-laminar delamination due to the poor impregnation of fibre regions. Due to the inadequate bonding between the fibre and matrix, transfer of load was minimal and hence no fibre breakage was noticed.

In the case of DREF yarn, the core fibre was subjected to the significant compaction between friction drums and hence the area of carbon tows reduced to 1/3rd the carbon tows of powder coated yarns. Therefore, the sheath around the carbon fibre fails to impregnate the compacted tow as the compaction reduces the effective space required for the resin impregnation. However, in the case of spray coated yarns, the powder was equally spread as micron-sized particles over the entire tow length. During consolidation, these powder particles spread uniformly under heat and pressure and flow easily into the spaces effectively (within carbon tow structure) due to capillary action. For film stacked composites, SEM images clearly indicated the resin rich and resin poor areas due to the high viscosity and low MFI properties of PP film as well as the nature of layer-by-layer stacking. These factors do not support the good wetting of each fibre layer and hence poorer impregnation observed for UDC-F than for UDC-D and UDC-P, respectively.

The study conducted by Merter, Baüer, and Tanoğlu (2016) concluded that composite prepared from air jet commingled hybrid yarns had better impact properties over direct twisted hybrid yarns. Bar, Das, and Alagirusamy (2017) concluded that flax-PP laminates made from DREF spun hybrid yarns show better mechanical properties over the film stacking technique. However, in later studies, it is revealed that flax-PP laminates prepared from thermally bonded roving (TBR) show better tensile, flexural and impact properties than DREF yarn compressed composites (Bar, Alagirusamy, and Das 2019). This was mainly attributed to the structure of hybrid yarns used. TBR showed uniform flax fibre matrix distribution across yarn cross sections as well in composite structure, almost as inseparable state, while in DREF composites, flax fibres are randomly distributed in the PP matrices.

Schneeberger, Wong, and Ermanni (2017) demonstrated a novel class of preform materials known as "hybrid bi-component fibres" for processing thermoplastic composites. A typical cross-section of unidirectional preforms composed of bi-component fibres is depicted in Figure 7.2. The study compared the unit cells of randomly mixed commingled yarns with a unit cell of hybrid bicomponent fibres, as shown in Figure 7.3. The authors concluded that laminates made from commingled yarns will have a bundle of reinforcement fibres that need to be filled during consolidation. On the other hand, preforms made with bi-component fibres barely show any dry areas as they avoid any impregnation flows altogether during consolidation. These preforms merely require a coalescence of the molten thermoplastic sheaths and an extraction or collapsing of remaining voids. Therefore, the preforms made from bi-component fibres have the potential to be processed and formed on time

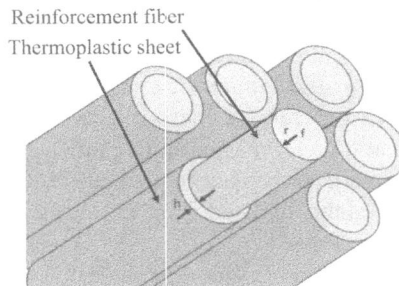

FIGURE 7.2   Cross-section of bi-component fibres (Schneeberger, Wong, and Ermanni 2017).

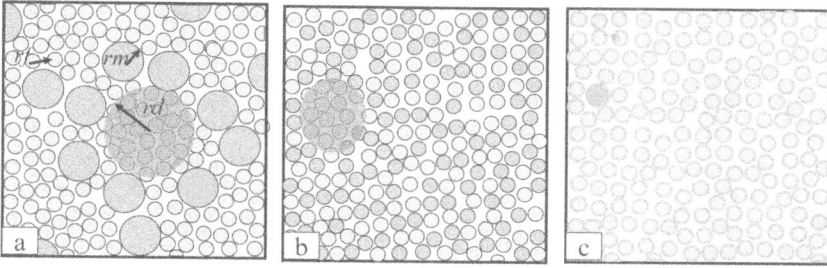

FIGURE 7.3 Unit cells of randomly distributed commingled yarn constituents with a relative matrix fibre diameter r_m/r_f of 3 in (a) and 1 in (b); and randomly distributed bicomponent fibres 7.3 (c) (Schneeberger, Wong, and Ermanni 2017).

scales like those of pre-consolidated intermediate materials. The authors also highlight the existing hybrid yarn structures with their processing features as shown in Figure 7.4.

Each hybrid yarn manufacturing technique has its own merits and demerits. Therefore, before choosing any technique one must have a general understanding of the method they choose. It is a well-known fact that one technique cannot process all forms of fibre or matrix-forming material. For example, the powder coating and solution coating techniques can only be used if matrix polymers are available in powder form. Similarly, commingling can only be employed if fibre and matrix material are in multi-filament form. Braiding also requires a matrix in the filament form and requires fibre either in filament or twisted yarn form. Fibres in either filament or staple form and a matrix in staple form are required for spinning core-sheath DREF hybrid yarns. Therefore, according to the form of fibre and matrix available one must choose the hybrid yarn technique.

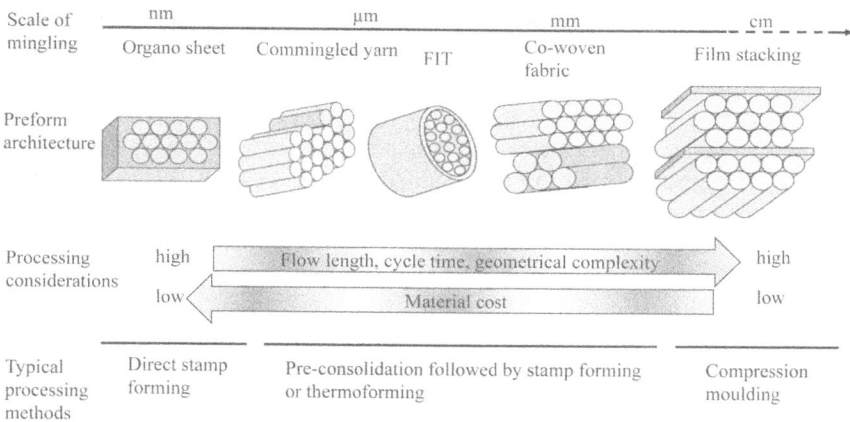

FIGURE 7.4 Schematic of various hybrid yarns structures with their processing features (Schneeberger, Wong, and Ermanni 2017).

7.4   EFFECT OF HYBRID YARN PROCESS PARAMETERS

The consolidation and resin impregnation quality of thermoplastic composites mainly depends on the viscosity of the polymeric resin. The high melt viscosity polymers pose a challenge in fully impregnating the laminates. To overcome viscosity associated problems, the best possible solution is to mix the reinforcement and matrix polymers well before advancing to the compression (Alagirusamy et al. 2006). In this regard, hybrid yarns play an important role as they consist of both fibre and matrix in the same yarn structure before the preforming and consolidation. The important feature of hybrid yarns is to reduce the melt flow distance during consolidation. It is possible only when fibre and resin materials are arranged adjacent to each other. When these hybrid yarns are subjected to heating under pressure, resin eventually melts and impregnates the surrounding fibres. The increased degree of fibre-matrix mixing decreases the melt flow distance. Therefore, the selection of a type of hybrid yarn and production technique becomes crucial as the laminate properties mainly depend on these factors. The quality of mixing is essentially dependent on the hybrid yarn processing technique and the variable process parameters associated with the respective technique (Alagirusamy et al. 2006; Jogur et al. 2018). The influence of these processes and process parameters is briefly discussed here.

Among various hybrid yarn formation techniques, commingling is the most promising one and it has the capability for a homogeneous distribution of fibre and matrix filaments over a cross section (Choi, Diestel, and Offermann 1999). An ideal commingled yarn consists of an individual matrix filament for a single fibre filament. Practically achieving such an ideal distribution is difficult due to many reasons. Yet commingled yarns with very short flow melt distance can be achieved, which allows fast and complete impregnation of fibre bundles. Any desired fibre to matrix volume ratio can be achieved just by adjusting the number of filaments.

The quality of mixing in commingling is mainly influenced by the three process factors, namely, air pressure, overfeed and delivery speed (Alagirusamy and Ogale 2004; Choi, Diestel, and Offermann 1999; Svensson, Shishoo, and Gilchrist 1998). The quality parameters of COM yarns, namely nip stability, nip frequency, nip regularity and degree of mixing, are mainly affected by the interaction of the aforementioned process variables. In general, the use of high air pressure with the low delivery speed at 1% overfeed offers the best quality commingled yarns (Alagirusamy et al. 2005; Mankodi and Patel 2009; Ogale and Alagirusamy 2008). Air pressure needs to be selected carefully to obtain good fibre distribution without causing any fibre damages. The process is known for its versatility and produces flexible and drapable yarns, which are important factors in forming complex shaped composite structures. The formation of nip/knots across the yarn structure introduces cohesion between the filaments and thereby provides stability to the yarn. Almost any known textile preforming techniques can therefore be chosen and processed with much ease and effortlessly (Alagirusamy et al. 2006).

Another method of producing hybrid yarns is co-wrapping, in which thermoplastic fibres are wrapped around by core reinforcing fibres. This wrapping process brings the individual filaments together and thus increases the structural integrity of the prepegs. In addition, wrapping also provides better protection to reinforcing fibres during textile preforming than that of commingling. However, the increase in wrapping density also increases the bending rigidity of the hybrid yarns. Therefore, warping density needs to be chosen carefully so that yarn can have enough stability and flexibility. In comparison to commingling, co-wrap spun yarns have poor fibre matrix distribution, which leads to poor impregnation and needs high processing temperature and pressure. The combination of commingling and co-wrapping provides better fibre-matrix distribution and good protection of fibres (Alagirusamy and Ogale 2004; Svensson, Shishoo, and Gilchrist 1998).

Friction spinning produces core-sheath hybrid yarns in which short thermoplastic fibres (staple fibres) are spun around a core of reinforcing fibres. A DREF friction spinning machine is commonly used to produce core-spun hybrid yarns. This technique has many merits over other hybrid yarn making methods. Firstly, any natural and synthetic fibres available in continuous or short-staple fibre form (sliver) can be used as core material, while fibres must be in the sliver form to be used as sheath material. Secondly, a wide range of core-sheath (fibre-matrix) ratio is tailored just by changing the feed input parameters. Thirdly, core filaments used are un-twisted and completely oriented, therefore maximum strength potential is realized. The technique also offers very high-speed production with minimum or no damage to the core yarn (Bar, Alagirusamy, and Das 2018, 2019; Bar, Das, and Alagirusamy 2017; Goud et al. 2019).

In the powder coating process, matrix in the form of powder is coated over the surface of the continuous fibre bundle. The important benefactor of this technique is the production of the twist free hybrid yarn structure. The process neither adds any twist nor any form of strain into the core reinforcing fibres, which in turn is beneficial in retaining maximum tensile properties. The process is environmentally friendly as no solvent is required. Another significance of this process is that the viscosity of matrix powder does not affect the quality of the composite formed. It is also said that powder coated yarns are effective in obtaining better mechanical properties at normal processing conditions as compared to commingled yarns. It may be due to the fact that resin flows better along the fibre in powder coated yarns and hence, impregnates the complete fibre bundle even at lower pressure. On the other hand, commingled yarns need severe consolidation conditions to meet similar mechanical properties of powder-coated yarns (Goud et al. 2018, 2019, 2020).

7.5 PERFORMING BEHAVIOUR

Textile preforming is an essential stage performed before composite fabrication. The method allows to tailor the architectures to suit the desired mechanical and other important properties in the final product. Preforming can be done with hybrid yarns and with parent yarns as basic raw material. However, the preforming operation may become difficult and challenging when hybrid yarns are employed as raw material.

Therefore, it is essential to have an understanding of the challenges and problems associated with different hybrid yarns. Preforming behaviour of different hybrid yarns is discussed below.

The braidability of nylon powder-coated carbon-nylon hybrid yarns were compared with commingled yarns and un-coated yarns. The number of interruptions as a function of braider speed was used to study the braidability of towpregs for a certain period. The powder coated yarns performed poorly in braiding with frequent stoppages and fibre damage due to the high coefficient of friction and bending rigidity over commingled yarns and un-coated yarns, respectively. The braidability of commingled yarns can be enhanced further by wrapping a thin nylon filament over the surface of COM yarns. However, increasing the wrapping density in powder coated yarns did not show much effect on the frictional coefficient. Since the braiding process involves greater fibre to fiber contact, a great amount of care must be needed while processing powder coated yarns to avoid frequent machine stoppages, fibre damages and entanglement of adjacent yarns (Ramasamy, Wang, and Muzzy 1996). To the best of the author's knowledge the same problems can also be seen in weaving and knitting as these operations put yarns under different strain levels, bending and curvature movements.

The spray-coated yarns can be heat-treated, where matrix powder is partially melted to improve the adhesion between fibre surface and powder. The heat-set yarns in turn minimize the powder slip off (loss) in subsequent textile preforming and thus improve the weavability and quality of preforms formed. However, the heat set yarns increase the flexural rigidity (due to partially molten powder) of the towpregs and may possess some challenge in textile preforming as well as in forming complex shaped composite structures (Goud et al. 2019).

The commingled yarns also have some limitations when it comes to textile preforming. The nips/knots formed across COM yarn structure impart cohesion between the filaments and thus increase the yarn stability, which again enhances preforming efficiency. However, a COM yarn having poor structural stability (due to the poor nip stability) may de-mingle when subjected to different strain levels during preforming. Such de-mingled yarns, if it becomes part of the fabric produced, may possess serious problems in the consolidation process. As a result, resin rich and resin starved areas may occur as individual yarn components (fibre and resin) get separated during preforming. To overcome this problem COM yarns are either warped with continuous thermoplastic filaments (co-wrapping) or the complete yarn is heat-treated to improve the yarn stability (Alagirusamy and Ogale 2004; Alagirusamy et al. 2006).

The preforming ability of DREF friction spun yarns is also poor due to the poor adhesion between core and sheath materials. During textile preforming, the sheath material is found flaring off the core surface, mainly due to abrasion at the healds and reed. To prevent such problems, a high level of twist cab be inserted to enhance the cohesion but at the cost of a reduction in production speed and yarn strength. In contrast, a new approach was used by Bar, Das, and Alagirusamy (2017) and Goud et al. (2020), in which DREF spun hybrid yarns were heat-treated on flat heated plates to enhance the core-sheath adhesion. The thermal treatment slightly melts the polymeric resin, and thus increases the bonding and prevents dislodging off sheath from the core surface. These heat-treated DREF yarns are compact, have better weave ability and thus make preforming easier.

Preforming efficiency of hybrid yarns has been enhanced using different approaches, namely co-wrapping, heat setting etc. in the past studies. The hybrid yarns can also be treated with synthetic size paste to enhance the preforming efficiency (Choi, Diestel, and Offermann 1999). It is always beneficial to have hybrid yarns having better preforming ability as they significantly reduce frequent machine stoppages, yarn breakages and defects in the produced fabric.

7.6 DAMAGE OF REINFORCING FIBRES

Reinforcing fibres are the load bearing component in FRP (fibre reinforcement plastic) composites. Therefore, any damage or deviations in the fibre structures directly affects the resultant composite properties. There are certain instances where fibres may experience severe to moderate damages and the orientation of fibre may be seriously affected. Hybrid yarn formation methods can sometime cause such damages in the reinforcement fibres. Therefore, process parameters for any hybrid yarn technique must be chosen in such a way that no harm is done to reinforcing filaments. Some studies have been reported to determine the effect of the hybrid yarn formation technique and its process parameters on different type of fibre reinforcement.

Choi, Diestel, and Offermann (1999) studied the effect of air jet texturing and intermingling nozzle and air pressure used during commingling on the carbon reinforcement damage and fibre distribution across hybrid yarn cross-section. The study concluded that irrespective of nozzle type the increase in air pressure from 0.4 MPa to 0.6 MPa caused decrease in the relative tensile strength and linear density of carbon fibre. The increased air pressure resulted in fibre damage with a loss of fibre. Therefore, commingling need to be done at suitable low pressure. Commingling process parameters were kept constant to determine the effect of nozzle types. The air jet texturing nozzle showed lower fibre damage than that of the intermingling nozzle. The nozzle offers gentle fibre treatment and hence low fibre damage. However, the fibre distribution in the yarn cross section was not comparable with the intermingling nozzle. The high fibre damage associated with the intermingling nozzle was attributed to the different working principles of the nozzle. When intermingled nozzles were used the intense air stream strikes the carbon fibre at a perpendicular direction. Since carbon fibres are highly sensitive to mechanical stresses perpendicular to their longitudinal direction, the use of the intermingling nozzle severely affects the reinforcement and results in high fibre damage. Figure 7.5 shows the effect of nozzle type and air pressure on the resultant strength and linear density of carbon fibre.

The commingling of brittle fibres like glass and carbon fibres is therefore quite challenging as the fibres may get damaged. Offerman et al. (2002) demonstrated the difficulties associated with the processing of glass fibres in textile operations. Since glass fibres have low bending and transversal compression properties, they do possess challenges in textile processing. Improperly designed machine parts as well as presence of deflection areas might impart high friction and bending force on glass fibres. These factors in turn caused fibre damage and eventually lead to reduced mechanical properties in final composite.

In another study it was found that glass fibres were more susceptible to damage in the case hybrid yarns made from friction spinning as compared to the yarns made from ring twisting, air interlacing and pneumatic texturing (Krucinska and Klata 2000).

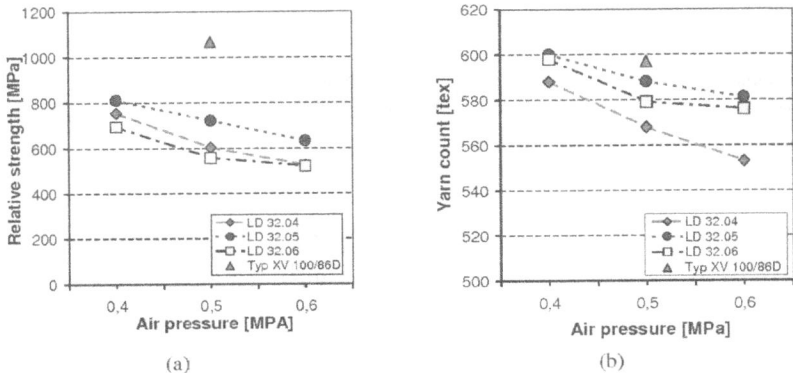

FIGURE 7.5   The effect of air pressure and nozzle type on the (a) relative strength and (b) yarn count (Choi, Diestel, and Offermann 1999).

7.7   FRICTION, FLEXIBILITY AND COMPRESSIVE PROPERTIES

From a textile processing point of view, towpreg properties such as friction, bending rigidity and compression resilience play an important role in the smooth passage and running of preforming processes and machineries. These factors mainly affect the weaving, knitting and braiding processes (Ramasamy, Wang, and Muzzy 1996). Fibrous material having high friction co-efficient, low flexibility and poor compression properties are used in textile operations. When such fibres are made to run at high speed in contact with metallic surface or with other adjacent fibrous strands, following two consequences may happen. Tension on fibrous strand might increase beyond the breaking strength of fibre, eventually lead to fibre breakage. The surface of the fibre may get damaged due to abrasion which causes either breakage of the fibre strand or degradation of fibre and resulting fabric quality. The mixing/deposition of matrix material with fibre reinforcement causes an increase in the bulkiness of resulting hybrid yarn. This in turn enhances the bulkiness of preform structure and subsequently causes the following problems.

- The bulkier preforms show poor formability while moulding for complex shaped composites. Due to this, it becomes difficult to produce fine quality products.
- The bulky preforms make it difficult to predict and control the fibre orientation in a laminate. Mould parts must move to a greater distance during compression to consolidate the laminates, which adds more deviations in fibre orientation. Therefore, the lateral properties of prepegs are very vital in composite fabrication (Alagirusamy et al. 2006).

7.8   SUMMARY AND FUTURE PROSPECTS

It is essential to have basic understanding of the influence of basic raw material properties and processing parameters on the structure and properties of flexible towpregs. Raw material parameters include filament denier, number of filaments, cross-sectional shape of fibres, filament rigidity and frictional characteristics of filaments. The process parameters include air pressure, feed ratio, take-up speed, and winding tension. The nozzle design, nozzle type, venturi, baffle settings, yarn entry angle and nozzle guide setting also influence flexible towpregs properties. Each hybrid yarn type has different fibre distribution behaviour and hence shows variation in impregnation and consolidation quality. The difference in fibre distribution is attributed to the structure of the yarns, which further depends on the technology of hybrid yarn production and its process parameters. The processing parameters should be chosen wisely to obtain good quality towpregs with no fibre damages and structural deviations. The quality hybrid yarns can be processed easily in subsequent textile preforming operations.

Preforming efficiency of hybrid yarns can be enhanced by various techniques like twisting, co-wrapping, thermal treatment, and sizing etc. These methods are effective in preventing frequent machine stoppages due to fibre breakages. This in turn yields good quality preform fabrics to be used for composite consolidation. The chapter also discusses effect of friction, flexibility and compression properties of hybrid yarns on composite properties.

For future research detailed comparison studies between different hybrid yarn techniques focusing on fibre distribution, preforming behaviour, fibre damage and impregnation quality can be performed. Simulation techniques can be used to understand and explore the mechanism of consolidation, index of blend quality (fibre: matrix distribution) and effect of process parameters on yarn properties. For example, numerical investigations can be carried out to understand the flow behaviour of mixing or spraying nozzles. Based on the fluid flow knowledge, nozzle designs can be optimized to improve process efficiency. Very limited publications are available on yarn structure analysis using image processing techniques. Detailed studies can be conducted to evaluate the yarn structure effect on the towpregs properties. The frictional, flexural and compressional properties of different hybrid yarns can be studied for different preforming techniques. This study gives an idea about the processability of yarns in weaving, braiding, and knitting etc. The studies can be explored to understand the effect of different hybrid yarns on different preforming architectures. Most importantly, focus need to be put on exploring the use of hybrid yarn techniques in the production of natural fibre reinforced thermoplastic composites. Very limited literature has been found in this area. Therefore, huge potential exists in the development of natural fibre-based hybrid yarns for sustainable composites. More attention is needed in developing state of the art modern hybrid yarn manufacturing machines and for commercialization of hybrid yarns and their composites in different sectors.

REFERENCES

Alagirusamy, R., R. Fangueiro, V. Ogale, and N. Padaki. 2006. "Hybrid Yarns and Textile Preforming for Thermoplastic Composites." *Textile Progress* 38 (4): 1–71.

Alagirusamy, R., Vinayak Ogale, Abhijeet Vaidya, and P. M. V. Subbarao. 2005. "Effect of Jet Design on Commingling of Glass/Nylon Filaments." *Journal of Thermoplastic Composite Materials* 18 (3): 255–68. doi:10.1177/0892705705049555

Alagirusamy Ramasamy, and Vinayak Ogale. 2005. "Development and Characterization of GF/ PET, GF/Nylon, and GF/PP Commingled Yarns for Thermoplastic Composites." *Journal of Thermoplastic Composite Materials* 18 (3): 269–85 doi:10.1177/0892705705049557

Alagirusamy Ramasamy, and Vinayak Ogale. 2004. "Commingled and Air Jet-Textured Hybrid Yarns for Thermoplastic Composites." *Journal of Industrial Textiles* 33 (4): 223–43. doi:10.1177/1528083704044360

Angell, Jr. Richard G., Jr. Michael J. Michno, John M. Konrad, and Kenneth E. Hobbs. 1986. Hot-melt prepreg tow process. 4,804,509, issued December 17, 1986. https://patents. google.com/patent/US4804509

Bar, Mahadev, R. Alagirusamy, and Apurba Das. 2018. "Properties of Fl Ax-Polypropylene Composites Made Through Hybrid Yarn and Fi Lm Stacking Methods." *Composite Structures* 197 (April). Elsevier: 63–71. doi:10.1016/j.compstruct.2018.04.078

Bar, Mahadev, R. Alagirusamy, and Apurba Das. 2019. "Influence of Friction Spun Yarn and Thermally Bonded Roving Structures on the Mechanical Properties of Flax/Polypropylene Composites." *Industrial Crops & Products* 135 (July 2018). Elsevier: 81–90. doi:10.1016/ j.indcrop.2019.04.025

Bar, Mahadev, Apurba Das, and R. Alagirusamy. 2017. "Studies on Flax-Polypropylene Based Low-Twist Hybrid Yarns for Thermoplastic Composite Reinforcement." doi:10.1177/ 0731684417693428

Choi, B.-D., Olaf Dr Ing Diestel, and Peter Offermann. 1999. "Commingled CF/PEEK Hybrid Yarns for Use in Textile Reinforced High Performance Rotors." In *International Conference on Composite Materials.*

Goud, Vijay, Ramasamy Alagirusamy, Apurba Das, and Dinesh Kalyanasundaram. 2018. "Dry Electrostatic Spray Coated Towpregs for Thermoplastic Composites." *Fibers and Polymers* 19 (2): 364–74. doi:10.1007/s12221-018-7470-7

Goud, Vijay, Ramasamy Alagirusamy, Apurba Das, and Dinesh Kalyanasundaram. 2019. "Influence of Various Forms of Polypropylene Matrix (Fiber, Powder and Film States) on the Flexural Strength of Carbon-Polypropylene Composites." *Composites Part B: Engineering* 166 (November 2018). Elsevier: 56–64. doi:10.1016/j.compositesb.2018.11.135

Goud, Vijay, Dilpreet Singh, Alagirusamy Ramasamy, Apurba Das, and Dinesh Kalyanasundaram. 2020. "Investigation of the Mechanical Performance of Carbon/ Polypropylene 2D and 3D Woven Composites Manufactured Through Multi-Step Impregnation Processes." *Composites Part A: Applied Science and Manufacturing* 130 (November 2019). Elsevier: 105733. doi:10.1016/j.compositesa.2019.105733

Krucinska, I., E. Klata, W. Ankudowicz, and H. Dopierala. 2000. "Preliminary Studies on the Manufacturing of Hybrid Yarns Designed for Thermoplastic Composites." *Fibers and Textiles in Eastern Europe* 8 (2): 61–5.

Jogur, Ganesh, Ashraf Nawaz Khan, Apurba Das, Puneet Mahajan, and R. Alagirusamy. 2018. "Impact Properties of Thermoplastic Composites." *Textile Progress* 50 (3). Taylor & Francis: 109–83. doi:10.1080/00405167.2018.1563369

Lauke, B., U. Bunzel, and K. Schneider. 1998. "Effect of Hybrid Yarn Structure on the Delamination Behaviour of Thermoplastic Composites." *Composites Part A: Applied Science and Manufacturing* 29 (11): 1397–1409. doi:10.1016/S1359-835X(98)00059-1

Mankodi, Hireni, and Pravin Patel. 2009. "Study the Effect of Commingling Parameters on Glass/Polypropylene Hybrid Yarns Properties." *Autex Research Journal* 9 (3): 70–3.

Merter, N. Emrah, Gülnur Baüer, and Metin Tanoğlu. 2016. "Effects of Hybrid Yarn Preparation Technique and Fiber Sizing on the Mechanical Properties of Continuous Glass Fiber-Reinforced Polypropylene Composites." *Journal of Composite Materials* 50 (12): 1697–1706. doi:10.1177/0021998315595710

Ogale, Vinayak, and R. Alagirusamy. 2008. "Properties of GF/PP Commingled Yarn Composites." *Journal of Thermoplastic Composite Materials* 21 (6): 511–23. doi:10.1177/ 0892705708091281

Offerman, P., O. Diestel, E. Mader, and T. Hubner. 2002. "Development of Commingling Hybrid Yarns for Thermoplastic Composites." *Technische Textilien* 45 (1): E12–4.

Ramasamy, A., Youjiang Wang, and John Muzzy. 1996. "Braided Thermoplastic Composites from Powder-Coated Towpregs. Part I: Towpreg Characterization." *Polymer Composites* 17 (3). John Wiley & Sons, Ltd: 497–504. doi:10.1002/pc.10639

Schneeberger, Christoph, Joanna C. H. Wong, and Paolo Ermanni. 2017. "Hybrid Bicomponent Fibres for Thermoplastic Composite Preforms." *Composites Part A: Applied Science and Manufacturing* 103. Elsevier Ltd: 69–73. doi:10.1016/j.compositesa.2017.09.008

Svensson, N., R. Shishoo, and M. Gilchrist. 1998. "Manufacturing of Thermoplastic Composites from Commingled Yarns - A Review." *Journal of Thermoplastic Composite Materials.* doi:10.1177/089270579801100102

Wu, G. M. 2000. "Processing and Properties of Solution Impregnated Carbon Fiber Reinforced Polyethersulfone Polymer Composites." 21 (2): 223–30.

8 Textile Preforming of Towpregs

Ganesh Jogur[a], Ashraf Nawaz Khan[a], and Naveen Padaki[b]

[a]Indian Institute of Technology
New Delhi, India
[b]Central Silk Technological Research Institute, Central
Silk Board, Ministry of Textiles, Government of India
Bengaluru, India

CONTENTS

8.1 INTRODUCTION

Lightweight fibre reinforced plastic (FRP) composites are of significant importance in the field of defense, aerospace, sports, marine, automobiles, and other technical textile sectors as well. These engineered materials offer unique features over the conventional metals in terms of high specific strength, stiffness, improved fatigue life and non-corrosion behaviour. In earlier days, an expensive and slow production technique called "Prepreg system" was used to manufacture FRP composites for aerospace applications. The prepreg system is an assembly of dry fibres that are pre-impregnated with matrix forming material and cured partially to maintain the stability of structure. These individual prepregs are then cut into desired shapes, stacked in required orientation into desired number of plies to meet preferred thickness and then subsequently cured in an autoclave to complete curing and consolidation process (Potluri, Sharif, and Jetavat 2008). In recent years, to overcome aforementioned problems, a new

DOI: 10.1201/9781003049715-8

approach has been adopted to fabricate FRP composites. In this new approach, reinforcing fibres are first arranged in a stable and strong architecture known as "textile preforms" prior to infusion with matrix forming material. Weaving, knitting, braiding and stitching are the well-known textile operations that are used to manufacture textile preforms for composite applications. These preforms are then impregnated with matrix material to prepare composite materials. However, this new approach is only suitable for thermoset composites and creates difficulty in processing thermoplastic composites. Thermoset resins are mainly available in liquid form and thus it is easy to introduce them into the core of fibre architecture. The low viscous nature of these resin materials helps in fabricating FRP composites using simple consolidation methods like film stacking and melt impregnation. However, these simple methods are not feasible to process highly viscous thermoplastic polymeric resins.

In case of thermoplastic systems, it is difficult to impregnate the highly viscous molten thermoplastic polymer resin into a close and firmly built textile preform and fibre/yarn bundle. Such an attempt leads to resin rich and resin poor areas causing poor fibre to matrix adhesion, which is one of the main reasons for increased void in the thermoplastic composite materials. High pressure injection moulding and heavier compression moulding machines need to be employed to avoid the above mentioned problems. However, these techniques require high capital cost and ultimately increase the final product price (Jogur et al. 2018).

The best way to eliminate the viscosity related problems in thermoplastic processing is to develop textile preforms made with flexible towpregs. According to Alagirusamy et al. (Alagirusamy et al. 2006), "thermoplastic prepregs having sufficient flexibility to go through textile preforming without getting damaged significantly" are known as flexible towpregs or thermoplastic hybrid yarns. These hybrid yarns can be produced by pre-impregnating through coating or mixing polymeric resin with reinforcing fibers well before textile preforming operations. These hybrid yarns offer homogeneous distribution of both fiber and resin in the yarn structure, which reduces the resin melt flow distance during compaction. Therefore, hybrid yarn preforms when heated and compacted for composite consolidation, the matrix forming material easily melts around the surrounding fiber and impregnates it immediately. This phenomenon results in reducing the heat and mass transfer issues associated with resin during the processing of thermoplastic composites. The hybrid yarn performs require less consolidation time, mound pressure and temperature in comparison to co-twisted, prepreg and co-woven preforms. This technique thereby reduces the capital expenses associated with the composite consolidation process. Commingling, powder coating, hot melt, slurry, emulsion, solution, film and surface polymerization techniques are ways of achieving reduced melt-flow distance; amongst these, commingling and powder-coating techniques can produce prepregs with good fiber-matrix distribution and considerable flexibility, which is a critical requirement in textile preforming.

8.2 TEXTILE PREFORMING

Technological advancement in the field of textile engineering has become a benefactor for the production of composites structures having high mechanical performance. Textile preforming is one such operation, which helps in fabricating various micro-structures of fiber preforms for the development of textile structural composites. These fibrous performs can be tailored to meet the specific needs for mechanical performance as well as adequate structural integrity of composite structures. These preforms must meet the key criteria to consider them in the composite sectors. They must have the capability for in-plane multiaxial reinforcement, through-thickness reinforcement and must be able to form into net or near-net shape structures. The preforming operation is economical, offers homogeneous distribution of fiber and matrix, provides good control over fiber placement and possesses ease of fiber handling. Textile preforming converts discrete or continuous length of fiber/yarn bundle into one, two, or three dimensional flexible, stable structures using either interlacing, interloping, twisting or inter-twining mechanisms. Textile preforming, because of its versatility, flexibility and simplicity, is considered as the backbone of the composite's structures.

The performance of composite structure mainly depends on the textile preforming as it influences matrix infiltration, consolidation and translation efficiency of the fibre properties into the final composite product. Textile preforming offers complete freedom to the composite designer in the selection of fibre geometry, architecture, fibre orientation and method of fibre placement to tailor the structural performance of the composites. Additionally, preforming permits to tailor made the preform structures as per the properties required in the compsoite by customising the pore sizes, its distribution, fibre volume fraction, fibre orientation and structural integrity. The technique also facilitates the production of net or near net shape structural composite parts (Alagirusamy et al. 2006; Jogur et al. 2018).

8.3 TYPES OF TEXTILE PREFORMS FOR TOWPREGS

Scardino (1989) classified textile preforms into different hierarchy levels like discrete, continuous, planar interlaced (2D), and fully integrated (3D) structures. This classification is based on considering the different criteria such as the manufacturing process used, structural integrity, fibre orientation and continuity, as well as geometric features of preform architecture (Alagirusamy et al. 2006). Fukuta and Aoki (1986) have also classified textile preforms considering the method of manufacturing, dimension of the structure, fabric architecture, yarn dimensions and directions within the preform. Khokar (2001) classified textile preforms based on shedding, thereby explaining the 3D preform weaving through shedding of yarns in both fabric length (horizontal) and through-thickness direction (vertical), thus allowing the insertion of weft in both directions. In another approach, Solden and Hills (1998) classified the woven fabrics by the weaving process, which comprises the warp binder, warp interlinked, in-plane interlacing yarns and stuffing warp yarns to produce an integrated structure (Alagirusamy et al. 2006).

8.4 WOVEN TEXTILE PREFORMS FROM FLEXIBLE TOWPREGS

Weaving is an ancient technology, which involves the conversion of different linear density yarns into a flexible sheet of material known as fabric. Weaving technology is versatile and facilitates the manufacturing of UD, 2D and 3D textile preforms as well as multi-axial preforms. The main features of woven preforms are the good drapability, flexibility, warmth keeping ability, low manufacturing cost and complex shape formation ability. These preforms find many applications in technical fields such as the aerospace industry, personal protective clothing and the marine industry (Alagirusamy et al. 2006). The flexible towpregs (hybrid yarns) can be used to construct different woven preforms in different configurations using weaving processes.

The neat (non-mixed) yarns can be woven into different structures with the available weaving technology. In contrast, the hybrid yarns are bulky in nature, sometimes rigid (in case of powder coating and melt-impregnation techniques) and hence, possess some difficulties during weaving operations. The bulky and stiff yarn shows poor flexibility and hence results in frequent yarn breakages and loom stoppages (Ramasamy, Wang, and Muzzy 1996a). The intermingled yarns may de-mingle due the strain generated during the shedding mechanism and may collide with the adjusting yarns and result in the entanglement of several yarn bundles and finally lead to yarn breakages. Further, the de-mingled yarn may cause the formation of more voids in the subsequent consolidation process of the composites (Alagirusamy and Ogale 2004). Therefore, it is essential to produce stable hybrid yarns having sufficient flexibility before going to weaving operations. This will surely lead in minimizing the yarn breakages, loom stoppages and ultimately delivers defect-less woven fabrics. Several researchers have explored the use of the woven preforms made of flexible towpregs in the composite applications. An attempt has been made to review the work done by those researchers in the same area.

8.4.1 Unidirectional Woven Textile Preforms from Flexible Towpregs

Uni-weaves are the one directional woven structures in which the majority of reinforcing yarns or towpregs are laid closely and arranged parallel to each other in one direction only. Uni-weaves may have their primary yarns either in $0°$ (warp UD) or in $90°$ (weft UD) directions. However, the structure needs some form of mechanism to hold the layers of unidirectional yarns together to form a fabric. Some known mechanisms are: i) a polymeric resin or film can be used to bind the individual layers, ii) very few weft/course yarns can be introduced to interlock the unidirectional yarns or iii) a separate set of yarns can be used to stitch all the individual unidirectional laminae (Cunningham and Pritchard 2005; Ryo Okada 2011). In weaving, a negligible amount of filling yarns can be inserted in the other directions to hold the primary yarns and thus to provide structural stability to the fabric. Stable uni-weaves prevent the handling problems in the subsequent operations. These structures provide the highest fibre volume per unit area as compared to any other weave. These weaves also have continuous straight yarns without any crimp, which leads to the highest fibre mechanical properties translation in the composite. Since the majority of primary yarns are oriented in one direction, the UD fabric shows the highest tensile strength and modulus in the yarn-oriented direction. Uni-weaves possess some limitations, such as intra- and inter-laminar weakness due to lack of their in-plane and out-plane yarn interlocking. This architecture is best suitable for filament-wound and angle-ply tape lay-up structures (Baucom and Zikry 2003; Jogur et al. 2018).

Uni-weave structures can be constructed using flexible towpregs (commingled, powder coated, friction spun, co-wrapped hybrid yarns etc). Several studies have been reported on the development of uni-weaves from the hybrid yarns and on the importance of these preforms (dry impregnated) against the traditional prepreg system. To understand and explore the importance of flexible towpregs and their preforms Ye et al. (1995) investigated the relationship between impregnation mechanisms and consolidation quality of carbon fibre/PEEK composites. The UD composites were made from commingled (COM) yarn fabric. The authors observed a longer processing time and higher applied pressure while consolidating the commingled yarn fabrics as compared to the sheet made of prepregs or powder/sheath fibre bundles. The authors concluded that the use of hybrid yarns having the highest degree of blend regularity in textile preforms would bring down the overall processing time and applied pressure. Quick pre-heating of the towpreg preforms (TP) before consolidation not only reduces the processing time, temperature and applied pressure but also helps to achieve composites with high performance properties. Choi, Diestel, and Offermann (1999) developed carbon/PEEK unidirectional composites from commingling and side-by-side (SBS) techniques and analysed their consolidation and impregnation quality. Microscopic examinations revealed that the COM composites had a homogeneous fibre to matrix distribution as compared to SBS composites and hence, significantly improved the impregnation quality in comparison to SBS variant. The COM composites also had shown good tensile strength in a 90° direction as compared to SBS. However, a reduction in tensile strength at 0° direction (longitudinal to fibre axis) was observed due to severe damage of carbon fibres during the commingling process at high air pressure.

Madär et al. (2007) suggested that the online commingling process will offer more merits over the air jet textured commingling process. The authors compared both hybrid yarns in terms of blend quality, thermal shrinkage and cross section geometry. Online spinning of commingled hybrid yarn involves simultaneous spinning and commingling of GF/PP fibres. The tensile strength for air jet textured hybrid yarns was lower as compared with online commingled hybrid yarns. The unidirectional composites made from online COM hybrid yarns showed no thermal shrinkage and improved tensile strength and transverse mechanical properties due to no fibre damage and homogeneous distribution of GF/PP. The hybrid yarns with no thermal shrinkage enhance the blend quality and tensile strength, further helping in production of complex shaped textile structures having high mechanical performance.

Selver et al. (2016) compared low velocity impact and compression after impact (CAI) performance of GF/PP commingled hybrid yarn unidirectional composites (0°/90° layer sequence) with their thermoset counterparts. Low velocity impact tests at 15J, 25J and 35J revealed that thermoplastic composites made from air jet textured commingled yarn showed better impact energy absorption (73–80%) due to their plastic deformation and high strain to failure behaviour as compared to thermoset composites (39–41%). CAI tests indicated that thermoplastic composites had higher residual strength and damage tolerance against thermoset composites.

Bar, Alagirusamy, and Das (2019) investigated the influence of flax/PP friction spun (FS) and thermally bonded roving (TBR) towpregs on the mechanical properties of UD composite structures. The UD composites made from TBR towpregs showed higher tensile strength and modulus, flexural strength and modulus than their FS composite counterpart when flax content was kept constant. The tensile and flexural strength of FS and TBR composites were improved with increase in flax content. However, the izod impact strength was decreased in FS composites with increase in flax content, but a reverse trend was observed with TBR composites. The difference in impact strength values were correlated with the nature of fibre-matrix distribution in the composite structure. In a similar study (Bar, Alagirusamy, and Das 2018) by the same authors, the mechanical properties of the UD composites made from FS hybrid yarn and the film stacking method were compared. At constant flax fibre content, FS composites exhibited excellent properties over film stacked composites. The FS hybrid yarn showed good fibre-matrix distribution over the film stacking method and hence improved results were observed with low void content.

The FS hybrid yarns are characterized by core sheath structures, meaning the flax reinforcing fibre is in the core (twistless) surrounded by the matrix component PP fibres. Since flax and PP are in close contact during consolidation, the PP fibre melts and impregnates the core fibre bundle as well as neighbouring fibre bundles for a very short time. However, in film stacking method, PP resin needs to travel a high melt flow distance to reach the fibre bundle. Moreover, under high moulding pressure the preform structure becomes more compact, the circular yarn becomes elliptical and this makes it difficult for the resin to impregnate into the porous structure created by the interlacing points. The failure of resin to impregnate the fibre bundle and perform leads to the formation of resin rich and resin starved areas and thereby increases the void content in laminates. Therefore, the film stacking method leads to the reduction in mechanical properties as compared to the FS hybrid yarns (Bar, Alagirusamy, and Das 2018).

Zhang and Miao (2010) compared the mechanical properties of flax/PP and hemp/PP UD composites made from conventional ring spun hybrid yarns to the wrap spun hybrid yarns. The wrap spun hybrid yarn that has twistless natural fibre in the core wrapped by a continuous PP filament has demonstrated excellent flexural modulus over twisted ring spun hybrid yarn. The improved fibre orientation in wrap yarn structure, and better fibre spreading and packing in laminate manufacturing enhanced the wrap spun composite properties.

8.4.2 2D Woven Preforms Based on Flexible Towpregs

Two dimensional woven structures are formed by interlacing two or more sets of yarns or fibre tows or flexible towpregs. In general, the woven fabrics consist of warp and weft sets of yarns. These yarns run alternatively over and under each other and are interlaced perpendicularly. The interlocking mechanism of these architectures help to solve the problem of intra-laminar failure associated with uni-weaves. However, these structures possess lower inter-laminar strength in comparison to uni-weaves. The main reason for the reduction in inter-laminar strength is the low fibre volume fraction per unit area of the 2D structures in comparison to uni-weaves. The other reason is that in the 2D structures matrix may not be able to impregnate into the structure effectively due to the restriction imposed by crossover points or inter-lacing or inter-looping points. Most importantly, the inter-laminar strength is limited by the matrix strength due to the absence of through-thickness fibre reinforcement in the structure. Since the inter-laminar strength mainly depends on the effective bonding between the fibre surface and the matrix material, the overall reduction in fibre volume fraction and poor matrix impregnation causes decrease in an interlaminar strength (Jogur et al. 2018).

The yarn interlacement inevitably yields the crimp (waviness) along the fabric length and width. The crimp formation limits the conformability, resulting in the poor in-plane shear resistance, reduced tensile and compressive properties as compared to the UD or cross ply laminates. The increment in the weft/fill density further increases the waviness in the structure, which in turn acts as a weak link and results in the significant reduction in the tensile strength and stiffness value and decreased in-plane modulus of woven fabric reinforced composites. When 2D woven composites are subjected to uniaxial or high impact tests, the presence of crimp in the fabric structure disturbs the propagation of stress waves to the boundary yarn layers and results in large normal and shear strain to concentrate. The strain concentration eventually leads to early damage initiation in the form of fiber/matrix interface debonding and finally causes localized damage failure in the composite. The crimp formation also promotes local fiber buckling under compression. Therefore, it's essential to have low crimp or optimized crimp woven fabrics, as the reinforcing efficiency of woven fabrics strongly depends on the percentage of crimp in the structure. In comparison to knit and braid structures, a higher yarn packing density (total yarns per unit area) with respect to thickness can be achieved in the woven fabrics. The basic 2D woven structures are plain, twill, satin & sateen and leno weaves.

Wakeman et al. (1998) studied the effect of pre-heating of balanced twill weave preforms made with GF/PP towpregs on the processing parameters and mechanical properties of the composites produced. The study claims that the preheat temperature has greater effect on the laminate flexural properties and porosity. A compaction time of 54 second was required to consolidate, cool and reduce the void content. A two-fold increment in laminate stiffness was found in contrast with equivalent glass matt reinforced thermoplastic composites. The homogeneous distribution of commingled yarn reduced the moulding pressure by a factor of 10.

Leal et al. (2016) introduced a novel technique for the production of polyolefin plastomer (POP) reinforced fully thermoplastic hybrid yarn UHMWPE composites. The authors used nano scale plasma polymer deposition methods to enhance the interface adhesion between UHMWPE and the POP matrix. The activated fibre is then coated with a layer of matrix film by the "over jacket extrusion" technique to produce thermoplastic hybrid yarns. These novel hybrid yarns have demonstrated improved interfacial strength, and maintained tensile strength and crystalline structure. The over jacket extrusion technique used, and cross section of hybrid yarn produced are demonstrated in Figure 8.1. The 4/1 twill woven fabrics were developed using OE hybrid yarns and then stacked with the layers of commercially available Dyneema® fabric in an alternate fashion. The produced composite from said composition had a V_f of 0.54 and density of 0.93 g/cm^3 and significantly better tensile properties as compared to self-reinforced PP or UHMWPE single polymer composites. The specific tensile properties of these composites were also better than the literature values of woven E-glass/epoxy composites. The 4/1 twill weave offers good conformability as it has low interlacements and also facilitates the good resin impregnation into the structure. The surface modified OE hybrid yarn further helps in easy resin impregnation by reducing the resin flow distance during consolidation.

Baghaei, Skrifvars, and Berglin (2015) reported the effect of 8 harness satin and basket weave on the mechanical properties of unidirectional composites made from

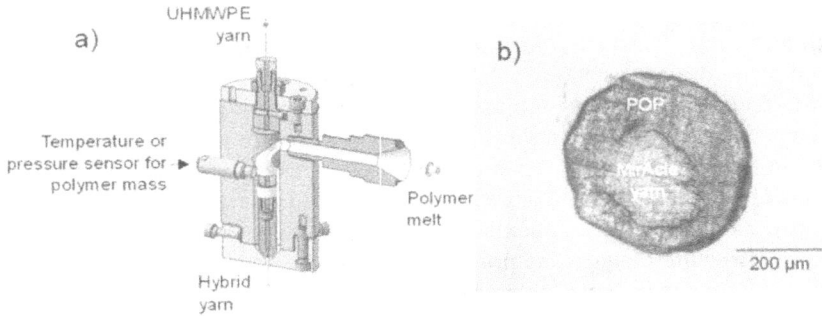

FIGURE 8.1   (a) Schematic of wire coating die used for the OE of the UHMWPE yarns and (b) cross-sectional view of a MirAcle®–POP hybrid yarn. (Adapted from Leal et al. 2016.)

prepregs of PLA (warp) and PLA/hemp wrapped spun hybrid yarns. Even though the chosen woven structures (satin and basket) are different from UD weave, yet the authors developed preforms that have the similar characteristics as UD fabric in weft direction. This method uses very fine warp yarns to considerably reduce the crimp; however, yields perform with poor structural stability. Authors investigated the effect of weave patterns on the mechanical and thermo-mechanical properties, amount of porosity and water absorption of the composites. The study claims that PLA/hemp UD composites made from satin weave showed excellent tensile, flexural and impact strength in comparison to the composites made from basket weave preforms and winded hybrid yarn (side-by-side). The enhancement in mechanical properties was correlated with the good impregnation quality, low void content and reduced fibre misalignments of the wrap spun hybrid yarns. The mechanical properties of winded hybrid yarn laminate were lower as compared to woven composites. However, the moisture absorption trend was opposite where the winded hybrid yarn composites absorbed less water than satin and basket weave composites. The satin weave has lower interlacing points, which reduce the crimp percentage and keep the yarn continuous straight. The structure has good porosity, which enables easy penetration of matrix resin into the core structure, eventually resulting in reducing the void content and enhanced mechanical properties.

8.4.3   3D Woven Preforms

To fulfill solutions to all the drawbacks associated with the UD and 2D structures, fully integrated structures have been developed and are popularly known as 3D structures. These structures belong to the third category of the basic textile architectures. The 3D structures consist of reinforcing fibres oriented in X, Y and Z directions, i.e., longitudinal and cross wise directions (in-planes) and in through thickness direction (out of plane). In the 3D textile structures, individual in-plane layers are first formed and then each layer is connected by the introduction of an additional series of yarn known as binders in the thickness direction of the fabric. These 3D structures have many meritorious properties over other woven structures. 3D weaving can produce complex net or near net shape textile preforms. The process is less expensive, and any complex shape preforms can be easily manufactured for 3D composites. It is possible to tailor the through thickness properties of 3D structures for any specific application. The composites made from these structures offer higher impact resistance, ballistic damage resistance, and delamination resistance. The 3D woven composite offers higher interlaminar fracture toughness and strain to failure properties (Ko and Du 1988; Gu and Zhili 2002). Apart from the benefactor properties of 3D weaving and its structures, there are some issues that are impeding the use of 3D preforms in composites (Mouritz et al. 1999).

In general, the 3D woven structures are categorized as through-thickness (TT) and layer-to-layer (LTL) interlock structures based on how deep the binder yarn penetrates through the fabric. Those structures in which the binder yarns penetrate all the way through the entire thickness of the fabric are termed as 3D through-thickness interlock structures. In contrast, the structures in which the binder yarns interlock only the adjacent fabric layers are known as 3D layer-to-layer structures. These weaves are further classified based on the interlacing angle between binder yarn and weft yarn. In the angle interlock (AI) structure the interlacing angle between binder and weft yarns lies in any value except 90°. The orthogonal interlock (ORT) occurs when the interlacing angle is equal to 90°. In addition, the weave pattern used during the weaving process can further classify the 3D preforms. For example, in the case of ORT weave, the pattern in which the Z binder sweeps the top and bottom surfaces of the fabric layers decides the 3D weave pattern. This pattern could be of plain, twill or satin, which could directly affect the features of 3D structures such as unit cell size, degree of crimp, elastic response and damage/delamination resistance (El-Dessouky and Saleh 2018).

Goud et al. (2020) investigated the mechanical performance of CF/PP 2D and 3D composites manufactured from two different pre-impregnation techniques: DREF friction spinning and powder coating. Among 3D woven structures, effects of 3D angle interlock (3DWA) and 3D orthogonal (3DWO) are also studied. 2D woven composites showed higher tensile strength followed by 3DWA and 3DWO. The reason was attributed to the lowest crimp % as well as a greater number of warp threads in 2D woven plain fabrics (when GSM is constant). The 2D woven fabrics have the lowest warp crimp % (1%) followed by 3DWA (7%) and 3DWO (30%) structures. In flexural testing, 2D woven fabrics showed quick failure whereas 3D composites underwent slow material separation. 3DWO composites having through thickness binder and closure wrapping in the structure absorbed excellent energy, thus showing excellent notch impact properties as compared to 2D composites.

The author also compared impregnation quality and composite properties of DREF and powder coated towpregs. The DREF hybrid yarn showed high melt flow distance as compared to powder coated, due to the transformation of DREF yarn cross section from elliptical to circular. This transformation results in highly compacted yarn structure and hence offers more resistance to the impregnation of PP molten liquid. The compact core structure combined with circular cross section of DREF yarn yields resin rich and resin poor areas in the composite manufactured. Therefore, composites made from DREF yarn showed poor tensile and flexural properties. In contrast, the powder coating process offers towpregs of homogeneously distributed fiber and matrix and thus low melt flow distance during consolidation.

Therefore, powder coated yarns showed good impregnation quality, less porosity and better mechanical properties as compared to the DREF hybrid yarns.

Torun, Mountasir, and Hoffmann (2013) introduced new production principles and automated production techniques for the development of complex 3D woven textile spacer fabrics with integrated stiffeners. Technological advancement was carried out for the face to face and terry weaving techniques based on the modification and adding modules to the double rapier loom. GF/PP commingled woven spacer preforms with integrated stiffeners were fabricated using the developed 3D production technique.

Mountasir et al. (2011) reports that spacer fabrics made from terry pile weaving consist of woven outer layers connected by additional pile yarns. The mechanical properties of these preforms, however, mainly depend on the pile yarns, which have limited bending stiffness. These structures cannot be used for complex shaped and highly strained composites. On the other hand, woven spacer fabrics constructed from woven outer layers connected by woven cross links instead of pile yarns offer excellent mechanical properties and thus find applications in multiple fields. Moreover, woven spacer fabrics offer high productivity and greatest flexibility in adapting their material structure to the applied loads. The authors developed multilayer woven fabrics in different weft densities to study their mechanical properties. The results revealed that tensile strength, flexural stiffness, and compaction properties improved in composites constructed with three-layer fabrics having no yarn crimping. The enhanced properties were the result of lower yarn damage and non-crimped layers of the structure. However, composites made with crimped yarns showed better impact energy absorption as compared to the three-layer fabrics.

8.5   KNITTED PREFORMS FROM FLEXIBLE TOWPREGS

Knitted structures are the commonly utilized preforms in the composite applications due to the ease of manufacturability. These structures are mainly classified as weft knit and warp knit. In weft knit structures, the yarns are interconnected in the form of a loop in horizontal course-wise directions. In contrast, loops are joined together in the vertical wale directions in the case of warp knit structures. Among these two structures weft knit preforms are widely preferred owing to their tailorability, ease of manufacturing, extensibility and formability (Horrocks and Anand 2000). The knitted fabrics are considered to be 3D due to non-planar configuration of the loops in the structure. These knitted preforms are highly extensible, thick, distribute stress better across the structure and have less flexural rigidity. Knit preform properties are greatly influenced by the fibre strength, modulus, knitted structure, stitch density, number of knitted fabric plies, pre-stretch parameters, inlays and other knitting parameters (Padaki, Alagirusamy, and Sugun 2006).

Laying-in, plating, open work and plush-pile constructions are the four known techniques used to modify the physical appearance and properties of knit preforms. Amongst these techniques, laying in technique has been explored and proven successful in both warp and weft knitting (Horrocks and Anand 2000). Isotropic and poor mechanical properties of knitted fabrics have led to structural modifications with inlay yarns in horizontal (weft), vertical (wale) and diagonal directions to develop preforms suitable for the composite applications. An inlaid preform consists of a ground structure made by intermeshing of loops. During knitting, these loops hold the inlaid (non-knitted) yarns in position, which eventually results the in modification of one or more fabric properties. The inlaid modification may result in altering the strength, stability, weight and handle of fabric preforms (Niklas 1998; Sarlin, Pentti, and Jarvela 1990).

8.5.1 UNI-KNITTED PREFORMS

In weft knitting, an inlay yarn can be inserted in any one direction to produce monoaxial structure known as "UD knitted preforms". Here, the inlay yarns are trapped inside double needle bed fabrics. For example, in 1 × 1 rib knit structure, UD tensile can be enhanced in the weft direction by inserting inlay yarns across the entire width of the fabric (Van Vuure and Frank 2003). The composites made from such preforms exhibited better tensile strength compared to standard woven preform counterparts (Padaki et al. 2000; Rao, Nahar, and Alagirusamy 2000). Two known principles commonly followed to insert weft inlays in warp knitting are the magazine return weft insertion and single-end magazine weft insertion. Weft inserted warp knit preforms have also been explored in the composite applications (Ko, Krauland, and Scardino 1982). Most generally used monoaxial knit structures are illustrated in Figure 8.2.

8.5.2 BI-AXIAL KNITTED STRUCTURES

Bi-axial knit structures involve the incorporation of two sets of same or different yarns inside the structures in two directions. The process introduces weft, warp or

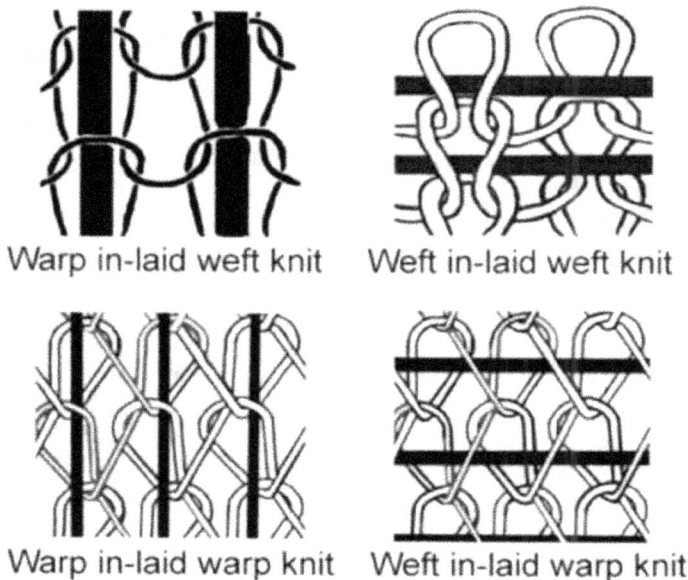

Warp in-laid weft knit   Weft in-laid weft knit

Warp in-laid warp knit   Weft in-laid warp knit

FIGURE 8.2   UD knitted structures. (Adapted from Raz 1988; David 2001.)

| Weft knit (In-lays 0/90) | Warp knit (In-lays +45/-45) | Warp knit (In-lays 0/90) |

FIGURE 8.3 Bi-axial knitted structures. (Adapted from Raz 1988; David 2001.)

diagonal yarns inside the warp and weft knitted fabrics to combine the benefits of woven and knitted fabrics together (Anand 1996). Co-we-knits are the unique structures produced by the modifying warp knitting technique, which combines weaving and knitting processes. In this technique weft yarns are laid in front of warp threads to simulate woven and knitted structures. During warp knitting, warp lay yarns are introduced either by miss lapping or by using the fall plate technique. Figure 8.3 depicts the bi-axial structures such as warp and weft inlaid weft knit, warp and weft in-laid warp knits, and diagonally inlaid knit structures [Raz 1988; David 2001.]

8.5.3 MULTIAXIAL-MULTILAYER KNITTED STRUCTURES (3D KNITS)

3D knitted composites are arguably the least studied and least understood of the other class of 3D textile composite structures, namely woven, braided, and stitched. A limited study has been reported on the mechanical properties and applications of 3D knitted composites as compared to their 2D counterparts. The 3D knit fabrics are generally classified as sandwich, non-crimp and near net shape structures. The composites made through these structures show different properties and applications. 3D knitted sandwich preforms are produced on double bed Raschel machines by knitting top and bottom layers simultaneously on each needle bed. During knitting, the core of through-thickness yarns known as piles are formed by interconnecting the fabric skins. In other words, a series of yarns are periodically interchanged between the two sets of needles to create a pile system. The relative orientation and density of pile yarns can be tailored as per the requirement just by controlling the level of yarn crossover between the two fabric layers.

3D warp knitted non-crimp preforms are formed by layers of non-crimp fabric (made through UD yarns by tow placement technique) stacked in desired orientation, then interconnected by binding yarns inserted in a through-thickness direction by warp knitting needles. Bi-, tri-, and quad-axial preforms of glass, carbon, and Kevlar® have been produced using polyester and Kevlar® binders. The amount of binders used is normally below 5% of the total fiber fraction to minimize the amount of damage to the non-crimp fabric (Hogg, Ahmadnia, and Guild 1993; Dexter and Hasko 1996; Bibo and Hogg 1997).

Development of fully integrated or near net shape 3D composites is still in an early stage despite having valuable composite performances. The high capital cost of knitting machinery as well as software development act as barriers in the progress

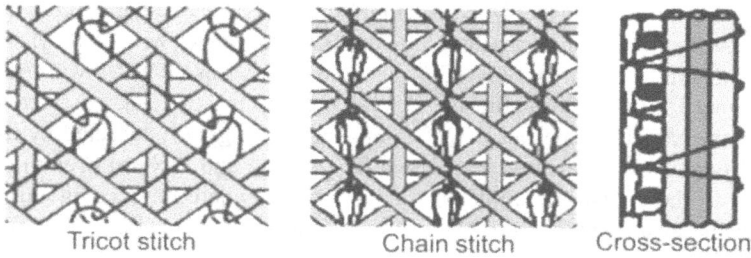

Tricot stitch   Chain stitch   Cross-section

FIGURE 8.4 3D Multiaxial–multilayer knitted structure. (Adapted from Raz 1988; David 2001.)

of 3D knit integrated composites. Two bed knitting machines along with additional needle beds can be used for the fabrication of 3D multi-layer integrated fabrics. The additional needle beds and yarn guides are required to facilitate the formation of different knit fabric layers and to integrate the fabric layers by means of transferring some part of yarns (interchanging yarn layers to lock fabric layers). These structures have found applications in the form of jet engine vanes, I-beams (Sheffer and Dias 1998), T-shape connectors (King and Greaves 1996), a rudder tip fairing for a mid-size jet engine aircraft and even a medical prosthesis (Leong, Falzon, and Bannister 1998). Multi-axial multilayer 3D knitted structures are illustrated in Figure 8.4.

Mechanical properties of thermoplastic flat knitted composites not only depend on the knit architecture, but also are affected by the hybrid towpregs used and their orientation inside the structure. To understand the effect of these parameters Abounaim, Diestel, and Cherif (2010) the developed GF/PP commingled hybrid yarn-based 3D multi-layer flat knit spacer fabrics curved in warp directions with four different angles, 0°, 90°, ±90°, and 360°, using single stage manufacturing. The study introduced two knitting techniques, one for integrating reinforcing yarns as bi-axial inlays (warp and weft direction), and the other one to introduce reinforcing yarns as tuck stitches to produce 3D multi-layer spacer fabrics. These 3D spacer fabrics were curved in warp directions by varying the length of surface layers while simultaneously knitting the connecting layers on both needle beds. 3D thermoplastic composites were also produced using U-shaped (0° curved) 3D spacer fabrics. Results revealed that the tensile strength of the reinforcement yarns after being pulled out from the 3D spacer fabrics and the strength of the 2D knit fabric were better when the reinforcement yarn integrated as bi-axial inlays. However, these properties were medium in the case of tuck stitch integration. Further, tensile and flexural properties of 2D knit composites were found to be excellent in both warp and weft directions for bi-axial inlays integration, while moderate properties are found for tuck stitch integration. In contrast, enhanced impact properties were recorded for 3D tuck knit composites in course directions, i.e., fiber-oriented directions, while optimum properties in both warp and weft directions for bi-axial inlays. Therefore, lightweight complex shaped thermoplastic composites with superior mechanical properties can be expected from these 3D multi-layer curvilinear spacer fabrics.

Abounaim, Diestel, and Cherif (2011) further explored the study and developed GF/PP hybrid yarn multi-layer TP composites to understand the effect of orientation of integrated reinforcement. Different knit structures were manufactured by integrating the load bearing yarns individually as warp yarns (WA), weft yarns (W), warp yarns and tuck stitches (WAT), warp and weft yarns (WWA), warp/weft and tuck stitches (WWAT), knit stitches (K) and tuck stitches (T). The study concluded that the mechanical properties of 2D composites were greatly affected by the type of reinforcement yarn arrangement used. The tensile and flexural properties were excellent in the course direction for the composites where reinforcing yarns were laid as weft yarns and weft with other yarn arrangement types (W, WWA, WWAT). Similarly, superior properties are found in the wale direction when warp yarns and warp with other yarn arrangement types (WA, WAT, WWA, WWAT) are used. Composite structures with bi-axial inlays showed improved tensile and flexural in both course and wale direction (WWA, WWAT). A great impact property (impact strength 230–240 kJ/m^2 and energy absorption 7–8 J) can be seen in the course direction for weft yarns (W) and tuck stitches (T). Similarly, impact strength of around 175 kJ/m^2 and energy absorption of around 7 J was reported in the wale direction for warp yarns. The authors finally concluded and recommended combining bi-axial inlays (offers good tensile and flexural properties in wale and course direction) and tuck stitch reinforcement integration (provides excellent impact energy absorption) in multi-layered knit structures for superior mechanical and impact properties. Nevertheless, unidirectional knit structures (W and WA) could be arranged asymmetrically (0°/90°) to obtain multi-directional textile preforms.

Abounaim and Cherif (2012) developed innovative GF/PP hybrid yarn-based 2D and 3D flat knitted spacer fabrics with integrated functional yarns for the creation of sensor networks. The authors, with the aim of monitoring the structural health of the end products, developed a flat knitting technology further to integrate the functional yarns into the 3D spacer fabric structures to create sensor networks. This innovation not only offers seamless, easy integration of a wide spectrum of functional materials (yarns) into the multi-layered knit structures but also forms multi-layered 3D spacer fabrics in one step process. The developed knit structures find promising applications in the area of lightweight composite structures, textile reinforced concretes, architectural designs, energy sectors, protective textiles, industrial textiles and geotextiles.

Hufenbach et al. (2011) compared the glass/PP commingled hybrid yarn-based multi-layered flatbed weft knitted (MKF) composite with the woven GF/PP composite and a NCF-GF/PP composite (Figure 8.5). Four MKF fabric layers ([0/90//90/0]$_2$)

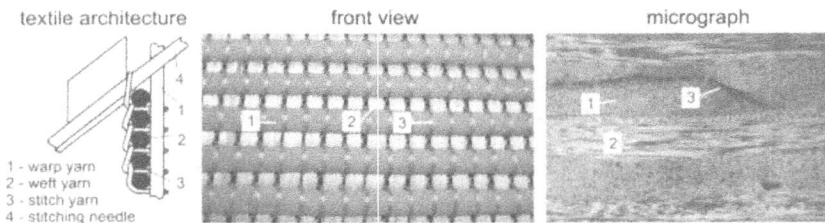

FIGURE 8.5 Textile architecture of multi-layered flatbed weft knitted fabrics (MKFs). (Adapted from Hufenbach et al. 2011.)

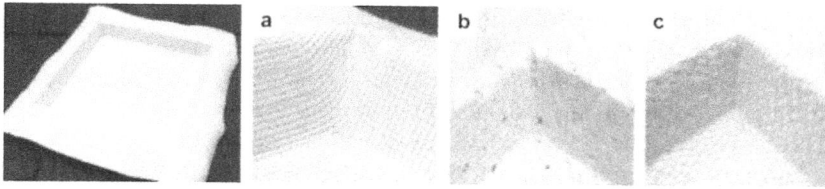

FIGURE 8.6 Textile preforming for a composite component with complex geometry made of (a) GF/PP-MKF, (b) GF/PP-NCF and (c) woven GF/PP. (Adapted from Hufenbach et al. 2011.)

of areal density 1058 g/m$_2$ are used to manufacture MKF-GF/PP composites. Non-crimp fabrics were arranged in [0/90//90/0]2 fashion and have an areal density of 1028 g/m2. Four woven layers of [(0/90)]$_4$ orientation with an areal density of 899 g/m$_2$ are used to fabricate woven composites. These MKF, NCF and woven GF/PP composites were subjected to tensile tests under different temperatures and test velocities. Charpy impact tests, open hole tension tests, and dynamic mechanical analysis were also carried out. The results concluded that MKF-GF/PP had highest energy absorption in out of plane direction or where three-dimensional stress occurs. The enhancement in load bearing capacity was attributed to the 3D stitching yarns inserted in z direction. These preforms can be used to form automotive parts with complex geometry due to their high energy absorption together with excellent drape behaviour. Figure 8.5 depicts MKF fabric structure with composite cross section, while Figure 8.6 shows textile preforming for MKF, NCF (b) and woven GF/PP.

8.6 FLEXIBLE PREFORMS BASED ON BRAIDING

Braiding is a versatile technique most commonly used in the composite industries since the 1980s. In the braiding process the continuous filaments/yarns are inter-twined to obtain the desired braid structure (Mouritz et al. 1999). This technique functions in two ways – it acts as hybrid yarn formation technique as well as a textile performing technique. In the former technique, two types of braided hybrid yarns can be constructed, such as core-sheath braided hybrid yarn and intermingled braided hybrid yarn. The core-sheath yarns can be produced in a braiding process by introducing reinforcing fibres such as carbon, glass, Kevlar, basalt, UHMWPE, ceramic, and metals as a core material, while supplying matrix forming materials such as PE, PP, PET, PEEK etc as a sheath forming material. In contrast, the intermingled hybrid yarn can be manufactured by arranging the fibre reinforcement and matrix filaments in clockwise and anticlockwise directions respectively, in the braiding machine carrier and then simultaneously feeding them to the braiding process. The braiding process intermingles both yarn bundles and yields intermingled braided hybrid yarns. In both the techniques it is possible to obtain any fibre to matrix volume fraction just by increasing or decreasing the number of fibre and matrix filaments fed. For higher V_f fractions the number of reinforcing fibres can be increased. Similarly, the number of matrix filaments can be increased to obtain more V_m fraction. It is also possible to change the braid angle by changing the delivery speed while keeping the braiding speed constant.

The braiding hybrid yarns/preforms are generally used in the production of high-performance composite by filament winding, pultrusion and tape layup techniques. The process is capable of producing complex preforms, different cross-sectional shapes, tapers, bends and bifurcations. Three-dimensional braiding was the first textile process used for the production of multidirectional near net shaped preforms for high damage resistance composites (Alagirusamy et al. 2006). Braided preforms are classified as two dimensional (2D) including tubular, flat bi-axial and tri-axial braids and three dimensional (3D) structures, including tubular, rectangular, and layer-interlock braids. The major difference between these two types is that 3D braids consist of reinforcement between layers or in through-thickness directions. In other words, 3D braided preforms are reinforced in Z-direction (out plane) also. The 3D braiding forms integrated structure either by simultaneous intertwining or orthogonal interlacement of yarns or by position displacement of yarns (Ramasamy, Wang, and Muzzy 1996a).

Alagirusamy et al. carried out a series of experiments to explore the possibility of using the braiding process as a textile performing technique. In the first part of the experiments the author characterized the flexibility, friction coefficient and lateral compressibility of nylon powder coated towpregs and commingled towpregs. In the study the flexural stiffness, friction coefficient and bulk factor property were found to be higher for powder coated towpregs as compared to the commingled towpregs and uncoated carbon tow (Ramasamy, Wang, and Muzzy 1996a).

In the second part of the study the damage inflicted by the braiding operation on the powder coated towpregs and commingled towpregs was studied. The study concluded that the powder coated towpregs experienced severe damage due to their high flexural rigidity and friction coefficient properties. Frequent interruptions were observed during braiding, which led to the increased inefficiency of the braiding process. In contrast, flexible and low friction coefficient commingled towpregs were braided better than powder coated tow pregs. Further, wrapping of these towpregs with fine filaments enhanced the braiding process (Ramasamy, Wang, and Muzzy 1996b). In the final part of the study the consolidation conditions of both towpregs were studied. The towpregs were first braided into bi-axial structures with a ± 45° braid angle and then consolidated to study the influence of consolidation conditions on the laminate properties. The results confirmed that the commingled towpregs required more severe conditions than the powder coated towpregs to arrive at comparable properties. The main reason for such difference in consolidation conditions was attributed to the difference in relative dispersion and deposition of resin. The powder coated towpregs had the well dispersed resin powder particles in between and above the carbon fibre surface. However, in the commingled yarns the nylon filaments were not well distributed. The laminates made with powder coated towpregs demonstrated better mechanical properties in braid axis direction at same consolidation condition, while commingled laminates performed poorly due to the presence of void content above the acceptance limit. The fracture analysis reveals significant fibre damage and the major failure mode of the laminates in the ± 45° test configuration is shear failure between the fibres and the resin. The fibre interlacement due to braiding helps to share the load in the post failure region (Ramasamy, Wang, and Muzzy 1996c).

Carbon/nylon schappe hybrid yarns were manufactured and braided at various angles to obtain laminates. The consolidation quality of schappe braided laminates was independent of the cover factor from 40–100%. The mechanical properties were studied in both axial and hoop directions. The tensile strength and modulus obtained at various braid angles fit in well with reported value from the literature. A bell-shaped curve was observed for the results, where a maximum value was found when the braid angle approaches 0°. Fracture analysis of the broken tensile samples elucidates that each laminate showed clear breaks exactly in line with the braid angle. Some consolidated yarns were also pulled out from the break line (Laberge-Lebel and Hoa 2007).

In another study, the commingled E-glass/polyamide 6 (PA6) hybrid yarns were used to produce 1 × 1 diamond braid structures at angles of 30°, 45° and 60°. Multi-layered braided preforms were produced using different angled fabrics (shown in Figure 8.7). Tubular beams were then produced from the braided commingled

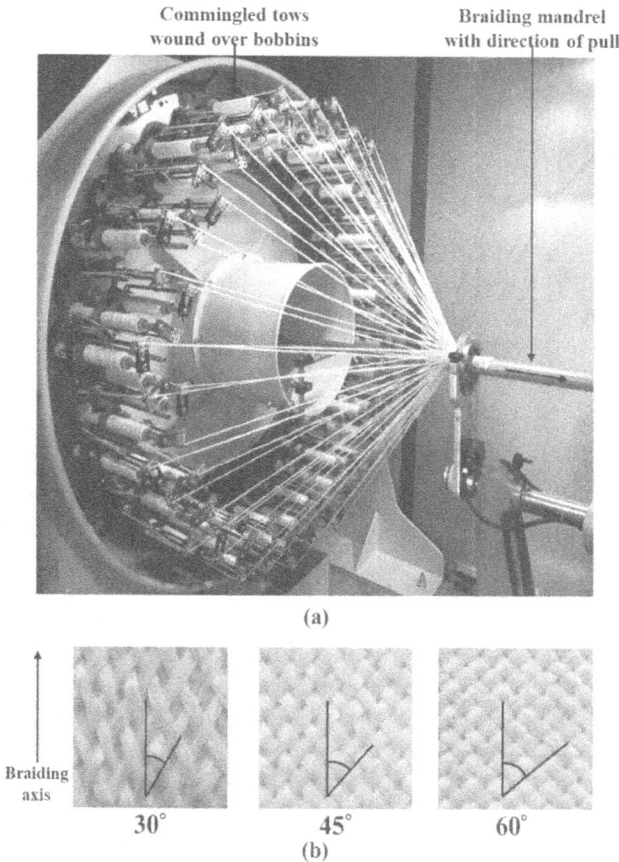

(a)

(b)

FIGURE 8.7 (a) Braiding of commingled tows; (b) flattened 1 × 1 diamond preforms with braid angles of 30°, 45° and 60°. (Adapted from Singh et al. 2021.)

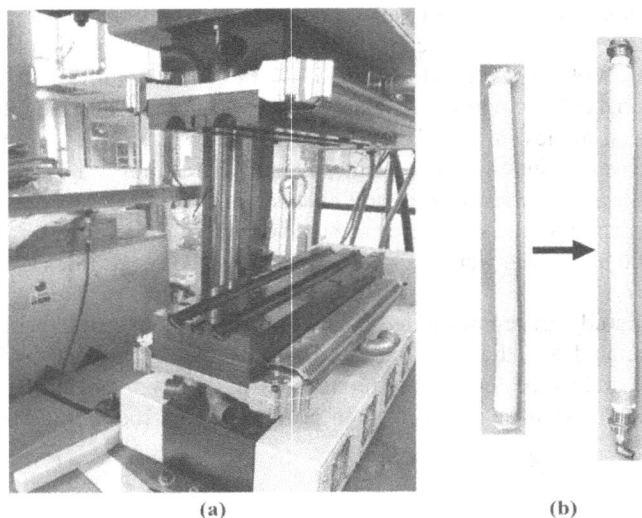

FIGURE 8.8 (a) RVM tool used for moulding of braided beams and (b) preparation of bladder-preform assembly from braided fabric. (Singh et al. 2021.)

preforms using bladder molding process (Figure 8.8). The static three-point flexure properties of the developed beams were investigated. The study explained the flexural deformation by two principal deformation modes such as localized cruising and global flexure. The localized crushing was found to decrease significantly with increasing the braid angle. The global flexural deformation was dominated when the braid angle was increased. The flexural stiffness and peak load were increased with the increments in braid angle (Singh et al. 2021).

8.7 SUMMARY AND FUTURE PROSPECTS

Textile preforming using thermoplastic hybrid yarns (flexible towpregs) offers high-quality composites with tailor-made properties as per the requirement of the application. The preforming technique overcomes the problems associated with the prepreg system. Polymeric resin is introduced already in the form of filaments/fibers/powder into the hybrid yarn. Therefore, viscosity related problems would be less or zero when towpregs-based preforms are used in the consolidation process. Textile preforming is economical, versatile, flexible and simple, known as the backbone of the composite structures. Textile preforms are generally classified as woven, knitted, braided and non-woven. These structures are further categorized as uni-directional, two-directional and three directional. Hybrid yarns, namely, commingled, powder-coated, friction spun, co-wrapped, Kemafil, side-by-side etc., can be used to construct textile preforms in different architecture. Preforming using these hybrid yarns poses some challenges, namely frequent loom stoppages, filament breakages, depreciation in the volume of matrix etc. Therefore, hybrid yarns must be stable enough to withstand the various strains generated during preforming. Multiple techniques such as co-wrapping, sizing, and twisting of hybrid yarns can be used to improve yarn stability. Hybrid yarn-based UD, 2D and 3D preforms can be constructed to meet desired specifications. The uni-directional preforms offer excellent tensile strength and modulus in fiber-oriented directions while possessing weak inter- and intra-laminar properties due to the lack of its in-plane and out-plane yarn locking. In comparison to UD weaves, 2D weaves offer better intra-laminar strength due to frequent yarn interlacement within the structure. However, the intra-laminar strength is low compared to UD weaves mainly because of the reduction in fiber volume fraction and poor matrix infiltration associated with the 2D weaves. The limitations of UD and 2D weaves can be overcome using 3D textile preforms, having yarns in X, Y and Z directions. The composite made from 3D structures offers higher inter-laminar fracture toughness and strain to failure properties, showing higher impact resistance, ballistic damage resistance and delamination resistance.

Extensive studies have been reported on the use of textile preforms in composite applications. However, most of these studies were focused mainly on developing preform structures without the use of hybrid yarns. Only a few studies are available on hybrid yarns-based preforming structures. Therefore, the current book chapter focused mainly on hybrid yarn-based textile preform structures and their composite properties. Woven textile preforms have been explored widely over knitted and braided counterparts. Solid and hollow 3D woven structures are also being used nowadays. Knitted structures are also gaining commercial interest due to their drapability and flexibility with good mechanical properties. Very few articles are available for hybrid yarn-based braided thermoplastic composites.

Textile preforming by hybrid yarns has an excellent scope in the coming days. Since limited studies are available, future studies can be pointed towards this area. Researchers can focus on exploring braided preforms using different types of hybrid yarns in the future. Similarly, more studies need to be focused on solid and hollow structures of knitted and woven preforms. The structural analysis of the composites made from hybrid-yarn-based preforms can be conducted to understand the effect of fabric structures.

REFERENCES

Abounaim, M. D., O. Diestel, G. Hoffmann, and C. Cherif. 2010. "Thermoplastic Composites from Curvilinear 3D Multi-Layer Spacer Fabrics." *Journal of Reinforced Plastics and Composites* 29(24): 3554–65.

Abounaim, M. D., O. Diestel, and C. Cherif. 2011. High Performance Thermoplastic Composite from Flat Knitted Multi-Layer Textile Preform Using Hybrid Yarn." *Composites Science and Technology* 71(4): 511–19.

Abounaim, M. D., and C. Cherif. 2012. "Flat-Knitted Innovative Three-Dimensional Spacer Fabrics: A Competitive Solution for Lightweight Composite Applications." *Textile Research Journal* 82(3): 288–98.

Alagirusamy, R., R. Fangueiro, V. Ogale, and N. Padaki. 2006. "Hybrid Yarns and Textile Preforming For Thermoplastic Composites." *Textile Progress* 38 (4): 1–71.

Alagirusamy, Ramasamy, and Vinayak Ogale. 2004. "Commingled and Air Jet-Textured Hybrid Yarns for Thermoplastic Composites." *Journal of Industrial Textiles* 33 (4): 223–43. doi:10.1177/1528083704044360

Anand, S. C. 1996. "Warp Knitted Structures in Composites." In *Proceedings of Seventh European Conference on Composite Materials*, London, 407–13.

Baghaei, Behnaz, Mikael Skrifvars, and Lena Berglin. 2015. "Characterization of Thermoplastic Natural Fibre Composites Made from Woven Hybrid Yarn Prepregs with Different Weave Pattern." *Composites Part A: Applied Science and Manufacturing* 76. Elsevier: 154–61. doi:10.1016/j.compositesa.2015.05.029

Bar, Mahadev, R. Alagirusamy, and Apurba Das. 2018. "Properties of Flax-Polypropylene Composites Made through Hybrid Yarn and Film Stacking Methods." *Composite Structures*. doi:10.1016/j.compstruct.2018.04.078

Bar, Mahadev, R. Alagirusamy, and Apurba Das. 2019. "Industrial Crops & Products Influence of Friction Spun Yarn and Thermally Bonded Roving Structures on the Mechanical Properties of Flax/Polypropylene Composites." *Industrial Crops & Products* 135 (July 2018). Elsevier: 81–90. doi:10.1016/j.indcrop.2019.04.025

Baucom, J. N., and M. A. Zikry. 2003. "Evolution of Failure Mechanisms in 2D and 3D Woven Composite Systems under Quasi-Static Perforation." *Journal of Composite Materials* 37 (18): 1651–74. doi:10.1177/0021998303035178

Bibo, G. A., P. J. Hogg, and M. Kemp. 1997. "Mechanical Characterisation of Glass-and Carbon-Fibre-Reinforced Composites Made with Non-Crimp Fabrics." *Composites Science and Technology* 57 (9–10): 1221–41.

Choi, B. D., O. Diestel, and P. Offermann. 1999, July. "Commingled CF/PEEK Hybrid Yarns for Use in Textile Reinforced High Performance Rotors." In *12th International Conference on Composite Materials (ICCM)*, Paris, 796–806.

Cunningham, D. V., and L. E. Pritchard. 2005. Quasi-Unidirectional Fabric for Ballistic Applications. US 6861378 B2. *United States Patent*, issued 2005.

Dexter, H. B., and G. H. Hasko 1996. "Mechanical Properties and Damage Tolerance of Multiaxial Warp-Knit Composites." *Composite Science and Technology* 56 (3): 367–80.

El-Dessouky, Hassan M., and Mohamed N. Saleh. 2018. "3D Woven Composites: From Weaving to Manufacturing." In *Recent Developments in the Field of Carbon Fibers*. InTech. doi:10.5772/intechopen.74311

Fukuta K. and E. Aoki. 1986. "3D Fabrics for Structural Composites." In *Proceedings of the 15th Textile Research Symposium*. Philadelphia, PA.

Goud, Vijay, Dilpreet Singh, Alagirusamy Ramasamy, Apurba Das, and Dinesh Kalyanasundaram. 2020. "Investigation of the Mechanical Performance of Carbon/Polypropylene 2D and 3D Woven Composites Manufactured through Multi-Step Impregnation Processes." *Composites Part A: Applied Science and Manufacturing* 130 (November 2019). Elsevier: 105733. doi:10.1016/j.compositesa.2019.105733

Gu, Huang, and Zhong Zhili. 2002. "Tensile Behavior of 3D Woven Composites by Using Different Fabric Structures." *Materials and Design* 23 (7): 671–74. doi:10.1016/s0261-3069(02)00053-5

Hogg, P. J., A. Ahmadnia, and F. J. Guild 1993. "The Mechanical Properties of Non-Crimped Fabric-Based Composites." *Composites* 24 (5): 423–32.

Horrocks, A. R., and S. C. Anand. 2000. *Handbook of Technical Textiles: Technical Textile Applications (Vol. 2)*. UK: CRC Press/Woodhead Publishing.

Hufenbach, W., R. Böhm, M. Thieme, A. Winkler, E. Mäder, J. Rausch, and M. Schade. 2011. "Polypropylene/Glass Fibre 3D-Textile Reinforced Composites for Automotive Applications." *Materials & Design* 32 (3). Elsevier: 1468–76. doi:10.1016/J.MATDES.2010.08.049

Jogur, Ganesh, Ashraf Nawaz Khan, Apurba Das, Puneet Mahajan, and R. Alagirusamy. 2018. "Impact Properties of Thermoplastic Composites." *Textile Progress* 50 (3). Taylor & Francis: 109–83. doi:10.1080/00405167.2018.1563369

Khokar, N. 2001. "3D-Weaving: Theory and Practice." *Journal of the Textile Institute* 92 (2): 193–207. doi:10.1080/00405000108659570

King J. E., R. P. Greaves, and H. Low. 1996. "Composite Materials in Aeroengine Gas Turbine: Performance Potential vs Commercial Constraint?" In *Keynote Lecture, Presented at Sixth European Conference on Composite Materials*, London, 14–16.

Ko, F. K., and G. W. Du. 1988. *Textile Preforming. In Handbook of Composites*. Boston, MA: Springer, 397–424.

Ko, F. K., K. Krauland, and F. Scardino. 1982. "Weft Insertion Warp Knit for Hybrid Composites, Progress in Science and Engineering of Composites." In Hayashi et al. (Eds.), *ICCM-V, Fourth International Conference on Composites*, 982–87.

Laberge-Lebel, Louis, and Suong Van Hoa. 2007. "Manufacturing of Braided Thermoplastic Composites with Carbon/Nylon Commingled Fibers." *Journal of Composite Materials* 41 (9): 1101–21. doi:10.1177/0021998306067273

Leal, A. Andres, Joshi C. Veeramachaneni, Felix A. Reifler, Martin Amberg, Dominik Stapf, Gion A. Barandun, Dirk Hegemann, and Rudolf Hufenus. 2016. "Novel Approach for the Development of Ultra-Light, Fully-Thermoplastic Composites." *Materials and Design* 93 (March). Elsevier Ltd: 334–42. doi:10.1016/j.matdes.2015.12.125

Leong K. H., P. J. Falzon, M. K. Bannister, and I. Herszberg. 1998. "An Investigation of the Mechanical Performance of Milano Rib Weft-Knit Glass/Epoxy Composites." *Composite Science and Technology* 58: 239–51.

Mäder, Edith, Christina Rothe, Harald Brünig, and Thomas Leopold. 2007. "Online Spinning of Commingled Yarns-Equipment and Yarn Modification by Tailored Fibre Surfaces." *Key Engineering Materials* 334–335: 229–32. doi:10.4028/www.scientific.net/kem.334-335.229

Mountasir, A., G. Hoffmann, and C. Cherif. 2011. "Development of Weaving Technology for Manufacturing Three-Dimensional Spacer Fabrics with High-Performance Yarns for Thermoplastic Composite Applications: An Analysis of Two-Dimensional Mechanical Properties." *Textile Research Journal* 81(13): 1354–66.

Mouritz, A. P., M. K. Bannister, P. J. Falzon, and K. H. Leong. 1999. "Review of Applications for Advanced Three-Dimensional Fibre Textile Composites." *Composites Part A: Applied Science and Manufacturing* 30 (12): 1445–61. doi:10.1016/S1359-835X(99)00034-2

Niklas, Svensson. 1998. "Textile Structures for Load-Carrying Composites." *Textile Magazine* 2: 6–13.

Padaki Naveen, V., A. Vani, V. Prakasha, C. J. Divakar, T. Ananth Krishnan, and R. M. V. G. K. Rao. 2000. "Studies on Mechanical Behavior of Knitted Glass-Epoxy Composites." *Journal of Reinforced Plastics and Composites* 19 (5): 396–402.

Padaki, Naveen V., R. Alagirusamy, and B. S. Sugun. 2006. "Knitted Preforms for Composite Applications." *Journal of Industrial Textiles* 35 (4): 295–321. doi:10.1177/1528083706060784

Potluri, Prasad, T. Sharif, and D. Jetavat. 2008. "Robotic Approach to Textile Preforming for Composites." *Indian Journal of Fibre and Textile Research* 33 (3): 333–38.

Ramasamy, A., Youjiang Wang, and John Muzzy. 1996a. "Braided Thermoplastic Composites from Powder-Coated Towpregs. Part I: Towpreg Characterization." *Polymer Composites* 17 (3). John Wiley & Sons, Ltd: 497–504. doi:10.1002/pc.10639

Ramasamy, A., Youjiang Wang, and John Muzzy. 1996b. "Braided Thermoplastic Composites from Powder-Coated Towpregs. Part II: Braiding Characteristics of Towpregs." *Polymer Composites* 17 (3). John Wiley & Sons, Ltd: 505–14. doi:10.1002/pc.10640

Ramasamy, A., Youjiang Wang, and John Muzzy. 1996c. "Braided Thermoplastic Composites from Powder-Coated Towpregs. Part III: Consolidation and Mechanical Properties." *Polymer Composites* 17 (3). John Wiley & Sons, Ltd: 515–22. doi:10.1002/pc.10641

Rao, R. M. V. G. K., G. Anand Nahar, and R. Alagirusamy. 2000. "Mechanical Properties of Glass Composites Based on Knitted Preforms with Inlays." *Indian Journal of Fibre and Textile Research* 25 (1): 115–20.

Raz, Samuel. 1988. "Karl Mayer Guide to Technical Textiles." *Karl Mayer Textilmaschinenfabtik*.

Okada, Ryo. 2011. Quasi-Unidirectional Fabrics for Structural Members Having Same. US8017532B2, issued 2011.

Sarlin, J., P. Pentti, and P. K. Jarvela. 1990. "A New Type of Knitted Reinforcing Fabrics and Its Application." In *Proceedings of Fourth European Conference on Composite Materials*, London, 621–23.

Scardino, F. 1989. "An Introduction to Textile Structures and Their Behavior. Textile Structural Composites." In Chou, T. W., and Ko, F. K. (Eds.), *Composite Materials Series* Vol. 3, Elsevier Science, New York, Chapter 1, 1-24. ISBN 0-444-42992-1.

Selver, Erdem, Prasad Potluri, Paul Hogg, and Costas Soutis. 2016. "Impact Damage Tolerance of Thermoset Composites Reinforced with Hybrid Commingled Yarns." *Composites Part B: Engineering* 91. Elsevier Ltd: 522–38. doi:10.1016/j.compositesb.2015.12.035

Sheffer, E., and T. Dias. 1998. "Knitting Novel 3-D Solid Structures with Multiple Needle Bars." In *UMIST Textile Conferences—Textile Engineered for Performance*, Manchester.

Singh, Anubhav, Neil Reynolds, Elspeth M. Keating, Alastair E. Barnett, Steve K. Barbour, and Darren J. Hughes. 2021. "The Effect of Braid Angle on the Flexural Performance of Structural Braided Thermoplastic Composite Beams." *Composite Structures* 261 (October 2020). Elsevier Ltd: 113314. doi:10.1016/j.compstruct.2020.113314

Soden, J. A., and B. J. Hill. 1998. "Conventional Weaving of Shaped Preforms for Engineering Composites." *Composites Part A: Applied Science and Manufacturing* 29(7): 757–62.

Spencer, David J. 2001. *Knitting Technology*. Woodhead Publishing Limited. CRC Press, UK.

Torun, Ahmet R., Adil Mountasir, and Gerald Hoffmann. 2013. "Production Principles and Technological Development of Novel Woven Spacer Preforms and Integrated Stiffener Structures," 275–85. doi:10.1007/s10443-012-9281-8

Van Vuure, Aart W., and K. Ko Frank. 2003. "Net Shape Knitting for Complex Composite Preforms." *Textile Research Journal* 73 (1): 1–10.

Wakeman, M. D., T. A. Cain, C. D. Rudd, R. Brooks, and A. C. Long. 1998. "Compression Moulding of Glass and Polypropylene Composites for Optimised Macro-and Micro-Mechanical Properties—1 Commingled Glass and Polypropylene." *Composites Science and Technology* 58 (12): 1879–98.

Ye, Lin, Klaus Friedrich, Joachim Kästel, and Yiu Wing Mai. 1995. "Consolidation of Unidirectional CF/PEEK Composites from Commingled Yarn Prepreg." *Composites Science and Technology* 54 (4): 349–58. doi:10.1016/0266-3538(95)00061-5

Zhang, Lu, and Menghe Miao. 2010. "Commingled Natural Fibre/Polypropylene Wrap Spun Yarns for Structured Thermoplastic Composites." *Composites Science and Technology* 70 (1). Elsevier Ltd: 130–35. doi:10.1016/j.compscitech.2009.09.016

9 Modelling the Consolidation of Flexible Towpregs

Ashraf Nawaz Khan[a], Puneet Mahajan[b],
and R. Alagirusamy[a]
[a]Department of Textile and Fibre Engineering
Indian Institute of Technology Delhi
New Delhi, India
[b]Department of Applied Mechanics
Indian Institute of Technology Delhi
New Delhi, India

CONTENTS

DOI: 10.1201/9781003049715-9

9.1   INTRODUCTION

The growing demand in the fields ranging from the automotive to aerospace industry increases the research interest in the thermoplastic composites. The thermoplastic material has superior inherent properties such as high impact strength, high chemical inertness, easy processability, and a short cycle time for the forming process due to the use of the semi-finished products at a relatively low cost. These properties make thermoplastic material more attractive for research and development (Ye, Klinkmuller, and Friedrich 1992; Scalea and Green 2000; Hattum, Nunes, and Bernardo 2005; Christmann, Medina, and Mitschang 2017). However, the uniform consolidation of the thermoplastic composite is the main challenge due to the high melt viscosity of the thermoplastic polymer matrix. Advanced manufacturing techniques have been developed to overcome the processability challenges associated with the polymer matrix. Several manufacturing techniques have been developed to produce the thermoplastic

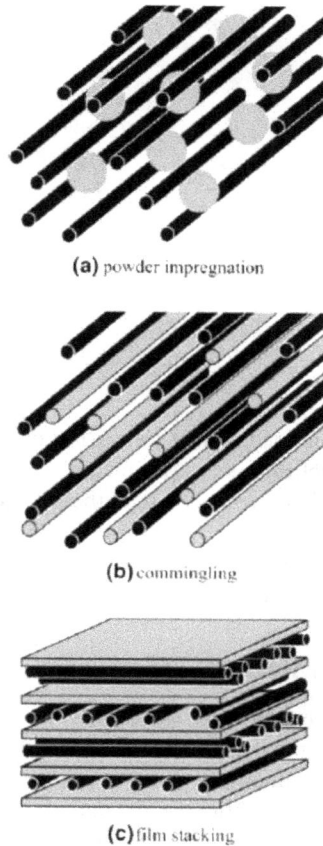

(a) powder impregnation

(b) commingling

(c) film stacking

FIGURE 9.1 Different impregnation techniques. (Reproduced with permission from Wysocki et al. 2005.)

composite such as solution processing, melt-impregnation, film stacking, filament bundle coating, hybrid yarn technique, and different methods of pre-impregnation (shown in Figure 9.1). Each technique has its merit and demerit and has a unique flow pattern of the molten polymer on consolidation as shown in Figure 9.2.

To overcome problems with the thermoset composite, the thermoplastic matrix evolved as a new variant to substitute the non-recyclable thermoset material with superior impact properties. But the high viscosity of the thermoplastic polymer matrix complicates the manufacturing of the laminates. The hybrid yarn-based manufacturing technique gives flexible towpregs, which reduces the melt flow distance at the sub-micron level. The low melt flow distance eventually improves the impregnation quality. The flexible towpreg is used to produce preform for the reinforcement purpose through the weaving, which was mainly concerned in the traditional techniques such as melt impregnation, slurry deposition, etc. (Steggall-Murphy et al. 2010; Jogur et al. 2018; Gupta et al. 2021; Khan et al. 2022). These preforms are used to produce laminates at optimized consolidation conditions. The quality of the

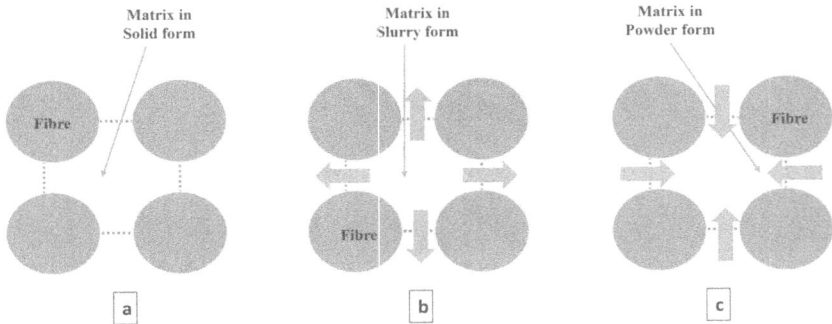

FIGURE 9.2 Illustration of the molten resin flow. (Adapted from Padaki and Drzal 1997.)

consolidation determines the mechanical utilization of its constituent components. Various mathematical modelling and experimental research work have been done on the manufacturing techniques to understand their consolidation mechanisms and influence of the processing factors on the quality of the impregnation. In the subsequent sections, the development of the mathematical consolidation models will be discussed to elaborate on the flow mechanics behind the impregnation process.

9.2 IMPREGNATION MODELS

Various consolidation models have been developed to estimate the rate of impregnation, void content, and mechanical properties as a function of the material parameters, processing conditions, and geometrical conditions of the prepreg. The consolidation model helps in optimizing the condition for the manufacturing of the composite laminates. It requires a deep understanding of the impregnation mechanics to develop the analytical model to predict the consolidation behavior of the laminate. Once the mathematical equation develops, it requires verification with the help of the experimental results. On successful validation, it becomes easy to predict the consolidation quality with respect to the different processing conditions. The consolidation process involves a combination of the various phenomena that can be categorized into different groups as shown in Figure 9.3. The consolidation model includes different submodels such as the heat and mass transfer model, fiber bed permeability model, etc.

There are various hybrid yarn-based techniques to fabricate the thermoplastic composites and each technique gives different fiber and matrix arrangements. The positioning of the fiber and matrix within the towpreg influences the interface property and impregnation mechanism during the consolidation process. The better impregnation leads to the formation of better adhesive bonding between the fiber and matrix constituents. This ultimately leads to better load transfer from the matrix to the reinforcing parts in the resultant thermoplastic composite laminates. It is important to understand the governing principle of the impregnation phenomena. This requires a comprehensive understanding of the flow mechanics of the molten polymer matrix at the micro and macro levels within the preform during the consolidation process. There are several works on developing the impregnation model to understand the

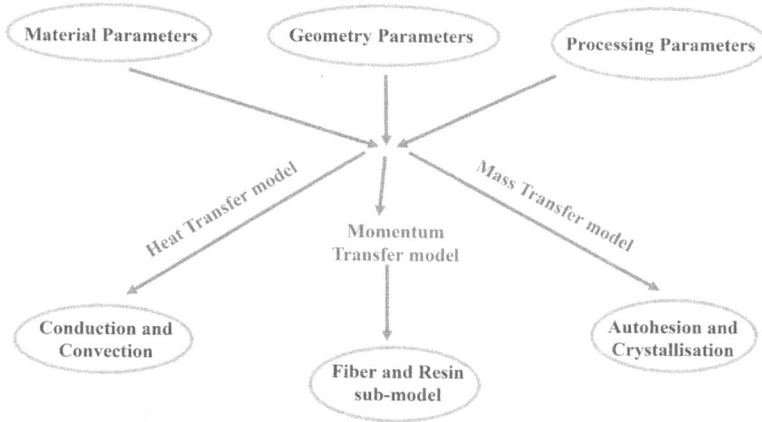

FIGURE 9.3   Model layout for the consolidation phenomena. (Adapted from Padaki and Drzal 1997.)

behavior of the consolidation process in terms of the mechanics of the flow at higher temperature and pressure, with several assumptions and hypotheses. The impregnation model helps in simulating the time cycle for the full consolidation at different processing conditions (van Hattum, Nunes, and Bernardo 2005; Alagirusamy et al. 2006; Steggall-Murphy et al. 2010; Jogur et al. 2018). The consolidation process involves three major steps as shown in Figure 9.4. The first one is the heating of the preform till the matrix phase gets melted, the second is the compression stage which forces the molten matrix to impregnate within the fibrous porous structure, and the third is the cooling of the preform once the saturation level is reached. The different steps can be modelled separately to simulate the whole consolidation process. The rate of heating and cooling, the holding temperature, pressure, and time play important role in determining the quality of the laminates.

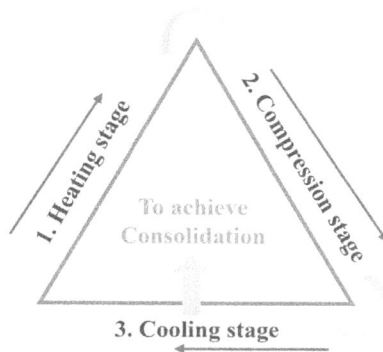

FIGURE 9.4   Major steps involve during consolidation. (Adapted from Padaki and Drzal 1997.)

The consolidation model is usually divided into three sub-models such as (Gil 2002):

   a. **Compaction Sub-model:** This sub-model deals with the deformation of the reinforcement during applied pressure.
   b. **Permeability Sub-model:** This sub-model predicts the permeability of the fiber assembly at higher pressure.
   c. **Impregnation Sub-model:** This sub-model predicts the impregnation of the molten matrix into the fiber bundle at higher temperature and pressure.

9.3 PROPOSED MODELS AND BASIC EQUATIONS FOR THE CONSOLIDATION PHENOMENA OF FLEXIBLE TOWPREGS

The consolidation of the composite laminate is a complex phenomenon that includes heat and mass transfer, the flow of the molten polymer through the complicated porous medium, deformation of the fiber bed under the applied pressure, and shear forces due to the high concentrated viscous matrix, including other unknown phenomena. The incorporation of all practical complexity in developing the mathematical model introduces difficulty in deriving the equations and their solutions. Since different parameters are inter-connected with their unexplored inter-dependency and furthermore their anisotropic nature makes it difficult to model the mathematical expressions. Therefore, it requires several assumptions and hypotheses to simplify the problem and derive the model. Once the problem is framed into mathematical terms, then it is possible to replace the assumptions with the existing reality. Although these assumptions lead to the diversion of the solution in comparison to the experimental one, through the correction factor these errors can be minimized.

    The governing mechanism of the consolidation process varies with different towpreg techniques such as commingled-based hybrid yarn, micro-braiding hybrid yarn, powder-coated towpregs, etc. This happens basically due to the change in the fiber and matrix configuration. The shape of the cross-section of the fiber bundle and the melt flow distance governs the flow of the molten matrix during the compaction and consolidation. The effect of the aspect ratio of the fiber bundle can be seen in Figure 9.5.

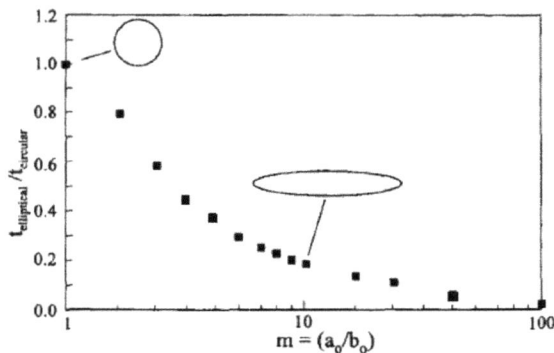

FIGURE 9.5 Time for full impregnation with respect to the aspect ratio of the fiber bundle. (Reproduced with permission from Cutolo and Savadori 1994.)

The boundary condition of the flow gets modified with the change in the arrangements of the fiber and matrix within the fiber bundle. Therefore, different approaches have been adopted to develop an impregnation model for different towpregs.

9.3.1 Impregnation Model for Commingled Hybrid Yarn-Based Laminate

For the commingled carbon/PEEK thermoplastic composite, Ye et al. (1995) explained the mechanics of the impregnation process during the consolidation. The consolidation behavior of the laminates was described as a function of the processing condition such as holding pressure and time using a mathematical model as discussed below.

The assumption made in this mathematical study of the consolidation process states that the Newtonian flow by the molten matrix through the porous medium (Figure 9.6) has negligible capillary effect, and there is no cross flow of the matrix from one bundle to another and across the cross-sectional plane.

The molten polymer impregnating the fiber bundle follows Darcy's law, which is very helpful in estimating the consolidation time. The governing differential equation for the rate of impregnation is given by Equation 9.1 (Ye, Klinkmuller, and Friedrich 1992; Ye et al. 1995).

$$\frac{dz}{dt} = \frac{K_P}{\mu} \times \frac{dp}{dz} \tag{9.1}$$

Where dp/dz is the pressure gradient, μ is the viscosity, K_p is the permeability of the fiber bed.

The permeability of the fiber bed is one of the important aspects of understanding the flow of the molten polymer matrix during impregnation. The permeability of the fiber bed can be measured through fiber volume fraction with the help of Carmen-Kozeny Equation 9.2 as:

$$K_P = \frac{d_f^2}{16k_0} \times \frac{\left(1 - V_f\right)^3}{V_f^2} \tag{9.2}$$

Where d_f is the fiber diameter, K_0 is the Carmen-Kozeny constant, v_f is the fiber volume fraction.

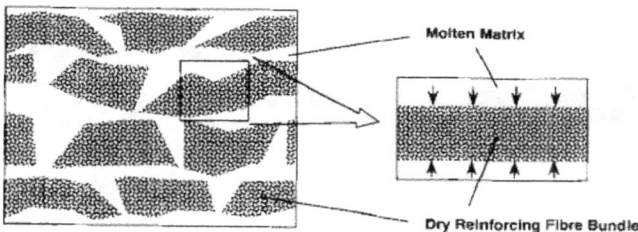

FIGURE 9.6 Depiction of the impregnation of the commingled towpregs. (Reproduced with permission from Ye, Klinkmuller, and Friedrich 1992.)

The processing time t required to reach the penetration depth z can be evaluated using Darcy's law by putting K_p (Equation 9.2) in Equation 9.1, on integration and simplification with appropriate boundary conditions, we get Equation 9.3 (Ye et al. 1995).

$$t = \frac{8\mu k_0 z^2}{d_f^2 p_a} \times \frac{V_f^2}{\left(1 - V_f\right)^3} \qquad (9.3)$$

Where p_a is the applied pressure that remains constant during the consolidation stage.

9.3.1.1　Fully Unmingled Model

In the fully unmingled model, it was assumed that the yarn was fully unmingled with the symmetrical distribution of the matrix. The total area of the matrix pools A_m was distributed as two equal matrix pools above and below the reinforcing fiber bundle due to the geometrical symmetry as shown in Figure 9.7.

The cross-section A_b of the fully consolidated yarn is measured using the density of the filaments ρ_t and weight per unit length of the filaments W as shown below.

$$A_b = W/\rho_t$$

The height h of the matrix pool can be evaluated as

$$h = V_m h_0$$

The height of the full consolidated yarn can be obtained as

$$h_0 = A_b/b$$

Where b is the width of the bundle which remains constant throughout the length and during the impregnation process, V_m is the volume fraction of the matrix polymer (Ye et al. 1995).

Once the impregnation process starts, the free space available inside the fiber bundle gets filled by the molten polymer. At the desired level of the impregnation

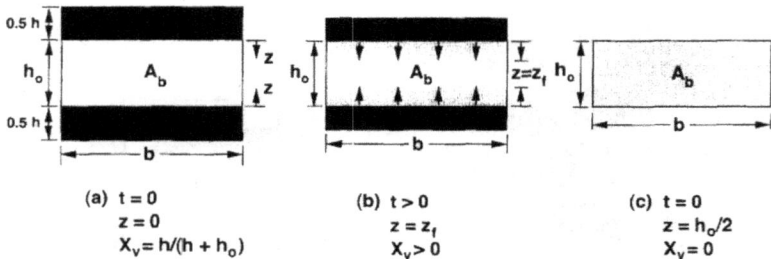

FIGURE 9.7　Illustration of fully unmingled yarn-based laminates. (Reproduced with permission from Ye, Klinkmuller, and Friedrich 1992.)

$z(t) = z_f$, the area of the remaining free space A_v in the fiber can be measured using Equation 9.4.

$$A_V = A_m - 2V_m b z_f \tag{9.4}$$

Therefore, the void content X_v (Equation 9.5) of the composite can be evaluated as a function of the penetration depth z_f during the impregnation process (Ye et al. 1995).

$$X_V = \frac{A_V}{A_b + A_V} = \frac{A_m - 2V_m b z_f}{A_b + \left(A_m - 2V_m b z_f\right)} \tag{9.5}$$

9.3.1.2 Partial Commingled Model

In practice, neither fully commingled nor fully unmingled yarns are available. Therefore, all the related models have some disagreement with the experimental results. To develop a more realistic model, the partial commingled model has been proposed to achieve better results. D_{com} indicates a degree of commingling that is $D_{com} = h_c/h_0$ where h_c is the height of the partially commingled yarn as shown in Figure 9.8.

$D_{com} = 0$ refers to fully unmingled yarn
$D_{com} = 1$ refers to fully commingled yarn

In the partial commingled model, the void content X_{vi} at the initial stage is measured as in Equation 9.6 (Ye et al. 1995).

$$X_{vi} = \frac{A_m - V_m b h_c}{A_b + \left(A_m - V_m b h_c\right)} = \frac{V_m \left(h_0 - h_c\right)}{h_0 + V_m \left(h_0 - h_c\right)} \tag{9.6}$$

When the impregnation proceeds, the free space inside the fiber bundle start filling with the molten matrix. The amount of the void content X_v at any stage of the

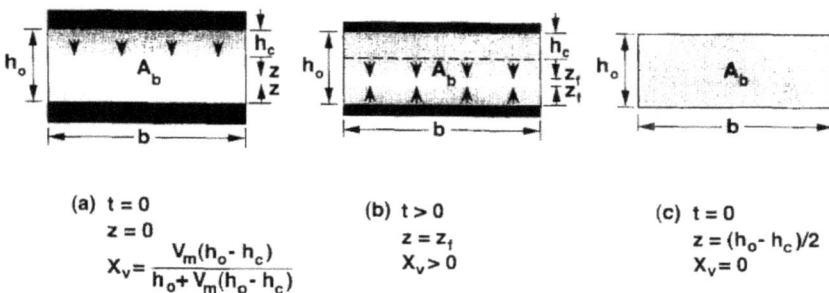

(a) $t = 0$
$z = 0$
$X_v = \dfrac{V_m(h_o - h_c)}{h_o + V_m(h_o - h_c)}$

(b) $t > 0$
$z = z_f$
$X_v > 0$

(c) $t = 0$
$z = (h_o - h_c)/2$
$X_v = 0$

FIGURE 9.8 Illustration of partially comingled yarn-based laminates. (Reproduced with permission from Ye, Klinkmuller, and Friedrich 1992.)

impregnation is the function of the penetration $z(t) = z_f$ and can be calculated through Equation 9.7.

$$X_V = \frac{A_m - V_m b \left(h_c + 2z_f \right)}{A_b + A_m - V_m b \left(h_c + 2z_f \right)} \qquad (9.7)$$

The above equation is a generalized equation which is valid for the fully unmingled model too. If $D_{com} = 0$ i.e. $h_c = 0$ then X_{vi} becomes zero in the equation which gives the condition of the full consolidation with zero voids.

Based on the assumption of the identical representative elements, the behavior of the consolidation procedure of the small control volume can be generalized throughout the laminates. If N number of the commingled yarn bundles, the thickness of the laminates can be calculated as the function of the voids as shown by Equation 9.8 (Ye et al. 1995).

$$H = \frac{NA_b \left(1 + X_v \right)}{B} = H_0 \left(1 + X_v \right) \qquad (9.8)$$

Where H is the actual height of the laminates, B is the width of the mold, X_V is the contents of the void.

9.3.2 EQUIVALENT PISTON AND SPRING MECHANICAL MODEL ANALOGOUS TO THE DIFFERENT CONSOLIDATION STAGES

In a study, Van West, Pipes, and Advani (1991) took a different approach to develop an impregnation model for the commingled-based thermoplastic composite material. The equivalent piston and spring mechanical models were considered in different consolidation stages (Figure 9.9) for the derivation of the mathematical model as

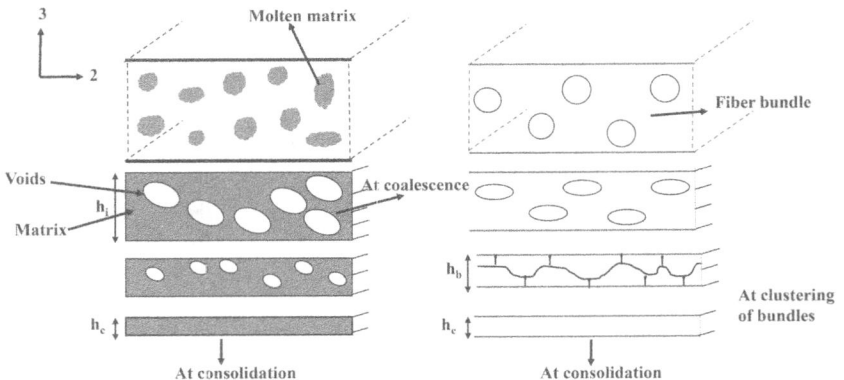

FIGURE 9.9 Steps for consolidation for (a) the matrix and (b) fiber respectively. (Adapted from Van West, Pipes, and Advani 1991.)

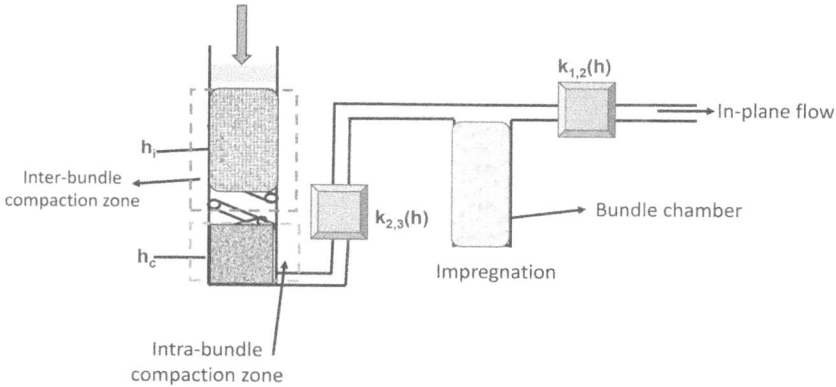

FIGURE 9.10 Equivalent mechanical model for consolidation. (Adapted from Van West, Pipes, and Advani 1991.)

discussed below. The model is helpful in determining the effect of the processing parameters on the consolidation behavior.

Basically, there are various factors which influence the rate and the quality of consolidation of the laminates. The processing parameters, material properties, and the geometry of the fiber network which includes weaving construction, hybrid yarn configuration, etc. have a direct influence on the rate of the consolidation and the compaction process (Van West, Pipes, and Advani 1991).

The consolidation mechanism of the laminate was represented by the equivalent mechanical model with the help of piston and spring arrangements as shown in Figure 9.10. Each process of the consolidation process was conceptualized by different elements of the equivalent mechanical model. The inter bundle compaction elasticity was represented by the spring mechanism which resists the downward motion of the piston on applied P_0. Initially, the load was not carried by the pool of the molten matrix. When the piston got displaced down to h_i height, the molten matrix started forming coalescences and began to bear the load. Further downward displacement of the piston leads to the matrix flow through the orifice that represents fiber bundle permeability and gets into a chamber that represents fiber bundles. When the piston was displaced to the height h_b, another spring representing intra-bundle compaction was added to the fiber bed elastic resistance. Furthermore, when the piston was displaced to h_c height, a molten polymer was passed through an orifice that represents transverse permeability, $k_{2,3}(h)$. When the molten polymer filled the chamber, the matrix flowed out through the second orifice which represents in-plane permeability $k_{1,2}(h)$ of the fiber bed. In this way, during the consolidation process, the matrix flows into the chamber representing the fiber/bundle impregnation while the flow outside represents the in-plane matrix flow (Van West, Pipes, and Advani 1991).

9.3.2.1 Compaction Sub-Model

The applied consolidation pressure is opposed by the elastic resistance of the fiber assembly during compaction through the hydrostatic pressure of the molten matrix during the impregnation process. A slight modification in the Hou model can be done to establish the relationship between the pressure of fiber bed and the thickness of the laminates as Equation 9.9 (Van West, Pipes, and Advani 1991).

$$P_{fb} = A_S \frac{(h_0 - h)}{\left[1 - \left(\dfrac{h_0 - h}{h_0 - h_\infty} \right) \right]^n}$$

(9.9)

Where P_{fb} is the fiber assembly pressure, A_s is the fiber assembly elastic constant, h_0 is the initial laminate thickness, h_∞ is the minimum possible laminate thickness, h is the instantaneous laminate thickness, n is an exponent.

9.3.2.2 Impregnation Phenomena

During the compaction, the fiber bundle takes the shape of ellipse. This complicates the derivation process for the impregnation phenomena. A numerical study is used to identify the equivalence factor to replace the elliptical cross-section with the circular cross-section. This was just to employ one-dimensional flow in a polar coordinate system in spite of the two-dimensional flow with an undefined flow front in the elliptical cross-section. The radius of the equivalent circle which takes the same time for the full consolidation as taken by the elliptical cross-section can be derived as Equation 9.10 (Van West, Pipes, and Advani 1991):

$$r_{eq} = \sqrt{2} \frac{a_0 b_0}{\sqrt{a_0^2 + b_0^2}}$$

(9.10)

Where a_0 and b_0 are the half major and minor axes of the ellipse.

The various assumption made for the impregnation model includes a) uniform flow of molten matrix across the cross-sectional plane, b) molten matrix has very low Reynolds number and follows Darcy's law during flow through the porous medium, c) the permeability of the fibrous medium depends on the fiber volume fraction, d) the thickness of the laminates and the applied pressure remains constant during the consolidation and many more.

The boundary condition for the flow front is molten matrix pressure at the periphery of the fiber bundle and ambient pressure within the fiber bundle (Van West, Pipes, and Advani 1991).

The expression of the relative velocity of the constituents through the porous medium during compaction and impregnation can be evaluated through Darcy's law as:

$$V_r - U_r = -\frac{K}{\mu} \frac{dP}{dr}$$

(9.11)

During the compaction, the relative fiber bundle velocity varies linearly from 0 at the center to the U_0 at the outer radius r_0.

The governing differential equation for the flow is given by Equation 9.12 (Van West, Pipes, and Advani 1991).

$$\frac{1}{r}\frac{d}{dr}\left(r\frac{dP}{dr}\right) = \frac{2\mu}{k}(1+\alpha)\frac{U_0}{r_0} = \text{Constant} \tag{9.12}$$

Where r is the radial location, V_r is the relative velocity of the matrix, U_r is the local relative velocity of the fiber, k is the permeability constant, μ is the viscosity of the matrix, U_0 is the fiber velocity at the outer fiber bundle boundary, α is the solid to liquid volume ratio.

The pressure gradient inside the fluids can be evaluated as Equation 9.13 (Van West, Pipes, and Advani 1991).

$$\frac{dP}{dr} = \frac{Cr}{2} + \frac{(P_r - P_v) - C/4\left(r_0^2 - r_i^2\right)}{r \ln r_0/r_i} \tag{9.13}$$

Where C is:

$$C = 2\mu(1+\alpha)U_0/Kr_0$$

Where r_0 is the circular bundle outer radius, r_i is the flow front radius within the fiber bundle, P_r is the molten matrix pressure, P_v is the pressure of the enclosed voids.

The pressure exerted by the voids (Equation 9.14) is obtained from the ideal gas equations under isothermal conditions (Van West, Pipes, and Advani 1991).

$$P_v = P_a \frac{\text{initial volume}}{\text{current volume}} = P_a \frac{r_{initial}^2}{r_t^2}\left(\frac{1 - v_{f\,initial}}{1 - v_f}\right) \tag{9.14}$$

The consolidation behavior of the single bundle is assumed to be identical throughout the bundle. The unit cell geometry used in the model has a definite relation with the thickness of the laminates as shown in Figure 9.11. The unit cell dimension is used to calculate the thickness of the laminate during the different stages of the consolidation process.

The thickness of the laminates can be obtained from the cross-sectional areas of the fiber, matrix, and voids as shown below (Van West, Pipes, and Advani 1991):

$$h = \frac{n_l}{s_b}\left[A_f + A_r + \pi a_i b_i\left(1 - v_f\right)\right] \tag{9.15}$$

Where A_f and A_r are the unit cell cross-sectional areas of the fiber and the resin, a_i and b_i is the major and minor half-length of the void's region, v_f is the fiber to non-fiber volume ratio and S_b is the center-to-center fiber spacing.

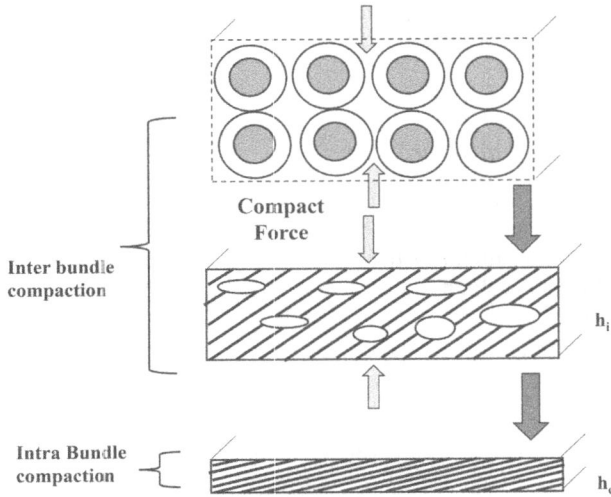

FIGURE 9.11 Depiction of inter/intra-bundle compaction. (Adapted from Van West, Pipes, and Advani 1991.)

9.3.3 THE EQUATION FOR THE RATE OF THE IMPREGNATION

Steggall-Murphy et al. (2010) developed a mathematical model to estimate the rate of impregnation as a function of holding pressure and dwell time during the consolidation process. The development of the model requires the unit cell, material property, and processing parameters with proper assumptions to simplify the model. Since there is a complex microstructure of the fiber assembly through which the molten polymer flows, it is assumed constant tow permeability throughout the laminate which makes it easy to derive the concerned equations. The derivation of the impregnation model is done based on the representative unit cell (as shown in Figure 9.12) by assuming similar physics throughout the laminates. The model is developed for the cross-ply six-layer unidirectional laminates (Steggall-Murphy et al. 2010). The unit cell represents two tows with one tow gap as shown in Figure 9.12.

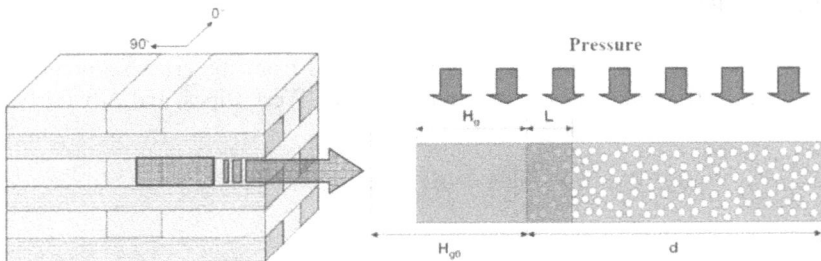

FIGURE 9.12 Unit cell with the tow arrangement. (Reproduced with permission from Steggall-Murphy et al. 2010.)

The unit cell is made up of half of the 0° tow gap and half of 0° tow where $Hg0$ is the initial value of the tow gap while H_g is the tow gap which decreases with the impregnation. As per Darcy's law, the average fluid velocity during impregnation can be expressed as:

$$u_{avg} = \frac{KP_r}{\mu L} = \emptyset \frac{dL}{dt}$$

Where K is the tow permeability, μ is the viscosity of the molten polymer, ϕ is the porosity of the fiber bundle, P_r is the molten resin pressure, L is the depth of the resin front impregnation at a given time t.

During the consolidation process, the applied pressure is shared by both the molten matrix and fiber bed. However, in the model, it is assumed that there is no tow deformation during the compression to simplify the derivation. Therefore, this eliminates the stress development in the fiber tow.

The differential equation representing impregnation of the resin into tow with respect to the time is expressed as Equation 9.16 (Steggall-Murphy et al. 2010):

$$\frac{dL}{dt} = \frac{K}{\mu\emptyset} \frac{P_a}{v_r} \frac{1}{L} \tag{9.16}$$

Where P_a is the applied pressure, v_r, v_f, and v_v is the volume fraction of the resin, fiber, and voids. The volume of the constituents such as resin, fiber, and voids at any time can be expressed as (Steggall-Murphy et al. 2010):

$$V_r = H_g + \phi L, \ V_f = d(1-\phi), \text{ and } V_v = (d - L)\phi$$

The governing differential equation for the flow front velocity is expressed as Equation 9.17 (Steggall-Murphy et al. 2010).

$$\frac{dL}{dt} = \frac{k}{\mu\emptyset} \frac{P_a}{L} \left(\frac{H_g + d}{H_g + L\emptyset} \right) \tag{9.17}$$

On solving the Equation 9.17 (Steggall-Murphy et al. 2010), the resin front position L_r within the tow during impregnation at time t_r is given by:

$$\frac{H_{go}(H_{go} + d)}{\emptyset^2} \ln \left\{ \frac{H_{go} + d}{H_{g0} + d - L_r\emptyset} \right\} - \frac{H_{go}}{\emptyset} L_r = \frac{kBt_r^2}{\mu\emptyset 2} \tag{9.18}$$

H_{g0} is the initial tow gap when the L was 0 and the width of the tow is assumed to be constant as shown in Figure 9.12.

9.3.4 IMPREGNATION MODELS FOR THE POWDER-COATED TOWPREGS

The powder towpreg technique is another technique where reinforcing tow is coated with the partially melted polymer matrix to produce flexible powder impregnated towpregs. The fiber and matrix melt flow distance (MFD) can be reduced to the sub-micron level through this technique. The low MFD helps in better impregnation quality during the consolidation period. This technique is a fast and cost-effective mode of production. The sprinkled powder size affects the towpreg structure, which influences the impregnation mechanism. The fiber volume fraction of the preform is the measure of the tortuosity present in the preform. The preform with the higher fiber volume fraction has less space available for the flow of the molten polymer matrix. Therefore, higher pressure is required for the flow of the molten resin across the fibrous assembly during the consolidation as shown in Figure 9.13 (Padaki and Drzal 1999; Bowman et al. 2018).

Padaki and Drzal (1997) studied the impregnation mechanism in the powder impregnated tapes during the consolidation stage. The mathematical expression was derived to describe the rate of heat transfer through the laminate during the consolidation. A three-dimensional squeeze flow model was used to describe the molten matrix flow in the layup. It is very crucial to understand the impregnation behavior of the powder impregnated tapes (Figure 9.14) during the consolidation. Since powder particles are available very close to the reinforcing fiber, the molten resin is required to travel a very small distance only.

9.3.4.1 Load Distribution

On applying the load, both the fiber and the matrix take the load. The amount of load sustained by the constituents depends upon the fiber volume fraction. The equation for the effective stress generation within the fibrous assembly can be obtained from the momentum balance equation in the transverse direction to the laminate plane.

During the consolidation stage, the applied pressure is taken by the molten matrix and then transferred to the fiber network. The molten matrix squeezes out of the

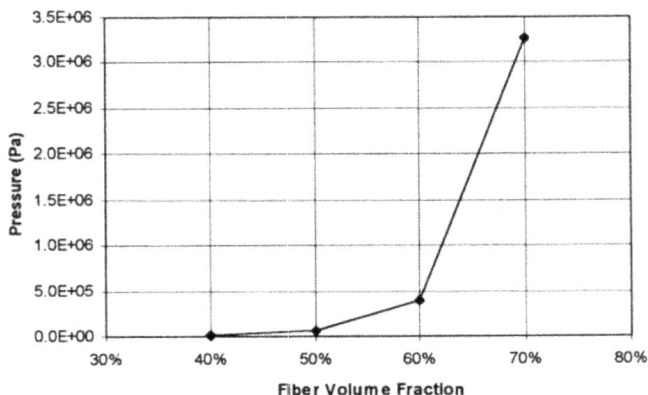

FIGURE 9.13 Inter-relationship: fiber deformation pressure vs. fiber volume fraction. (Reproduced with permission from Padaki and Drzal 1999.)

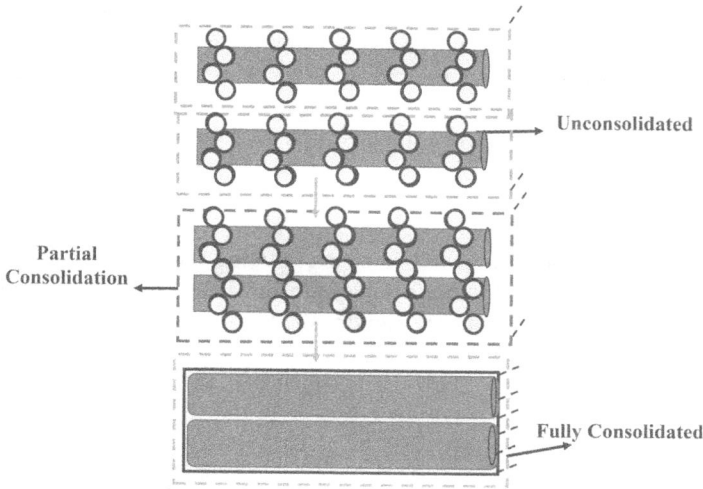

FIGURE 9.14   Consolidation steps for powder impregnated tape. (Adapted from Padaki and Drzal 1997.)

mold due to the pressure gradient and therefore the fiber volume fraction changes. On further compression, the fiber assembly starts deforming against the compressive load, which results in stress generation within the fiber network. The improved Equation 9.19 for the average effective stress is given by Padaki and Drzal 1997:

$$\sigma = 3\pi \frac{E}{\beta^4} \frac{\sqrt{\dfrac{V_f}{V_0}} - 1}{\left(\sqrt{\dfrac{V_a}{V_f}} - 1\right)^4} \qquad (9.19)$$

Where σ is the average effective stress generated in the fiber network, E is the bending stiffness of the fiber, β is the ratio of the span length and the span height of the fiber beam network, V_0 is the original (or initial) fiber volume fraction when σ is zero while V_a is the maximum available fiber volume fraction when the σ tends to infinity (that is maximum fiber volume fraction), V_f is the fiber volume fraction. The maximum fiber volume fraction attainable for the square and hexagonal packed array of the unidirectional laminates is 78.5% and 90.7%.

9.3.5   CONSOLIDATION MODEL FOR THE POWDER IMPREGNATED PREPREG COMPOSITES

It has been observed that the impregnation mechanism in the powder impregnated prepregs has some difference in comparison to others in terms of flow direction from inside to outside and distance between fiber and matrix. The input parameters such as material property (density, viscosity, etc.), geometry condition (fiber shape and size, powder size, etc.), and the processing condition (temperature, pressure, and time) are required to develop the impregnation model (Padaki and Drzal 1997). The assumption for the model includes a flat rectangular layup for the analysis purpose, and the thickness is very small in comparison to the other two dimensions and uniform distribution of the powder inside.

9.3.5.1 Heat Transfer Model

For the heat transfer analysis, the heat is supplied to the unconsolidated slab through the conduction mode to achieve uniform temperature, T as shown in Figure 9.15.

The fundamental differential equation for the heat transfer mode is expressed as Equation 9.20 (Padaki and Drzal 1997):

$$\frac{\partial T}{\partial t} = \alpha \frac{\partial^2 T}{\partial x^2} \tag{9.20}$$

Where α is the thermal diffusivity of the slab, x is the location within the slab.

Where h is the heat conductivity, k_c is the thermal conductivity of the composite laminates.

On applying the appropriate boundary conditions, the solution of Equation 9.20 (Padaki and Drzal 1997) can be found as:

$$\frac{T_\infty - T}{T_\infty - T_i} = \sum_{n=1}^{\infty} exp\left(-\frac{\lambda_n^2 \alpha t}{h^2}\right) cos\left(\frac{\lambda_n x}{L}\right) \tag{9.21}$$

Where λ_n are the eigenvalues, v is the fiber volume fraction.

The thermal properties of the laminates can be measured through the rule of the mixture relationships for the concern constituents, namely fibers, matrix, and voids content (Padaki and Drzal 1997).

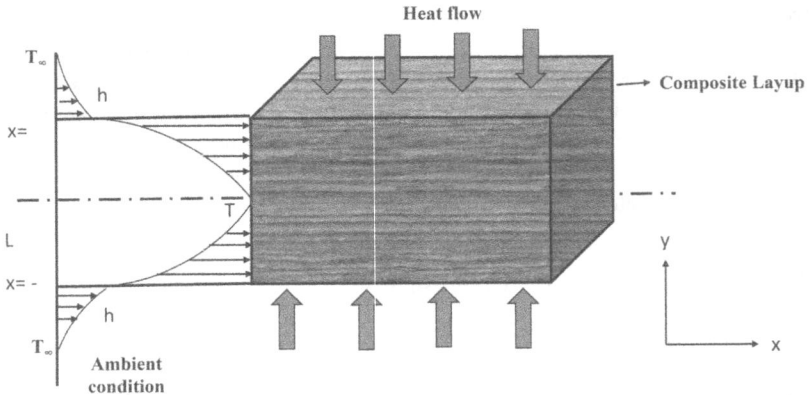

FIGURE 9.15   Illustration of heat transfer. (Adapted from Padaki and Drzal 1997.)

9.3.5.2 Momentum Transfer

The consolidation process for the powder impregnated prepreg is different from the other prepreg material such as commingling hybrid yarn, etc. This happens due to the distribution of the powder, i.e., each fiber is coated with the powder. During the consolidation stage, the tow comes close together and squeezes out the powder polymer between the fibers and achieves void-free consolidation. Therefore, the traditional approach for model development needs to change.

For the modelling of the flow, the spherical powder particle is taken as the equivalent cubical structure using mathematical transformation as shown in Figure 9.16. A creeping flow model is used for the analysis purpose. A constant viscosity and incompressible fluid are assumed to simplify the problem. The in-plane flow of the molten matrix occurs equally in two directions.

Simplifying the continuity and pressure gradient equation with the given boundary condition results in Equation 9.22 (Padaki and Drzal 1997):

$$P_r - P_{r0} = \frac{3\mu v_w}{h}\left[\frac{\left(a^2 - x^2\right)}{h^2} + \frac{\left(b^2 - y^2\right)}{h^2} - \frac{2z}{h}\left(1 - \frac{z}{h}\right)\right] \qquad (9.22)$$

Where v_w is the velocity of one surface with respect to the stationary surface.

The expression of the pressure distribution for the molten resin is given by Equation 9.22. On integrating Equation 9.22 over the cross-sectional plane across the flow direction with respect to time, we get the total force F on the molten resin as shown in Equation 9.23 (Padaki and Drzal 1997):

$$F = \frac{8\mu a^4}{t}\left(\frac{1}{h^2} - \frac{1}{h_0^2}\right) \qquad (9.23)$$

Where a is the side of the cube, μ is the viscosity, h_0 is the initial spacing and h is the spacing at any time t.

In another study for the powder-coated towpreg, Ye, Klinkmuller, and Friedrich (1992) formulated the equation to estimate the time for the impregnation, void content during the impregnation, and thickness of the resultant composite laminate for the powder-coated glass/PP composite system.

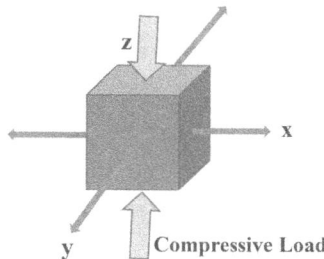

FIGURE 9.16 Depiction for creeping flow model. (Adapted from Padaki and Drzal 1997.)

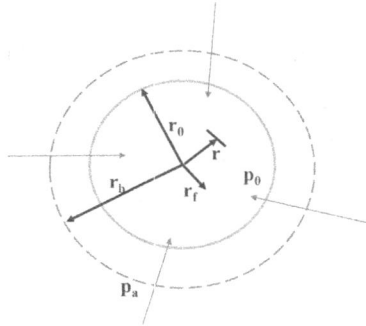

FIGURE 9.17 Radial impregnation for the fiber bundle. (Adapted from Ye, Klinkmuller, and Friedrich 1992.)

The time taken by the molten matrix to reach the desire radial position (Figure 9.17) is given by Equation 9.24 (Ye, Klinkmuller, and Friedrich 1992):

$$t = \frac{\mu r_0^2}{4k_p\left(p_a - p_0\right)}\left[2\left(\frac{r_f}{r_0}\right)^2 In\left(\frac{r_f}{r_0}\right) + 1 - \left(\left(\frac{r_f}{r_0}\right)^2\right)\right] \tag{9.24}$$

Where K_p is the permeability, p_0 is the void pressure within the tow, p_a is the molten resin pressure surrounding the fiber bundle, μ is the viscosity, r_0 is the radius of the fiber-tow border, r_f is the radius of the resin front.

The degree of the impregnation (Figure 9.18) can be expressed as:

$$D_{imp} = \frac{\text{area impregnated}}{\text{fully impregnated tow area}}$$

During the impregnation, the free space inside the bundle can be measured through Equation 9.25 (Ye, Klinkmuller, and Friedrich 1992).

$$V_v = \left(1 - V_f - V_{imp}\right)\pi r_f^2 = \pi r_0^2\left(1 - D_{imp}\right)\left(1 - V_f - V_{imp}\right) \tag{9.25}$$

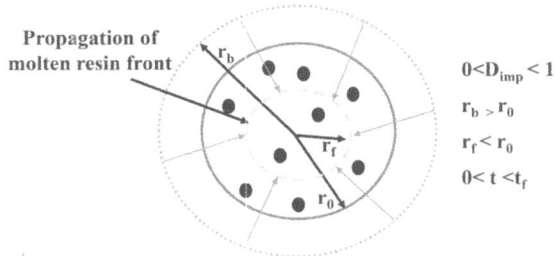

FIGURE 9.18 Impregnation within the fiber bundle. (Adapted from Ye, Klinkmuller, and Friedrich 1992.)

The void content during the impregnation is measured through Equation 9.26.

$$X_v = \frac{V_v}{\pi r_b^2} \tag{9.26}$$

Based on the idea of the representative bundle, the same impregnation behavior can be considered for the whole laminate. If there are N bundles of the commingled yarn, then the thickness of the laminate H can be calculated by Equation 9.27 (Ye, Klinkmuller, and Friedrich 1992):

$$H = \frac{NA_b(1+X_v)}{B} = H_0(1+X_v) \tag{9.27}$$

Where B is the width of the laminates.

9.3.6 EVALUATE THE DEGREE OF COALESCENCE AND THE TEMPERATURE WITH RESPECT TO THE TIME

In another study, Ramani and Hoyle (1995b) investigated the heat transfer mechanism for the flexible powder-coated towpreg. An experiment-based analytical study was performed to evaluate the degree of coalescence and the temperature with respect to the time. The temperature is the dominating factor to affect the behavior of molten polymer material. A coalescence model has been formulated based on the processing parameters to measure the degree of coalescence of the powder and its influence on the flexibility of the towpreg. The flexibility of the towpreg can be predicted through the degree of coalescence of the powder particles (DOC). DOC is the ratio of the contact radius of the coalescing particles to the initial powder particle radii. The magnitude of the DOC varies from 0 (uncoalesced) to 1 (completely coalesced particles). The quality of consolidation can be examined through image analysis and the mechanical property of the laminates such as flexural properties.

9.3.6.1 Coalescence Sub-Model

The coalescence model is developed to predict the degree of coalescence (x/r) during the heating. The coalescence forms above the melting point of the matrix polymer and follows the relation given by Equation 9.28 (Ramani and Hoyle 1995b).

$$\left(\frac{X}{r}\right)^m = \frac{F(T)}{r^n} t \tag{9.28}$$

Where x is the contact radius of the two coalescing particles, r is the particle radius, t is the time, m and n are the constants which characterize the type of flow during coalescence, and $F(T)$ is the ratio of surface tension to the viscosity.

The general expression for the surface tension of the polymer as a function of the temperature is expressed through Equation 9.29. At higher temperatures, the wettability/flowability of the molten polymer increases due to the reduction in the surface tension.

$$v = v_0 \left(1 - \frac{T}{T_c} \right) \tag{9.29}$$

v and v_0 are the surface tension and reference surface tension of the polymer respectively and T_c is the critical temperature.

The viscosity of the polymer following the Arrhenius type relation is described through Equation 9.30.

$$\eta = \eta_0 \, exp^{\frac{Q}{Rt}} \tag{9.30}$$

Where η and η_0 are the viscosity and reference viscosity, Q is the activation energy of the flow, and R is the universal gas constant.

9.3.6.2   Heat Transfer Sub-Model

A control volume for the tow is chosen for heat transfer analysis through the energy balance equation. The balance of the energy equations helps in deriving the ordinary governing differential equation. The tow temperature is an important aspect in determining the viscosity and surface tension as shown in Equations 9.29 and 9.30. It has been observed that the rate of energy transfer through the change in the crystallinity and coalescence is negligible in comparison to the total rate of change in the energy since the phase change of the polymer matrix occurs during the coating process. Therefore, the governing differential Equation 9.31 (Ramani and Hoyle 1995b) for the rate of the temperature change can be obtained with the help of energy balance equation through substituting the expression suggested by Chiang for the specific heat C_p.

$$\frac{dT}{dt} = \frac{\sigma P_t \epsilon_t \left(T_{wall}^4 - T^4 \right)}{\rho A_c \left(m_f P_1 + m_r P_2 \right)} \tag{9.31}$$

Where p_1 and p_2 are,

$$p_1 = f_1 + f_2 \left(2T - f_3 \right)$$

$$p_2 = c_1 + c_2 \left(2T - c_5 \right) + c_3 e^{-C4(T - c_5)^2} \left(1 - 2c_4 T^2 + 2c_5 T \right)$$

where σ is the Stefan-Boltzman constant, ρ is the equivalent tow density, ε_t is the emissivity of the tow, P_t is the tow perimeter, T_{wall} is the wall temperature of the coating heater, m_f and m_r are the mass fractions of the fiber and resin respectively, A_c is the cross-sectional area of the tow, P is the pressure, f and c are constants used for the fiber and the resin. The governing differential equation can be solved through the fourth order Runga-Kutta methods. The residence time of the small control volume can be calculated using the tow velocity and length of the heater and used in the simulation analysis (Ramani and Hoyle 1995b).

The contact angle of the molten droplet with the fiber is the measure of the surface tension. The droplet remains stable up to the threshold value of the spreading coefficient. The spreading coefficient increases with an increase in the temperature and crosses the threshold value at a certain temperature. At this stage, instability generated inside the droplet leads to the breakdown of the droplet and spreading occurs. This is the temperature limit when the coating of the tow becomes possible. The consolidation process is dominant with the longitudinal flow of the resin in comparison to the transverse flow. The maximum voids are removed during the local longitudinal flow of the molten polymer. The consolidation process generally comprises of fiber bed compaction and polymer flow within the tow (Ramani and Hoyle 1995b).

In a similar study, Ramani and Hoyle (1995a) developed a model to predict the temperature and moisture content of the tow through coupled heat and mass flow analysis for the powder slurry technique. The assumptions made in the developed model includes (a) a rectangular cross-section of the tow, (b) the heat and mass transfer in the orthogonal direction of the tow axis, and (c) the small thickness of the tow in comparison to the width, giving higher temperature and moisture gradient across the thickness. This led to the one-dimensional heat and mass transfer analysis in the thickness direction. The boundary condition for the heat and mass analysis is shown in Figure 9.19.

For the heat and mass transfer analysis, a control volume can be chosen to perform the energy and mass balance within the tow. The tow is modelled as unidirectional fiber with small capillaries in the third direction. For the prediction of the heating and drying of the tow, Luikov's coupled equation is used for the capillary-porous media. The rate of the temperature change within the tow directly depends upon the conductance

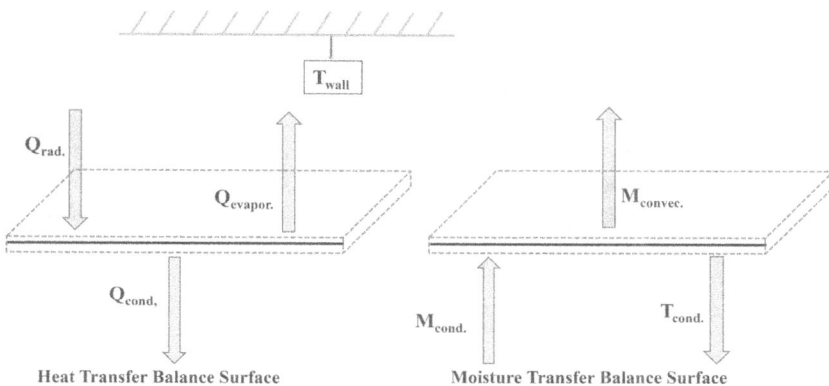

FIGURE 9.19 Heat and mass transfer balance surface analysis. (Adapted from Ramani and Hoyle 1995a.)

of the tow and the presence of the heat sink or sources. The coupled heat and mass transfer differential Equation 9.32 are given by (Ramani and Hoyle 1995a).

$$c\rho_0 \frac{\partial T}{\partial t} = \frac{\partial}{\partial x}\left(k \frac{\partial T}{\partial x} \right) + \rho_0 \varepsilon h_{fg} \frac{\partial u}{\partial t} \qquad (9.32)$$

The coupled Equation 9.32 can be solved to measure the temperature and moisture content within the tow with the help of the effective thermo-physical parameters such as specific heat of the towpreg c, the density of the fiber tow ρ_0, transverse thermal conductivity of the fiber tow k. When the temperature of the tow increases, the moisture content within tow decreases due to evaporation.

The heat transfer occurs through conduction, convection, and radiation mode. The heat balance Equation 9.33 is derived as (Ramani and Hoyle 1995a).

$$-k \frac{\partial Tl_s}{\partial x} - (1-\varepsilon) h_{fg} \beta_u \rho_0 \left(u l_s - u_{inf} \right) + \sigma \varepsilon_{tow} \left(T_{wall}^4 - Tl_s^4 \right) = 0 \qquad (9.33)$$

Where T is the temperature of the tow, T_{wall} is the wall temperature of the heater, ε is a phase conversion factor, h_{fg} is the latent heat of the vapourization, β_u is the convection coefficient of the mass transfer, u is the moisture content, u_{inf} is the moisture content of the ambient air, σ is the Stefan-Boltzman constant, ε_{tow} is the emissivity of the fiber tow. In the heat balance Equation 9.33 (Ramani and Hoyle 1995a), the first term represents the conduction within the fiber bundle, the second term is heat transfer through convection, and the last term is the heating of the tow through the radiation.

The mass balance can be performed at tow surfaces likewise in the case of the energy balance. During the process, the moisture comes to the tow surface through moisture conduction and is removed by the means of mass convection. The mass balance Equation 9.34 is given as (Ramani and Hoyle 1995a):

$$\alpha_m \left[\frac{\partial u}{\partial x} l_s + \delta \frac{\partial T}{\partial x} l_s \right] + \beta_u \left(u l_s - u_{inf} \right) = 0 \qquad (9.34)$$

Where α_m is the total mass transfer coefficients of moisture diffusion, β_u is the convection coefficient of the mass transfer, u is the moisture content, u_{inf} is the moisture content of the ambient air. The first two terms in Equation 9.34 represent the moisture conduction due to the moisture gradient and the temperature gradient, and the third one represent mass convections. The measurement of the heat and mass transfer Biot (Bi) number has its significance in determining the external and internal resistance to the heat and mass transfer. The explicit finite-difference technique can be used to solve the coupled heat and mass transfer equations (Ramani and Hoyle 1995a).

9.3.7 PREDICTION OF THE IMPREGNATION TIME

Phillips, Akyüz, and Månson (1998) studied a model to predict the rate of the impregnation of the woven fiber-reinforced thermoplastic laminate considering the isothermal processing condition. The three major processes involved during the consolidation are intimate contact, autohesion, and fiber impregnation.

9.3.7.1 Autohesion

The effect of the autohesion can be studied through the double cantilever beam to estimate critical energy during crack propagation. The critical fracture toughness can be measured through Equation 9.35 (Phillips, Akyüz, and Månson 1998).

$$G_{IC} = \frac{3F\delta}{2Ba} \tag{9.35}$$

Where F is the load, δ is the displacement, B is the sample width, a is the crack length.

9.3.7.2 Intimate Contact

On applying pressure to the laminate, the load is shared by both the fiber and matrix constituents. For the pressure evolution measurements, the pressure-time response (Equation 9.36) can be estimated with the help of the Maxwell equation.

$$P(t) = P_f + \left(P_{app} - P_f\right)exp\left(-\frac{t}{\tau_0}\right) \tag{9.36}$$

Where P_{app} is the applied pressure, P_f is the asymptotic value of the pressure, τ_0 is the relaxation time (Figure 9.20).

Based on the pressure relaxation analysis, the viscous deformation of the rectangular resin element can be expressed as Equation 9.37.

$$b(t) = b_0 \left[1 + \frac{5\phi P_{app}t}{\eta}\left(1 + \frac{w_0}{b_0}\right)\left(\frac{a_0}{b_0}\right)^2\right]^{\frac{1}{5}} \tag{9.37}$$

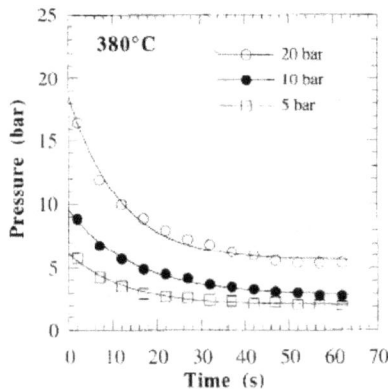

FIGURE 9.20 Pressure-time graph obtained during pressure relaxation tests. (Reproduced with permission from Phillips, Akyüz, and Månson 1998.)

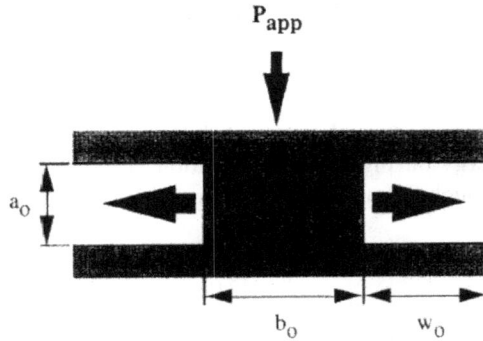

FIGURE 9.21   Resin bridge. (Reproduced with permission from Phillips, Akyüz, and Månson 1998.)

Equation 9.37 is the extension of the Lee-Springer model, which is considered pressure distribution of the fiber bed and resin. Equation 9.37 (Phillips, Akyüz, and Månson 1998) is valid for the unidirectional and woven preforms for the thermoplastic composites. Where $b(t)$ is the width of the rectangular element as a function of the time, η is the viscosity, w_0, a_0, b_0 are the initial dimensions of the resin rectangular elements, ϕ is a constant, P_{app} is the applied pressure as shown in Figure 9.21.

Although the plies are bonded together, the macro void present between the plies requires pressure to form a full resin bridge to achieve full intimate contact.

9.3.7.3   Fiber Impregnation

For the fiber impregnation, the fiber matrix system is heated to form the molten matrix which starts impregnating the fiber bundles. The unidirectional molten polymer flow (Figure 9.22) follows Darcy's law with laminar motion. The derivation done to evaluate the rate of the impregnation follows the steps discussed below.

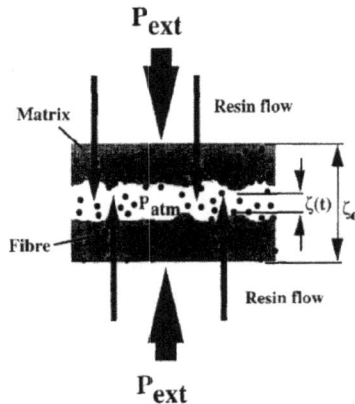

FIGURE 9.22   Resin flow during fiber impregnation. (Reproduced with permission from Phillips, Akyüz, and Månson 1998.)

Putting the values such as $\Delta P = P_{ext} - P_{atm}$, $dx = \zeta - \zeta_0$ to the differential equation obtained through Darcy's law for the laminar flow which on simplification through the Viete relationship takes the form of Equation 9.38 (Phillips, Akyüz, and Månson 1998) through initial condition $t = 0$, $\zeta = \zeta_i$. The fiber impregnation is represented as a function of time by Equation 9.38 at constant temperature and pressure.

$$\zeta(t) = \zeta_0 - \sqrt{\zeta_0^2 - 2\left[\zeta_0\zeta_i - \frac{\zeta_i^2}{2} - \frac{4k_p}{\eta}(P_{ext} - P_{atm})t\right]} \qquad (9.38)$$

Where ζ_0 is the thickness of the fiber, ζ is the thickness of the non-impregnation zone at t time.

Furthermore, the impregnation time can be measured using Equation 9.39 (Phillips, Akyüz, and Månson 1998).

$$t_{imp} = \left(\frac{\zeta_0\zeta_i}{4} - \frac{\zeta_i^2}{8}\right)\frac{\eta}{K_p\left(\phi P_{app} - P_{atm}\right)} \qquad (9.39)$$

The main advantages of the powder impregnation process are the low melt flow distance, the feasible processability of the high molecular weight polymer matrix powder, and the high flexibility and homogeneous distribution of the fiber and matrix parts which lead to excellent consolidation (Itoi and Pipes 1990).

9.3.8 THE RESIN IMPREGNATION FOR THE MICRO-BRAIDED YARN-BASED COMPOSITE

Kobayashi, Tanaka, and Morimoto (2012) studied the resin impregnation process (as depicted in Figure 9.23) in the micro-braided yarn based on Darcy's law and continuity equations. Following Darcy's law and the continuity equation with appropriate boundary condition, equations for the pressure distribution and velocity

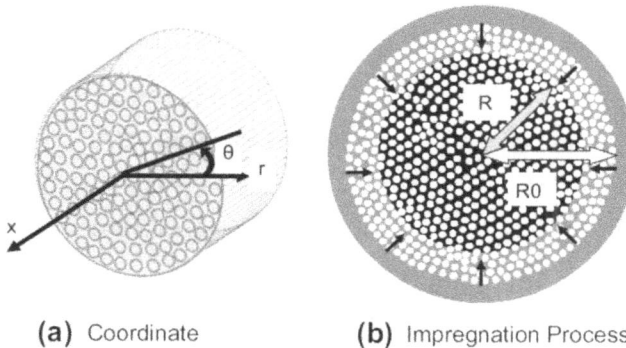

(a) Coordinate   (b) Impregnation Process

FIGURE 9.23 Impregnation process. (Reproduced with permission from Kobayashi, Tanaka, and Morimoto 2012.)

distribution can be expressed by Equations 9.40 and 9.41 (Kobayashi, Tanaka, and Morimoto 2012).

$$P = \frac{P_m}{In(R_0/R)} In\ r - \frac{P_m}{In(R_0/R)} In\ R = \frac{P_m}{In(R_0/R)} In\left(\frac{r}{R}\right) \qquad (9.40)$$

$$u_r = -\frac{\pi}{\mu}\cdot\frac{\partial P}{\partial r} = -\frac{k}{\mu}\cdot\frac{\partial}{\partial r}\left\{\frac{P_m}{In(R_0/R)} In\left(\frac{r}{R}\right)\right\} = -\frac{K}{\mu}\cdot\frac{P_m}{In(R_0/R)}\cdot\frac{1}{r} \qquad (9.41)$$

Impregnation ratio I is expressed as Equation 9.42.

$$I = \frac{\pi R_0^2 - \pi R^2}{\pi R_0^2} = 1 - \left(\frac{R}{R_0}\right)^2 \qquad (9.42)$$

The relation between impregnation ratio and integration of the resin pressure with respect to the time is expressed as Equation 9.43 (Kobayashi, Tanaka, and Morimoto 2012):

$$\int P_m\ dt = \frac{\mu R_0^2(1-V_f)}{4k}\left[(1-I)\{1-In(1-I)\}-(1-I_0)\{1-In(1-I_0)\}\right] \qquad (9.43)$$

Where r is the distance from the center of the resin flow front, u_r is the velocity of the fluids in the r-directions, k is the permeability, μ is the viscosity of the fluids, P_m is the resin pressure at R_0. If the resin pressure history and the I_0 can be determined from the experiment, then the impregnation time and ratio can be calculated through numerical methods. The graph (Figure 9.24) was drawn for the impregnation ratio and the holding time for validating the model to the experiments.

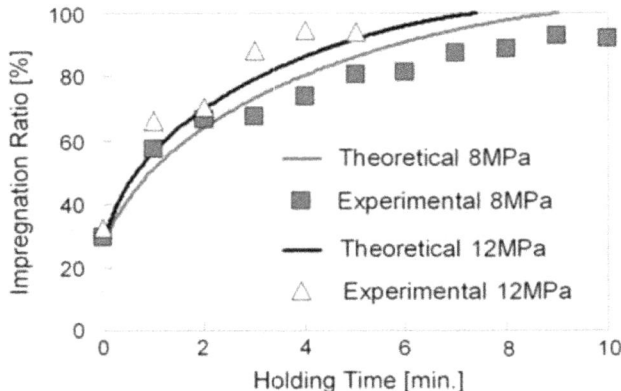

FIGURE 9.24 Impregnation ratio vs holding time. (Reproduced with permission from Kobayashi, Tanaka, and Morimoto 2012.)

9.4 INCORPORATION OF ADDITIONAL PARAMETERS IN THE IMPREGNATION MODELS

Different models have been developed for the impregnation process of the hybrid yarn-based polymer composites with time. The purpose of the model is to predict the behavior of the consolidation process for the material, processing, and geometrical parameters. But most of the models give accurate results in a certain condition only. This happens due to the assumption made for developing the general model for the consolidation process and due to the analysis of the single fiber system and assumption of the identical behavior despite considering inter-fiber interaction.

Many of the models skip the condition of the real situation such as the non-isothermal consolidation process, temperature gradient existing throughout the thickness, effect of the capillary force on the impregnation, the effect of the pressure exerted by the trapped gas, the three-dimensional flow of the molten matrix during impregnation within, and many more. In a study, Mehl and Rebenfeld (1993) studied the influence of the crystallization kinetics and morphology of the semicrystalline polymers. Van West et al. (1991) performed drapability simulation and developed a consolidation model to study the mechanical property of the composite. This study was done for the commingled yarn or co-woven fabric-based composite fabricated in an arbitrary shape. The drapability simulations help in knowing the fiber orientation, wrinkling, and bridging throughout the fabric. The drapability simulation helps in controlling the fiber orientation. This will be very useful in predicting the mechanical properties as well as optimizing the excellent quality of the laminate by combining the drapability model with the consolidation model.

For considering practical situations, Phillips, Akyüz, and Månson (1998) attached the pressure transducer and thermocouple inside the woven carbon reinforced polyetherimide composite for monitoring the real-time pressure and temperature variation during the consolidation process. The evolution of the intimate contact and fiber impregnation was studied through the microscopic analysis of the partially consolidated composite laminates to develop a comprehensive understanding. A double cantilever beam test was carried out to understand the influence of the autohesion on bond strength at the ply interface. Helmus et al. (2016) considered the consolidation mechanism which involves air evacuation through a partially impregnated microstructure. In this section, such a model has been discussed, which took these important aspects into consideration while developing the model. The model development requires a better understanding of the impregnation mechanics during the advanced manufacturing process as shown in Figure 9.25.

The intra-ply slip mechanisms in prepregs lead to a characteristic pattern of dry zones and resin-rich boundaries that might affect the impregnation in a later phase (Baumard et al. 2018). The fiber and matrix arrangement gets distorted when the faults exist in the close packing arrangement. The different types of fiber arrangement and corresponding unit cell are shown in Figure 9.26.

The high melt viscosity of the thermoplastic composite poses a problem of the resin impregnation inside the fiber bundle (Larock, Hahn, and Evans 1989). The variation of the temperature within the laminate sometimes results in the incomplete curing of the matrix and resin-rich and deficient zone within the laminate as shown in Figure 9.27.

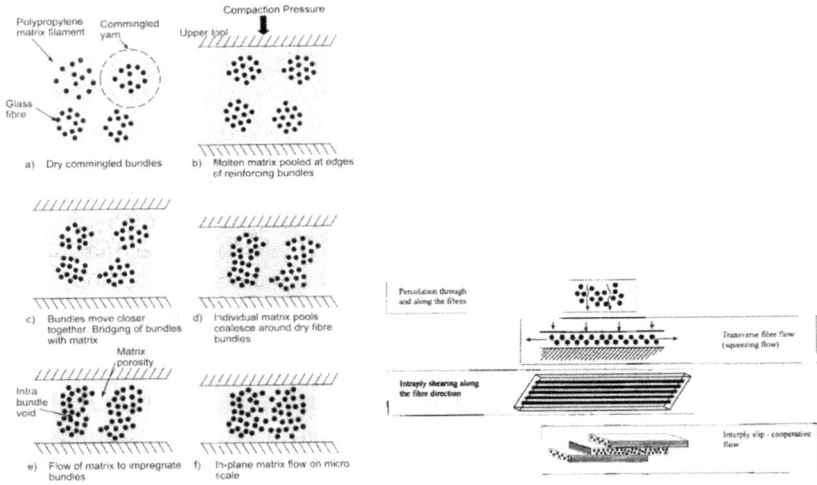

FIGURE 9.25 Steps in consolidation process and flow mechanism in composite manufacturing. (Reproduced with permission from Cutolo and Savadori (1994) and from Wakeman et al. 1998.)

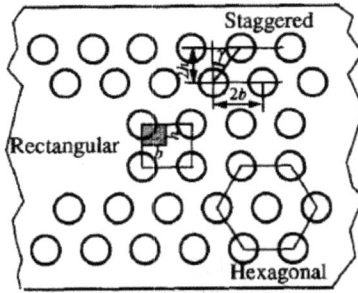

FIGURE 9.26 Different fiber arrangements and corresponding unit cells. (Reproduced from Guo and Derby 1994.)

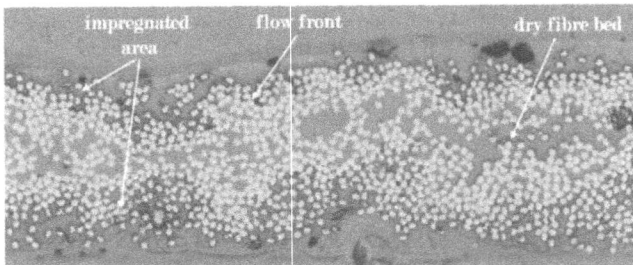

FIGURE 9.27 Microscopic image for uncured composite. (Reproduced with permission from Helmus et al. 2015.)

Various theoretical models have been developed for the prediction of permeability in the fibrous porous structure. Most of the models are in good agreement at low fiber volume fractions but show deviations of the results at high fiber volume fractions. The most widely used model applicable for the fibrous composite structure to predict the permeability is the capillary-based model, such as Carmon-Kozeny equations. However, there are also discrepancies beyond a certain porosity range. But this problem can be overcome through the experimental measurement of the Kozeny constant for a good prediction. The lack of an analytical model for the prediction of the permeability of the fibrous structure system encourages the researchers to investigate the numerical model for the solutions (Phelan, Leung, and Parnas 1994).

9.4.1 CONSIDERATION OF THE CAPILLARY PRESSURE AND TRAPPED GAS PRESSURE

The model developed by Bernet et al. (1999) includes the prediction of the resin front flow with respect to the time, void content, and the effect of the capillary. This model assumes that the impregnation process occurs at the same time and the impregnation of the single commingled yarn (Figures 9.28 and 9.29) represents the whole laminate consolidation process due to the similar representative structure.

In reality, the fiber and matrix do not have uniform distribution inside the commingled hybrid yarn structure. The fiber bundle is surrounded by the molten matrix pools during the consolidation. When the pressure is applied, the hydrostatic force generates at the fiber bundle boundary surface through the molten matrix and the impregnation process starts.

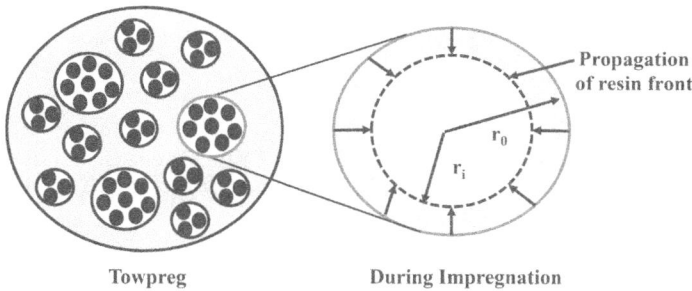

FIGURE 9.28 The towpreg cross-section during the consolidation. (Adapted from Bernet et al. 1999.)

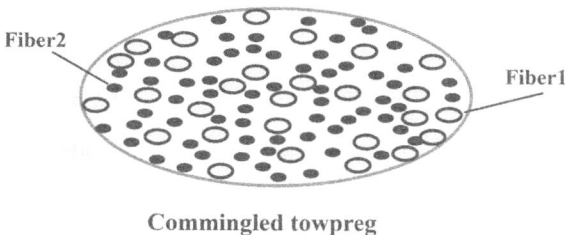

FIGURE 9.29 Commingled towpreg. (Adapted from Bernet et al. 1999.)

9.4.1.1 Governing Law for the Impregnations

The model is derived for the impregnation across the cross-sectional plane of the dry fiber bundle. The representative volume element ΔV is taken for the analysis purpose with fiber and matrix having an initial radius of the fiber bundle r_0 and at constant applied pressure P_a. The flow through the control volume follows Darcy's principle given by Equation 9.44.

$$(1-v_f)(u_l - u_s) = \frac{-k_p}{\eta}\nabla \tag{9.44}$$

Where v_f is the fiber volume fraction of the bundle, u_l and u_s are the local velocities of the resin and the fibers, K_p is the permeability tensor which depends upon fiber volume fraction, η is the viscosity and P is the local pressure, t is the time. The continuity equations for the fiber and the molten matrix phase within the control volume is used to solve Equation 9.44. The solution of the above non-linear equation requires knowledge of the numerical analysis and permeability tensor. The permeability of the fibrous structure is a function of fiber volume fraction and the stress and strain behavior of the fiber bed in the molten matrix (Bernet et al. 1999).

Once the fiber bed gets compressed, the fluctuation in the local pressure does not change the fiber volume fraction. This employs the $u_s = 0$ which means constant fiber volume fraction. On simplifications and integration with respect to radial distance, Equation 9.45 (Bernet et al. 1999) will be:

$$r_i In\left(\frac{r_i}{r_0}\right)\dot{r}_i = \frac{-k_p}{\eta(1-v_f)}\left(P_g(r_i) - P_c - P_a\right) \tag{9.45}$$

Where P_a is applied pressure, P_g is the gas pressure, P_c is the capillary pressure.

The position of the resin front in the interval of time Δt can be obtained through the integration of Equation 9.45, which is expressed as Equation 9.46 (Bernet et al. 1999).

$$\Delta t = \frac{\eta(1-v_f)}{K_p\left(P_a + P_c - P_g(r_i)\right)}\left[\frac{r_i^2}{2}In\left(\frac{r_i}{r_0}\right) - \frac{R_i^2}{2}In\left(\frac{R_i}{r_0}\right) - \frac{r_i^2}{4} + \frac{R_i^2}{4}\right] \tag{9.46}$$

9.4.1.2 Void Content during Impregnation

During the initial stage of the impregnation, the voids inside the dry fiber bundles are regarded as open pores. Once the impregnation process starts, the cross-section of the fiber bundle starts decreasing due to the occupying of the unfilled region inside the tow. Furthermore, the formation of the resin bridges may result in the entrapment of the voids inside the laminate. The resin bridges result in the non-uniform distribution of the resin and temperature difference inside. This leads to the change in the viscosity of the molten resin, variation in the permeability, and the quality of the commingled yarns. The voids inside the laminate create void pressure inside the reinforcement, and the pressure exerted by the entrapped gas can be calculated with the help of the ideal gas equations. The voids inside laminates lead to the longitudinal flow of the resin which converts the two-dimensional flow into the three-dimensional flow which further complicates the flow equations. The volume fraction of the voids within commingled yarn can be measured through Equation 9.47 (Bernet et al. 1999).

$$X_v = \frac{\sum_{i=1}^{n} \pi N_a^i r_i^{i2} \left(1 - v_f\right)}{A_t + \sum_{i=1}^{n} \pi N_a^i r_i^{i2} \left(1 - v_f\right)}$$

(9.47)

Where A_t is the cross-sectional area of the fully commingled yarn, r_0 is the initial bundle boundary radius, N represents bundle having size group i (varies from 0 to n).

9.4.1.3 Capillary Pressure

The expression for the capillary pressure (Equation 9.48) is derived with the help of the Young Laplace equations (Bernet et al. 1999):

$$P_c = \frac{2\gamma_{lv} \cos \theta}{D_f} \frac{v_f}{1 - v_f}$$

(9.48)

Where γ_{lv} is the surface tension of the liquid, θ is the contact angle between solid and liquid, D_e is the equivalent hydraulic channel diameter, D_f is the fiber diameter.

The overall pressure applied during the consolidation is represented as:

$$P_a = P_f + P_r + P_g - P_c$$

(9.49)

The voids pressure during the consolidation varies by assuming the pores as cylindrical of the constant length is expressed as:

$$P_g = P_0 \left(r_i \geq r_c\right)$$

(9.50)

$$P_g = P_0 \left(\frac{r_c}{r_i}\right)^2 \left(r_i < r_c\right)$$

(9.51)

Where r_c is the unimpregnated bundle radius below which pore is closed, r_i is the position of the resin front, P_0 is the ambient pressure.

In another work, Centea and Hubert (2012) explained the effect of process and material parameters on the impregnation mechanisms. The dynamics of the tow impregnation depend upon its constituent elements and production methods. The gas trapped and void formation during laminate production always have a complex effect on the impregnation phenomena depending upon their interaction with the surrounding material and the processing conditions.

9.4.2 MODEL FOR THE THREE-DIMENSIONAL FLOW AND FIBER BED DEFORMATION

In another study, Gutowski, Morigaki, and Cai (1987b) formulated an impregnation model by addressing more of the realistic situations such as fiber deformation during the consolidation, non-linear elastic behavior of the fiber bed, and time-dependent viscosity factor of the molten polymer, etc. This model was developed with the help of Darcy's law for the three-dimensional flow of the molten matrix through an anisotropic porous medium. The model assumes (a) one-dimensional deformation of the fiber network, (b) negligible spatial variation of viscosity of the polymer, (c) incompressible resin, and (d) fiber density is considered invariable etc. The molten polymers get squeezed out of the mold due to the pressure gradient. The pressure is transferred to the fiber tow by the compression, which eventually leads to the compaction of the laminates.

A flat laminate is taken for the analysis (as shown in Figure 9.30). This analysis is valid for both the woven as well as cross-ply composite laminates. In the analysis, x-direction represents the direction of the fiber, and y- and z-axes represent transverse directions. The analysis is done for the one-dimensional consolidation process with the three-dimensional flow of the resin. The fiber movement is restricted in the x- and y-directions but can be deformed in the z-direction (Gutowski, Morigaki, and Cai 1987b).

The new variable ξ is introduced in the z-direction to account for deformation in the fiber networks and w is the displacement in the z-directions, $\xi = w + z$. The thickness of the elements is $(\partial\xi/\partial z)dz$ as shown in Figure 9.31 (Gutowski, Morigaki, and Cai 1987b).

The resin flow rate q_i (Gutowski, Morigaki, and Cai 1987b) through the porous medium given by the Scheidegger is:

$$q_x = \frac{Q_x}{A_x}, \quad q_y = \frac{Q_y}{A_y}, \quad q_z = \left(1 - V_f\right)\left(u_z - \frac{\partial\xi}{\partial t}\right) \tag{9.52}$$

where q_x, q_y, and q_z are the volume flow rates through the porous material, Q_i is the total volume flow rate through the cross-sectional area A_i in the i-direction (i is x-, y-,

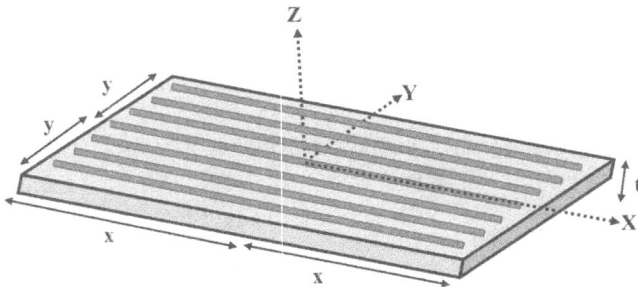

FIGURE 9.30 Flat laminate representation for the analysis. (Adapted from Gutowski, Morigaki, and Cai 1987b.)

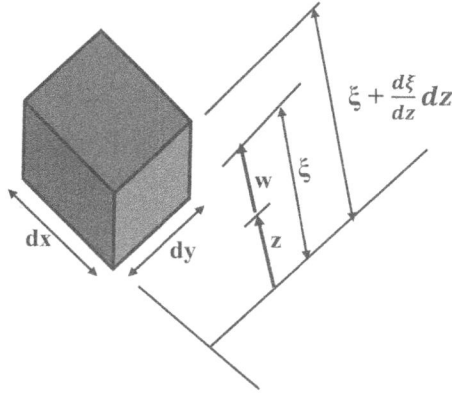

FIGURE 9.31 Representation of small element for the analysis. (Adapted from Gutowski, Morigaki, and Cai 1987b.)

and z-directions), u_z is the velocity of the resin in the z-direction. The fiber bed velocity was considered due to the deformation of the fiber network in the z-direction while there was no in-plane movement of the fiber. The fundamental law of the momentum, continuity, and the resin flow can be used to derive the model (Gutowski, Morigaki, and Cai 1987b).

9.4.2.1 Momentum Balance Equation

The momentum balance equation in the z-direction is considered for the control of the volume flow region. Here, the inertial force is assumed to be very small, and the applied load is shared by both fiber and matrix phases (Gutowski, Morigaki, and Cai 1987b; Steggall-Murphy et al. 2010). The applied pressure can be shown (Equation 9.53) in terms of the average pressure.

$$p = \overline{p_f} V_f + \overline{p_r} \tag{9.53}$$

Where p is the applied pressure, p_f and p_r is the average fiber and the resin pressure and V_f is the area fraction of the fiber. The average pressure p_r is usually modified during the flow through the porous media.

9.4.2.2 Fiber Continuity Equation

The volume of the fiber in the control volume flow (refer to the deformable element in Figure 9.31) region will always be constant and follows Equation 9.54 (Gutowski, Morigaki, and Cai 1987b):

$$V_0 = V_f \frac{\partial \xi}{\partial z} \tag{9.54}$$

Where V_0 is the initial volume fraction.

9.4.2.3 Resin Continuity Equation

Since the fiber bed gets deformed in the z-direction on applying pressure, the molten matrix gets squeezed out due to the pressure gradient. The flow of the matrix can be predicted with the principle of resin continuity. Through the principle of resin continuity, the governing differential equation for the three-dimensional flow of the molten polymer during the consolidation is obtained as Equation 9.55 (Gutowski, Morigaki, and Cai 1987b).

$$\frac{\partial}{\partial x}\left(q_x \frac{\partial \xi}{\partial z}\right) + \frac{\partial}{\partial y}\left(q_y \frac{\partial \xi}{\partial z}\right) + \frac{\partial}{\partial z}(q_z) + \frac{\partial}{\partial t}[1 - V_f]\frac{\partial \xi}{\partial z} = 0 \tag{9.55}$$

9.4.2.4 Darcy's Law

The volume flow rate of the molten matrix through the transversely isotropic porous material can be explained by the help of Darcy's law as represented by Equation 9.56.

$$\begin{Bmatrix} q_x \\ q_y \\ q_z \end{Bmatrix} = - \begin{bmatrix} S_{xx}/\mu & 0 & 0 \\ 0 & S_{zz}/\mu & 0 \\ 0 & 0 & S_{zz}/\mu \end{bmatrix} \begin{Bmatrix} \partial p_r/\partial x \\ \partial p_r/\partial y \\ \partial p_r/\partial \xi \end{Bmatrix} \tag{9.56}$$

Where μ is the viscosity of the molten polymer, S_{ii} is the permeabilities of the fiber network.

On combining the equations, the three-dimensional pressure distribution within the molten polymer can be expressed as Equation 9.57.

$$\frac{S_{xx}}{V_f}\frac{\partial^2 pr}{\partial x^2} + \frac{S_{zz}}{V_f}\frac{\partial^2 pr}{\partial y^2} + \frac{1}{V_0^2}\frac{\partial}{\partial z}\left(V_f S_{zz}\frac{\partial pr}{\partial z}\right) = \mu\frac{\partial}{\partial t}\left(\frac{1 - V_f}{V_f}\right) \tag{9.57}$$

The above differential Equation 9.57 can be solved through the numerical methods to obtain pressure distribution, local fiber volume fraction, the volume of the resin squeeze out of the laminates, and other variables (Gutowski, Morigaki, and Cai 1987b).

9.4.3 CONSIDERATION OF FIBER DEFORMATION

The stiffness of the tow (given by Equation 9.58) can be measured by considering fiber as a beam with multiple contact points. The multiple contact points are due to the fiber misalignment and waviness inside the preform. The beam length was found to be directly proportional to the thickness of the laminates. The misalignment and the crossover of the fiber leads to a decrease in the fiber volume fraction (Gutowski, Morigaki, and Cai 1987b).

$$\sigma = A_s \frac{\left(\frac{V_f}{V_0} - 1\right)}{\left(\frac{1}{V_f} - \frac{1}{V_a}\right)^4} \tag{9.58}$$

V_0 is the original fiber volume fraction, V_a is the available fiber volume fraction, A_s is the spring constant.

9.4.3.1 Permeability of the Fiber Assembly

To address the practical issues, it is important to know the dependency of the permeability in the compressed stage of the laminates. Carmen-Kozeny's theory is one of the efforts to establish the relation of the permeabilities with the variation of the fiber volume fraction. The study can be done to conceptualize the flow through the porous media as a flow through the capillaries having the average hydraulic radius. Thus, an expression is developed as Equation 9.59.

$$s_{ii} = \frac{r_f^2}{4k_{ii}} \frac{\left(1 - V_f\right)^3}{V_f^2} \tag{9.59}$$

Where r_f is the fiber radius, k_{ii} is the Kozeny constant.

In other literature, the effect of the fiber volume fraction was studied on the fiber bed distortion on application of compression load during the consolidation process. Gutowski et al. (1987a) conducted a similar study on the effective stress generation in the fibers against the content of the fiber volume fraction. It was found that a finite amount of the compressive load is always required for the stress generation within the fiber bed. At low fiber volume fractions, the contribution of the fiber bed in the load bearing is small in comparison to the molten matrix while it becomes significant for the higher fiber volume fraction fibrous network.

The tow actually has different fiber arrangements such as quasi-hexagonal and quasi-quadratic fiber arrangements as shown in Figure 9.32. The tow transverse

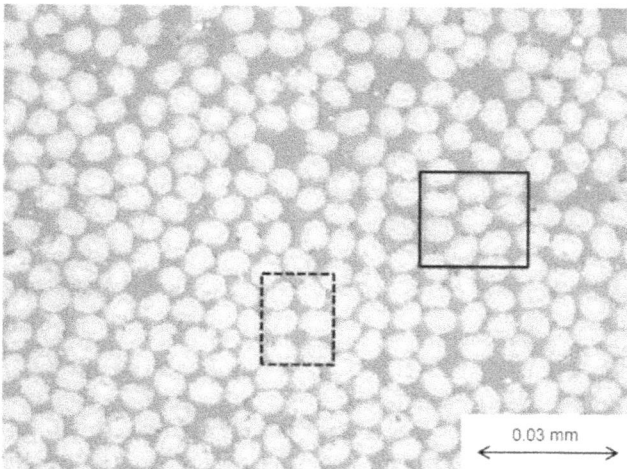

FIGURE 9.32 Quasi-hexagonal and quasi-quadratic fiber arrangements. (Reproduced with permission from Centea and Hubert 2012.)

permeability relations for the quadratic and hexagonal packing proposed by Gebart are given by Equation 9.60 (Centea and Hubert 2012).

$$K_{quad} = \frac{16}{9\pi\sqrt{2}} \left[\sqrt{\frac{\pi/4}{V_f}} - 1 \right]^{5/2} R^2, \quad K_{hex} = \frac{16}{9\pi\sqrt{6}} \left[\sqrt{\frac{\pi(2\sqrt{3})}{V_f}} - 1 \right]^{5/2} R^2 \qquad (9.60)$$

Where K is the permeability, R is the fiber radius, V_f is the tow fiber volume fraction.

In another study, Yang and Colton (1994) carried out a study to measure the permeability of the laminates and elastic property of the fiber bed network during compression as these properties depend upon the microstructure of the fibrous structure. Therefore, the experiments were done on the tape prepregs, commingled tow prepregs, and powder coated tow prepregs to understand and verify the theoretical results.

For the one dimensional flow in the axial direction, the expression for the axial permeability of the fibrous assembly can be expressed as Equation 9.61.

$$S_{xx} \frac{\partial^2 p_r}{\partial x^2} + \mu \frac{\dot{v}_f}{v_f} = 0 \qquad (9.61)$$

On integrating Equation 9.61 with appropriate boundary conditions, we have Equation 9.62 for the resin pressure distributions.

$$P_r = \frac{\mu}{2S_{xx}} \frac{\dot{v}_f}{v_f} \left(L^2 - x^2 \right) \qquad (9.62)$$

The total pressure is supported by both fiber assembly as well as molten resin. On solving equations 9.61 and 9.62 (Yang and Colton 1994), the expression for the axial permeability is given by:

$$\frac{s_{xx}}{r_f^2} = \frac{\mu}{3} \frac{L^2}{r_f^2} \left[-\frac{\dot{h}}{h} \right] \frac{1}{P_r} \qquad (9.63)$$

Where P_r is the resin pressure, S_{xx} is the axial permeabilities, L is the half axial distance of the mold, μ is the resin viscosity, v_f is the fiber volume fraction, r_f is the fiber radius, and h is the thickness of the composite.

9.4.4 Polymer Bridge Length for the Different Fiber Arrangements

Various models have been developed based on the assumption of the isothermal consolidation of the laminate, which is far from reality. In general, there is a temperature difference that exists across the laminate which makes the consolidation process non-isothermal. Nunes et al. (2001) devised the finite element program to validate the isothermal and non-isothermal consolidation models. The study was done for the constant press closing speed of the mold. The isothermal model was in good agreement for the low press closing speed for long preheating time while these limitations did not exist for the non-isothermal model and gave good validation for the wide range of the works.

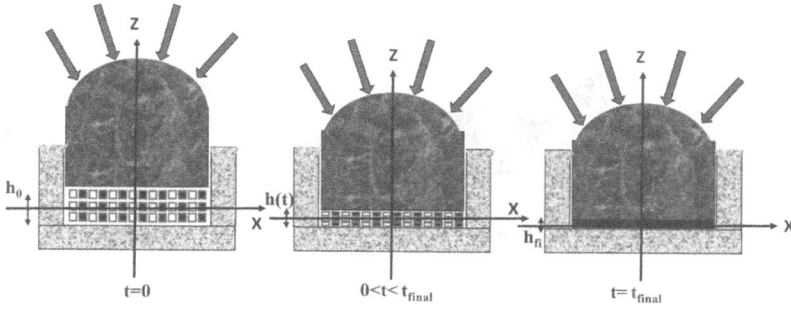

FIGURE 9.33 Preform under compression load during consolidation. (Adapted from Nunes et al. 2001.)

From Figure 9.33, the variation in the thickness of the laminates as a function of fiber volume fraction during the consolidation is shown by Equation 9.64.

$$v_f = \frac{h_{fi} v_{fi}}{h(t)}; \quad v_p = \frac{h_{fi}(1 - v_{fi})}{h(t)}; \quad v_v = \frac{h(t) - h_{fi}}{h(t)} \tag{9.64}$$

Where h_{fi} is the final thickness of the laminates, v_{fi} is the final fiber volume faction

The final polymer bridge length L (shown in Figure 9.34) can be estimated through equations 9.65 and 9.66 (Nunes et al. 2001) for the different fiber arrangements (Figure 9.35).

$$\text{Hexagonal arrangement } L(t) = \frac{8(r_p)^3 h_{fi} v_{fi}}{3(r_f)^2 h(t) \left(1 - \dfrac{h_{fi} v_{fi}}{h(t)}\right)} \tag{9.65}$$

$$\text{Square arrangement } L(t) = \frac{4(r_p)^3 h_{fi} v_{fi}}{3(r_f)^2 h(t) \left(1 - \dfrac{h_{fi} v_{fi}}{h(t)}\right)} \tag{9.66}$$

The hexagonal close packing of the circular fibers gave better results.

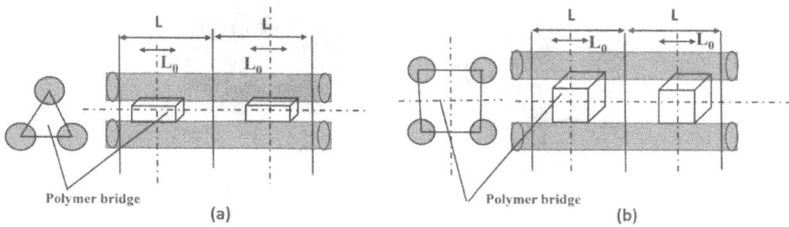

FIGURE 9.34 Polymer bridge at the beginning of the consolidation for the hexagonal and square arrangements. (Adapted from Nunes et al. 2001.)

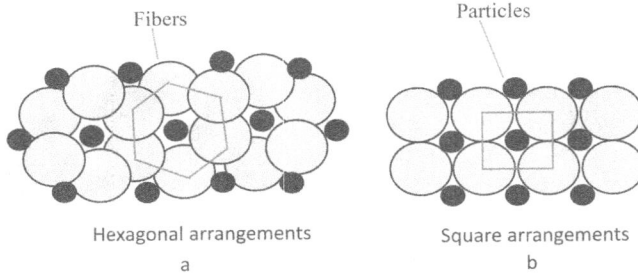

FIGURE 9.35   Different fibers arrangements. (Adapted from Nunes et al. 2001.)

9.4.5 SHEAR RATE

Haffner et al. (1998) investigated the micro-impregnation through finite element modelling for the thermoplastic preform. There are various finite element methods to predict the flow of the resin surrounding the reinforcement fiber during the consolidation. Here, the hexagonal and square arrangement (Figure 9.36) of fibers was considered for simulating the impregnation process for the laminates. The shear rate was predicted for the preform made with commingled and powder-coated processes during the consolidation process.

A triangular and square unit cell has been selected for the simulation process as shown in Figure 9.36. To validate the simulation results of the two models with the experimental, a constant initial fiber volume fraction was chosen, which is 37.5%. The fiber spacing 'a' is given in terms of the fiber volume fraction as in Equation 9.67 (Haffner et al. 1998).

$$a = D_f \cdot \left(\sqrt{\frac{\varnothing max}{\varnothing fiber}} - 1 \right) \tag{9.67}$$

Where ϕ_{max} is the maximum fiber volume fraction, which is 78.5% for the square packing and 90.7% for the hexagonal model, D_f is the diameter of the fiber.

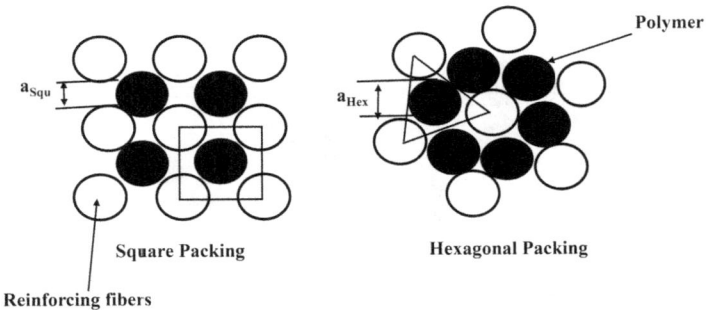

FIGURE 9.36   Square and hexagonal packing. (Adapted from Haffner et al. 1998.)

In general, the fiber and matrix phase does not distributed homogeneously within the towpreg which can be confirmed by the SEM images in both cases i.e, commingled hybrid yarn and powder-coated towpreg. This happens due to the agglomeration of the fibers inside the bundle. The idealized initial fiber volume fraction differs from the actual fiber volume fraction due to the occupation of the extra fibers in place of the matrix, which increases the fiber volume fraction. The average radial displacement toward the center can be measured simply from the geometry (Haffner et al. 1998).

$$\Delta r_{squ} = \frac{\Delta a_{squ}}{\sqrt{2}} \text{ for square fiber packing}$$

$$\Delta r_{Hex} = \frac{\Delta a_{Hex}}{\sqrt{3}} \text{ for hexagonal fiber packing} \qquad (9.68)$$

Where Δr is the change in the radial distance and Δa is the change in the fiber spacing.

The change in the radial displacement and the fiber spacing has been incorporated into the FE model to make the simulation process more realistic. Initially in the FE model, the uniform distribution of the fiber and matrix was considered throughout (Haffner et al. 1998).

The maximum shear rate is formulated as Equation 9.69.

$$\dot{\gamma}_{Max} = 2 \cdot \frac{\Delta v_{Deform}}{\Delta a_{Final}} \qquad (9.69)$$

Where γ_{Max} is the maximum shear rate, Δv_{Deform} is the deformation velocity, Δa_{Final} is the final fiber spacing.

9.4.6 CONSTITUTIVE RELATION FOR AN ANISOTROPIC VISCOUS FLOW

Pipes et al. (1991) developed a constitutive relation for an anisotropic viscous flow in an oriented fiber assembly as shown in Figure 9.37. The compliance matrix was represented in terms of three constants that are given by the relation of the viscosity of the resin, fiber volume fraction, and fiber aspect ratio.

Viscous compliance matrix (Pipes et al. 1991) for the transversely isotropic and incompressible fluid can be expressed as below.

$$
\begin{bmatrix} \dot{\varepsilon}_1 \\ \dot{\varepsilon}_2 \\ \dot{\varepsilon}_3 \\ \dot{\varepsilon}_4 \\ \dot{\varepsilon}_5 \\ \dot{\varepsilon}_6 \end{bmatrix} =
\begin{bmatrix}
\beta_{11} & -\beta_{11/2} & -\beta_{11/2} & 0 & 0 & 0 \\
0 & \beta_{22} & (\beta_{11} - 2\beta_{22})/2 & 0 & 0 & 0 \\
0 & 0 & \beta_{22} & 0 & 0 & 0 \\
0 & 0 & 0 & 4\beta_{22} - \beta_{11} & 0 & 0 \\
0 & 0 & 0 & 0 & \beta_{66} & 0 \\
0 & 0 & 0 & 0 & 0 & \beta_{66}
\end{bmatrix}
\begin{bmatrix} \sigma_1 \\ \sigma_2 \\ \sigma_3 \\ \sigma_4 \\ \sigma_5 \\ \sigma_6 \end{bmatrix}
$$

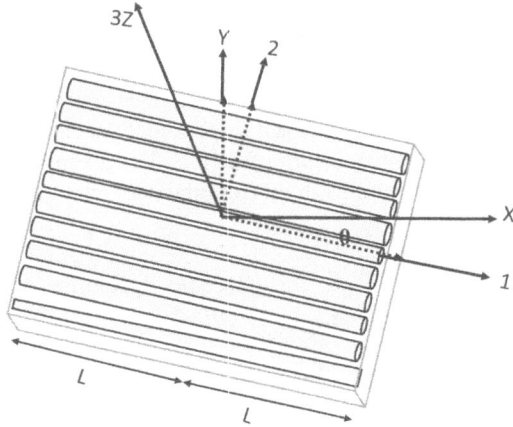

FIGURE 9.37 Unidirectional fiber assembly (UD) fiber assembly. (Adapted from Pipes et al. (1991).)

Where β_{ij} is a viscous compliance matrix constant, ε_i is normal strain rate components, σ_i is normal stress components. The anisotropic viscosity terms are shown in Table 9.1.

Where 'B' and 'F' terms for hexagonal arrangements are $(2\sqrt{3}/\pi)^{1/2}$ and $\pi/2\sqrt{3}$ respectively while for square arrangements, 'B' and 'F' are $2/\sqrt{\pi}$ and $\pi/4$ respectively.

The fiber volume fraction f can be expressed for both fiber arrays (Figure 9.38) (Pipes et al. 1991).

$$f = \frac{1}{B^2}\left(\frac{D}{S}\right)^2 \tag{9.70}$$

$$B^2 = \frac{2\sqrt{3}}{\pi} \quad (\text{hexagonal array})$$

$$B^2 = \frac{4}{\pi} \quad (\text{square array})$$

TABLE 9.1
Different Anisotropic Viscosity Terms

	η_{11}/η	η_{12}/η	η_{22}/η
In Viscosity			
Corresponding Viscous Compliance	$\left(\beta_{11}\,\eta\right)^{-1}$	$\left(\beta_{66}\,\eta\right)^{-1}$	$\left(\beta_{22}\,\eta\right)^{-1}$
In General Form	$\dfrac{f}{2}\left[\dfrac{B\sqrt{f}}{1-B\sqrt{f}}\right](L/D)^2$	$\dfrac{1}{2}\left[\dfrac{2-B\sqrt{f}}{1-B\sqrt{f}}\right]$	$4\left[1-B^2 f\right]^{-2}$

Source: Adapted from Pipes et al. (1991).

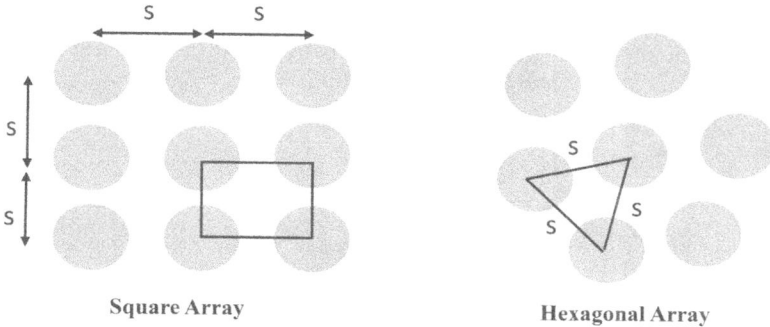

FIGURE 9.38   Fiber arrangements. (Adapted from Pipes et al. 1991.)

Where D is the diameter of the fiber, S is the spacing between the fibers, B is the constant.

9.4.7   VISCOUS FLOW IN THE LONGITUDINAL AND TRANSVERSE DIRECTION

Cai and Berdichevsky (1993) studied two-dimensional viscous flow in longitudinal and transverse directions using Navier-Stokes's equations. The idealized packing structure and quasi-random structure were investigated, and numerical simulation was performed on the available finite element package for the prediction of the permeability of the fiber assembly. The variation inside the fiber bundle should be included to minimize the uncertainty in the estimation of the permeability of the fiber bed even at the same fiber volume fraction. The ultimate fiber volume fraction (V_a) is the effective parameter that can be used with the average fiber volume fraction to predict the permeability of the fiber bed.

The permeability of the fiber network is expressed as Equation 9.71.

$$K = \frac{\varphi}{A} \frac{\mu}{(\Delta P/L)} \tag{9.71}$$

Where Q is the total amount of the flow, A is the total cross-sectional area including fibers and fluids, L length and height of the representative cell used in the numerical methods, Δp is the pressure of the fluids, μ is the viscosity, K is the permeability, r_f is the fiber radius, where V is velocity components.

For the transverse flow with a small Reynolds number, the Navier-Stokes equations become:

$$\frac{1}{\mu} \frac{\partial P}{\partial y} = \nabla^2 V_y, \ \frac{1}{\mu} \frac{\partial P}{\partial z} = \nabla^2 V_z, \ \nabla^2 = \frac{\partial^2}{\partial y^2} + \frac{\partial^2}{\partial z^2} \tag{9.72}$$

The permeability can be calculated from the average flow rate. The proposed formula for the longitudinal and transverse permeabilities is given by Equations 9.73 and 9.74 respectively (Cai and Berdichevsky 1993).

$$k_x^* = 0.211(V_a - 0.605) \times \left(0.907 \frac{V_f}{V_a} \right)^{-1.81} \left(1 - 0.907 \frac{V_f}{V_a} \right)^{2.66} +$$

$$0.292(0.907 - V_a)(V_f)^{-1.57}(1 - V_f)^{1.55} \tag{9.73}$$

$$k_z^* = \frac{k_z}{r_f^2} = \frac{1}{4k_z} \frac{\left(\sqrt{V_a/V_f} - 1 \right)^3}{(V_a/V_f + 1)} \tag{9.74}$$

Where V_f is the fiber volume fraction, V_a is the ultimate fiber volume fraction.

On the other hand, Connor et al. (1995) discussed the influence of the contact angle on spreading the molten polymer droplets on the fiber surface. The contact angle was greatly influenced by factors such as polarity of the fiber and surface treatment of the fiber. There is an important role of the surface energy and contact angle in spreading of the molten polymer droplet on the fiber surface and this varies from fiber/polymer system to system. The fiber bridging occurs spontaneously when the contact angle is less than 90° while permanent fiber spacing occurs for the contact angle greater than 90° without applying external load, which results in non-consolidation conditions (Figure 9.39).

The pressure exerted by the air in the trapped location can be evaluated with the help of ideal gas law Equation 9.75 (Åkermo and Åström 1998).

$$P_g = \frac{nRT}{V_g} \tag{9.75}$$

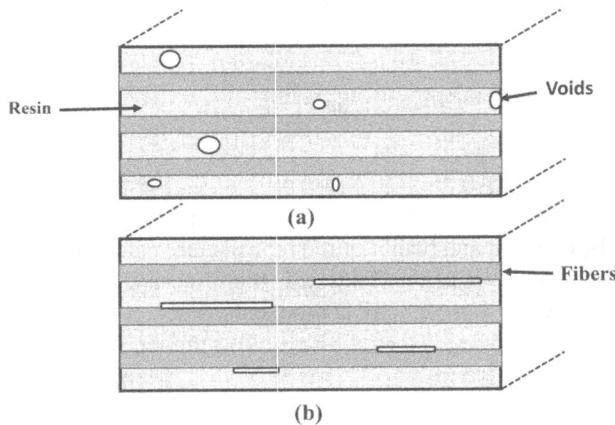

FIGURE 9.39 The topology of the voids as a function of the contact angle (a) $\theta < 90°$ and (b) $\theta > 90°$. (Adapted from Connor et al. 1995.)

Where R is the molar gas constant, N is an instantaneous number of the mole of gas in the cell, V_g and T is the instantaneous volume and temperature of the gas.

9.5 FIBER-MATRIX INTERFACE ADHESION

The mechanical property of the composite material is mainly governed by the adhesive force acting in the fiber and matrix interface zone. The principal function of the matrix in the composite system is to transfer the load to the reinforcing parts. Therefore, the adhesion between the two constituent parts plays an important role in utilizing the strength of the material. The adhesion force between the fiber and matrix can be modified by several means such as chemical treatment, through the change in the surface morphology of the fibers, and by the manufacturing techniques. The inhomogeneous distribution of the fiber and matrix and uncured laminates exhibits poor adhesion force. These problems especially come in the manufacturing of the thermoplastic composite. The high viscosity of the thermoplastic matrix leads to poor consolidation and poses the challenge of the interface related property. To overcome these problems, a short melt flow distance is required for the homogeneous distribution of the two phases. These are fulfilled by the hybrid yarn structure, such as commingled yarn, micro braiding, DREF spinning, co-wrapping, powder coating towpregs, etc. Each hybrid yarn has a unique structure and different degrees of pre-mingling of the matrix parts with the reinforcing fibers. The hybrid yarn-based composite exhibits strong adhesion force due to excellent consolidation of the resultant laminates (Alagirusamy et al. 2006; Jogur et al. 2018; Bar, Alagirusamy, and Das 2018; Bar, Das, and Alagirusamy 2019; Goud et al. 2019; 2020). Several works have been performed to understand the influence of the hybrid yarn structure on the adhesion between the fiber and matrix parts. A better interface strength is desirable for high inter-laminar shear strength, fatigue strength, etc. while a weak interface is required for low damage tolerance during the drop weight impact test to dissipate the energy through the delamination.

Micro-braiding is a hybrid yarn technique for the manufacturing of the long continuous fiber-reinforced composites which is generally used for complex textile preform. Sakaguchi et al. (2000) produced glass/PA6-nylon-based composite laminate and examined the effect of the micro-braiding structure on the interface adhesion property and consolidation quality of the laminates. The correlation between the impregnation quality and mechanical property of the resultant laminates has been done to understand their interdependency. The laminate having a better distribution of the matrix and least void content tends to exhibit higher mechanical property due to the effective utilization of the reinforcing parts.

Four different micro-braiding yarn structures were used to study the impregnation phenomena. Different micro-braided yarn structures (as shown in Figure 9.40) with glass as a reinforcing fiber and PA-6 fiber bundle as a matrix filament having constant fiber volume fraction were developed. The different configurations of the reinforcing and matrix filaments in the respective structure leads to different impregnation distances. The influence of the impregnation distance was investigated on

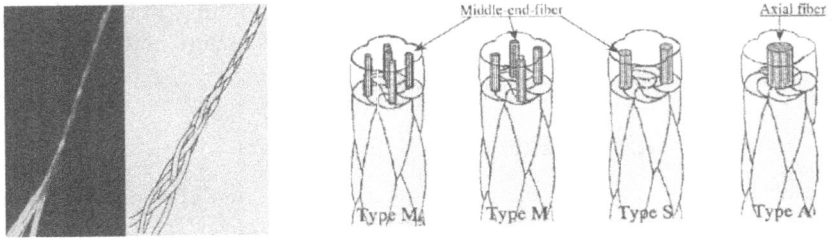

FIGURE 9.40 Fiber configuration in micro-braiding structure. (Reproduced with permission from Sakaguchi et al. 2000.)

the impregnation quality (Figure 9.41) and the mechanical behavior of the resultant laminates. The unidirectional laminate was manufactured through compression molding to study the consolidation process of the laminates (Sakaguchi et al. 2000).

The void content indicates the degree of impregnation. The better the impregnation, the lower will be the voids content. Generally, the duration of impregnation is measured till the laminate acquires the constant void contents. The least void content was measured for the hybrid yarn structure which has the smallest melt flow distance among the four following the trend $A > S > M > M_s$, while in the case of flexural property, the laminate with the least void content showed the highest flexural property followed by M, S, and A respectively as shown in Figure 9.42. From the graph (Figure 9.43), it can be easily observed that voids start decreasing with an increase in the holding time. This happens due to the decrease in the viscosity and the surface tension of the molten matrix polymer at a higher temperature and full time to impregnate further inside the fiber region. All four laminates followed the same trend of decreasing voids with an increase in the holding time (Sakaguchi et al. 2000).

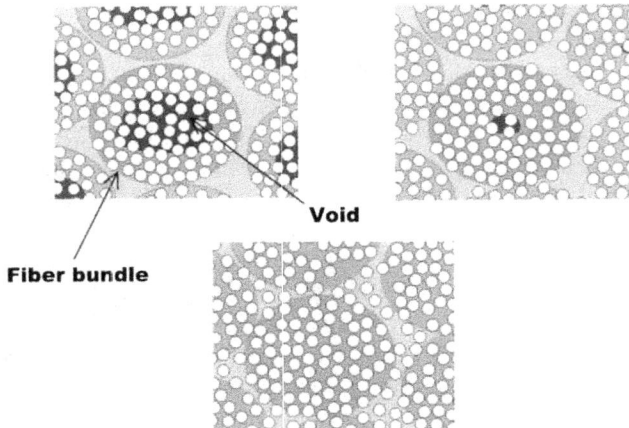

FIGURE 9.41 Illustration of micro-braided composite and fiber/matrix interface. (Reproduced with permission from Sakaguchi et al. 2000.)

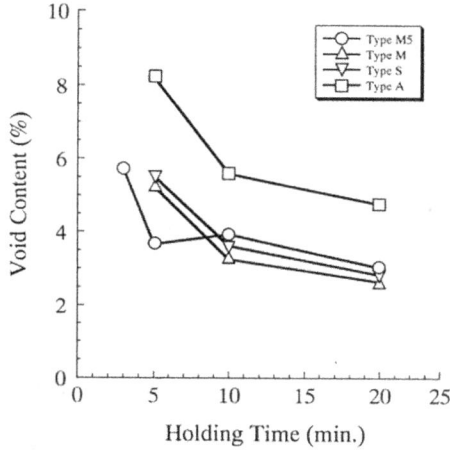

FIGURE 9.42 Graph for axial and transverse bending strength vs. holding time respectively. (Reproduced with permission from Sakaguchi et al. 2000.)

The mechanical property such as flexural strength was evaluated with respect to the holding time. The flexural properties in the axial and transverse directions were found to increase with the increase in the holding time. This can be directly correlated with the improvement in the fiber/matrix adhesion leading to decrease in the void content and an increase in the flexural property of the final laminates (Sakaguchi et al. 2000).

Li et al. (2017) studied the effect of the rheology of the matrix on the tensile property of the composite laminate. The carbon fiber and two different grades of the PEEK were taken for the study. The two PEEK had different values of viscosity, PEEK1 having higher MFI than the PEEK2. It was observed that the final tensile strength of the laminate was higher for the resin having a higher MFI value (PEEK1 > PEEK2). The powder impregnation technique was used for the production of the

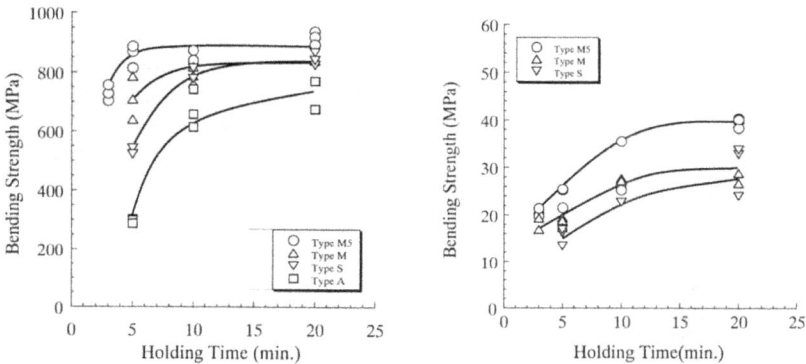

FIGURE 9.43 Void content vs. holding time. (Reproduced with permission from Sakaguchi et al. 2000.)

composite materials. It was observed that the fiber distribution has a strong dependency on the mobility of the molten resin matrix. The fiber distribution was found to be more concentrated for the PEEK2 resin having a higher viscosity. In another study, Shishoo and Bernhardsson (2000) scrutinized the influence of the consolidation parameters on the interface property of glass/PP commingled-based laminate. Based on the experiment, it was concluded that the holding temperature has a dominant effect on the consolidation quality and the mechanical property of the resultant laminate followed by the holding time and pressure, respectively. The interaction of the processing parameters has been studied on the response of the short beam flexural strength with respect to the holding temperature and the time. The pressure was not considered due to the insignificant influence it had on the laminate property. It was reported that the strength of the laminates increased more at the higher holding temperature in comparison to the lower holding temperature with respect to the holding time. Consequently, it can be said that the higher temperature results in higher strength in a shorter time which is basically due to the increase in the rate of the impregnation process. The increase in the strength due to the prolonged holding time can be explained in terms of the decrease in the void content and improvement in the interfacial fiber matrix adhesions.

The impregnation quality of the laminate can be quantified with the help of the inter-laminar shear strength test i.e., higher the ILSS means high quality of the laminate and strong fiber/matrix adhesion. For the full consolidation, it is suggested to keep higher temperature for a long holding time at lower processing pressure. This helps the molten polymer matrix to enter space within the inter- and intra-fiber bundle by removing micro-voids present well inside the bundle (Shishoo and Bernhardsson 2000).

9.6 EFFECT OF TOWPREG STRUCTURE ON THE CONSOLIDATION QUALITY OF THE LAMINATES

To overcome the problem of the long melt flow distance, the concept of the hybrid yarn technique has evolved to achieve good fiber and matrix impregnation on consolidation. Practically, the laminate shows far less strength in comparison to its theoretical value based on the rule of mixture. This shows the poor utilization of the reinforcing fiber strength during the application of the load. There is always scope to improve the property of the laminate if there is good impregnation during the compaction and the consolidation process. The quality of the consolidation of the laminates depends upon several parameters such as material property, processing parameters, and structure of the hybrid yarn. Several studies have been conducted to explore the effect of the processing conditions (such as holding temperature, pressure, and time), type of the towpregs (commingled, powder coated, DREF, etc.), and material conditions (such as viscosity, MFI, shape factor, etc.) on the consolidation behavior of the laminates, which has been discussed in this section.

The consolidation quality of the laminate can be evaluated through the measurement of the void content and improvement in the mechanical property. The measurement of the void content can be done through various ways such as image analysis, which includes SEM analysis, micro-CT-based analysis, optical microscopic image, etc. The apparent void percentage can be calculated through the prescribed ASTM standard. The theoretical and actual density will be required for the measurement of the voids present in the laminates. The higher void content within the laminate shows the inferior quality of the laminate. Another way of describing the quality of the consolidation is in terms of mechanical properties such as tensile, flexural, ILSS, and many more. There is an inverse relationship between the void content and the mechanical property of the composite laminates. The composite laminate having the same material constituents exhibits different mechanical properties for the varying consolidation quality.

Various studies have been conducted to extract the influence of the consolidation conditions on the behavior of the laminates. Greco et al. (2011) explored the impregnation mechanism for the glass/polyester commingled-based laminate during the consolidation. The impregnation of the molten polymer was observed on two scales: one is micro-scale impregnation where molten polymer flows within the fiber bundle, and the other is macro-scale impregnation where molten polymer flows between the tow. This study reveals that the impregnation process is governed by the material property and quality of the pre-mingled hybrid yarn. In fact, the material property such as viscosity decreases with the temperature. The degree of impregnation of the molten polymer improves at higher temperatures due to an increase in the melt flow index, while the degree of the pre-mingling of the fiber and matrix helps in reducing the melt flow distance, which ultimately helps in getting void proof laminates. The study of Bullions, Loos, and McGrath (1999) on the dry powder prepreg laminate shows the relative positions of the fiber and matrix with excellent quality of the impregnation even for the high viscous thermoplastic matrix. The same trend was found by Novais et al. (2004) for the glass/PP dry powder towpreg-based composite laminate where the composite was consolidated through the compression molding technique. The resultant laminate exhibited higher flexural strength on testing.

Lin Ye et al. (1995) observed the change in the apparent void content from 6 to 0% resulted in a 6-fold increase in the ultimate strength of the composite. There is a strong dependency of the mechanical property on the quality of the consolidation. The quality of the composite laminate depends upon how well impregnation occurs. The processing condition, such as compression pressure, holding time, and temperature has a direct influence on the consolidation quality of the laminates. It has been observed that void content starts decreasing with an increase in temperature, pressure, and holding time. The increase in these parameters leads to the better impregnation of the molten matrix and hence improves the adhesion between the fiber and matrix which imparts excellent mechanical property in the resultant composite laminates. There are several ways of measurement of the quality of the consolidation. The mechanical property and the voids measurement through the density measurement is one of the assessments to understand the quality of the laminate. The theoretical density of the fully consolidated laminate can be calculated according to the ASTM-792 given below.

$$\rho_t = \frac{\rho_f \rho_m}{W_f \rho_m + W_m \rho_f} \tag{9.76}$$

Where ρ_t is the theoretical density of the composite laminates, ρ_c is the actual laminate density, ρ_f is the density of the fiber, ρ_m is the density of the matrix, W_f and W_m are the weight fraction of the fiber and matrix, respectively.

The state of commingled yarn also has a great influence on the impregnation phenomena. For the higher degree of the commingling D_{com}, it takes less time to reach the full consolidation. The distributions of the fiber and matrix have a great influence on the consolidation phenomena of the laminates. Ye and Friedrich (1997) experimented to study the interaction of the void content and the processing parameters in CF/PEEK laminates made through commingled yarn and powder sheath bundle techniques. In the study, it was observed that to attain full consolidation stage, commingled yarn-based laminates required higher holding time and applied pressure in comparison to the powder-coated towpreg-based composite laminates. The effect of the void content on the transverse flexural property of the respective laminate was plotted as obtained in the experimental result in Figure 9.44.

Ye, Friedrich, and Kastel (1995) studied the effects of the processing parameters (such as pressure, time, and temperature) on the impregnation quality in terms of the void contents and subsequently, the influence of the void content was also studied in terms of the mechanical property. Their study explained the consolidation quality based on the void measurement and their functional relation with the mechanical property. The ASTM standard method was used for the calculation of the void content based on the density measurement. The void content can also be measured through microscopic image analysis.

In the experiment, it was found that the void content was decreasing with an increase in holding time as well as with higher temperature and pressure. At the higher temperature, the viscosity of the matrix decreases which makes it possible to impregnate more of the unfilled region within the laminates. The increase in the applied pressure creates a higher-pressure gradient which results in higher force to push the molten matrix inside the dry fiber bundle, while the long holding time enables the molten matrix to travel long-distance during the impregnation. On the other hand, the influence of the impregnation quality was also reported for the transverse flexural strength and the transverse elastic modulus response of the laminate. The mechanical properties can be seen to decline with an increase in the

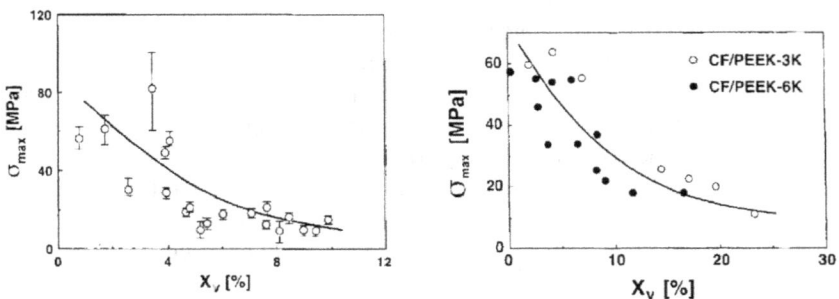

FIGURE 9.44 Effect of the voids on the transverse flexural property of the CF/PEEK commingled yarn and powder/sheath-based composite respectively. (Reproduced with permission from Ye and Friedrich 1997.)

void content. The voids generated inside the laminate give weaker interface strength and create a difficulty in the load transferring mechanism which in turn gives lower material strength (Ye, Friedrich, and Kastel 1995). In another study, the laminate consolidated at higher temperature and pressure gave a higher value of the yield and ultimate stress. This is an indication of the better consolidation of the laminate (Ye, Klinkmuller, and Friedrich 1992).

9.7 EFFECT OF MOLDING PARAMETERS ON THE MICROSTRUCTURES AND MECHANICAL PROPERTIES OF THE COMPOSITE MATERIALS

The microstructure of the laminate is influenced by the molding parameters such as temperature, pressure, and holding time. In the derivation of the consolidation model, it is commonly assumed for the microstructure of the fiber bundle to be intact and fiber arrangement to be uniform to simplify the mathematical equations. But actually the fiber arrangement and shape of the bundle get distorted due to high pressure and irregularity in the processing conditions. Further distortion in the fiber bundle occurs due to the melting of the constituent polymer matrix and due to the shear force generated during the impregnation of the molten matrix in the three-dimensional flow. These usually lead to the irregular distortion in the shape of the fiber bundle inside the laminate after the consolidation. And therefore, it has its repercussion on the mechanical property of the laminate in comparison to the ideal microstructure based on assumed theoretical properties.

There have been various studies conducted experimentally and theoretically to understand the influence of the consolidation parameters on the mechanical response of the laminate. These studies help in reaching the full consolidation of the void-free laminate to utilize its maximum strength. The anisotropic behavior and indifferent chemical and thermo-mechanical nature of fiber and matrix limits the predictability of the simulations tools for reaching the optimum processing consolidation condition. However, a better understanding at the micro level helps in validating the experimental results by incorporating the minute micro-mechanical details. In this section, the mechanical response of the different laminates has been discussed to understand the effect of the molding parameters observed in various works.

Zhang and Peijs (2010) studied the influence of the consolidation temperature and pressure on the mechanical property of the self-reinforced poly (ethylene terephthalate) (PET) composites (Figure 9.45).

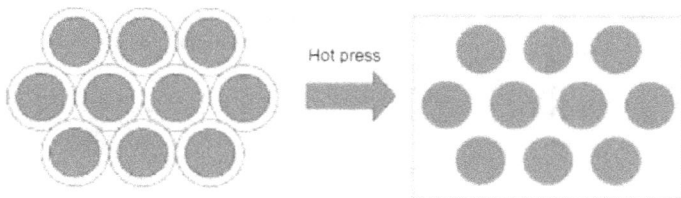

FIGURE 9.45 Illustration of laminate formation. (Reproduced with permission from Zhang and Peijs 2010.)

In this experiment, the graph (as shown in Figure 9.46) has been plotted for the thickness and the density of the laminate as a function of the temperature at a different level of the holding pressure. The thickness of the laminates decreases with the increase in the compaction temperature. This occurs mainly due to the impregnation of the unfilled space within the preform by removing the voids. The thickness of the laminates tends to decrease with the increase in the pressure level. This can be understood that at high-pressure levels two phenomena occur: one is molten polymer matrix forces to fill the unoccupied region well inside the preform and the other is due to the squeezing of the molten matrix outside the mold. For the density, it can be easily seen that it increases with an increase in the compaction temperature and compaction pressure. Since the same quantity of the material occupied lesser volume inside the resultant laminate, this led to an increase in the density of the final laminates (Zhang and Peijs 2010).

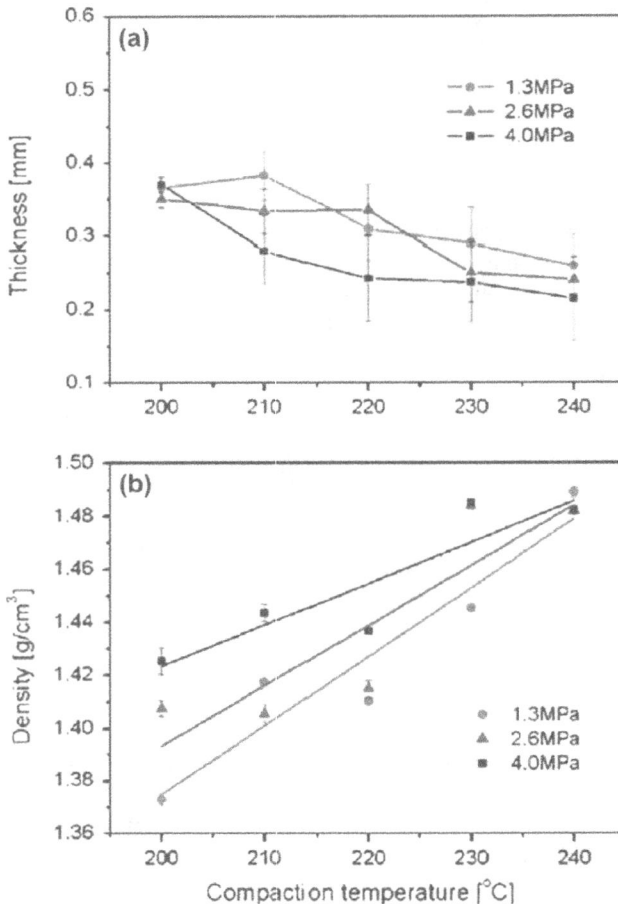

FIGURE 9.46 Influence of the processing temperature on the thickness of the composite at different pressures. (Reproduced with permission from Zhang and Peijs 2010.)

Unlike other polymer composites, the self-reinforcing PET composite shows some diversion in the mechanical response with the consolidation parameters. This happens due to the unique nature of its composition. The fiber and matrix of the self-reinforcing laminate have the same polymer unit with different molecular weights having much less melting point difference. It was reported that the longitudinal strength and modulus of the laminate increased with an increase in the consolidation temperature. This happened due to the better impregnation of the matrix into the fiber bundles. But after a certain temperature, there was a drop in the strength of the laminate. This primarily happened due to the thermal softening (drop in the tenacity) of the reinforcing parts of the core fiber. The partial melting of some portion of the reinforcing fiber was reported which eventually led to the drop in the tensile strength of the laminate in the fiber direction but the reverse phenomena was noticed in the transverse direction. It was expected to have lower strength and modulus. But at a higher temperature, due to good impregnation and little diffusion of the core yarn in the transverse direction, there is increase in strength in the transverse direction as explained in Figure 9.47 (Zhang and Peijs 2010).

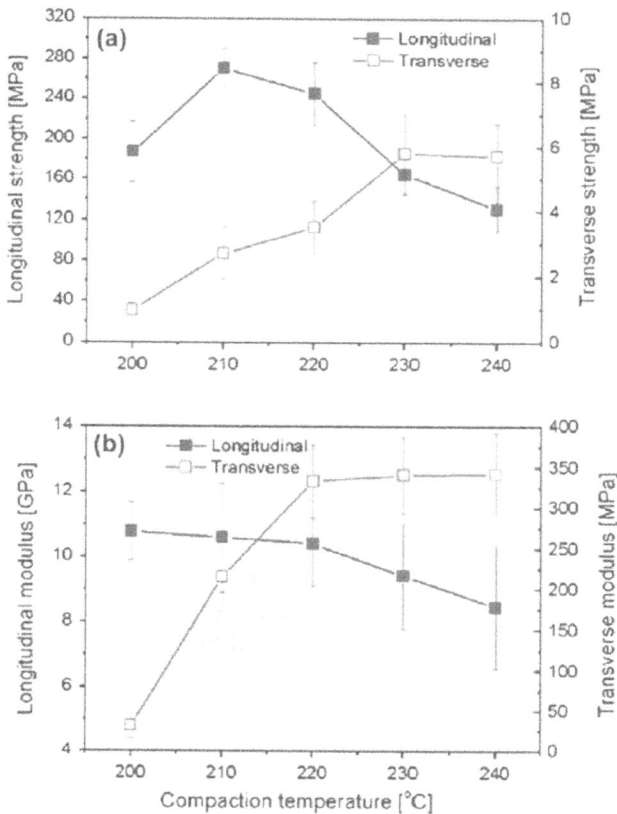

FIGURE 9.47 Influence of the processing temperature on the tensile property of the composite. (Reproduced with permission from Zhang and Peijs 2010.)

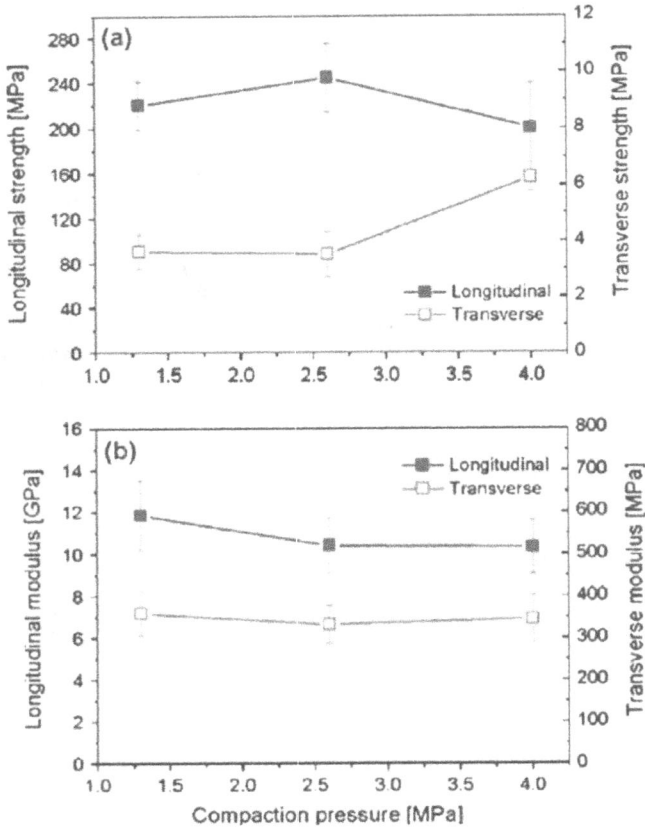

FIGURE 9.48 Influence of the processing pressure on the tensile property of the composite. (Reproduced with permission from Zhang and Peijs 2010.)

At higher compaction pressure, there was drop in the longitudinal strength which was due to the excessive squeeze out of the matrix. This led to insufficient matrix for the load transfer, but the reverse trend was found for the compaction pressure and the reason was better impregnation (Figure 9.48). During the study, an interesting behavioral change occurred, that is, the increase in the consolidation temperature lead to the transformation of the failure mechanism from the ductile yarn pull-out and debonding dominant failure mode to the brittle failure mode. This happened mainly due to the enhanced interface bonding strength between the fiber and matrix (Zhang and Peijs 2010).

A similar study was performed by Wu et al. (2013) based on the single polymer PLA-based composite made with a 5-directional braiding technique where the composite was processed at different temperatures (Figure 9.49). The flexural properties of the laminates were studied at different temperatures and the performance curves were plotted for the flexural strength and modulus against each holding

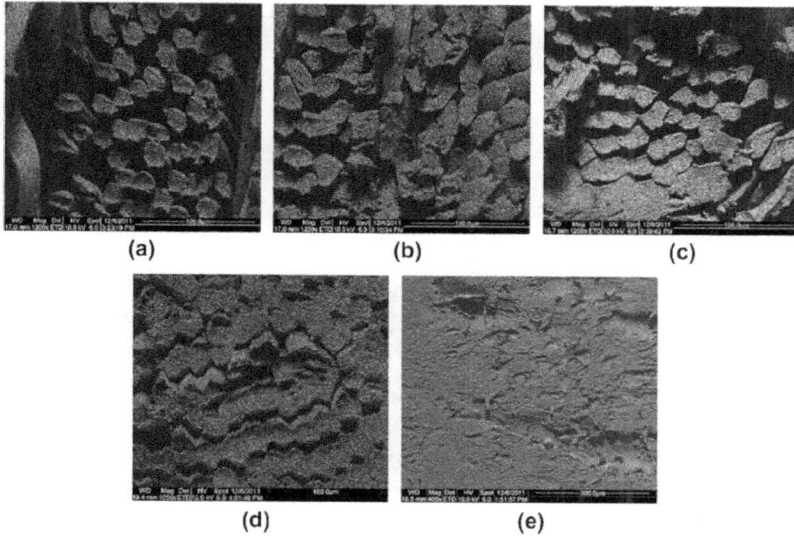

FIGURE 9.49 Microscopic image of the cross-section at different consolidation temperature. (Reproduced with permission from Wu et al. 2013.)

temperature as shown in Figures 9.50 and 9.51. The flexural strength and modulus increased with the increase in the temperature. The increase in temperature leads to the better fusion of the fiber and matrix which ultimately imparts higher flexural property by transforming ductile failure to brittle failure. The better impregnation of the matrix can be seen in the SEM images' examination of the cross-section at a different consolidated temperatures as shown in Figure 9.51.

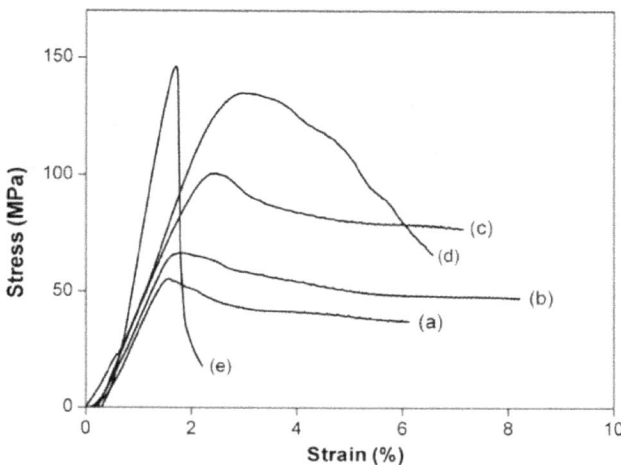

FIGURE 9.50 Flexural behavior of the laminates at different consolidation temperatures. (Reproduced with permission from Wu et al. 2013.)

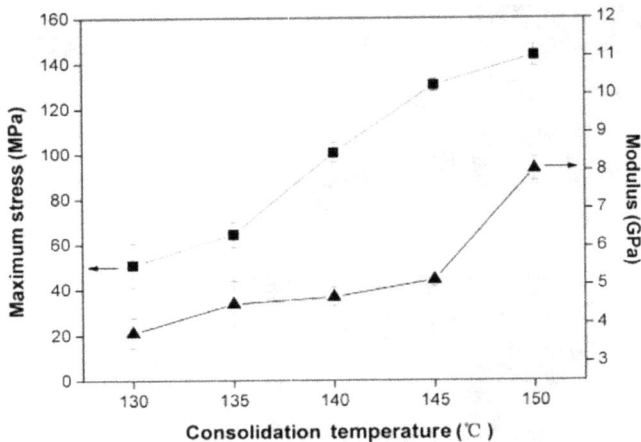

FIGURE 9.51 Effect of the consolidation temperature on the maximum flexural strength and modulus. (Reproduced with permission from Wu et al. 2013.)

Wu, Lin, and Tsai (2016) examined the effect of the consolidation parameters on the mechanical property of the self-reinforced poly (ethylene terephthalate) composite made of double-covered unmingled yarn made with the co-wrap spinning technique. The quality of the composite is a function of the level of impregnation, void content, and the degradation of the fiber and matrix components. For the temperature-sensitive material, it is recommended to optimize the level of temperature and the holding time to avoid overheating/degradation conditions for the same level of the consolidation. There is an inverse relationship between these two factors, that is, reducing the holding time will require an increase in the consolidation temperature to achieve the same magnitude of the impregnation.

Different behavior of the laminate was observed when consolidation was done at the following temperatures: 235°C, 245°C, and 250°C. The laminate consolidated at 245°C has shown the highest tensile strength in comparison to others. This was attributed to the better impregnation at a higher temperature level than the 235°C, while in the case of the laminate prepared at 250°C, a sudden drop in the tensile strength was observed due to the partial melting of the reinforcing fiber which resulted in the loss of strength. But in contrary to the tensile strength, the laminate prepared at 250°C exhibited the highest tensile modulus among the three. This was attributed to the good impregnation and good structural integrity of the matrix into the fiber bundle. The laminate consolidated at the higher temperature showed brittle failure with low failure strain (Wu, Lin, and Tsai 2016). Yang and Colton (1994) consolidated the preform comprising of commingled towpregs, powder-coated towpregs, and tape prepregs. The laminate made from the commingled-based yarn showed the highest compressive elastic modulus at corresponding fiber volume fraction followed by the powder-coated laminate and then tape prepreg laminate. The higher value of the compressive elastic modulus can be related to the higher consolidation quality of the laminate. In another study, Klinkmuller et al. (1994) examined the influence of the pressure and temperature on the shear strength of the composite laminate. The variation of the shear strength was studied as a function of the pressure at a different consolidated temperature. The shear strength of the laminate was reported to increase with an increase in the applied pressure and the temperature. The improvement in the shear strength of the laminate can be correlated with the reduction of the void for the corresponding pressure and temperature. And the better load transferability led to an increase in the shear strength of the composite laminate.

9.8   SUMMARY

In this chapter, different impregnation mechanisms have been discussed for different heterogeneous fibrous networks. Evolution of the fundamental mathematical equations from very basic to complex models has been illustrated. Some simplified equation took various assumptions such as isotropic homogeneous flow medium, ideal behavior of the molten resin, void free laminate and many more. While a few equations have been developed in a very realistic approach such as the non-Newtonian time and temperature dependent fluid behavior during the consolidation period and effect of the air or real gas entrapment during the impregnation phenomena etc. The effect of the processing parameters has been discussed on the interface quality and its mechanical properties of the final laminates. Moreover, this chapter presents an overview of the consolidation process which will be helpful in understanding analytically the matrix flow behavior at high temperature and pressure in an anisotropic medium.

REFERENCES

Åkermo, Malin, and B. Tomas Åström. 1998. "Modelling Face-Core Bonding in Sandwich Manufacturing: Thermoplastic Faces and Rigid Closed-Cell Foam Core." *Composites Part A: Applied Science and Manufacturing* 29 (5–6): 485–94. https://doi.org/10.1016/S1359-835X(98)00001-3

Alagirusamy, R., R. Fangueiro, V. Ogale, and N. Padaki. 2006. "Hybrid Yarns and Textile Preforming for Thermoplastic Composites." *Textile Progress* 38 (4): 1–71. https://doi.org/10.1533/tepr.2006.0004

Bar, Mahadev, R. Alagirusamy, and Apurba Das. 2018. "Properties of Flax-Polypropylene Composites Made through Hybrid Yarn and Film Stacking Methods." *Composite Structures* 197 (August): 63–71. https://doi.org/10.1016/j.compstruct.2018.04.078

Bar, Mahadev, Apurba Das, and Ramasamy Alagirusamy. 2019. "Influence of Flax/Polypropylene Distribution in Twistless Thermally Bonded Rovings on Their Composite Properties." *Polymer Composites* 40 (11): 4300–10. https://doi.org/10.1002/pc.25291

Baumard, T., G. Menary, O. De Almeida, F. Schmidt, and J. Bikard. 2018. "Characterisation and modelling of the temperature and rate dependent shear behaviour of a non- consolidated powder impregnated fabric." In *ECCM18-18th European Conference on Composite Materials* (June):24–28.

Bernet, N., V. Michaud, P. E. Bourban, and J. A. E. Manson. 1999. "An Impregnation Model for the Consolidation of Thermoplastic Composite Made Form Commingled Yarns." *Journal of Composite Materials* 33 (8): 751–72.

Bowman, Sean, Qiuran Jiang, Hafeezullah Memon, Yiping Qiu, Wanshuang Liu, and Yi Wei. 2018. "Effects of Styrene-Acrylic Sizing on the Mechanical Properties of Carbon Fiber Thermoplastic Towpregs and Their Composites." *Molecules* 23 (3): 547. https://doi.org/10.3390/molecules23030547

Bullions, T. A., A. C. Loos, and J. E. McGrath. 1999. "Advanced composites manufactured via dry powder prepregging." In *Proceedings of the International Conference on Composite Materials* 12:1–10.

Cai, Zhong, and Alexander L. Berdichevsky. 1993. "Numerical Simulation on the Permeability Variations of a Fiber Assembly." *Polymer Composites* 14 (6): 529–39. https://doi.org/10.1002/pc.750140611

Centea, T., and P. Hubert. 2012. "Modelling the Effect of Material Properties and Process Parameters on Tow Impregnation in Out-of-Autoclave Prepregs." *Composites Part A: Applied Science and Manufacturing* 43 (9): 1505–13. https://doi.org/10.1016/j.compositesa.2012.03.028

Christmann, Marcel, Luisa Medina, and Peter Mitschang. 2017. "Effect of Inhomogeneous Temperature Distribution on the Impregnation Process of the Continuous Compression Molding Technology." *Journal of Thermoplastic Composite Materials* 30 (9): 1285–1302. https://doi.org/10.1177/0892705716632855

Connor, M., S. Toll, J. -A. E. Månson, and A. G. Gibson. 1995. "A Model for the Consolidation of Aligned Thermoplastic Powder Impregnated Composites." *Journal of Thermoplastic Composite Materials* 8 (2): 138–62. https://doi.org/10.1177/089270579500800201

Cutolo, D., and A. Savadori. (1994). "Processing of Product Forms for the Large-Scale Manufacturing of Advanced Thermoplastic Composites." *Polymers for Advanced Technologies* 5 (9): 545–53. https://doi.org/10.1002/pat.1994.220050913

Gil, Rafael Garcia. 2002. "Consolidation modelling for textile composites," *FRC 2002: 9th International Fibre Reinforced Composites*, 362–69.

Goud, Vijay, Ramasamy Alagirusamy, Apurba Das, and Dinesh Kalyanasundaram. 2019. "Influence of Various Forms of Polypropylene Matrix (Fiber, Powder and Film States) on the Flexural Strength of Carbon-Polypropylene Composites." *Composites Part B: Engineering* 166 (June): 56–64. https://doi.org/10.1016/j.compositesb.2018.11.135

Goud, Vijay, Dilpreet Singh, Alagirusamy Ramasamy, Apurba Das, and Dinesh Kalyanasundaram. 2020. "Investigation of the Mechanical Performance of Carbon/Polypropylene 2D and 3D Woven Composites Manufactured through Multi-Step Impregnation Processes." *Composites Part A: Applied Science and Manufacturing* 130 (March): 105733. https://doi.org/10.1016/j.compositesa.2019.105733

Greco, Antonio, Alessandra Strafella, Carola La Tegola, and Alfonso Maffezzoli. 2011. "Assessment of the Relevance of Sintering in Thermoplastic Commingled Yarn Consolidation." *Polymer Composites* 32 (4): 657–64. https://doi.org/10.1002/pc.21080

Guo, Z. X., and B. Derby. 1994. "Theoretical Model for Solid-State Consolidation of Long-Fibre Reinforced Metal-Matrix Composites." *Acta Metallurgica Et Materialia* 42(2): 461–73. https://doi.org/10.1016/0956-7151(94)90501-0

Gupta, Mohit, Harpreet Singh, Ashraf Nawaz Khan, Puneet Mahajan, R. T. Durai Prabhakaran, and R. Alagirusamy. 2021. "An Improved Orthotropic Elasto-Plastic Damage Model for Plain Woven Composites." *Thin-Walled Structures* 162 (May): 107598. https://doi.org/10.1016/j.tws.2021.107598

Gutowski, T. G., Z. Cai, S. Bauer, D. Boucher, J. Kingery, and S. Wineman. 1987a. "Consolidation Experiments for Laminate Composites." *Journal of Composite Materials* 21 (7): 650–69. https://doi.org/10.1177/002199838702100705

Gutowski, Timothy G., Tadahiko Morigaki, and Zhong Cai. 1987b. "The Consolidation of Laminate Composites." *Journal of Composite Materials* 21 (2): 172–88. https://doi.org/10.1177/002199838702100207

Haffner, S. M., K. Friedrich, P. J. Hogg, and J. J. C. Busfield. 1998. "Finite Element Assisted Modelling of the Microscopic Impregnation Process in Thermoplastic Preforms." *Applied Composite Materials*, 5(4), 237–55. https://doi.org/10.1023/A:1008850332291

Hattum, F. W. J. van, J. P. Nunes, and C. A. Bernardo. 2005. "A Theoretical and Experimental Study of New Towpreg-Based Long Fibre Thermoplastic Composites." *Composites Part A: Applied Science and Manufacturing* 36 (1): 25–32. https://doi.org/10.1016/j.compositesa.2004.06.031

Helmus, Rhena, Timotei Centea, Pascal Hubert, and Roland Hinterhölzl. 2016. "Out-of-Autoclave Prepreg Consolidation: Coupled Air Evacuation and Prepreg Impregnation Modeling." *Journal of Composite Materials* 50 (10): 1403–13. https://doi.org/10.1177/0021998315592005

Helmus, R., R. Hinterhölzl, and P. Hubert. 2015. "A Stochastic Approach to Model Material Variation Determining Tow Impregnation in Out-of-Autoclave Prepreg Consolidation." *Composites Part A: Applied Science and Manufacturing* 77: 293–300. https://doi.org/10.1016/j.compositesa.2015.03.021

Itoi, Masaaki, and R. Byron Pipes. 1990. "PAN and Pitch-Based Carbon Fiber-Reinforced Polyethernitrile Composites." *Journal of Thermoplastic Composite Materials* 3 (3): 172–89. https://doi.org/10.1177/089270579000300301

Jogur, Ganesh, Ashraf Nawaz Khan, Apurba Das, Puneet Mahajan, and R. Alagirusamy. 2018. "Impact Properties of Thermoplastic Composites." *Textile Progress* 50 (3): 109–83. https://doi.org/10.1080/00405167.2018.1563369

Khan, Ashraf Nawaz, Vijay Goud, Ramasamy Alagirusamy, Puneet Mahajan, and Apurba Das. 2022. "Optimization Study on Wet Electrostatic Powder Coating Process to Manufacture UHMWPE/LDPE Towpregs." *Journal of Industrial Textiles* (January): 15280837211070995. https://doi.org/10.1177/15280837211070995

Klinkmuller, V., M. -K. Um, M. Steffens, K. Friedrich, and B. -S. Kim. 1994. "A New Model for Impregnation Mechanisms in Different GF/PP Commingled Yarns." *Applied Composite Materials* 1 (5): 351–71. https://doi.org/10.1007/BF00568041

Kobayashi, Satoshi, Atsushi Tanaka, and Tetsuya Morimoto. 2012. "Analytical Prediction of Resin Impregnation Behavior during Processing of Unidirectional Fiber Reinforced Thermoplastic Composites Considering Pressure Fluctuation." *Advanced Composite Materials* 21 (5–6): 425–32. https://doi.org/10.1080/09243046.2012.740773

Larock, J. A., H. T. Hahn, and D. J. Evans. 1989. "Pultrusion Processes for Thermoplastic Composites." *Journal of Thermoplastic Composite Materials* 2 (3): 216–29. https://doi.org/10.1177/089270578900200304

Li, Xuekuan, Yan Zhao, Kai Wang, Junlin Chen, and Wenkuo Lu. 2017. "Study on the rheology and tensile property of ccf300/peek composites via powder impregnation method." In *21st International Conference on Composite Materials* (August):1–8.

Ye, Lin, and Klaus Friedrich. 1997. "Processing of CF/PEEK Thermoplastic Composites from Flexible Preforms." *Advanced Composite Materials* 6 (2): 83–97. https://doi.org/10.1163/156855197X00012

Mehl, Nathan A., and Ludwig Rebenfeld. 1993. "Computer Simulation of Crystallization Kinetics and Morphology in Fiber-Reinforced Thermoplastic Composites. I. Two-Dimensional Case." *Journal of Polymer Science Part B: Polymer Physics* 31 (12): 1677–86. https://doi.org/10.1002/polb.1993.090311203

Novais, F, J. P. Nunes, J. F. Silva, F. van Hattum, P. Vieira, and R. Fangueiro. 2004. "Production of gf/pp woven fabrics from powder coated thermoplastic matrix towpregs." In *Proceedings of 11th European Conference on Composite Materials-ECCM* (Vol. 11) (May):1–6.

Nunes, J. P., A. M. Brito, A. S. Pouzada, and C. A. Bernardo. 2001. "Isothermal and Non-Isothermal Consolidation of Carbon Fiber Towpregs." *Polymer Composites* 22 (1): 71–9. https://doi.org/10.1002/pc.10518

Padaki, S., and L. T. Drzal. 1997. "A Consolidation Model for Polymer Powder Impregnated Tapes." *Journal of Composite Materials* 31 (21): 2202–27.

Padaki, S., and L. T. Drzal. 1999. "A Simulation Study on the Effects of Particle Size on the Consolidation of Polymer Powder Impregnated Tapes." *Composites Part A: Applied Science and Manufacturing* 30 (3): 325–37. https://doi.org/10.1016/S1359-835X(98)00115-8

Phelan, F. R., Y. Leung, and R. S. Parnas. 1994. "Modeling of Microscale Flow in Unidirectional Fibrous Porous Media." *Journal of Thermoplastic Composite Materials* 7 (3): 208–18. https://doi.org/10.1177/089270579400700303

Phillips, Richard, Devrim A. Akyüz, and Jan-Anders E. Månson. 1998. "Prediction of the Consolidation of Woven Fibre-Reinforced Thermoplastic Composites. Part I. Isothermal Case." *Composites Part A: Applied Science and Manufacturing* 29 (4): 395–402. https://doi.org/10.1016/S1359-835X(97)00099-7

Pipes, R. Byron, J. W. S. Hearle, A. J. Beaussart, A. M. Sastry, and R. K. Okine. 1991. "A Constitutive Relation for the Viscous Flow of an Oriented Fiber Assembly." *Journal of Composite Materials* 25 (9): 1204–17. https://doi.org/10.1177/002199839102500907

Ramani, Karthik, and Chris Hoyle. 1995a. "Processing of Thermoplastic Composites Using a Powder Slurry Technique. I. Impregnation and Preheating." *Materials and Manufacturing Processes* 10 (6): 1169–82. https://doi.org/10.1080/10426919508935100

Ramani, Karthik, and Chris Hoyle. 1995b. "Processing of Thermoplastic Composites Using a Powder Slurry Technique. II. Coating and Consolidation." *Materials and Manufacturing Processes* 10 (6): 1183–1200. https://doi.org/10.1080/10426919508935101

Sakaguchi, M., A. Nakai, H. Hamada, and N. Takeda. 2000. "The Mechanical Properties of Unidirectional Thermoplastic Composites Manufactured by a Micro-Braiding Technique." *Composites Science and Technology* 60 (5): 717–22. https://doi.org/10.1016/S0266-3538(99)00175-X

Scalea, Francesco Lanza di, and Robert E. Green Jr. 2000. "A Hybrid Non-Contact Ultrasonic System for Sensing Bond Quality in Tow-Placed Thermoplastic Composites." *Journal of Composite Materials* 34 (21): 1860–80.

Shishoo, Roshan, and Jonas Bernhardsson. 2000. "Effect of Processing Parameters on Consolidation Quality of GF//PP Commingled Yarn Based Composites." *Journal of Thermoplastic Composite Materials* 13: 292–313.

Steggall-Murphy, Claire, Pavel Simacek, Suresh G. Advani, Shridhar Yarlagadda, and Shawn Walsh. 2010. "A Model for Thermoplastic Melt Impregnation of Fiber Bundles during Consolidation of Powder-Impregnated Continuous Fiber Composites." *Composites Part A: Applied Science and Manufacturing* 41 (1): 93–100. https://doi.org/10.1016/j.compositesa.2009.09.026

Van West, B. P., R. Byron Pipes, and S. G. Advani. 1991a. "The Consolidation of Commingled Thermoplastic Fabrics." *Polymer Composites* 12 (6): 417–27. https://doi.org/10.1002/pc.750120607

Van West, B. P., R. Byron Pipes, M. Keefe, and S. G. Advani. 1991b. "The Draping and Consolidation of Commingled Fabrics." *Composites Manufacturing* 2 (1): 10–22. https://doi.org/10.1016/0956-7143(91)90154-9

Wakeman, M. D., T. A. Cain, C. D. Rudd, R. Brooks,, and A. C. Long. 1998. "Compression Moulding of Glass and Polypropylene Composites for Optimised Macro-and Micro-Mechanical Properties—1 Commingled Glass and Polypropylene." *Composites Science and Technology* 58 (12): 1879–1898. https://doi.org/10.1016/S0266-3538(98)00011-6

Wu, Chang Mou, Po Chung Lin, and Cheng Te Tsai. 2016. "Fabrication and Mechanical Properties of Self-Reinforced Polyester Composites by Double Covered Uncommingled Yarn." *Polymer Composites* 37 (12): 3331–40. https://doi.org/10.1002/pc.23531

Wu, Ning, Yunxing Liang, Kegang Zhang, Wenzheng Xu, and Li Chen. 2013. "Preparation and Bending Properties of Three Dimensional Braided Single Poly (Lactic Acid) Composite." *Composites Part B: Engineering* 52 (September): 106–13. https://doi.org/10.1016/j.compositesb.2013.02.047

Wysocki, M., R. Larsson, and S. Toll. 2005. "Hydrostatic Consolidation of Commingled Fibre Composites." *Composites Science and Technology* 65 (10): 1507–19. https://doi.org/10.1016/j.compscitech.2005.01.002

Yang, Heechun, and Jonathan S. Colton. 1994. "Microstructure-Based Processing Parameters of Thermoplastic Composite Materials. Part I: Theoretical Models." *Polymer Composites* 15 (1): 34–41. https://doi.org/10.1002/pc.750150106

Ye, L., V. Klinkmuller, and K. Friedrich. 1992. "Impregnation and Consolidation in Composites Made of GF/PP Powder Impregnated Bundles." *Journal of Thermoplastic Composite Materials* 5 (1): 32–48. https://doi.org/10.1177/089270579200500103

Ye, Lin, Klaus Friedrich, Joachim Kästel, and Yiu-Wing Mai. 1995a. "Consolidation of Unidirectional CF/PEEK Composites from Commingled Yarn Prepreg." *Composites Science and Technology* 54 (4): 349–58. https://doi.org/10.1016/0266-3538(95)00061-5

Ye, Lin, Klaus Friedrich, and Joachim Kastel. 1995b. "Consolidation of GF/PP Commingled Yarn Composites." *Applied Composite Materials* 1 (6): 415–29. https://doi.org/10.1007/BF00706502

Zhang, Jian Min, and Ton Peijs. 2010. "Self-Reinforced Poly(Ethylene Terephthalate) Composites by Hot Consolidation of Bi-Component PET Yarns." *Composites Part A: Applied Science and Manufacturing* 41 (8): 964–72. https://doi.org/10.1016/j.compositesa.2010.03.012

10 Manufacture of Composites with Flexible Towpregs

V. Balakumaran[a], R. Alagirusamy[b],
and Dinesh Kalyanasundaram[ac]
[a]Indian Institute of Technology
New Delhi, India
[b]Indian Institute of Technology
New Delhi, India
[c]All India Institute of Medical Science
New Delhi, India

CONTENTS

DOI: 10.1201/9781003049715-10

10.1　INTRODUCTION

Thermoplastic composites (TPC) find extensive use in automobile, aeronautical, aerospace, biomedical, and impact applications (Hamada and Ramakrishna 1995; Ramakrishna et al. 2001; Klaus Friedrich and Almajid 2013; Jogur et al. 2018). TPCs are preferred for the high strength, toughness, impact strength, strength retention at elevated temperatures, thermal stability, solvent resistance, low moisture absorption, low coefficient of thermal expansion, repairability, weldability, and recyclability (Cogswell 1992; Vodicka 1996; Ahmed et al. 2006; Ramakrishna et al. 2009; Zhang et al. 2018). TPC production cycles are shorter, with cycle time in minutes compared to thermosets, whose production cycle (curing cycle) runs in hours (Loos and Springer 1983; Colton et al. 1992). The thermoplastics are fully reacted high-molecular weight resins in an amorphous or semi-crystalline state (Bigg et al. 1988). Unlike thermosets, thermoplastics are biocompatible (Evans and Gregson 1998). Thermoplastic is solid at room temperature; it is heated above the melting point to achieve melt flow to form a composite during the manufacturing process. It is a physical phenomenon, and no chemical reaction occurs. In contrast, thermosets are liquid in the form of low-molecular weight monomers. During the curing cycle, thermosets undergo extensive cross-linking reaction between the monomers, started by an initiator molecule to build highly cross-linked high molecular weight molecules (Loos and Springer 1983). They become stiff and gain strength with an increase in molecular weight upon curing. Thermosets are brittle and lack toughness, due to the presence of extensive cross-links in the fully cured resin (Bigg et al. 1988; Iyer and Drzal 1990). Thermosets release harmful low-molecular-weight volatile organic compounds during the curing cycle (Evans and Gregson 1998).

Due to the above discussed points, the use of TPC is favored over thermosets. However, thermoplastics present a different set of challenges. The thermoplastic resin is fully cured and solid under room temperature. Thus, thermoplastics cannot be processed into a composite by liquid impregnation techniques like layup process, resin infusion method, vacuum-assisted resin transfer molding, which are followed for thermoset composite manufacturing. Thermoplastics have high viscosity (10^2 –10^4 Pa.s whereas thermosets < 100 Pa.s), and flow is achieved when heated above the melting point with pressure application. Due to the high viscosity of thermoplastics, it cannot flow over large distances into the spaces between the reinforcement fibers. Poor melt flow behavior of thermoplastics hinders complete wetting of the reinforcement fiber by the thermoplastic. The incomplete wetting of the reinforcement fiber affects composite quality. Due to this reason, the film stack method of producing composites is not prepared for manufacturing thermoplastic composite even though it is straightforward (Goud et al. 2019). In the film stack method, the reinforcement in the form of woven fabric and thermoplastic matrix in the form of a film are placed alternatively as laminates, one over the other inside a mold. After the laminates are laid, they are pressed under heat and pressure to melt the matrix and impregnate the woven reinforcement fabric. Since the viscosity is high, the flow is not enough to achieve complete wetting and impregnation (Gibson and Manson 1992), leading to poor quality composite. The solution to this problem is to get the thermoplastic resin in close vicinity of the reinforcement fiber. By doing so, the distance the thermoplastic resin must flow will be reduced. Thus, melt flow will be adequate to wet out the reinforcement fibers thoroughly and achieve total impregnation to form void-free composites. The solution to achieve this comes in the name of flexible towpregs of thermoplastics.

Flexible towpregs for manufacturing TPC refers to the structural hybridized textile preforms (mixing matrix and reinforcement in the fibrous form), which are in a semi-impregnated state. A semi-impregnated state refers to the thermoplastic resin present in the vicinity of the reinforcement fibers without wetting (fiber and matrix separated). Hybrid yarn, hybrid fabric, and powder-impregnated towpregs are different forms of flexible prepregs discussed in earlier chapters. One of the hybrid yarns is a commingled yarn, where the matrix thermoplastic is mixed with the reinforcement fiber in the fibrous form, at the yarn level (Alagirusamy et al. 2010). The mixing at the yarn level is achieved by spinning, wrapping, braiding, and air-jet texturing methods (Alagirusamy et al. 2010). The hybrid fabric is a co-woven or flat/circular braided or knitted textile fabric structure (Alagirusamy et al. 2010). The thermoplastic matrix yarn and reinforcement yarn are used in different combinations to produce these hybrid fabric textile preforms. In powder impregnated tows, the thermoplastic matrix in the form of powder is deposited on the reinforcement fiber. The powder matrix is coated on the reinforcement yarn by dry and wet powder coating methods, followed by sintering of the coated powder to fuse it with the reinforcement fiber (Iyer and Drzal 1990; Ho et al. 2011). Complete wetting of the reinforcement fibers does not happen during the sintering process (Iyer and Drzal 1990). In all forms of the flexible towpregs, the wetting of the reinforcement fibers does not occur.

Several other methods of producing thermoplastic towpregs are melt impregnation and solution impregnation, which provide pre-impregnated tows; the thermoplastic polymer wets the fibers (Iyer and Drzal 1990; Ramasamy, Wang, and Muzzy 1996a). The tows produced by these two methods are stiff, and the technique by itself is complicated. In solution impregnation, the thermoplastic polymer is dissolved in an organic solvent and the reinforcement fiber is dipped into the solution to impregnate the fiber with the polymer (Iyer and Drzal 1990). Upon impregnation, the solvent must be removed by evaporation or heating under vacuum (Iyer and Drzal 1990). Removal of solvent is challenging, and improper removal leads to void formation during the processing of the towpregs into composite (Wu and Schultz 2000). There are not many solvents that can dissolve thermoplastics at room temperature and many are dissolvable in volatile solvents at high temperatures (Talley et al. 2017). In melt impregnation, the thermoplastic polymer is melted by heating the polymer above the melting point in an impregnator. The reinforcement fiber is dipped into the melt polymer to achieve impregnation (Iyer and Drzal 1990). The viscosity of the matrix is high, and achieving uniform impregnation is challenged even in melt state during melt impregnation. Thus, plasticizer chemicals are added to the polymer to lower the inherent viscosity (Gibson and Manson 1992). These compatibilizer chemicals react with the polymer affecting mechanical and long-term properties. Due to the above discussed reasons, flexible towpregs are preferred for the manufacture of TPC.

Flexible towpregs should undergo a further processing step called consolidation to fabricate them into a composite. The towpregs are arranged in layers to be made into a composite. The tow height is non-uniform and the surface of the towpreg is rough (Campbell 2014). This roughness creates air gaps between the layers. The matrix polymer is present on the yarn's surface only, and complete wetting of fiber is not observed at the towpreg stage. Thus, consolidation of laminate with heat and applied pressure must be carried to produce useful composites. Consolidation involves three steps, intimate fiber contact, autohesion, and fiber impregnation (Van West, Pipes, and Advani 1991; Hou et al. 1998; Alagirusamy et al. 2010). First, intimate fiber contact happens with the applied mechanical pressure. Here, the separated laminate layer of towpregs is pressed together. The matrix polymer around the fiber at the polymer-polymer interface undergoes deformation to allow the intimate contact of fibers (Iyer and Drzal 1990).

Second step in consolidation of thermoplastic composite is autohesion due to melt flow. When the temperature is raised above the glass transition temperature (Tg) for amorphous thermoplastic and melting point (Tm) for semi-crystalline thermoplastics, melt flow occurs. The matrix polymer flows into the interstitial space between the individual dry tows in the same layer (Iyer and Drzal 1990). Then it is followed by the flow of the melt polymer into the inter-layer space between laminates of towpregs. This removes the air and creates a resin-rich inter-layer, which allows the bonding of laminates together; the polymer-polymer interface disappears due to autohesion. The matrix polymer's melt flow distance in a commingled towpreg is 20–40 µm (Svensson, Shishoo, and Gilchrist 1998). The polymer molecule at the polymer-polymer interface undergoes diffusion across the interface to entangle with the neighboring chains of the polymer. The contact time decides the extent of polymer entanglement, thus deciding the interfacial bond strength (Iyer and Drzal 1990; Campbell 2014). The fiber tows are still dry internally, and this is observed under a microscope examination by Ye et al. (1995), Ye and Friedrich (1997), and Hou et al. (1998).

Third step in consolidation of thermoplsatic composite is fiber impregnation, achieved due to applied pressure and heat over an extended time. The pressure pushes the melt polymer layer present surrounding the fiber into the inter-fiber space (Ye and Friedrich 1997; Hou et al. 1998), thus, removing air and further reducing void content to form a good composite. The completion of the consolidation process is identified as total uniform impregnation of fibers forming a monolithic phase without any distinct interface. The assumed cross-section and consolidation mechanism for a commingled yarn is shown in Figure 10.1.

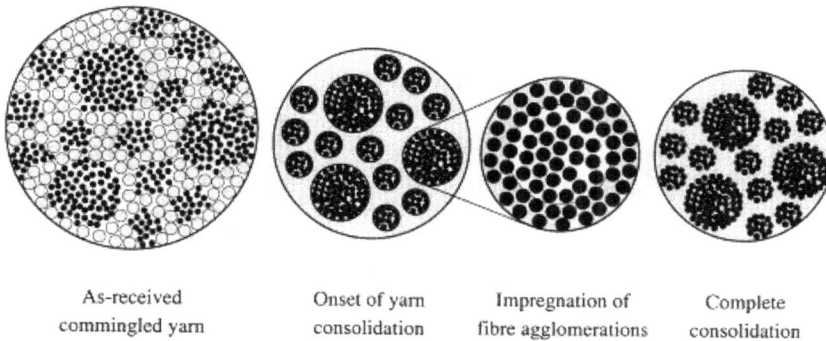

As-received   Onset of yarn   Impregnation of   Complete
commingled yarn   consolidation   fibre agglomerations   consolidation

FIGURE 10.1   Commingled yarn cross-section and the consolidation mechanism. (Reproduced with permission from Bernet et al. 2002.)

Several manufacturing methods can achieve the consolidation of TPC. The most common is the compression moulding method due to the simplicity of the process. The process is carried using a heated press that applies heat and pressure over a fixed time. The process achieves consolidation over time, and the material is cooled under pressure to achieve a good quality composite. However, the compression molding process suffers limitations. Thus, there is a need to develop new methods to address the shortcoming of the compression molding process. Several methods like autoclave molding, filament winding, thermal forming, automated tape placement, continuous consolidation, and pultrusion were developed. These methods helped produce TPC in complex shapes, hollow structures, and large parts at low labor costs made possible by automation. The above-listed methods will be discussed concerning the manufacture of TPC using flexible towpregs in the subsequent sections. The merits and critical process parameters that affect composite quality in each method are discussed.

10.2   COMPRESSION MOLDING

Compression molding of thermoplastic composites is a low-flow process of achieving thermoplastic composite consolidation (Hou et al. 1998). The laminate sequence of prepreg preforms is stacked inside a matched two-part cavity mold and pressed under heat and pressure. The applied heat melts the thermoplastic polymer. Upon melting, with pressure, the molten polymer is squeezed and forced into the gaps between the reinforcement fiber to form the composite. No shear forces are involved in driving the molten polymer into the voids, unlike in injection molding (Hou et al. 1998). The compression molding process is a high-pressure process where the consolidation pressure ranges a few MPa (0–5 MPa) (Fujihara et al. 2004). The process yields flat consolidated laminate sheets of composites, where the laminate thickness is far lower than the length of the composite. The produced laminate sheet does not require any trimming. These consolidated flat laminates may be subjected to a thermal forming process to achieve complex shapes. The process of thermal forming will be explained in the subsequent section. The compression molding method is primarily used in research and development to produce flat panels for testing and profiles for aerospace applications (Chang et al. 1991). The composite produced has a fiber volume fraction typically up to 60% by volume (Muzzy, Wu, and Colton 1990; Chang et al. 1991).

10.2.1 MACHINE AND PROCESS DESCRIPTION

The compression molding process is performed on a heated hydraulic hot press, as shown in Figure 10.2. The hot press consists of two heated platens on which the two-part flat cavity closed mold is placed. The platens are enclosed inside a vacuum chamber that is evacuated by a vacuum pump. The vacuum helps achieve composite panels with the least void content (Yu and Davis 1993). The two platens are heated electrically using electrical heaters that are embedded inside. The heaters have a total power of 3 kW with six heaters per platen for one square foot area. However,

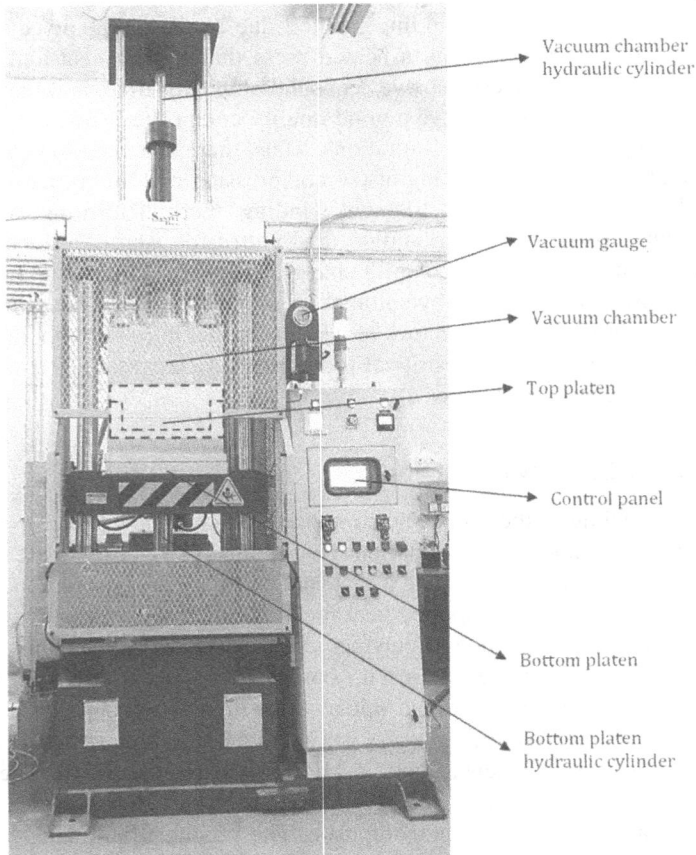

Vacuum chamber
hydraulic cylinder

Vacuum gauge

Vacuum chamber

Top platen

Control panel

Bottom platen

Bottom platen
hydraulic cylinder

FIGURE 10.2 A vacuum compression molding hot press used for producing thermoplastic composites in the laboratory.

the heater capacity and the number of heaters shall vary based on the platen area, temperature to be maintained, and heating rate required. The platen temperature is maintained as per the value set by the user. The machine turns the heaters on and off to maintain the temperature automatically controlled by a PID temperature controller. Each platen heating is controlled independently by separate PID controllers, and different temperatures can be maintained in each platen as set by the user. The user set temperature is generally a few degrees above the polymer's melting point, which is the processing temperature of the thermoplastic polymer. As a general practice in compression molding, the top platen is set at a temperature of 10°C or 5°C above the bottom platen to help in consolidation. For the controlled cooling of the laminate after consolidation, the platens are placed with coolant channels. Inside these coolant channels, cold water or compressed air as the coolant is circulated to take away heat from the platen. Depending on temperature and process cycle requirements, the coolant is selected.

One of the two platens is movable, either the top or the bottom platen. The moving platen is connected to a hydraulic cylinder that drives the platen up and down. A servo-hydraulic control system controls the hydraulic cylinder's movement. The other stationary platen is attached to a load cell, which measures the pressure between the mold when the platens close. This load value helps maintain constant pressure on the mold throughout the consolidation cycle. The machine maintains consolidation pressure at the set value automatically based on a closed-loop control algorithm with feedback from the load cell. A PLC controller in the machine controls the whole process of consolidation. The user sets the process parameters of consolidation time, consolidation pressure, cooling rate, breathing cycles, travel rate, and breather cycles in the PLC controller. Once the process parameters are set, the user initiates the cycle after placing the laminate stack in the mold to start the process. Upon starting the process, the PLC controller performs several tasks like mold closing, breather cycles, vacuum build-up, consolidation, vacuum vent, and mold cooling, followed by the opening of the platens. The mold closing is when the moving platen is moved up to close the mold, and upon closing the mold, breather cycles are carried. The breather cycle process releases the pressure in the mold after mold closure by the movement of the platen up and down. Breathing is performed for few counts to allow the removal of any entrapped air bubbles between the laminates. After the breather cycle, consolidation begins with the build of vacuum in the chamber at the consolidation temperature and pressure maintained in the mold for the time set by the user. Upon consolidation, the controller automatically vents out the vacuum slowly for a set time, followed by a cooling cycle where water or compressed air is circulated to lower the mold temperature. The cooling happens under pressure before the mold opening. Upon cooling, the mold is opened, and the process cycle is complete. The user is indicated for removing the composite from the mold and preparing the mold for the next cycle.

The user prepares the mold before each consolidation cycle by applying a mold release agent to ensure a smooth surface for easy release of the composite from the mold (Krämer et al. 2020). The release agent is applied to all the surfaces that make contact with the composite material. After the application of mold release, a layer of release media is placed inside the mold. It further prevents sticking of the composite with the mold. The release media shall be a Teflon sheet, coated glass fabric, or

Consolidation cycle

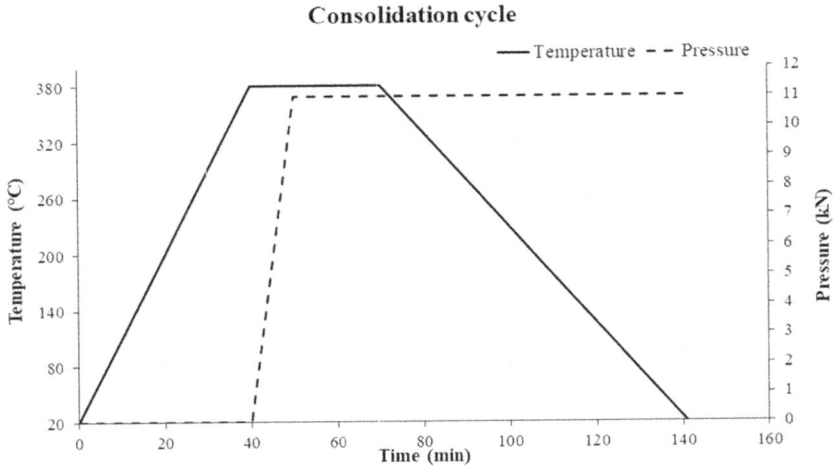

FIGURE 10.3 Consolidation cycle applied for producing carbon/PEEK laminate composite. (Adapted from Fujihara et al. 2003.)

polyimide film (UPILEX or Kapton film) (Texier et al. 1993; Ye et al. 1995), depending on the composite being prepared and the consolidation temperature. This step is followed by placing the laminate stack inside the mold to start the next consolidation cycle. The typical consolidation cycle is exemplified in Figure 10.3.

10.2.2 EFFECT OF PROCESS PARAMETERS ON THE QUALITY OF COMPOSITE

The process parameters that affect thermoplastic composite quality in compression molding are as follows:

1. Temperature: The processing temperature during the consolidation affects impregnation quality (Ramasamy, Wang, and Muzzy 1996b; Fujihara et al. 2004). The higher the processing temperature above the thermoplastic melting point, the melt viscosity will be lowered to allow for easy flow into the inter-fiber and intra-fiber spaces. However, the temperature cannot be raised indefinably, since the polymer will start to degrade when heated above the degradation temperature (Fujihara et al. 2004).
2. Time: The time through which the consolidation happens helps in achieving impregnation. The higher the consolidation time, the higher distances the thermoplastic polymer can flow to achieve complete impregnation.
3. Pressure: The pressure applied during consolidation is the driving force that drives the melt polymer into the inter-fiber and intra-fiber spaces. The higher consolidation pressure translates to a higher driving force applied to the laminate to achieve better consolidation. The holding time and applied pressure affect fiber impregnation (Ye and Friedrich 1997). The towpreg with high fiber content requires higher pressure for consolidation compared to towpreg with lower fiber content (Iyer and Drzal 1990).

4. Cooling rate: The cooling rate affects the rate of recrystallization of the thermoplastic polymer into a solid. It affects the interface properties between the fiber and matrix thermoplastic polymer, affecting the composite's mechanical properties. The cooling rate also creates waviness in the composite.
5. Preheating: Generally, before compression molding, the preforms are subjected to the preheating process. The preheat removes any volatile substances present on the towpreg and improves interfacial bond strength (Ramani and Hoyle 1995; Li et al. 2019).

10.2.3 Advantages

The following are the advantages of the compression molding process:

1. Compression molding can produce a composite with a very high fiber volume fraction of up to 70%.
2. Low cost of investment. The process requires only a heated press and a matched two-part cavity mold to produce the laminate composites.
3. High-temperature polymer with high viscosity can be effectively processed.
4. The process gives finished composites with a good surface finish, often requiring no post-manufacture trimming.
5. Low cost at high volume production rates.
6. The composite laminate can be reformed by a thermoforming process to produce complex shapes.

10.2.4 Limitations

1. The process is suited for only small components. The size of the heated press limits the size of the component.
2. High labor cost involved. The laminate stacks must be prepared manually.
3. Complex shapes cannot be made and are limited to only flat laminate panels predominantly in research and development.
4. The process is not suited for small volume parts production cycle with requirement of the order less than 100, since the initial investment in the mold and machine shall end up not economical.
5. Thick composite parts are not possible. The heat must be transferred from the two half molds into the laminate to achieve melting of thermoplastic resin and impregnations. If the part is thick, the heat transfer will not be sufficient to melt the thermoplastic resin.

10.3 THERMAL FORMING

The thermal forming process is specifically developed to produce thermoplastic fiber-reinforced composite parts. The method is not suited for thermoset composites. Complex two-dimensional (2D) and three-dimensional (3D) structural shapes required for helmets, aircraft parts, wing inserts, structure preforms, etc., are produced by thermoforming. The process uses heat and pressure to deform a pre-consolidated laminate sheet into an intricately shaped structural part. For the thermal forming process, pre-consolidated laminate sheets are preferred over loose un-consolidated ply stacks. The pre-consolidated flat sheets are free of void and porosity and do not slip during the forming process, preventing waviness and wrinkles in the final part (Friedrich, Hou, and Krebs 1997; Akkerman and Haanappel 2015). The pre-consolidated laminate sheets are produced by compression molding on a heated press (refer to Section 10.2). The pre-consolidated sheets are heated above the polymer's melting point (T_m) for semi-crystalline polymer and glass transition temperatures (T_g) for amorphous polymers. The pre-consolidated laminate on heating melts the polymer to allow deformation of the sheet to the required part shape. The process may seem simple up front and when observed. However, it is complicated due to the stiffness of the continuous fiber reinforcement in a thermoplastic matrix along the fiber direction. There are chances that the laminate sheet may fracture, buckle at the bending radii, and the required part shape may not be formed. Care is needed in designing a proper mold for effective forming, considering the constraints. The primary phenomenon and deformation modes that help in forming the part in the order of increasing complexity are a) resin percolation, b) transverse flow, c) inter-ply shear, d) intra-ply shear, and e) inter-laminar rotation. Thermal forming is predominantly possible due to the shearing within and between the laminate stack's individual layers to allow deformation. Resin percolation and transverse flow are phenomena during consolidation, and they also have a role in forming. Each phenomenon involved is shown in Figure 10.4.

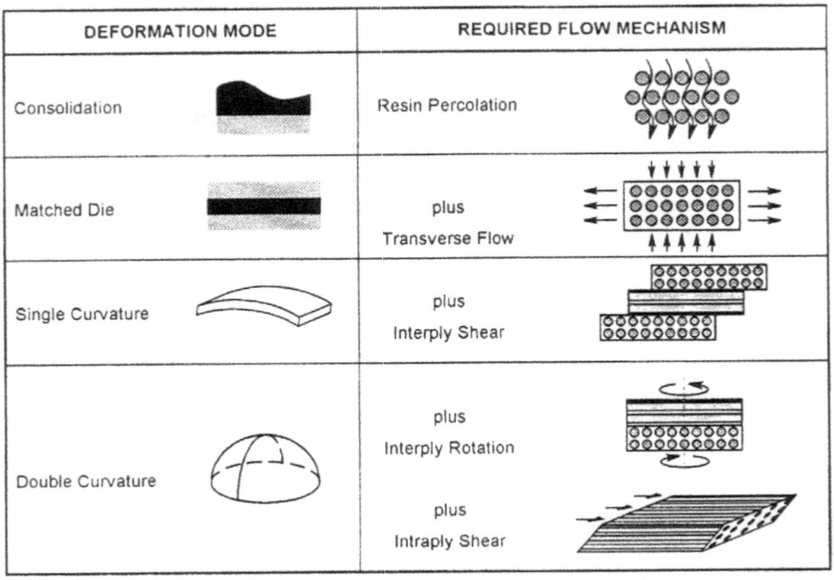

DEFORMATION MODE	REQUIRED FLOW MECHANISM
Consolidation	Resin Percolation
Matched Die	plus Transverse Flow
Single Curvature	plus Interply Shear
Double Curvature	plus Interply Rotation plus Intraply Shear

FIGURE 10.4   The deformation and flow mechanics involved in thermal forming. (Reproduced with permission from Friedrich, Hou, and Krebs 1997.)

The first and most basic phenomenon common to all thermoplastic composites fabrication methods and observed in thermal forming is resin percolation. The melt flow of thermoplastic polymers into and along the laminate reinforcement fibrous layers is resin percolation. This resin percolation helps in the bonding of the individual laminate plies to form a single sheet. The resin flow happens when a consolidation pressure is applied over the laminate stack. Second is the transverse flow phenomenon, where the fibers spread along the lateral direction perpendicular to the direction of the unidirectional fiber when pressure is applied. Transverse flow happens due to slight thickness variation and surface unevenness of the laminate stack or the mold's uneven surface. Unevenness creates localized pressure differences over the laminate stack during consolidation and forming inside a matched mold. Another situation where transverse flow occurs is when a shear force is exerted on the laminate due to a mismatch of thermal expansion coefficient between the mold and thermoplastic polymer matrix. However, transverse flow due to shear may have a possible adverse effect leading to waviness in the laminate structure affecting mechanical properties. Third, the inter-ply shear is the relative shear slip movement of two adjacent laminate layers at the resin-rich layer bonding the laminates. Fourth, the intra-ply shear is the deformation of the laminate layers along the plane of the laminate. There is no limit to shearing in unidirectional laminate, whereas in a woven fabric, the shearing is limited by woven interlock structure. The shearing deformation allows the forming of the part to produce curvatures. The fifth is the inter-laminate rotation, which is the relative rotation of the laminate due to shearing at the resin layer between the laminates. This effect helps in forming a thick laminate structure into a double curved part. The first two phenomena of resin peculation and transverse flow are more of resin and material-based phenomenon common to consolidation and thermal forming. The primary phenomenon helping in thermal forming are the shearing effect due to inter-ply shearing, intra-ply shearing, and inter-laminate rotation.

10.3.1   PROCESS DESCRIPTION

Several steps involved in the thermal forming process are explained. First, the process starts with the preparation of a pre-consolidated flat sheet by compression molding. The laminate sequence is cut to the required size and placed inside the hot press, and it is consolidated into a single composite flat panel without voids. The consolidation is carried under heat and pressure, with the temperature applied being above the melting point of the thermoplastic polymer. Second, the consolidated flat panel is cooled under pressure to a temperature below the glass transition temperature (T_g) and after cooling, the pressure is relieved. The third step is the reheating of the consolidated laminate to prepare it for thermal forming. The heating is carried to make the laminate pliable and allow shear-induced deformations during the thermal forming step. The laminate heating is carried either using an infrared (IR) heater or a convection oven or a heated press. The IR heaters are limited to use with only thin laminates. In the case of thick laminates, there are chances for a temperature gradient across the thickness. Since the heating is from the surface due to absorption of the IR radiation, there are chances for the surface to overheat if exposed for a longer time. The heating is rapid and limited to thin laminates only. On the other hand, the convection oven achieves uniform heating even with thick laminates, the heating time is in the range of a few minutes, and the rate of heating is lower than IR heaters. The other method is by using a heated hot press similar to a compression molding press. The heating of the pre-consolidated laminate sheets deconsolidates the sheet and allows for the deformation of the part during forming. The third step is the transfer of the heated sheet into the forming die. The transfer should be made rapidly and without any delay to prevent cooling. Fourth is the forming step where pressure is applied, and the laminate sheet is bent to the required part shape in the mold. After forming the part inside the mold, the pressure is released after the part cools and then the part is removed. The process is repeated for the next production cycle.

10.3.2 Types of Thermal Forming Process

The thermal forming process described in the earlier section is the general process flow, which can be performed in several ways. The forming can be performed using two matched open molds on a hot press machine called the stamp forming. However, the stamp forming method of thermoforming requires closely matched mold in the shape of the final part. The machining of the mold to the required shape with close tolerance is a challenge and the cost of producing such a large mold is high. Thus, newer thermal forming methods, namely hydroforming, bladder/diaphragm forming, and roll forming, were developed. Each of the mentioned thermal forming methods is discussed in the subsequent sub-sections.

10.3.2.1 Stamp Forming

As the name suggests, stamp forming is the process of stamping a flat consolidated sheet of the thermoplastic composite to the required shape in a mold using a heated molding press. The process is similar to stamping forming of sheet metal. The process starts with a stack of unconsolidated laminate plies. This is placed in the heated compression molding press to be pressed under heat and pressure to produce consolidated flat blanks. The flat consolidated blank panel is removed from the press after cooling for further steps. The consolidated blank is cut to the required two-dimensional shape and size required to produce the final part after thermal forming. The flat cut blank of the required size is heated to a temperature above the thermoplastic polymer's melting point to allow workability during the stamping step. The heating of the flat blank is carried using infrared heaters. The infrared heaters achieve the required heating at a very rapid rate, maintaining uniform temperature throughout. The heated blank is rapidly transferred into the heated stamping mold, which is at the stamping temperature of thermoplastic in the composite. The mold is closed, and pressure is applied rapidly with the pressure maintained till the part is cooled below the forming temperature of the thermoplastic in the laminate. The forming temperature is a few degrees higher than the melting point and glass transition temperature for semi-crystalline and amorphous thermoplastic. The mold is cooled rapidly by using water circulation into the cooling channels available in the mold.

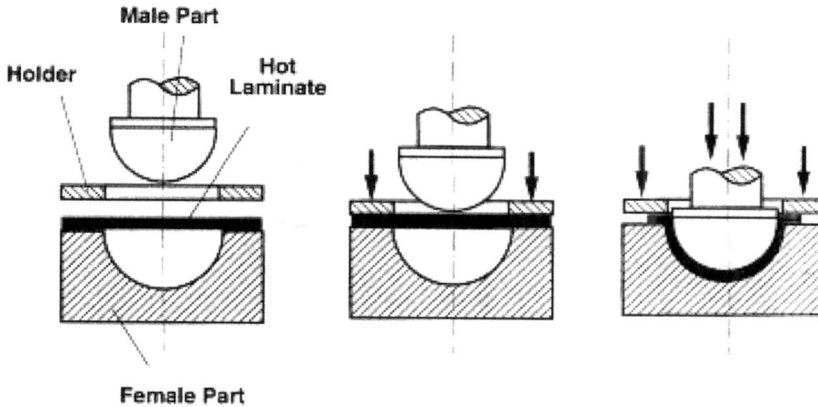

FIGURE 10.5 The stamp forming of a hot laminate in a two-part matched mold setup for a hemispherical shape. (Reproduced with permission from Hou 1997.)

After cooling of the part, the stamping mold is opened, and the formed part is removed and trimmed if required to get the final part. After part removal, the molds are heated to the required forming temperature to prepare for the next cycle. Electrical cartridge heaters are used to heat the molds.

The stamp forming process is carried out using matched two-part open molds. The hot flat cut laminate for forming is placed inside the bottom mold, and the mold is closed by the top mold to form the part in the required shape. The top mold consists of the punch, which forms the laminate into the required shape when the mold is closed. The bottom mold is the female part and the top mold is the male part. The setup used in stamp forming is shown in Figure 10.5 (Hou 1997). The hot laminate is clamped with the bottom half of the mold with the help of the holder, as seen in Figure 10.5. The clamping prevents the slipping and shifting of the laminate during the forming step. The clamping force is varied according to the process requirement (Hou 1997). The shape of the mold shall be different based on the requirement and not limited to only spherical parts. Complex 2D and 3D shaped parts can be produced by stamp forming (Friedrich, Hou, and Krebs 1997). An example of an angular stamp forming mold used to produce parts for brackets and several other applications is shown in Figure 10.6. The molds are placed on a heated hydraulic press similar to the one used for compression molding. The press is attached with a load cell that measures the load applied on the molds during the closure. It helps to maintain constant pressure, as set by the user during the process cycle. The press applies pressure via a hydraulic cylinder attached to the top mold. The speed of mold closure, also known as stamping velocity, is controlled during the process. The factors of temperature, stamping velocity, the pressure applied, the time duration of pressure applied, and cooling rate affect the composite quality (Hou 1997; McCool et al. 2012). It is critical to design and manufacture molds with a uniform gap to achieve uniform consolidation pressure over the part when formed (Hou 1997; Friedrich, Hou, and Krebs 1997). This design and production of the mold with close

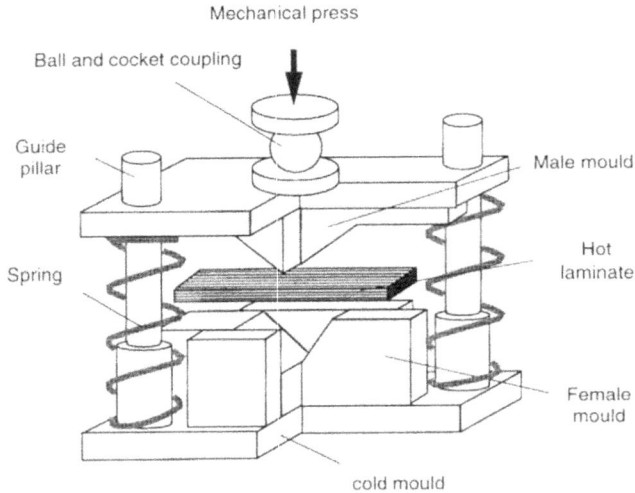

FIGURE 10.6   An example of an angular mold used for part production. (Reproduced with permission from Friedrich, Hou, and Krebs 1997.)

tolerance is an expensive and complicated process. To overcome the complexity of stamp forming, newer hydroforming and diaphragm forming methods are proposed and discussed in the next sections.

10.3.2.2   Hydroforming

The hydroforming process uses an incompressible hydraulic liquid (oil) in the bottom mold, and the top mold consists of the punch for the shape required. The blank is separated from making contact with the oil in the bottom mold using a diaphragm in between. The top mold has the shape of the part, functioning as the punch during the thermal hydroforming process. The process starts with the cutting of the required shape and size of the consolidated flat blank panel. The required cut blank is heated to the processing temperature in an infrared oven. After heating, the blank is transferred above the diaphragm present above the hydraulic liquid in the bottom mold. Swiftly, the top mold with the punch is lowered to form the part. While the top mold is lowered over the blank, the hydraulic oil in the bottom mold is maintained under pressure. After the top mold touches the blank, it is further lowered into the bottom mold, generating the compressive force over the blank against the bottom mold's oil. This compressive force applied over the blank forms the part into the required shape. The compressive force exerted on the blank is due to hydrostatic pressure (Wakeman and Manson 2005) and incompressibility (Ray 2005) of the hydraulic fluid. The hydrostatic pressure applied ensures there is uniform pressure applied over the blank (Zampaloni, Pourboghrat, and Yu 2004). This overcomes the requirement for fabricating a closely matched mold, which is expensive and complicated (Zampaloni, Pourboghrat, and Yu 2004). The typical setup used for thermal hydroforming is shown in Figure 10.7. The top mold with the rigid shaped punch and hydraulic fluid in the other mold is electrically heated to the processing temperature.

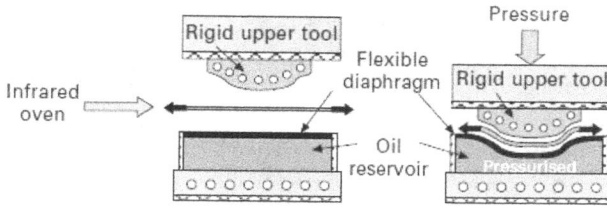

FIGURE 10.7 Thermal hydroforming of thermoplastic composites. (Reproduced with permission from Wakeman and Manson 2005.)

After the part is formed, the mold and hydraulic fluid are cooled to allow cooling of the part before removal from the mold. The mold is again heated to prepare for the next cycle. The method is suited for producing deep drawn parts like helmets and complex 3D shapes (Zampaloni, Pourboghrat, and Yu 2004). However, the process is limited to low-temperature thermoplastic resins. Since the method uses a rubber/elastomeric diaphragm to separate the part from the hydraulic fluid, the elastomer is not stable at higher temperatures.

10.3.2.3 Elastomer Block Forming

The elastomer block or rubber block thermal forming process is similar to the stamp and hydroforming of thermoplastic composites as shown in Figure 10.8. In this method, the top mold half is replaced with an elastomeric rubber block to apply the pressure. The bottom mold consists of the tool shaped to the part required. The tool and the bottom mold are electrically heated. Over the heated tool, the heated blank paned is placed. After the blank is placed, the top half with the elastomeric block is lowered over the blank and tool to apply the consolidation pressure by compression. During the application of consolidation pressure, the rubber deforms over the blank in the shape of the tool to apply a uniform pressure over the blank. The use of elastomer block ensures constant pressure applied over the blank. However, the process is suited for only low temperature melting thermoplastics only. Since the elastomer degrades at high temperature and thermoplastic with a melting point below, the elastomer block's degradation temperature only can be processed.

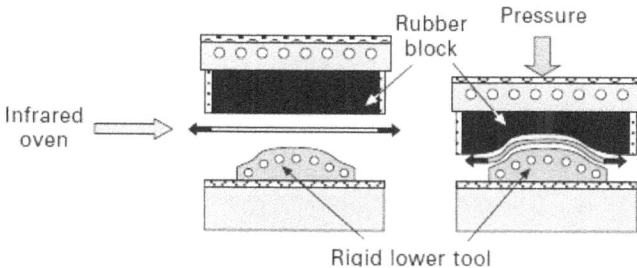

FIGURE 10.8 Elastomer block forming process. (Reproduced with permission from Wakeman and Manson 2005.)

10.3.2.4   Bladder/Diaphragm Forming

Diaphragm forming uses a diaphragm to apply the forming pressure over the heated blank during forming. During the diaphragm forming process, the blank sheet is placed between diaphragms and heated, followed by applying a positive pressure over the diaphragm to press and form the part. The positive pressure is applied using an autoclave or inside closed molds with a chamber to apply pressure using compressed gas. The diaphragm used shall be an elastomeric rubber (Friedrich, Hou, and Krebs 1997), or superplastic aluminum foil (Mallon, O'Brádaigh, and Pipes 1989; O'Brádaigh, Pipes, and Mallon 1991; Friedrich, Hou, and Krebs 1997), or polymeric films of polyimide and Kapton/Upilex (Mallon, O'Brádaigh, and Pipes 1989; Friedrich, Hou, and Krebs 1997; Wakeman and Manson 2005). The diaphragm used in the process needs to be flexible allowing for the diagram to take the shape of the required part. The diaphragm should not tear during the process when pressure is applied and should be mechanically stable at the processing temperature of the thermoplastic in the composite. The part can be performed in two different ways of diaphragm forming. First shall be the single diaphragm forming. In this type, the blank is placed above the part shaped tool over which the blank will be formed. The diaphragm is placed above the blank, and the tool and clamped secured. There is a line for vacuum placed below the diaphragm which is used to evacuate and create a negative pressure below the diaphragm. A line is provided above the diaphragm to apply an appositive pressure by forcing compressed gas into the press. The combination of vacuum and positive pressure applied crates a hydrostatic pressure above the part that helps form the part. The tool is heated to the processing temperature using electric heaters placed inside the tool. The typical setup of the single diaphragm process is shown in Figure 10.9. The other type of diaphragm forming is double diaphragm forming. Here two layers of the diaphragm are used in the process, one above and the other below the blank. The two diaphragms are taped in between a sealant tape or a vacuum ring, and a vacuum line is placed through to evacuate air between the diaphragms. The vacuum created between the diaphragms removes the entrapped air between the layers in the blank to help with uniform consolidation and prevents voids. A positive pressure is applied above the top diaphragm, this generates a hydrostatic pressure in relation to the bottom diaphragm, which is at

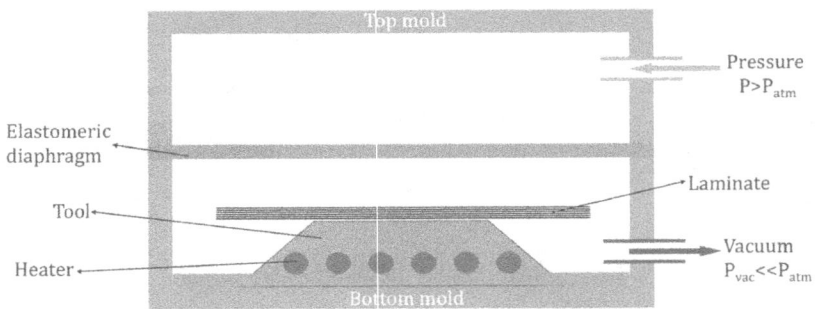

FIGURE 10.9   Single diaphragm forming method.

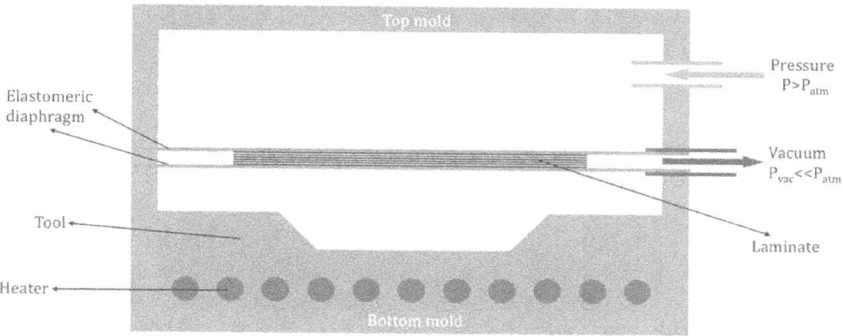

FIGURE 10.10 Typical setup for double diaphragm forming.

atmospheric pressure. The positive pressure applied forms the blank to the required shape against the heated tool placed below the bottom diaphragm and ensures consolidation. This pressure can be applied by placing the entire setup inside an autoclave or using a matched mold with a clamping plate similar to a single diaphragm forming. The typical setup of double diaphragm forming is shown in Figure 10.10. The tool below the blank and the diaphragm is electrically heated to the processing temperature of the thermoplastic. In double diaphragm forming, the entire setup can also be placed inside an autoclave and heater with the application of consolidation pressure. The diaphragm forming process is advantageous over rigid matched mold forming for producing composites from flexible preforms, since the diaphragm used stretches and deforms to overcome local stress and ensures uniform pressure over the flexible prepreg stacks (Mallon, O'Brádaigh, and Pipes 1989; Krebs, Friedrich, and Bhattacharyya 1998).

10.3.2.5 Roll Forming Process

The roll forming process starts with a constant cross-section consolidated planar sheet of continuous or semi-continuous (of a certain length) length. The sheets are deformed progressively by passing through a series of rollers with a matched profile cut on them. The process is more similar to that of roll forming of metal sheets. The process achieves simple open sections like c-section and hat profiles and complex 3D closed profiles. The closed sections are similar to that of tubes, square profiles, and rectangular profiles, which are complex and cannot be formed by other thermal forming methods (stamp/press forming). The required profiles are formed from a flat sheet by gradual and progressive deformation of the flat sheet when passed through several matched sets of rollers (Dykes, Mander, and Bhattacharyya 2000; Henninger and Friedrich 2004). An example of a series of matched rollers used is shown in Figure 10.11. During the process, the flat sheet undergoes transverse bending when passed through the rollers to form the part. Besides, undesirable longitudinal strains are also involved, limiting the degree of forming per stage of rollers (Henninger and Friedrich 2004). Thus, the forming is achieved in stages. The forming of the part is made possible by weave stretching and shearing at the inter-ply and intra-ply levels

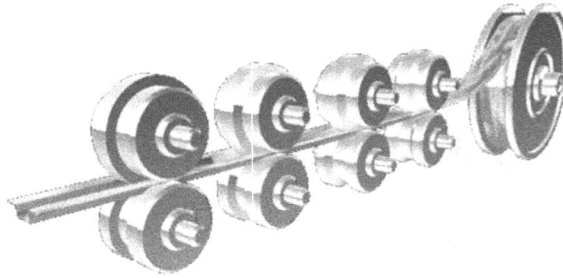

FIGURE 10.11 An example of rollers used in the roll forming of thermoplastic composites. (Reproduced with permission from Henninger and Friedrich 2004.)

(Dykes and Logue 2002; Henninger and Friedrich 2002; 2004). The formed parts produced are always symmetric over the central axis and are shown in Figure 10.12. The figure also explains the roll scheme to show the steps in forming the required shape.

The roll forming process is most suited for processing thermoplastic composites. The process requires a wide processing temperature range over which the flat blank can bend and form (Henninger and Friedrich 2004). The thermoplastic flat blanks used for roll forming are fully consolidated sheets of discontinuous lengths. The blanks are dried before processing to remove any absorbed moisture. The blanks are heated to the processing temperature above the melting point, rapidly using an infrared heating oven placed before the roll forming rollers. After heating, the blank is slightly cooled to the initial inlet temperature. Forming is achieved by a series of rollers depending on the shape that is to be produced. The forming rollers take up some heat during the process, which is prevented by coating them with Teflon sheets. After the shaping is completed, the shape is set by spraying water to cool the profile at the final step. The setting is achieved when the part is cooled below the crystallization temperature of the thermoplastic polymer in the composite.

Roll Schedule B

Roll schedule designed for CFRT's

Roll Station	α	β
1	30	30
2	60	60
3	80	80
4 5 6	90	90

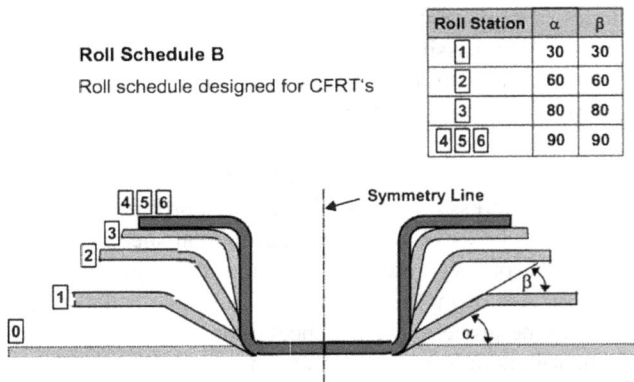

FIGURE 10.12 Roll scheme for forming a top hat profile part. (Reproduced with permission from Henninger and Friedrich 2004.)

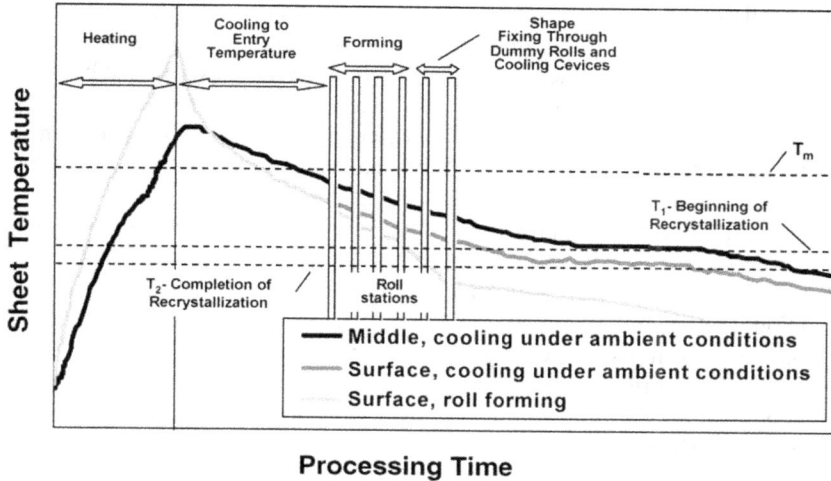

Processing Time

FIGURE 10.13 Temperature profile for roll forming process. (Reproduced with permission from Henninger and Friedrich 2004.)

Figure 10.13 shows the typical temperature profile during the roll forming of the carbon fiber reinforced sheet. The movement of the bank through the rollers is made by the pulling force applied on the blank by the drive of the roller. The speed of the rollers decides the pulling speed. The rollers are driven by a motor and chain drive. The process parameter of the initial inlet temperature and pulling speed affects composite quality (Henninger and Friedrich 2004). The process parameters control voids, delamination, and wrinkles that may affect composite strength.

10.3.3 Effect of Process Parameters

The process parameters namely, temperature, dwell time, holding pressure, stamping pressure, stamping velocity, and preheat time, influence the composite quality. The holing pressure is the consolidation pressure applied over the flat blank after stamping to ensure successful part formation. The stamping pressure is the pressure applied during the initial stamping to form the part. The stamping velocity is the speed at which the molds are closed during the forming process. Preheat time is the time during which the flat blank is heated to the processing temperature. The individual parameters effects are explained below:

1. Temperature: The higher the temperature, the lower the melt viscosity of thermoplastic and easier forming. It is supported by Zheng et al., where findings reported that for a CF/PEEK composite the flexural strength increased with increases in process temperature from 360°C to 400°C (Zheng et al. 2019). When it was increased further to 420°C, the strength dropped by 15MPa due to matrix degradation observed as dark coloration on the samples (Zheng et al. 2019).

2. Dwell time: With the increase in dwell time, the void content decreased from 2.46% to 0.2% when the time was increased from 5 minutes to 20 minutes for a CF/PEEK laminate produced by thermal forming as reported by Jiang et al. (2019). The void size reduced with an increase in dwell time. With the increase in time, the edges were void-free, and minute voids were seen in the laminate center (Jiang et al. 2019). The void content affects consolidation and in turn, the strength of the composites.

3. Holding pressure: The pressure applied after forming cannot be high as it influences the thickness of the part, shape, and consolidation. Excess pressure created wrinkles, shrinkage, and reduction in thickness of the part (Friedrich, Hou, and Krebs 1997). At lower pressure, the thickness of the part will be higher, affecting consolidation quality and geometry of the formed part (Friedrich, Hou, and Krebs 1997; Hou 1997). Thus, the holding pressure should be at the optimum level and is a critical process parameter for successful part forming.

4. Stamping pressure: The process parameter of stamp pressure and steaming velocity has a combined effect. However, the stamping pressure affects the melt flow of the matrix in the laminate and the bend forming of the part inside the mold (Hou, Friedrich, and Scherer 1994). The stamping pressure required for part forming is decided by the fiber volume fraction in the laminate, laminate thickness, and bend angle (Hou, Friedrich, and Scherer 1994). Stamping pressure that is too high leads to matrix migration at the bends with reduction of thickness at the bends, and low pressure leads to breakage or buckling of the fibers at the bends (Hou, Friedrich, and Scherer 1994).

5. Stamping velocity: Lower stamping velocity will lead to a lower temperature at the laminate. The lower temperature leads to higher viscosity of the matrix and leads to a situation where inter-ply slip is not allowed due to buckling and fiber breakage (Hou, Friedrich, and Scherer 1994).

6. Preheat duration: The time required to heat the blank to the processing temperature plays a vital role. The heating is by contact heating or by non-contact heating by an infrared oven or air ovens. The laminate layers at the outside reach the processing temperature fast, and the middle layer takes a longer time to reach the processing temperature (Friedrich, Hou, and Krebs 1997). Due to this, a thicker blank requires more preheating duration. When the temperature is not ideal, then the part will not be formed. Thus, preheat duration should be determined with care, keeping in mind the blank's thickness (Friedrich, Hou, and Krebs 1997).

10.3.4 ADVANTAGES

1. The part can be produced with reasonably complex 2D and 3D shapes.
2. The process produces high strength composites.
3. Small and large parts can be produced.
4. High fiber volume fraction in the composite is possible.
5. The process is quick and the consolidation pressure is applied only for short duration.
6. Large volume production cycles are possible.

10.3.5 LIMITATIONS

1. Thick parts cannot be produced.
2. The size of the press limits the process and large parts are not possible.
3. Achieving distortion and warpage free part is tricky.
4. Not suited for low volume production runs since the tooling cost involved is high.

10.4 AUTOCLAVE MOLDING

The autoclave molding process of consolidation of thermoplastic composites is similar to that of the diaphragm forming process, more like the single diaphragm forming. However, in autoclave molding, unconsolidated prepreg sheets are stacked as laminate and consolidated, whereas in the diaphragm forming, the part is formed from a consolidated blank. During autoclave molding, the required heat and pressure for consolidation are applied by the autoclave. The process is an open molding technique where the laminate is stacked on the mold and vacuum bagged. The mold with the laminate and vacuum bag is transferred into the autoclave for consolidation. The autoclave follows a user-defined preset consolidation cycle to produce thermoplastic composites. The laminate is made from flexible prepreg sheets, referred to as semipreg sheets (Fernández, Blas, and Frövel 2003). The flexible prepreg sheets are woven textile preforms available as a 3D and 2D woven fabric preform sheet or unidirectional sheet. The autoclave molding method is more suited to producing laminates of complex shapes that cannot be produced by compression molding and thermal forming methods. The process is repeatable and produces good quality composites.

10.4.1 MACHINE AND PROCESS DESCRIPTION

The autoclave is a pressure vessel that is used to apply uniform heat and pressure. The pressure vessel is a cylindrical tube made of strong metal capable of withstanding high temperature and pressure. The autoclave consists of a door opening at one end through which the molds are placed inside and removed. A typical diagram of the autoclave used for producing thermoplastic composites is shown in Figure 10.14. The autoclave consists of two sections, namely the central working section and an outer ducting section. The central section is where the mold with the laminate stack is placed for consolidation (Monaghan, Brogan, and Oosthuizen 1991). The outer ducting section consists of band heaters used for heating the central working area by heating the air inside (Monaghan, Brogan, and Oosthuizen 1991). The heaters used are generally electric resistance heaters (Monaghan, Brogan, and Oosthuizen 1991). The air is circulated by a fan revolving inside the inner section of the autoclave. The fan moves air from the central section into the outer ducting section, where the band heaters heat the air, and hot air is circulated back into the central section. Thermocouples monitor the air temperature inside the central section, and another measures the temperature at the autoclave's metal wall. Additional thermocouples can be placed on the molds at a different location inside to monitor the real-time temperature. The walls of the autoclave pressure vessel are insulated to prevent the escape of heat

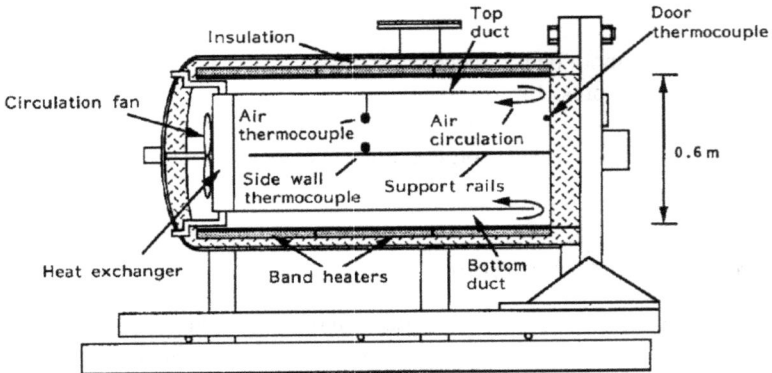

FIGURE 10.14 A diagram of autoclave used for producing thermoplastic composites. (Reproduced with permission from Monaghan, Brogan, and Oosthuizen 1991.)

to the atmosphere. The autoclave heating is controlled by a computer system that monitors temperature and runs a PID control to modulate the heating (Monaghan and Mallon 1990). The computer program ensures uniform heating inside the autoclave. The autoclave heating process is started after placing the laminate stack vacuum bagged on a mold inside the central section. The autoclave takes time to heat up to the set temperature. The computer program controls the heating and cooling rate set by the user (Monaghan and Mallon 1990). After the autoclave central section has achieved the set temperature, the pressure is applied. The pressure is built by pumping compressed nitrogen gas into the autoclave (Monaghan and Mallon 1990). Control values open and close the compressed nitrogen supply to build and regulate the set value pressure. The control values used are solenoid valves that are controlled by the computer (Monaghan and Mallon 1990). There are safety valves present on the autoclave to prevent over- pressurization and prevent catastrophic events during the consolidation cycle (Monaghan and Mallon 1990). After consolidation, the chamber is depressurized by slowly venting the chamber by opening the let-off valve. The rate of depressurization is controlled. Along with depressurization, the chamber is also cooled in a rate-controlled manner, all controlled by the computer. A chiller unit cools the hot gas that comes out of the autoclave before releasing it into the atmosphere. During the entire cycle of pressurization and depressurization, the autoclave door is closed airtight with a gasket that forms a seal with the pressure chamber.

The process of producing thermoplastic composites by autoclave molding starts with the prepreg sheets. The prepreg is woven flexible preforms available as a continuous sheet in the form of a roll. The first step in the process is cutting the prepreg sheet to the required shape to produce the laminate stack. For complicated shaped composite parts, the cut prepreg sheets required are also in complex shapes and thus, they are mostly cut using an automated preform cutting machine (Fernández, Blas, and Frövel 2003). After cutting the prepreg, the second step is the preparation of mold. The mold is coated with a mold release agent to help remove the consolidated composite after processing. The third step is laying the laminate stack over the mold. The cut preforms are stacked on the metal mold, and spot welds are created locally to hold the laminate stack together (Fernández, Blas, and Frövel 2003). The spot welds were created locally by melting the thermoplastic polymer in the prepreg sheet. The spot welds prevent dislocation of the laminate and also prevent wrinkles in the prepreg during stacking. The fourth step is the vacuum bagging of the laminate placed over the mold. Vacuum bagging is performed to build a vacuum over the laminate stack. This vacuum when applied removes any air voids present between the laminates to help with consolidation. The film used for vacuum bagging ranges from polyimide film and superplastic aluminum film to Milar film. The vacuum bagging film is selected based on the processing temperature required for the thermoplastic polymer. For PEEK polyimide film or superplastic, aluminum films are used (Lystrap and Andersen 1998). The vacuum bag is sealed airtight with a lining of the sealant between the mold and bagging film. High-temperature sealant tape was used for thermoplastic composite manufacturing since the processing temperature required is high for thermoplastics. After the bagging is complete, the bag is connected to a vacuum pump to evacuate the air void between the layers of the laminates. After the vacuum is built, the entire setup over the metal mold with the laminate and the vacuum bag is transferred into the autoclave for heating under pressure for consolidation. The consolidation cycle followed is typically the same as the consolidation cycle followed during compression molding, and this is decided based on the thermoplastic used. After the consolidation cycle is complete, the autoclave is opened after cooling, and the consolidated composite is removed from the mold.

10.4.2 Advantages

1. The process can achieve uniform consolidation over the entire part.
2. The process is suited for producing high-temperature thermoplastic composite parts.
3. The process is suited for producing complex 2D and 3D shaped laminate parts.
4. The process is suitable for producing large laminate parts.
5. The method is also suited for co-consolidating the composite to produce complex shapes and profiles like a double hat structure and elliptical profiles used in aircraft.

10.4.3 Limitations

1. The purchase of an autoclave requires for a large capital investment upfront.
2. The consolidation pressure that the autoclave can apply is limited by the design of the autoclave. Large pressure as possible in compression molding is not possible in autoclave molding
3. The process requires comprehensive labor-intensive prep work for stacking and vacuum bagging of the laminate. It leads to higher processing costs.
4. It is an open molding process, and thus adequate care is required during laminate stacking to prevent wrinkles in the stack of semipregs. It is because the thermoplastic prepreg sheet is not tacky.
5. The thermoplastic prepreg preforms lack tack, which leads to a need to spot weld the laminates to hold it in place.

10.5 FILAMENT WINDING

The filament winding process is one of the automated composite production methods developed during the late 80s to produce thermoset and thermoplastic composites rapidly. The method reduced production costs by reducing the labor cost drastically (Romagna, Ziegmann, and Flemming 1995). The process was automated using computer numeric control (CNC) to allow for large volume production runs (Nunes et al. 2005). The process suits the production of small and large parts at a low cost. This section shall focus only on the production of thermoplastic composites from flexible preforms by the filament winding process. In thermoplastic composite filament winding, predominantly flexible preforms in the form of towpregs are used, woven pre-impregnated preforms are not preferred (Romagna, Ziegmann, and Flemming 1995). The flexible preforms shall be a pre-impregnated preform like hybrid yarn, powder impregnated towpreg, and melt impregnated towpreg. However, apart from the pre-impregnated flexible preforms, there is a possibility to use an online impregnation setup with filament winding for manufacturing thermoplastic composites (Henninger and Friedrich 2002; Henninger, Hoffmann, and Friedrich 2002; McGregor et al. 2017). During the online impregnation with filament winding, the continuous fiber roving/tow of the reinforcement fibers are passed through an impregnation unit where the thermoplastic is coated in the melt form on roving/tow (Henninger, Hoffmann, and Friedrich 2002) or in the powder form (McGregor et al. 2017). During the melt form online impregnation, the tow is pulled through an impregnation die fed continuously with melt polymer from an extruder where the molten polymer coats the tow. During powder form online impregnation, the tow is coated with the powder by dispersion coating when a tow is pulled through a resin impregnation bath (McGregor et al. 2017) followed by heating to fuse the polymer with tow before further processing.

As the name suggests, the filament winding process involves the winding of the flexible thermoplastic towpregs over a rotation mandrel as layers to form a rigid consolidated composite part. The mandrel shape shall be symmetric or asymmetric over the central axis of the mandrel. The geometry of the mandrel decides the shape of the parts produced by filament winding. The shapes produced range from cylindrical, spherical, elliptical, square, rectangular, and triangular, to complex three-dimensional shapes that are a closed shape, and all shapes limited to a convex profile. During filament winding of thermoplastic composites, the pre-impregnated continuous towpreg is heated in a continuous process and consolidated over a rotating mandrel by on-line consolidation/in-situ consolidation (Werdermann et al. 1989; Wong, Blanco, and Ermanni 2018). The on-line consolidation involves heating of the flexible preforms rapidly above the thermoplastic melting point and is followed by application of the consolidation pressure for a short duration of time to achieve consolidation swiftly (Wagner and Colton 1994). The heating of the towpreg is carried using an infra-red heater, laser source, hot air guns, tubular oven, contact heater, induction heater, microwave energy, and ultrasound energy (Wagner and Colton 1994; Funck and Neitzel 1995; Romagna, Ziegmann, and Flemming 1995; Lionetto et al. 2016; Wong, Blanco, and Ermanni 2018). The next sub-sections shall explain the working of filament winding machine for producing thermoplastic fiber reinforced composite parts.

10.5.1 Machine and Process Description

The thermoplastic filament winding process starts with the pre-impregnated towpreg or an uncoated tow/roving in case of online consolidation. The towpreg is wound on a spool and loaded to the filament winding machine at the let-off or on a creel stand setup. The creel is used when more than one tow is used to achieve a wider tow deliver at the mandrel. It is achieved by placing individual tows close to each other to form a wider tape. The let-off on the filament winder and the creel stand has tension measurement units for each tow to measure the tow tension during the process. The tension is measured with the help of spool mounted load cell or a roller mounted load cell that the tow passes through before moving to the next station. The tension on the tow is created by the resistance force generated by the brake mounted on the spool mounts on the let-off or the creel stand. Each spool has individual brake. The tow tension is maintained at a constant value by forming a closed-loop control system with the load cell measuring tension and the spool mounted break. The load cell, break, and spool mount/creel stand form the let-off system of the filament winding machine. From here, the tow is passed to the preheat zone for further process.

The preheat zone of the machine consists of the oven used to heat the towpreg to the desired preheat temperature. Towpreg preheating is critical for right consolidation in filament winding. Since in filament winding, the consolidation happens in a short time by in-situ consolidation, preheating is critical for consolidation and achieving good composites. The preheating oven is generally a tubular or box oven with ceramic, infrared, or resistance heaters. The heaters inside the oven provide the heating, and the tow is heated to the required preheat temperature by non-contact heating methods. Generally, infrared radiation heaters are used. After the preheating oven, a non-contact temperature measurement device is used to monitor the preheat on the towpreg. The temperature measurement is made possible by an infrared camera or infrared sensor arrays. It forms a closed-loop control with the preheating oven to achieve the required heating. The temperature control unit controls the heater power according to the towpreg temperature recorded via the infrared sensor and adjusts the heater power in the oven. After the towpreg is preheated, it is sent to the nip point heating section known as the winder head and consolidation head. The consolidation head guides the tow to be wound on the mandrel to form the part. A typical thermoplastic filament winding process flow is shown in Figure 10.15.

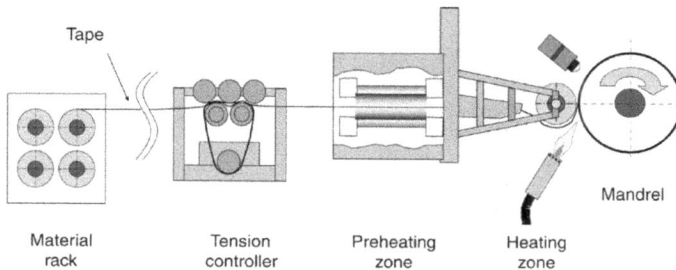

FIGURE 10.15 A schematic of the thermoplastic filament winding process. (Reproduced with permission from Mack and Schledjewski 2012.)

The consolidation head consists of a heater, consolidation roller, pressure application device, tow guide, and eyeball delivery head. In the consolidation head, the heater heats the towpreg to a temperature above the melting point of the thermoplastic polymer. The heating helps with impregnation and consolidation. The heating is made possible by air blowers, gas torch, laser source, infrared heater, induction heating, microwave energy, and ultrasound energy (Wagner and Colton 1994; Funck and Neitzel 1995; Romagna, Ziegmann, and Flemming 1995; Wong, Blanco, and Ermanni 2018). Here, a rapid heating method like laser heating, air blower, gas touch, or ultrasound energy is preferred. The towpreg must be heated above the melting point of the thermoplastic from the preheat temperature in a short duration. After heating, the tow is passed to the consolidation roller via the eyeball delivery head.

The eyeball delivery head ensures delivery of the towpreg without any twist to the consolidation roller over the mandrel. It also helps in guiding the yarn to achieve unique winding patterns on the mandrel. From the eyeball guide, the towpreg reaches the consolidation roller held against the mandrel. The consolidation roller applies the required pressure by pressing against the mandrel. The towpreg is pressed between the consolidation roller and mandrel to achieve consolidation. The pressure applied by the consolidation roller is controlled by an air cylinder to which the consolidation roller is attached via an arm setup. By controlling the air pressure applied to the air cylinder, the pressure exerted by the consolidation roller on the towpreg is controlled. Consolidation pressure at the required value is maintained throughout the process. The mentioned process steps achieve on-line consolidation, and a diagram depicting the consolidation head is shown in Figure 10.16. The guide roller on the consolidation head helps to guide the towpreg without lateral or vertical shift during the continuous winding process. The winding of the towpreg on the mandrel after on-line consolidation produces the composite over the mandrel. The mandrel controls the shape and dimension of the part produced.

The mandrel used for the thermoplastic filament winding process is mostly a metal mandrel made of steel or aluminum, and often it is heated (Munro 1988; Lauke and Friedrich 1993). In a few setups, the consolidation rollers shall also be heated based on the requirement. The mandrel and roller heating is made by resistive heaters or by the circulation of heated oil via channels in them (McGregor et al. 2017). Over the

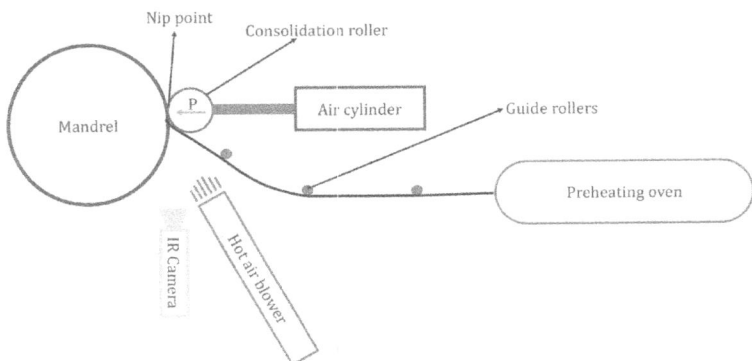

FIGURE 10.16   Schematic of an online consolidation head.

metal mandrel, the towpreg is wound; after the part is fully consolidated, the mandrel with the part is removed from the filament winder and transferred to a hydraulic puller. With the help of a hydraulic puller, the composite part is removed from the mandrel. The mandrel is coated again with a release agent to prepare for the next cycle. The metal mandrel helps with the production of open-ended cylindrical parts only. Closed-end cylindrical parts are produced using a dissolvable mandrel made of polyvinyl alcohol (PVA) or sand with a dissolvable binder and soluble plasters (Munro 1988). These mandrels can be dissolved in water after the part is produced to have a hollow closed-end part finally. The mandrel is fixed on a shaft or a rotating chuck mount that rotates over its axis and does not move horizontally or vertically. The mandrel is always fixed horizontally in all the machines (Werdermann et al. 1989), since the mandrel is easy to remove from the machine mounts in a horizontal position. Also, removing the composite from the mandrel is easier horizontally. The eyeball guide and consolidation unit that deliver towpreg over the mandrel need not travel against gravity when it is horizontal traversal (Wagner and Colton 1994). To achieve unique winding patterns on the mandrel, the eyeball delivery head mounted on the on-line consolidation unit moves along the horizontal over the width of the mandrel from left to right. The eyeball moves along with the entire carriage left and right, along with the preheating oven and on-line consolidation unit mounted over the carriage.

The degrees of freedom of movement possible in a conventional filament winding system is a maximum of six. The mandrel rotation around the shaft or central axis is the first axis referred to as the X-axis. The rotation of the mandrel can be clockwise or anti-clockwise. The second axis is the eyeball guide's movement along the horizontal parallel to the mandrel and is referred to as the Y-axis. The most straightforward filament winding system has a minimum of two axes, the X- and Y-axis. The two-axis filament wider can achieve cylindrical parts with open ends. The third axis is the distance between the eyeball guide and the rotating mandrel, referred to as the Z-axis. The fourth axis is the rotation of the eyeball guide around the horizontal; it is referred to as the U-axis. The fifth axis is the rotation of the eyeball guide around the vertical; it is referred to as the V-axis. The sixth axis is the height at which the eyeball guide is placed. The change in height of the eyeball guide is referred to as the W-axis. The possible axis of movement in filament winding is shown in Figure 10.17. The six-axis filament winder setup can produce closed-end cylindrical parts by following a polar winding pattern. A six-axis system is required for all complex shapes

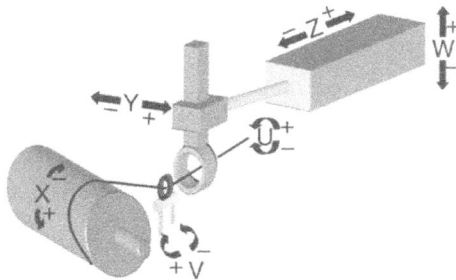

FIGURE 10.17   The six axes of the filament winding system. (Reproduced with permission from Mack and Schledjewski 2012.)

like the elliptical profile, T joints, elbows, and several others. For all the parts produced by filament winding, computer aided design (CAD)-based visualization and winding pattern generation are followed to achieve the optimum part strength and geometry (Munro 1988; Johansen, Lystrup, and Jensen 1998).

Apart from the above-mentioned type of filament winding system, a more advanced one is the use of a six-degrees-of-freedom robotic arm to deliver the towpreg at the mandrel. It offers more control over the eyeball delivery head, which helps achieve complex shapes like T-joints, elbows, and free form shapes. The whole winding process can be controlled by codes based on CNC control (Scholliers and Van Brussel 1994). The whole process is automated with a robotic winding system. The robot performs the winding and there is a separated rotating mandrel that is working in unison with the robot to achieve the winding patterns. There is no need for manual labor, and the only process requiring human intervention is the mandrel change and loading towpreg spools. Even the mandrel change can be automated to remove the human requirement with the process (Munro 1988). However, even the robotic filament winding can only achieve closed shapes with a convex profile. Still, there was a need to develop a more advanced system suited for producing open geometry parts and producing parts with concave profiles. This need has led to developing an automated tape/fiber placement method of manufacturing composites using towpreg.

10.5.2 WINDING PATTERNS

The filament winding process is carried out by following a few distinct winding patterns suited for producing different parts. The simplest and the most straightforward winding pattern is the hoop winding or the circumferential winding pattern. Here the towpreg is wound at an angle of roughly 90° with the mandrel. With each rotation of the mandrel, the towpreg is wound close to the previous winding without a gap to cover the mandrel area completely. It is similar to the unidirectional laminate structures for a filament winding process. The second is the helical winding pattern. In the helical winding pattern, the winding forms the required helical winding angle by the carriage movement and the angle shall range roughly about 20°–80°. The third is the polar winding pattern. In polar winding, the towpreg is wound from the polar end (i.e., from end to end) tangentially to form an end closed part. The polar winding is more suited for spherical and cylindrical parts with ends closed. The diagrammatic representation of the different winding patterns is shown in Figure 10.18.

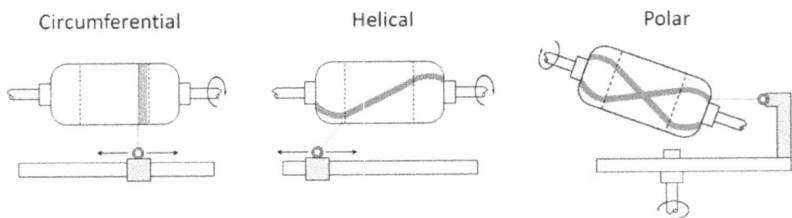

FIGURE 10.18 Different types of winding patterns possible in the filament winding process. (Reproduced with permission from Henriquez and Mertiny 2017.)

10.5.3 Advantages

1. The process is automated and does not require high labor costs.
2. Suited for large volume production runs.
3. Large scale parts also can be produced.
4. High production rates achievable.
5. Low-cost process and completely computer controlled.
6. Low-cost raw materials in the form of towpregs and rovings can be used.

10.5.4 Limitations

1. Limited to convex shaped closed geometries only. No open geometry is possible.
2. Complex 3D shapes are not possible. Always the part geometry needs to be symmetric over the central axis of the mandrel.
3. Achieving uniform resin content and consolidation shall be a challenge.

10.6 AUTOMATIC TAPE PLACEMENT

The automated tape placement (ATP) is one of the automated composite production methods. The ATP process is capable of processing thermoplastics and thermoset composites. However, this section focuses on the production of thermoplastic composites. The method is an additive manufacturing method, where layers of unidirectional prepreg tape are laid down as layers to produce a laminate structure over the surface of the mold. The ATP process uses a robot or a gantry with a placement head that consists of an online consolidation unit to produce composites. The layup process in ATP is similar to hand layup, but it is automated and controlled by CNC control (Lukaszewicz, Ward, and Potter 2012). The tool path generated for the CNC control is performed by CAD-based path generation and simulation programs. ATP provides repeatability, removing human errors and reducing production costs by up to 40% by eliminating the high labor cost involved in manual layup (Lukaszewicz, Ward, and Potter 2012). Initially, the ATP process was limited to producing composites over a flat mold. However later, with introducing a multi-axis robot with an on-line consolidation head having a tape heating system and pressure rollers, it was possible to work with intricate shapes (Lukaszewicz, Ward, and Potter 2012). The robot allows six degrees of freedom and the placement head also allows some degrees of freedom (Van Hoa, Duc Hoang, and Simpson 2017) and as a combination helped to layup over complex shaped molds. During the on-line consolidation in ATP, the tape is heated above the thermoplastic melting point before laying and is consolidated by a roller applying pressure. The heating of the tape is made possible by a laser or hot air gun/blow torch, or ultrasound is used (Rizzolo and Walczyk 2016; Yassin and Hojjati 2018; Rodriguez-Lence, Martin, and Fernandez Horcajo 2019). Predominantly, the laser heating or blow torch is used since it offers effective localized heating of the tape (Yassin and Hojjati 2018). Similar to ATP, another automated process for manufacturing thermoplastics is automated fiber placement (AFP). The only difference between ATP and AFP is the width of the preform processed.

The AFP process lays down towpreg, which is smaller in width, at about 3 and 12 mm for individual towpregs (Lukaszewicz, Ward, and Potter 2012). The AFP always lays down several bands of towpreg simultaneously to form a wider band than laying a single towpreg. The machine can lay about 32 towpregs simultaneously parallel to each other to help with faster material layup (Lukaszewicz, Ward, and Potter 2012; Yassin and Hojjati 2018). Whereas ATP process can lay unidirectional tapes with a maximum width of up to 300 mm and generally smaller tape with width between 50 and 76 mm are used (Lukaszewicz, Ward, and Potter 2012). Due to the greater width of the tape laid than AFP, the ATP achieved greater material deposition rates. The method is suited for producing large parts used for aircraft and aerospace applications (Lukaszewicz, Ward, and Potter 2012; Yassin and Hojjati 2018). The ATP process promoted the use of composite parts in aircraft, and it is a widely used method for manufacturing fuselage, wings, tail, and several other parts (Yassin and Hojjati 2018). The ATP process overcame the filament winding process's limitation by laying tape on an open mold with a concave shape and non-symmetric shapes. The ATP process is a low-pressure consolidation process with applied pressure in the range of only 0.5 MPa for carbon PEEK composites (Stokes-Griffin and Compston 2015).

10.6.1 Machine and Process Description

The ATP process is performed using an on-line/in-situ consolidation unit/head mounted on a multi-axis robot or a gantry, as shown in Figure 10.19. The robot or gantry offers up to six axes of movement and the head mounting shall also allow an

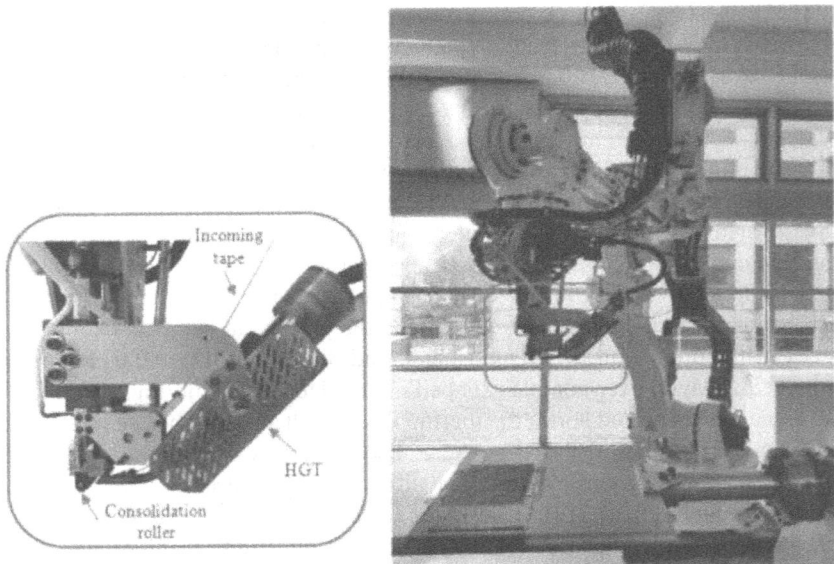

FIGURE 10.19 Typical ATP with on-line consolidation unit mounter on a robot. (Reproduced with permission from Oromiehie et al. 2019.)

additional degree of movement. The robot used in ATP is generally a KUKA industrial robot fixed on a movable platform (Denkena, Schmidt, and Weber 2016). The on-line consolidation head consists of a towpreg let-off/unwinder system, backing paper winder, cutting unit, heating system to heat the tape before laying, consolidation roller, guide roller, tension control, and IR camera to monitor the operation. To allow fast movement of the robot and achieve higher material delivery rates, the on-line consolidation head is designed as light as possible (Denkena, Schmidt, and Weber 2016). The unwinder system consists of the tape spool holder onto which the tape spool is loaded. The unwinder is driven by a servo motor that controls the unwinding speed to ensure constant let-off and maintain tension in the tape. The tape has a backing tape made of paper applied on one side of the prepreg tape, which needs to be removed before laying down the tape. Thus, the backing tape winder driven by a servo motor winds the backing paper to an empty spool separately. Some thermoplastic prepreg tape has a backing paper. However, predominantly thermoplastic prepreg tapes are not tacky and do not require a backing paper. The heating system is generally a laser beam or a hot air torch focused at the nip point to heat the tape above the melting point for the layup. Nip point is the point at which the tape contacts the mold in the molten state to be fused with and consolidated by a roller. The laser beam source can be a CO_2 laser, IR or near-infrared (NIR) laser, or ND: YAG lasers. The CO_2 laser was not preferred as the first choice for laser heating since it tends to burn away thermoplastics. Thus, IR and NIR lasers are preferred (Yassin and Hojjati 2018). When a hot air torch is used for heating, mostly nitrogen gas is used as the heating medium, since it prevents the oxidation of the thermoplastic (Qureshi et al. 2014). Laser heating provided localized heating and eliminated the need for convection heat transfer, and the heating rate was high (Yassin and Hojjati 2018). Thus, using the laser as the heat source helped achieve greater material delivery rates (Stokes-Griffin and Compston 2015; Yassin and Hojjati 2018). The consolidation roller applies vertical pressure over the freshly laid tape to consolidate it with the mold surface or weld and consolidate with the previously laid tape layer. Guide rollers help with the process by preventing tape from twisting and ensure complete flatness of tape. The tension control unit measures the tension of the tape and helps maintain uniform tension throughout the process by controlling the unwinder rate of the tape from the spool. The cutting unit consists of blades that help in cutting the tape to the required length after it has reached the end of layup. The IR camera monitors the temperature profile over the tape and at the nip point to achieve and maintain the required temperature.

The schematic diagram representing the process flow in ATP is shown in Figure 10.20. The first step starts with preparing the mold surface with a mold release agent and a release film layer. The release film is generally a polyimide film for thermoplastics (Venkatesan et al. 2020). The second step is the loading of the prepreg tape spool on the placement head. The prepreg tapes or towpregs are available as continuous length tapes with backing paper on one side wound over a spool made of polymer or cardboard core (Lukaszewicz, Ward, and Potter 2012). After loading of the spool, the tape is fed on the head via the guide rollers till the nip point (delivery) below the consolidation roller in the head. It marks the readiness of the ATP to start laying down tape over the mold. The third step is the tool path generation. The tool

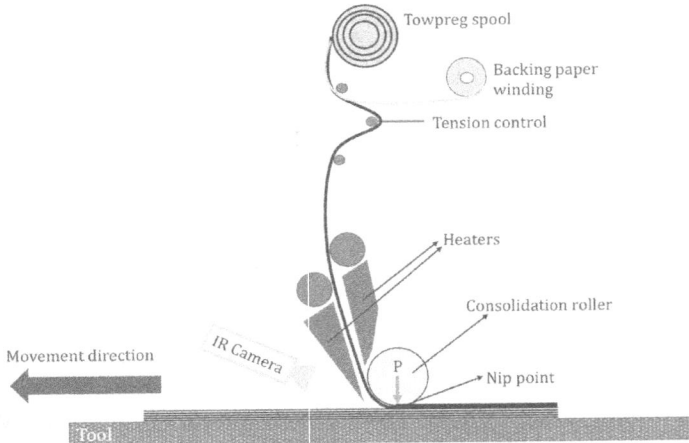

FIGURE 10.20 Schematic showing the process flow in the ATP process.

paths are based on G codes used for CNC control and software generates the code based on CAD drawings. When the CNC control code is available, the ATP process can be started. The fourth step is the start of the layup process, where the prepreg tape is fed to the nip point by the feeding mechanism. The nip point is the point below the consolidation roller where the tape touches the mold or the previous laid layer, where the localized heating of the tape is concentrated, creating a small melt pool. The fifth step is the layup of the tape at the nip point. The concentrated heating created a small melt pool, make the tape pliable and weldable. Due to this melt pool formation and with the application of pressure by the consolidation roller at the nip point, the tape bonds with the mold or the previous laid layer. The compression pressure applied by the consolidation roller compresses the tape to consolidate forming composites. The compression pressure applied by the compaction roller is measure and controlled accordingly to maintain a set value. A load cell connected to the roller shall measure the pressure applied. The sixth step is the movement of the head (i.e., the consolidation roller and heat source) past the tape segment (i.e., nip point) under consideration. The nip point and the variables involved in ATP at the nip point are shown in Figure 10.21. The lower ambient temperature creates a temperature gradient leading to heat transfer from the laid tape to the atmosphere to cool the tape. The cooling of the tape recrystallizes the thermoplastic polymer to a solid state. The removal of the compression pressure and cooling of the tape relieves stress from the tape to form a flat composite. However, mismatch in thermal expansion coefficient, temperature gradient, crystallization, and shrinkage lead to residual stress and distorts the part (Van Hoa, Duc Hoang, and Simpson 2017). After completing a layer of the tape, it is cut by the cutting blade to allow movement of the head to the next point in the layup process. The fourth and fifth steps are repeated to achieve multiple layers of the tape over the previously laid layer to produce a laminate structure. The finished part does not require any additional consolidation step like autoclaving, and the process produces near-net shape parts.

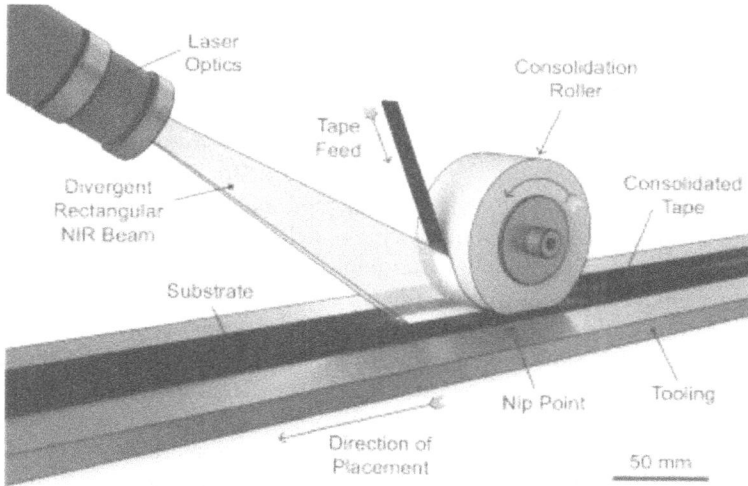

FIGURE 10.21 Schematic representation of the consolidation process of thermoplastics in the laser ATP process. (Reproduced with permission from Stokes-Griffin and Compston 2015.)

10.6.2 ADVANTAGES

1. Complex shapes can be produced.
2. The cost of production is low since labor cost is eliminated.
3. The production rates possible are high.
4. The tapes are consolidated in-situ and do not require autoclave consolidation post layup.

10.6.3 LIMITATIONS

1. The initial investment cost is exponentially high due to the cost of the robot or gantry and the on-line consolidation head cost.
2. The defects arising during the layup process affects the composite quality adversely.
3. The process is highly complicated and several process parameters such as heating intensity, heat transfer rate, lay-down speed, the thickness of laminate, and others affect composite quality. These parameters are not the same for all thermoplastic matrices and there is a need to perform optimization of process parameters for each thermoplastic matrix.
4. The surface finish achieved is not perfectly smooth.

10.7 PULTRUSION PROCESS

The pultrusion process, as the name suggests, involves pulling and extrusion processes combined to produce composites. The pultrusion process was initially used for producing thermoset composites for the ease of production, availability of resin in liquid state, and uniformity in impregnation due to easy wetting of the tows (Ramani, Borgaonkar, and Hoyle 1995; Michaeli and Jürss 1996). However, thermoplastic composites gained interest for the weldability, strength, and environmental friendliness. Therefore, the thermoplastic pultrusion process was developed to produce thermoplastic composite rods, sections, and profiles (Ramani, Borgaonkar, and Hoyle 1995; Michaeli and Jürss 1996; Novo et al. 2016). Thermoplastic pultrusion can be performed in several ways. The first method shall use dry unimpregnated roving as the starting material. The dry roving is pulled through an impregnation and pultrusion die supplied with molten thermoplastic from an extruder, where the dry tow is impregnated and consolidated when pulled through the pultrusion die (Novo et al. 2016). The production of pultruded composites from dry rovings and on-line impregnation using a die supplied with molten polymer is not preferred, since the high melt viscosity of the thermoplastic creates a need to pull the dry tow with high pulling force through the die. The high pulling force applied during pulling shall lead to fiber damage and composite breakage (Lapointe and Laberge Lebel 2019). The second type uses pre-impregnated towpregs produced by commingling, braiding, wrapped yarn, and powder coating to produce composites by pultrusion (Novo et al. 2016). Here, the towpreg is heated above the melting point and pulled through a pultrusion die to produce the composite. The third method is called pull braiding or braidtrusion; this is a combination of pultrusion and on-line braiding (Michaeli and Jürss 1996; Bechtold, Wiedmer, and Friedrich 2002). Pull braiding is more suited for producing hollow cylindrical parts like tubes by pultrusion (Bechtold et al. 1999). The most popular and the easiest is the production of pultruded composite using pre-impregnated towpregs (Novo et al. 2016).

10.7.1 Machine and Process Description

The pultrusion setup consists of a creel, guiding system, pre-heating oven, pultrusion die (pressurizing and consolidation die), cooling die, a pulling system, and a cutter. The schematic representation of the thermoplastic pultrusion process is shown in Figure 10.22. The creel holds several spools of towpreg and lets it unwind in a controlled manner when towpreg is pulled. The guiding system consists of a specific path for individual towpregs to pass through and ensure flatness, preventing entanglement

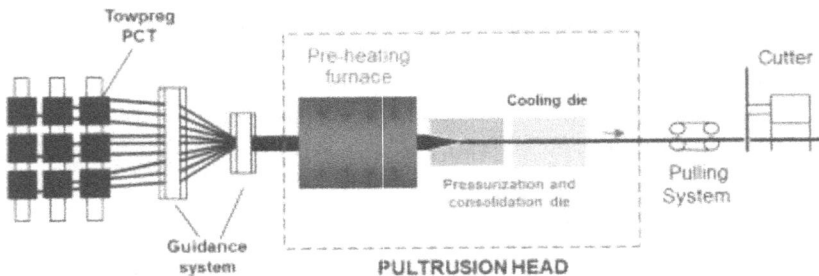

FIGURE 10.22 Schematic diagram of the thermoplastic pultrusion process. (Reproduced with permission from Novo et al. 2016.)

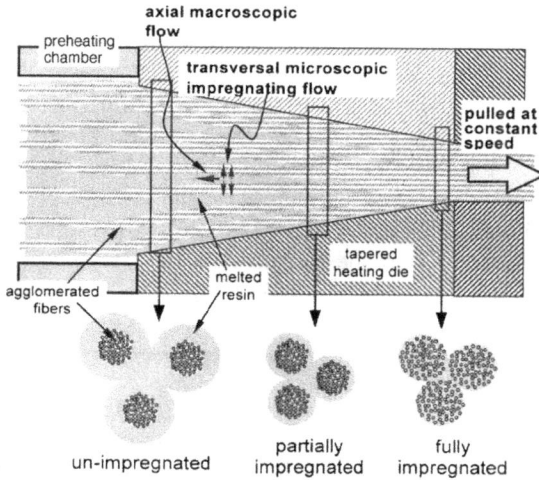

FIGURE 10.23   Impregnation and consolidation phenomena in a tapered die. (Reproduced with permission from Kim, Lee, and Friedrich 2001.)

and twisting of towpreg. The pre-heating oven consists of an IR radiation oven or a convection oven heated by resistive heaters or ceramic heaters to heat the towpreg to the required pre-heat temperature. For thermoplastics, pre-heating is critical to achieve a good quality composite. The pre-heating removes the adsorbed water from the surface of the reinforcement fiber and also helps achieve the required temperature above the melting point in the pultrusion die (Van De Velde and Kiekens 2001). Achieving the processing temperature in the pultrusion die alone without pre-heating will lead to extensive heating of the towpreg, possibly leading to degradation of the thermoplastic polymer. The pultrusion die consists of a tapered profile that is heated and used to apply the required consolidation pressure when the towpreg is pulled through it (Kim, Lee, and Friedrich 2001). The typical profile of a pultrusion die and the microscopic phenomenon in the impregnation and consolidation mechanism is explained in Figure 10.23.

The process starts with the loading of the towpreg spools on the creel. The creel consists of a tension control system maintaining constant tension during operation. From the creel, the towpreg passes through the guiding system before entering the pre-heating oven. The guiding system ensures that there is no twist or entanglement in the towpreg delivered. Through the guiding system, the towpreg enters the pre-heating oven. In the pre-heating oven, the towpreg is heated to the required pre-heat temperature by convective heating from resistance heaters or IR heaters placed inside the oven. After pre-heating, the towpreg enters the heated pultrusion die, where the towpreg is heated above the melting point of the thermoplastic polymer and consolidated. The pultrusion die, a taper profiled die, is shown in Figure 10.23 and helps in applying the consolidation pressure to achieve consolidation when the towpreg is pulled with force through the die. The speed of pulling and the temperature in the die decided the consolidation pressure generated in the die, and this affects composite quality (Michaeli and Jürss 1996; Kim, Lee, and Friedrich 2001). The higher the pulling speed, the higher is the consolidation pressure generated (Kim, Lee, and Friedrich 2001). The temperature affects the melt flow and the impregnation quality. The higher the temperature, the greater is the melt flow and higher is the consolidation. After consolidation, the towpreg exits the pultrusion die as consolidated thermoplastic composite profiles. The pultruded profiles are still pliable due to the heat in them. To set the shape of the profile, it needs to be cooled below the glass transition temperature (Tg). The cooling is performed using a cooling die where the profiles are cooled under pressure. The cooling is made by water being circulated in the die that exchanges heat from the pultruded profiles. The profiles exit the cooling the cooling die as fully consolidated profiles that are fit for use. The required pulling force is applied by a double belt pull placed further to the cooling die as shown in Figure 10.22. After the puller, there is a cutter that cuts the pultruded profiles to discrete length for easier transportation and handling.

The thermoplastic pultrusion process can be performed in another, slightly different process flow method suited for producing hollow profiles like cylinders and tubes (Bechtold et al. 1999). The method combines pultrusion and braiding on an on-line process, as shown in Figure 10.24. The pultrusion system is fed with circular braided towpreg preform to produce composites instead of flat towpreg tape. The process starts with a Teflon core over which layers of circular braided towpreg are added. Over the Teflon core, a braided layer of towpreg sleeve is placed. To this base braided layer, additional layers of circular braided towpreg are added from an on-line braiding unit forming the over braid layers (Michaeli and Jürss 1996; Bechtold et al. 1999; Bechtold, Wiedmer, and Friedrich 2002; Tatsuno et al. 2020). The Teflon core with the braid and over braid layers is pulled through the pultrusion die where the braided towpreg layers are heated above the melting point of the thermoplastic and consolidation pressure applied to produce a composite. After pultrusion, the tube with the core is cooled. Finally, the Teflon core is pulled out to provide the hollow pultruded composite tubes after the pultruded tube is cooled.

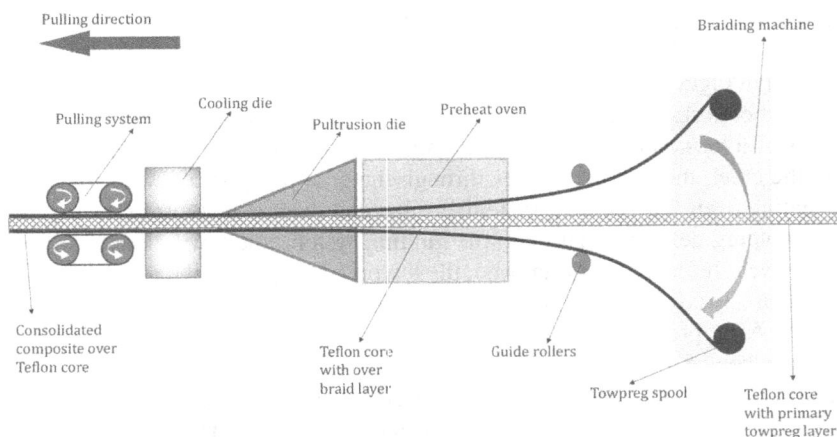

FIGURE 10.24 A schematic of the pull braiding line to produce hollow cylindrical thermoplastic composites.

10.7.2 ADVANTAGES

1. Able to achieve high fiber volume fractions.
2. A wide range of thermoplastics can be processed and even high-performance thermoplastics of PEEK, PEI, PAEK, PEKK, PE, PP, etc. can be processed.
3. It is a continuous production process, and high production volumes are possible.
4. Requires minimum human intervention during the operation of the process.

10.7.3 LIMITATIONS

1. The initial investment is high.
2. The process is not suited for small production cycles.
3. The surface finish is poor.
4. The shapes produced are limited to rods, circular uniform sections, or simple regular geometrical cross section profiles only. No complex 3D shapes can be manufactured.
5. The initial cost of the thermoplastic hybrid towpreg is high.

10.8 SUMMARY

The TPCs find wider use with the development of the newer manufacturing methods, as discussed in this chapter. The applications range from engineering, aerospace, aircraft, defense and military, and biomedical applications. Their use is not limited, and over the coming years, their use is expected to increase due to the advantages offered by TPCs. With the broader adaptation of TPCs, there is a need to achieve higher production rates and lower the labor costs. In the TPCs' manufacturing process, labor is the main contributing factor for the part's overall cost next to the raw material cost. The production cost is considerably reduced by automation of the process. The focus in the present-day scenario is the automation of manufacturing and the same is being brought into TPCs manufacturing by adopting methods like filament winding and automated tape placement. However, automation in composite manufacturing is not suited for all applications and require a hefty initial investment. Thus, the landscape requires research and development in adapting newer manufacturing methods like additive manufacturing to produce TPCs. Thus, there is scope for improvement of the present manufacturing methods and the development of new automated production methods for producing TPCs.

REFERENCES

Ahmed, T. J., D. Stavrov, H. E. N. Bersee, and A. Beukers. 2006. "Induction Welding of Thermoplastic Composites-an Overview." *Composites Part A: Applied Science and Manufacturing* 37 (10): 1638–1651. doi:10.1016/j.compositesa.2005.10.009

Akkerman, R., and S. P. Haanappel. 2015. "Thermoplastic Composites Manufacturing by Thermoforming." *Advances in Composites Manufacturing and Process Design*. Elsevier Ltd. doi:10.1016/B978-1-78242-307-2.00006-3

Alagirusamy, R., R. Fangueiro, V. Ongale, and N. Padaki. 2010. "Hybrid Yarn and Textile Preforming for Thermoplastic Composites." *Textile Progress* 5167 (January 2013): 37–41. doi:10.1533/tepr

Bechtold, G., K. Kameo, F. Langler, H. Hamada, and K. Friedrich. 1999. "Pultrusion of Braided Thermoplastic Commingled Yarn - Simulation of the Impregnation Process." In *Proceedings of the 5th International Conference on Flow Processes in Composite Materials*, Plymouth, UK, July 1999, 257–264.

Bechtold, G., S. Wiedmer, and K. Friedrich. 2002. "Pultrusion of Thermoplastic Composites – New Developments and Modelling Studies." *Journal of Thermoplastic Composite Materials* 15 (5): 443–465. doi:10.1177/0892705702015005202

Bernet, N., M. D. Wakeman, P. E. Bourban, and J. A. Månson. 2002. "An Integrated Cost and Consolidation Model for Commingled Yarn Based Composites." *Composites Part A: Applied Science and Manufacturing* 33 (4): 495–506. doi:10.1016/S1359-835X(01)00140-3

Bigg, D. M., D. F. Hiscock, J. R. Preston, and E. J. Bradbury. 1988. "High Performance Thermoplastic Matrix Composites." *Journal of Thermoplastic Composite Materials* 1 (April): 146–160.

Campbell, F. C. 2014. *Manifacturing Process for Advanced Composites. Igarss 2014.* doi: 10.1007/s13398-014-0173-7.2

Chang, Ike Y., James F. Pratte, E. I. Du, and Magnolia Run. 1991. "LDF™ Thermoplastic Composites Technology Fibers and Composites Development Center." *Journal of Thermoplastic Composite Materials* 4 (July): 227–252.

Colton, J., J. Muzzy, S. Birger, H. Yang, and L. Norpoth. 1992. "Processing Parameters for Consolidating PEEK/Carbon Fiber (APC-2) Composites." *Polymer* 13 (6): 421–426.

Denkena, Berend, Carsten Schmidt, and Patricc Weber. 2016. "Automated Fiber Placement Head for Manufacturing of Innovative Aerospace Stiffening Structures." *Procedia Manufacturing* 6: 96–104. Elsevier B.V. doi:10.1016/j.promfg.2016.11.013

Dykes, R. J., and T. B. Logue. 2002. "Roll Forming of Advanced Thermoplastic Composite Materials the Roll Forming Process." *SPE Automotive Composites Conference and Exposition.*

Dykes, R. J., S. J. Mander, and D. Bhattacharyya. 2000. "Roll Forming Continuous Fibre-Reinforced Thermoplastic Sheets: Experimental Analysis." *Composites Part A: Applied Science and Manufacturing* 31 (12): 1395–1407. doi:10.1016/S1359-835X(00)00076-2

Evans, S. L., and P. J. Gregson. 1998. "Composite Technology in Load-Bearing Orthopaedic Implants." *Biomaterials* 19 (15): 1329–1342. doi:10.1016/S0142-9612(97)00217-2

Henninger, F., and K. Friedrich. 2002. "Process Analysis of Roll Forming of Thermoplastic Composites." In *The 6th International Conference on Flow Processes in Composite Materials*, 63–72.

Fernández, I., F. Blas, and M. Frövel. 2003. "Autoclave Forming of Thermoplastic Composite Parts." In *Journal of Materials Processing Technology* 143–144:266–269. doi:10.1016/S0924-0136(03)00309-1

Cogswell, Frederic Neil. 1992. *Thermoplastic Aromatic Polymer Composites Materials.* Butterwort Heinemann.

Friedrich, K., M. Hou, and J. Krebs. 1997. "Chapter 4 Thermoforming of Continuous Fibre/Thermoplastic Composite Sheets." *Composite Materials Series* 11 (C): 91–162. doi:10.1016/S0927-0108(97)80006-9

Friedrich, Klaus, and Abdulhakim A. Almajid. 2013. "Manufacturing Aspects of Advanced Polymer Composites for Automotive Applications." *Applied Composite Materials* 20 (2): 107–128. doi:10.1007/s10443-012-9258-7

Fujihara, K., Zheng Ming Huang, S. Ramakrishna, and H. Hamada. 2004. "Influence of Processing Conditions on Bending Property of Continuous Carbon Fiber Reinforced PEEK Composites." *Composites Science and Technology* 64 (16): 2525–2534. doi: 10.1016/j.compscitech.2004.05.014

Fujihara, K., Zheng Ming Huang, S. Ramakrishna, K. Satknanantham, and H. Hamada. 2003. "Performance Study of Braided Carbon/PEEK Composite Compression Bone Plates." *Biomaterials* 24 (15): 2661–2667. doi:10.1016/S0142-9612(03)00065-6

Funck, R., and M. Neitzel. 1995. "Improved Thermoplastic Tape Winding Using Laser or Direct-Flame Heating." *Composites Manufacturing* 6 (3–4): 189–192. doi:10.1016/0956-7143(95)95010-V

Gibson, A. G., and J. A. Manson. 1992. "Impregnation Techniques for Thermoplastic Matrix Composites." *Composites Manufacturing* 3 (4): 223–233.

Goud, Vijay, Ramasamy Alagirusamy, Apurba Das, and Dinesh Kalyanasundaram. 2019. "Influence of Various Forms of Polypropylene Matrix (Fiber, Powder and Film States) on the Flexural Strength of Carbon-Polypropylene Composites." *Composites Part B: Engineering* 166: 56–64. Elsevier Ltd. doi:10.1016/j.compositesb.2018.11.135

Hamada, H., and S. Ramakrishna. 1995. "Scaling Effects in the Energy Absorption of Carbon-Fiber/PEEK Composite Tubes." *Composites Science and Technology* 55 (3): 211–221. doi:10.1016/0266-3538(95)00081-X

Henninger, F., and K. Friedrich. 2002. "Thermoplastic Filament Winding with Online-Impregnation. Part A: Process Technology and Operating Efficiency." *Composites Part A: Applied Science and Manufacturing* 33 (11): 1479–1486. doi:10.1016/S1359-835X(02)00135-5

Henninger, F., and K. Friedrich. 2004. "Production of Textile Reinforced Thermoplastic Profiles by Roll Forming." *Composites Part A: Applied Science and Manufacturing* 35 (5): 573–583. doi:10.1016/j.compositesa.2003.12.001

Henninger, F., J. Hoffmann, and K. Friedrich. 2002. "Thermoplastic Filament Winding with Online-Impregnation. Part B. Experimental Study of Processing Parameters." *Composites Part A: Applied Science and Manufacturing* 33 (12): 1684–1695. doi:10.1016/S1359-835X(02)00186-0

Henriquez, Raelvim Gonzalez, and Pierre Mertiny. 2017. "Filament Winding Applications." *Comprehensive Composite Materials II* 3: 556–577. Elsevier Ltd. doi:10.1016/B978-0-12-803581-8.10313-3

Ho, K. K. C., S. R. Shamsuddin, S. Riaz, S. Lamorinere, M. Q. Tran, A. Javaid, and A. Bismarck. 2011. "Wet Impregnation as Route to Unidirectional Carbon Fibre Reinforced Thermoplastic Composites Manufacturing." *Plastics, Rubber and Composites* 40 (2): 100–107. doi:10.1179/174328911X12988622801098

Hou, M. 1997. "Stamp Forming of Continuous Glass Fibre Reinforced Polypropylene." *Composites Part A: Applied Science and Manufacturing* 28 (8): 695–702. doi:10.1016/S1359-835X(97)00013-4

Hou, M., K. Friedrich, and R. Scherer. 1994. "Optimization of Stamp Forming of Thermoplastic Composite Bends." *Composite Structures* 27 (1–2): 157–167. doi:10.1016/0263-8223(94)90077-9

Hou, M., L. Ye, H. J. Lee, and Y. W. Mai. 1998. "Manufacture of a Carbon-Fabric-Reinforced Polyetherimide (CF/PEI) Composite Material." *Composites Science and Technology* 58 (2): 181–190. doi:10.1016/S0266-3538(97)00117-6

Iyer, Shridar R., and Lawrence T. Drzal. 1990. "Manufacture of Thermoplastic Composites." *Journal of Thermoplastic Composite Materials* 3 (October): 325–355.

Jiang, Wei, Zhigao Huang, Yunming Wang, Bing Zheng, and Huamin Zhou. 2019. "Voids Formation and Their Effects on Mechanical Properties in Thermoformed Carbon Fiber Fabric-Reinforced Composites." *Polymer Composites* 40 (S2): E1094–E1102. doi:10.1002/pc.24876

Jogur, Ganesh, Ashraf Nawaz Khan, Apurba Das, Puneet Mahajan, and R.Alagirusamy. 2018. "Impact Properties of Thermoplastic Composites." *Textile Progress* 50 (3): 109–183. Taylor & Francis. doi:10.1080/00405167.2018.1563369

Johansen, B. S., Aa Lystrup, and M. T. Jensen. 1998. "CADPATH: A Complete Program for the CAD-, CAE- and CAM-Winding of Advanced Fibre Composites." *Journal of Materials Processing Technology* 300 (3–4): 194–200. doi:10.1016/s0924-0136(97)00417-2

Kim, Dae Hwan, Woo Il Lee, and Klaus Friedrich. 2001. "A Model for a Thermoplastic Pultrusion Process Using Commingled Yarns." *Composites Science and Technology* 61 (8): 1065–1077. doi:10.1016/S0266-3538(00)00234-7

Krämer, E. T. M., W. J. B. Grouve, S. Koussios, L. L. Warnet, and R. Akkerman. 2020. "Real-Time Observation of Waviness Formation during C/PEEK Consolidation." *Composites Part A: Applied Science and Manufacturing* 133 (March): 105872. Elsevier. doi:10.1016/j.compositesa.2020.105872

Krebs, J., K. Friedrich, and D. Bhattacharyya. 1998. "A Direct Comparison of Matched-Die versus Diaphragm Forming." *Composites Part A: Applied Science and Manufacturing* 29 (1–2): 183–188. doi:10.1016/S1359-835X(97)82706-6

Lapointe, Felix, and Louis Laberge Lebel. 2019. "Fiber Damage and Impregnation during Multi-Die Vacuum Assisted Pultrusion of Carbon/PEEK Hybrid Yarns." *Polymer Composites* 40 (S2): E1015–E1028. doi:10.1002/pc.24788

Lauke, B., and K. Friedrich. 1993. "Evaluation of Processing Parameters of Thermoplastic Composites Fabricated by Filament Winding." *Composites Manufacturing* 4 (2): 93–101. doi:10.1016/0956-7143(93)90076-K

Li, Na, Junlin Chen, Hansong Liu, Anqi Dong, Kai Wang, and Yan Zhao. 2019. "Effect of Preheat Treatment on Carbon Fiber Surface Properties and Fiber/PEEK Interfacial Behavior." *Polymer Composites* 40 (S2): E1407–E1415. doi:10.1002/pc.25020

Lionetto, Francesca, Riccardo Dell'Anna, Francesco Montagna, and Alfonso Maffezzoli. 2016. "Modeling of Continuous Ultrasonic Impregnation and Consolidation of Thermoplastic Matrix Composites." *Composites Part A: Applied Science and Manufacturing* 82: 119–129. Elsevier Ltd. doi:10.1016/j.compositesa.2015.12.004

Loos, Alfred C., and George S. Springer. 1983. "Curing of Epoxy Matrix Composites." *Journal of Composite Materials* 17 (March): 135–169.

Lukaszewicz, Dirk H. J. A., Carwyn Ward, and Kevin D. Potter. 2012. "The Engineering Aspects of Automated Prepreg Layup: History, Present and Future." *Composites Part B: Engineering* 43 (3): 997–1009. Elsevier Ltd. doi:10.1016/j.compositesb.2011.12.003

Lystrap, Aage, and Tom L. Andersen. 1998. "Autoclave Consolidation of Fibre Composites with a High Temperature Thermoplastic Matrix." *Journal of Materials Processing Technology* 300 (3–4): 80–85. doi:10.1016/s0924-0136(97)00398-1

Mack, J., and R. Schledjewski. 2012. "Filament Winding Process in Thermoplastics." In *Manufacturing Techniques for Polymer Matrix Composites (PMCs)*. Woodhead Publishing Limited. doi:10.1533/9780857096258.2.182

Mallon, P. J., C. M. O'Brádaigh, and R. B. Pipes. 1989. "Polymeric Diaphragm Forming of Complex-Curvature Thermoplastic Composite Parts." *Composites* 20 (1): 48–56. doi:10.1016/0010-4361(89)90682-4

McCool, R., A. Murphy, R. Wilson, Z. Jiang, M. Price, J. Butterfield, and P. Hornsby. 2012. "Thermoforming Carbon Fibre-Reinforced Thermoplastic Composites." *Proceedings of the Institution of Mechanical Engineers, Part L: Journal of Materials: Design and Applications* 226 (2): 91–102. doi:10.1177/1464420712437318

McGregor, O. P. L., A. A. Somashekar, D. Bhattacharyya, O. P. L. McGregor, and M. Duhovic. 2017. "Pre-Impregnated Natural Fibre-Thermoplastic Composite Tape Manufacture Using a Novel Process." *Composites Part A: Applied Science and Manufacturing* 101: 59–71. Elsevier Ltd. doi:10.1016/j.compositesa.2017.05.025

Michaeli, W., and D. Jürss. 1996. "Thermoplastic Pull-Braiding: Pultrusion of Profiles with Braided Fibre Lay-up and Thermoplastic Matrix System (PP)." *Composites Part A: Applied Science and Manufacturing* 27 (1): 3–7. doi:10.1016/1359-835X(95)00004-L

Monaghan, M. R., and P. J. Mallon. 1990. "Development of A Computer Controlled Autoclave for Forming Thermoplastic Composites." *Composites Manufacturing* 1 (1): 8–14. doi: 10.1016/0956-7143(90)90269-3

Monaghan, P. F., M. T. Brogan, and P. H. Oosthuizen. 1991. "Heat Transfer in an Autoclave for Processing Thermoplastic Composites." *Composites Manufacturing* 2 (3–4): 233–242. doi:10.1016/0956-7143(91)90145-7

Munro, M. 1988. "Review of Manufacturing of Fiber Composite Components by Filament Winding." *Polymer Composites* 9 (5): 352–359. doi:10.1002/pc.750090508

Muzzy, John D., Xiang Wu, and Jonathan S. Colton. 1990. "Thermoforming of High Performance Thermoplastic Composites." *Polymer Composites* 11 (5): 280–285. doi: 10.1002/pc.750110505

Novo, P. J., J. F. Silva, J. P. Nunes, and A. T. Marques. 2016. "Pultrusion of Fibre Reinforced Thermoplastic Pre-Impregnated Materials." *Composites Part B: Engineering* 89: 328–339. Elsevier Ltd. doi:10.1016/j.compositesb.2015.12.026

Nunes, João P., Ferrie W. J. van Hattum, Carlos A. Brito, António M. Bernardo, António S. Pouzada, João F. Silva, and António T. Marques. 2005. "Production of Thermoplastic Towpregs and Towpreg-Based Composites." In: *Polymer Composites*. Springer, 189–213. doi:10.1007/0-387-26213-X_11

O'Brádaigh, Conchúr M., R. Byron Pipes, and Patrick J. Mallon. 1991. "Issues in Diaphragm Forming of Continuous Fiber Reinforced Thermoplastic Composites." *Polymer Composites* 12 (4): 246–256. doi:10.1002/pc.750120406

Oromiehie, Ebrahim, B. Gangadhara Prusty, Paul Compston, and Ginu Rajan. 2019. "Automated Fibre Placement Based Composite Structures: Review on the Defects, Impacts and Inspections Techniques." *Composite Structures* 224 (May). Elsevier: 110987. doi:10.1016/j.compstruct.2019.110987

Qureshi, Z., T. Swait, R. Scaife, and H. M. El-Dessouky. 2014. "In Situ Consolidation of Thermoplastic Prepreg Tape Using Automated Tape Placement Technology: Potential and Possibilities." *Composites Part B: Engineering* 66: 255–267. Elsevier Ltd. doi:10.1016/j.compositesb.2014.05.025

Ramakrishna, S., W.K. Tan, S.H. Teoh, and M. O. Lai. 2009. "Recycling of Carbon Fiber/Peek Composites." *Key Engineering Materials* 137: 1–8. doi:10.4028/www.scientific.net/kem.137.1

Ramakrishna, S., J. Mayer, E. Wintermantel, and Kam W. Leong. 2001. "Biomedical Applications of Polymer-Composite Materials: A Review." *Composites Science and Technology* 61 (9): 1189–1224. doi:10.1016/S0266-3538(00)00241-4

Ramani, Karthik, Harshad Borgaonkar, and Chris Hoyle. 1995. "Experiments on Compression Moulding and Pultrusion of Thermoplastic Powder Impregnated Towpregs." *Composites Manufacturing* 6 (1): 35–43. doi:10.1016/0956-7143(95)93711-R

Ramani, Karthik, and Chris Hoyle. 1995. "Processing of Thermoplastic Composites Using a Powder Slurry Technique. I. Impregnation and Preheating." *Materials and Manufacturing Processes* 10 (6): 1169–1182. doi:10.1080/10426919508935100

Ramasamy, A., Youjiang Wang, and John Muzzy. 1996a. "Braided Thermoplastic Composites from Powder-Coated Towpregs. Part I: Towpreg Characterization." *Polymer Composites* 17 (3): 497–504. doi:10.1002/pc.10639

Ramasamy, A., Youjiang Wang, and John Muzzy. 1996b. "Braided Thermoplastic Composites from Powder-Coated Towpregs. Part III: Consolidation and Mechanical Properties." *Polymer Composites* 17 (3): 515–522. doi:10.1002/pc.10639

Ray, Pinaki. 2005. *Computer Aided Optimization of Tube Hydroforming Processes*. Dublin City University.

Rizzolo, Robert H., and Daniel F. Walczyk. 2016. "Ultrasonic Consolidation of Thermoplastic Composite Prepreg for Automated Fiber Placement." *Journal of Thermoplastic Composite Materials* 29 (11): 1480–1497. doi:10.1177/0892705714565705

Rodriguez-Lence F, Martin MI, Fernandez Horcajo K. In-situ consolidation of integrated thermoplastic fuselage panels: The future in structural comercial aerocomposites. *ECCM 2018 - 18th European Conference on Composite Materials.*, 2018.

Romagna, J., G. Ziegmann, and M. Flemming. 1995. "Thermoplastic Filament Winding– An Experimental Investigation of the on-Line Consolidation of Poly(Ether Imide) Fit Preforms." *Composites Manufacturing* 6 (3–4): 205–210. doi:10.1016/0956-7143(95)95013-O

Sch_olliers, Johan, and Hendrik Van Brussel. 1994. "Computer-Integrated Filament Winding: Computer-Integrated Design, Robotic Filament Winding and Robotic Quality Control." *Composites Manufacturing* 5 (1): 15–23. doi:10.1016/0956-7143(94)90015-9

Stokes-Griffin, C. M., and P. Compston. 2015. "The Effect of Processing Temperature and Placement Rate on the Short Beam Strength of Carbon Fibre-PEEK Manufactured Using a Laser Tape Placement Process." *Composites Part A: Applied Science and Manufacturing* 78: 274–283. Elsevier Ltd. doi:10.1016/j.compositesa.2015.08.008

Svensson, N., R. Shishoo, and M. Gilchrist. 1998. "Manufacturing of Thermoplastic Composites from Commingled Yarns – A Review." *Journal of Thermoplastic Composite Materials* 11 (1): 22–56. doi:10.1177/089270579801100102

Talley, Samantha J., Christian L. Anderson Schoepe, Christopher J. Berger, Kaitlyn A. Leary, Samuel A. Snyder, and Robert B. Moore. 2017. "Mechanically Robust and Superhydrophobic Aerogels of Poly(Ether Ether Ketone)." *Polymer* 126: 437–445. Elsevier Ltd. doi:10.1016/j.polymer.2017.05.047

Tatsuno, D., T. Yoneyama, T. Kinari, E. Sakanishi, T. Ochiai, and Y. Taniichi. 2020. "Braid-Press Forming for Manufacturing Thermoplastic CFRP Tube." *International Journal of Material Forming.* International Journal of Material Forming, 1–10. doi:10.1007/s12289-020-01584-7

Texier, A., R. M. Davis, K. R. Lyon, A. Gungor, J. E. McGrath, H. Marand, and J. S. Riffle. 1993. "Fabrication of PEEK/Carbon Fibre Composites by Aqueous Suspension Prepregging." *Polymer* 34 (4): 896–906. doi:10.1016/0032-3861(93)90378-N

Van De Velde, Kathleen, and Paul Kiekens. 2001. "Thermoplastic Pultrusion of Natural Fibre Reinforced Composites." *Composite Structures* 54 (2–3): 355–360. doi:10.1016/S0263-8223(01)00110-6

Van Hoa, Suong, Minh Duc Hoang, and JeffSimpson. 2017. "Manufacturing Procedure to Make Flat Thermoplastic Composite Laminates by Automated Fibre Placement and Their Mechanical Properties." *Journal of Thermoplastic Composite Materials* 30 (12): 1693–1712. doi:10.1177/0892705716662516

Van West, B. P., R. Byron Pipes, and S. G. Advani. 1991. "The Consolidation of Commingled Thermoplastic Fabrics." *Polymer Composites* 12 (6): 417–427. doi:10.1002/pc.750120607

Venkatesan, Chadurvedi, Rajkumar Velu, Nahaad Vaheed, Felix Raspall, Tong Earn Tay, and Arlindo Silva. 2020. "Effect of Process Parameters on Polyamide-6 Carbon Fibre Prepreg Laminated by IR-Assisted Automated Fibre Placement." *International Journal of Advanced Manufacturing Technology* 108 (4). The International Journal of Advanced Manufacturing Technology: 1275–1284. doi:10.1007/s00170-020-05230-z

Vodicka, Roger. 1996. "Thermoplastics for Airframe Applications a Review of the Properties and Repair Methods for Thermoplastic Composites." *Defence Science and Technology Organisation* DSTO-TR-04.

Wagner, P., and J. Colton. 1994. "On-line Consolidation of Thermoplastic Towpreg Composites in Filament Winding." *Polymer Composites* 15 (6): 436–441. doi:10.1002/pc.750150608

Wakeman, M. D., and J. A. E. Manson. 2005. "Composites Manufacturing – Thermoplastics." In Design and Manufacture of Textile Composites, 197–241. Woodhead Publishing Limited. doi:10.1533/9781845690823.197

Werdermann, Cord, Klaus Friedrich, Mark Cirino, and R. Byron Pipes. 1989. "Design and Fabrication of an On-Line Consolidation Facility for Thermoplastic Composites." *Journal of Thermoplastic Composite Materials* 2 (4): 293–306. doi:10.1177/089270578900200404

Wong, Joanna C.H., Javier Molina Blanco, and Paolo Ermanni. 2018. "Filament Winding of Aramid/PA6 Commingled Yarns with In Situ Consolidation." *Journal of Thermoplastic Composite Materials* 31 (4): 465–482. doi:10.1177/0892705717706528

Wu, G. M., and J. M. Schultz. 2000. "Processing and Properties of Solution Impregnated Carbon Fiber Reinforced Polyethersulfone Composites." *Polymer Composites* 21 (2): 223–230. doi:10.1002/pc.10179

Yassin, Khaled, and Mehdi Hojjati. 2018. "Processing of Thermoplastic Matrix Composites through Automated Fiber Placement and Tape Laying Methods: A Review." *Journal of Thermoplastic Composite Materials* 31 (12): 1676–1725. doi:10.1177/0892705717738305

Ye, Lin, and Klaus Friedrich. 1997. "Processing of CF/PEEK Thermoplastic Composites from Flexible Preforms." *Advanced Composite Materials* 6 (2): 83–97. doi:10.1163/156855197X00012

Ye, Lin, Klaus Friedrich, Joachim Kästel, and Yiu Wing Mai. 1995. "Consolidation of Unidirectional CF/PEEK Composites from Commingled Yarn Prepreg." *Composites Science and Technology* 54 (4): 349–358. doi:10.1016/0266-3538(95)00061-5

Zampaloni, Michael A., Farhang Pourboghrat, and Woong Ryeol Yu. 2004. "Stamp Thermo-Hydroforming: A New Method for Processing Fiber-Reinforced Thermoplastic Composite Sheets." *Journal of Thermoplastic Composite Materials* 17 (1): 31–50. doi:10.1177/0892705704038219

Zhang, Yunhe, Wei Tao, Yu Zhang, Lin Tang, Junwei Gu, and Zhenhua Jiang. 2018. "Continuous Carbon Fiber/Crosslinkable Poly(Ether Ether Ketone) Laminated Composites with Outstanding Mechanical Properties, Robust Solvent Resistance and Excellent Thermal Stability." *Composites Science and Technology* 165 (March): 148–153. Elsevier. doi:10.1016/j.compscitech.2018.06.020

Zheng, Bing, Maoyuan Li, Tianzhengxiong Deng, Helezi Zhou, Zhigao Huang, Huamin Zhou, and Dequn Li. 2019. "Process–Structure–Property Relationships of Thermoformed Woven Carbon-Fiber-Reinforced Polyether-Ether-Ketone Composites." *Polymer Composites* 40 (10): 3823–3834. doi:10.1002/pc.25241

11 Flexible Towpreg Structure and Composite Properties

Ashraf Nawaz Khan, Ganesh Jogur,
and R. Alagirusamy
Department of Textile and Fibre Engineering
Indian Institute of Technology Delhi
New Delhi, India

CONTENTS

11.1 INTRODUCTION

For the thermoplastic composite, the main issue is the impregnation of the highly viscous resin into the fibrous reinforcement elements. There are various manufacturing techniques which are developed for the fabrication of the thermoplastic composite to achieve the least void contents. The void formation in the composite laminate mainly depends upon the impregnation quality of the matrix (Jogur et al. 2018). The void contents can be minimized with better impregnation. The melt flow distance travelled by the thermoplastic matrix during the consolidation process plays an important role in determining the interfacial properties and the fiber-matrix distribution. The melt flow distance for the highly viscous material must be as low as possible to

DOI: 10.1201/9781003049715-11

ensure better impregnation into the reinforcing fibers, giving the least amount of voids. There are numerous hybrid yarn techniques available such as commingled yarn technique, co-wrapping technique, DREF spinning, and many more, where the matrix is available in the form of fibers that are used to combine with the reinforcing elements (Alagirusamy et al. 2006; Khan et al. 2022). This is to minimize the matrix melt flow distance during the consolidation. There are other methods such as the dry powder-coated towpreg technique where the matrix in the powder form is diffused into the fiber tows. The powder combined with the filaments at micron levels results in homogenous fiber and matrix distribution. During the consolidation process, the resin moves in the longitudinal direction in the case of powder-coated towpregs on applying pressure, which is different from the transverse movements of the matrix in the commingled towpregs (Miller et al. 1998; Goud, Alagirusamy, et al. 2019). The mechanical properties of the fiber-reinforced composite are widely governed by the orientation of the reinforcing fiber, interface property, and void content. The thermoplastic composite made with different manufacturing techniques results in different fiber-matrix distribution as well as void content and its distribution in a longitudinal and transverse direction. The processing parameter influences the property of the laminate considerably during the manufacturing stage. The processing pressure, holding time, and the temperature during the consolidation control the void content and thereby the quality of the composites. The void content decreases with the increase in the consolidation pressure. But after a certain pressure limit, no further reduction in the void content is noticed. The increase in the consolidation temperature and holding time also results in the diminishing of the voids, but up to a certain temperature and holding time limit (Ramasamy, Wang, and Muzzy 1996; Alagirusamy et al. 2006).

11.2 VOIDS AND THEIR DISTRIBUTIONS IN THE LAMINATE MADE THROUGH HYBRID YARN TECHNIQUES

Among the hybrid yarn techniques, the commingled yarn technique is the commonly used one where the matrix in the form of the continuous filaments is mixed with the continuous reinforcing filaments. The continuous multifilament mingle at filament-to-filament level and form a hybrid yarn where matrix distribution occurs from the core to surface level. This results in inclusive impregnation of the matrix phase across the reinforcing fiber bundle. The main challenge in the commingled towpreg is the thorough mixing of the two types of filaments. The incomplete mixing may result in fiber-rich and resin-rich zones inside the composite. Therefore, to assure the guaranteed well mixing of the filaments, there must be a similar diameter of the filaments (Miller et al. 1998).

Dry powder coating technique is another famous technique where a matrix is available in the powder form for the production of the towpreg. There are various versions of the powder coating method which use different powder impregnation principles. The powder-coated towpreg structure (as shown in Figure 11.1) depends upon the degree of the filaments' exposure to the powder matrix. The initial dispersion of the powder inside the tow influences the melt flow distance during wet out phenomena. There are various studies done on the spreading of the matrix powder

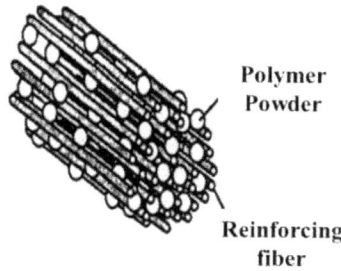

Polymer Powder

Reinforcing fiber

FIGURE 11.1 Powder impregnated towpregs. (Reproduced with permission from Miller et al. 1998.)

within the tow for good impregnation (Miller et al. 1998; Chen, Arakawa, and Xu 2015; Goud, Alagirusamy, et al. 2019; Goud et al. 2020). For the better laminate quality, the fiber should realign and the molten matrix should move to fill the available interstitial spaces to avoid any residual air entrapment during the consolidation process. The gradual increase of the processing pressure facilitates the ease in the fiber movements and the uniform mixing of the matrix and the reinforcing parts. The increase in the holding time, processing temperature, and pressure leads to the minimization of the void contents in the composite laminates, which is confirmed by the ultrasonic C-scan (Hugh et al. 1992).

There are various studies reported on the characterization of the voids inside the polymer composite. Boriek et al. (1990) developed a probabilistic model for the distribution of the voids in the polymer composites. The effect of the distribution of the voids was discussed with respect to the stress concentration in the composite laminates. These voids act as a zero-strength domain from which the crack initiates and propagates. However, the presence of the voids helps in relieving the stress during the curing process. But it is undesirable due to their adverse effect on the mechanical properties of the composite. In most of the composite fabrication processes, there is the degassing step to remove the entrapped air pockets using vibration or vacuum. Miller et al. (1998) developed a model for the powder-coated composites to understand the mechanism of the powder impregnation and consolidation step based on the geometric and flow assumptions. The rate of particle flow is considered as the main mechanism of the powder impregnation and the void formation and its distribution. The volume of the void was determined by Equation 11.1 (Miller et al. 1998) through the immersion technique as per ASTM-792.

$$V_V(t) = \frac{\left(h(t) - h_1\right)}{\left(h_0 - h_1\right)}(V_{V0} - V_{V1}) + V_{V1} \tag{11.1}$$

Where h_1, h_2 is the thickness of the laminate at the beginning and at the end of the experiments respectively. V_{v0} is the volume of the cavity when the upper part of the mold just touched the heater before consolidation, V_{v1} is the final void fraction of consolidated laminates after cooling.

The formation of the voids and its distribution within the composite laminate depends upon the production technique as well as the flow behavior of the resin in the molten stage. The flowability of the matrix depends upon the state of the matrix such as solid, liquid, etc. Generally, a void can be categorized on the multi-scale basis that is macro-, meso-, and micro-voids owing to the multi-scale nature of the fiber-reinforced composite. These voids form at different regions: micro-voids generally form within the tows, meso-voids are formed in between the tows, while macro-voids are formed in the broader region of the preforms. The void formation at different scales is controlled by the flow at the micro- and macro-levels (Park et al. 2011; Chen, Arakawa, and Xu 2015; Mehdikhani et al. 2019).

Composites prepared with prepreg technology have different mechanisms of void formation and its propagation. The void formation in the laying-up stage and curing stage are much discussed but in the impregnation stage, much less study has been performed in the prepreg technology. The phenomena of the void formation and its growth are yet to be explored in the curing stage. The majority of the studies have been done on the general study of the void content in the final composite laminate with respect to different manufacturing parameters. The intra-laminar and inter-laminar voids are main causes of concern in the prepreg technology, which is not a critical issue in the LCM technology. The air entrapment during the impregnation leads to the formation of the intra-laminar voids and formation of the inter-laminar voids happens during the lay-up stage. The release of the volatile material during the curing process and the moisture absorption in the resin are the main reasons for the void content in the laminate made with past generation of the matrix materials. But these issues are resolved in the modern grades of resin. Hence, the void formation mainly occurs through the mechanical air entrapment (Boey and Lye 1992a; 1992b; Centea, Grunenfelder, and Nutt 2015; Mehdikhani et al. 2019).

The characterization of the voids in the composite through different methodologies is also one of the crucial phenomena. This includes the destructive and non-destructive techniques. The voids determination methods are developed from the very basic density determination method to the advanced microscopic methods. The measurement of the exact void content in the polymer laminate is difficult to measure. The different determination methods give varying results from the actual one. However, there are various approaches through which the precise measurement of the volume of the void is possible with little error. The void development in the fiber-reinforced composite depends upon the manufacturing process which has varying thermodynamic and rheological phenomena. Mehdikhani et al. (2019) studied void formation in different manufacturing technologies. There are various techniques used in the characterization of the voids such as density determinations, ultrasonic testing, x-ray micro-CT and many more. Also, other methods such as x-ray radiography, shearography, thermography, etc. are tools to characterize the voids throughout the composite. All the tools are used according to the application and based on demands such as time factor, cost-effectiveness, 2D/3D analysis, research or industrial application, and many more. The voids can be categorized in various ways which include void content, void shape and size, and its distribution within the composite material (Stone and Clarke 1975; Purslow 1984; Jones, Polansky, and Berger 1988; Hung 1996; Desplentere et al. 2005; Schell et al. 2006; Podymova and Karabutov

2013; Karabutov and Podymova 2014; Madra, Hajj, and Benzeggagh 2014; Straumit, Lomov, and Wevers 2015; Sisodia et al. 2016; Mehdikhani et al. 2019; Gupta et al. 2021; Muralidhar and Arya 1993).

Ramasamy, Wang, and Muzzy (1996) manufactured composite laminates through powder-coating and commingling. The void content in the laminates which were fabricated at different processing conditions such as consolidated temperature, pressure, and holding time but at a constant cooling rate were studied. Two types of the void were observed: one is within the tow and the other is in between the tows. The void content in the powder-coated composite was in the range of 0.8 to 5.4% while in the case of commingled laminate, it varied from 2.8 to 5.3%. The powder-coated laminate showed a better fiber-matrix distribution in comparison to the commingled one. In another study, the hybrid yarn was prepared through commingling and co-wrapped technique. The laminate made through co-wrapped hybrid yarn demonstrated more void content, resin-rich, and fiber-rich zones, which acted adversely for the laminate mechanical properties. These irregular fiber and matrix distributions were resolved by an improved commingled hybrid yarn technique where a better impregnation was observed, which resulted in good fiber and matrix distribution and lesser void content in the laminate (Alagirusamy et al. 2006). Klinkmuller et al. (1994) conducted a microscopic analysis of the composite laminates and studied the void content variations with the processing temperature and pressure. The influence of the temperature and pressure was observed up to certain point beyond which no change in the void content was noted. The specific gravity of the composite can be measured as per the given Equation 11.2 (Klinkmuller et al. 1994). This specific gravity is used for the calculation of the absolute density of the composite laminate as in Equation 11.3 (Klinkmuller et al. 1994).

$$\gamma = \frac{w_1}{w_3 - w_2} \tag{11.2}$$

$$p = \gamma p_w \tag{11.3}$$

Where γ is the specific gravity, p is the density of the composite, p_w is the density of the distilled water, w_1 is the weight of the laminate in air, w_2 is the of the weight of laminate in the water, w_3 is the weight of the laminate saturated with the water.

Klinkmuller et al. (1994) studied the quality of the impregnation for the commingled yarn technique. It is observed that the degree of the impregnation depends upon the size of the fiber agglomerations (as shown in Figure 11.2) which is a function of the initial distribution of the fiber and matrix. The fiber agglomeration is determined through the mathematical model and observed through the SEM analysis. The area of the components is calculated as per Equation 11.4 (Klinkmuller et al. 1994).

$$A_i = \frac{Tex_i}{\rho_i . 10^3} \tag{11.4}$$

The impregnation time depends upon the size of the fiber agglomeration, and the penetration distance works as a function of the voids content in the laminate. When the

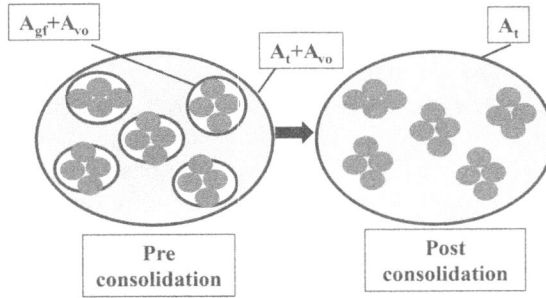

FIGURE 11.2   Cross-sectional view of the towpreg. (Adapted from Klinkmuller et al. 1994.)

voids between the fibers are included in the area determination, the total area of the component is modified to Equation 11.5 (Klinkmuller et al. 1994).

$$A_i = \frac{Tex_i}{\rho_i \cdot \rho_{pck,i} \cdot 10^3} \tag{11.5}$$

Tex_i is the weight of material i per unit length, ρ_i is the specific weight in g/cm³ of material i, $\rho_{pck.i}$ is the packing density of the material i.

The effect of the processing condition such as consolidation pressure, temperature, and holding time on fiber-matrix distribution and voids content inside the laminate was investigated. The major finding of this work is the relation between the mechanical property and the void content in the laminates. The mathematical model for the number of the fiber agglomerations per unit fiber bundle is calculated as per Equation 11.6 (Klinkmuller et al. 1994).

$$n = \frac{A_{gf}}{\pi \cdot r_0^2 \cdot \rho_{pck,gf}} \tag{11.6}$$

n is the number of fiber agglomerations per bundle, A_{gf} is the area of the fiber, r_0 is the radius of the fiber agglomerations, $\rho_{pck.gf}$ is the packing density of the fiber.

The mechanical property of the composite such as tensile, flexural, and shear increases with the decrease in the void contents. The reduction in the void content reduces the stress concentration and the weaker region inside the composite. Therefore, it has a high resistance to crack initiation and its propagation (Klinkmuller et al. 1994; Ramasamy, Wang, and Muzzy 1996; Alagirusamy et al. 2006). Ramasamy, Wang, and Muzzy (1996) conducted a study to determine the volume fraction of the void in two different laminates made with powder-coated and commingled yarn techniques as shown in Table 11.1. An optical microscope was used to capture high-resolution images of the cross-section of the composite. The threshold level of magnification of the microscope was kept to see the fiber, matrix, and void phase distinctly. The image analysis was done to determine the percentage of the area of the fiber, matrix, and void across the entire cross-sectional area, which ultimately helped in determining the void contents. The void content of the

TABLE 11.1

Voids Content in Powder-Coated and Commingled-Based Laminate

Consolidation Conditions			Fiber Volume (%)		Density(g/cc)		Void Content(%)	
			Powder-Coated	Commingled	Powder-Coated	Commingled	Powder-Coated	Commingled
T (Cel.)	P (kPa)	t (min)	Laminates	Laminates	Laminates	Laminates	Laminates	Laminates
260	345	5	59.7	56.2	1.464	1.373	2.1	5.3
260	345	30	61.2	56.5	1.467	1.399	1.7	4.6
260	690	5	61.9	58.8	1.523	1.437	1.1	3.2
260	690	30	62.2	60.2	1.528	1.447	0.8	2.8
232	345	5	55.1	53.9	1.388	1.205	5.4	19.1
232	345	30	59.4	54.2	1.41	1.221	4.5	18.2
232	690	5	60.5	53.7	1.463	1.258	2.2	15.5
232	690	30	60	53.9	1.449	1.26	1.6	15.4
245	520	18	58.2	–	1.486	–	1.7	–

Source: Ramasamy, Wang, and Muzzy (1996).

laminate was determined at different processing conditions. The void content of the two types of laminates was compared and it was found that powder-coated composite material has lesser void content in comparison to the commingled composite material. The lesser void content in the powder-coated laminate indicated better impregnation and consolidation of the composite during curing and hence better quality of the laminate.

Generally in the composite laminate, there are two types of voids: one is within the towpreg, which is of smaller size ranging from 12 to 20 microns while the other is between the towpreg, which is of larger size ranging from 80 to 120 microns. The lesser void content in the powder-coated composite indicates the better dispersion of the resin in comparison to the commingled composite (Ramasamy, Wang, and Muzzy 1996). The void content can be reduced by elevating consolidation temperature and pressure. The reduction in the melt viscosity of the resin at a higher temperature helps in filling the gap within the preform. The similar effect was also found at higher pressure. Higher pressure forces the melted resin into the tow and fills the gap. Hence, a significant reduction in the void content of the laminate was found with higher processing temperature and pressure as shown in Figure 11.3 (Ramasamy, Wang, and Muzzy 1996).

The mechanical property of the composite changes significantly with the processing condition as discussed. For the manufacturing of the composite laminate, there must be a parametric study to reach optimum value. As the composite material is non-isotropic, it is difficult to model its mechanical behavior. For the development of any model for the performance of the composite, there should be a complete understanding of the micro-features of the material such as interface bonding and interaction of the fiber and matrix on application of load.

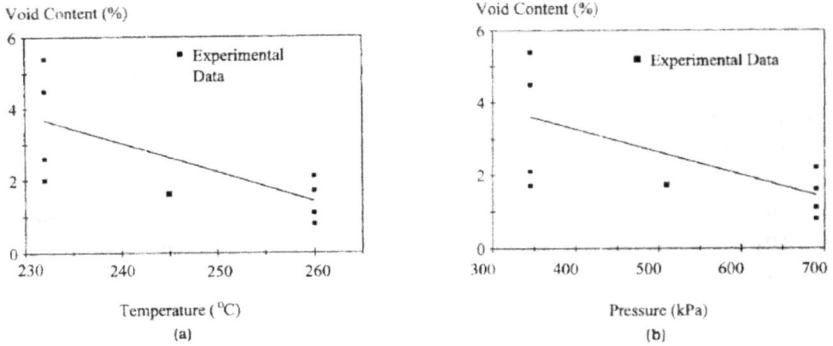

FIGURE 11.3 Influence of processing condition on void content. (Reproduced with permission from Ramasamy, Wang, and Muzzy 1996.)

11.3 FIBER DISTRIBUTIONS

The fiber distribution is one of the important factors which decide the performance of the composite material. As there are different hybrid yarn techniques for the manufacturing of the thermoplastic composite, each has different orientation and distribution of the fibers across the towpreg in the longitudinal as well as in the transverse direction. Hartness (1991) studied the fabrication of the composite laminate through powder-coated towpregs and commingled towpregs. For the commingled yarn technique, fiber and matrix should be in the form of continuous multifilament. The polymer which cannot be spun is of no use for the commingled yarn technique while it can be used in the fused powder technology. The fused powder technology reveals many advantages such as the elimination of the binder, handling issues, and this technique is also applicable for the polymer which cannot be spun. The fiber alignment in the powder-coated composite remains untwisted and parallel to the longitudinal direction while in the case of the commingled yarn technique, the fibers misalign due to the intermingling of the filaments and wave like structure forms along the length of the towpreg.

The fiber matrix distribution inside the composite is strongly dependent upon the hybrid yarn structure. The fiber and matrix distribution in the laminate made through different hybrid yarn techniques are shown in Figure 11.4. Lauke, Bunzel, and Schneider (1998) studied the effect of the hybrid yarn structure on the behavior of the composite laminate. The SEM images of the polished cross-section were taken to study the fiber/matrix distribution within the composite. Comparison of different hybrid yarn structures such as kemafil hybrid yarn (KEM), schappe hybrid yarns (SCH), commingling hybrid yarn (COM), friction spinning (FS), and side by side hybrid yarn (SBS) was made to understand the effect of the hybrid yarn structure on the interface property of the fiber-reinforced composite. Among these different hybrid yarns, SCH and COM composites depicted excellent mixing of the fiber and matrix phase. The degree of mixing of the fiber and resin phase depends upon the melt flow distance. The lower melt flow distance improves the fiber/matrix

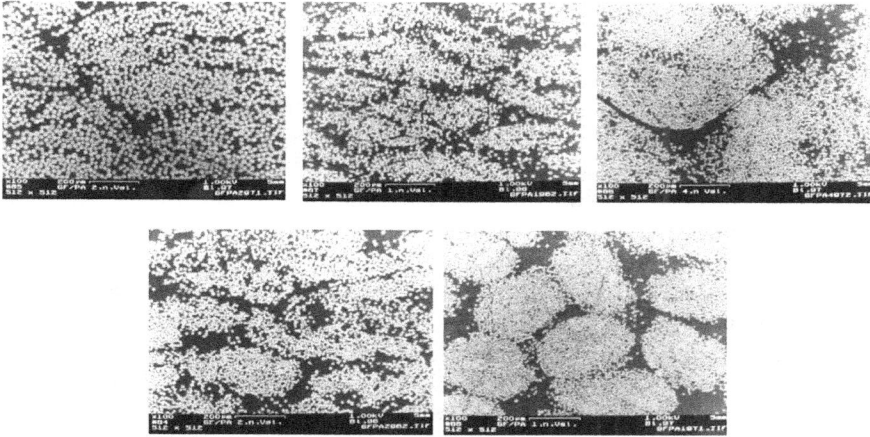

FIGURE 11.4 Cross-sectional view of the laminates manufactured through different tow-preg techniques: SCH, COM, KEM, SBS, and FS respectively. (Reproduced with permission from Lauke, Bunzel, and Schneider 1998.)

distribution during the consolidation stage. These different hybrid yarns are listed in terms of increasing melt flow distance: SCH, COM, KEM, SBS, and FS (Lauke, Bunzel, and Schneider 1998; Alagirusamy et al. 2006).

Ramani, Borgaonkar, and Hoyle (1995) studied the effect of the processing parameters on the impregnation and consolidation of the powder-coated composite laminate, which is different from the melt impregnated prepreg. The flow of the polymer matrix depends upon its state, such as powder, film, fiber, etc., while in the case of powder-coated towpregs it further depends upon the particle size, its distribution, and flow characteristics of the polymer. The quality of the impregnation and consolidation controls the formation of the voids within the laminates. The quality of the towpreg depends upon the interaction of powder particle and filaments in the tow during the powder coating process and the extent of heating and fusing of the powder onto the tow. The partially fused powder-coated towpreg is used to weave the complex preform structure due to the flexibility of the towpreg (Goud et al. 2018; Goud et al. 2019a; 2019b; Goud et al. 2020).

There are various factors which usually influence the consolidation process of the composite, such as capillary action, surface tension, and resistance to flow. There is a two-way flow of the molten polymer matrix during the consolidation process. The localized longitudinal flow of the molten matrix sets the initial level of the consolidation while the transverse flow is essential to complete the consolidation process. The localized flow of the polymer during consolidation is dependent upon various factors such as polymer particle size, fiber volume fraction, particle size distribution, and flowability of the molten polymer matrix. When the particle size exceeds the diameter of the filaments then the polymer sheath formation occurs. The temperature rate and displacement rate influence the consolidation quality of the composite laminate, which can be confirmed based on the void content and strength measurements

(Ramani, Borgaonkar, and Hoyle 1995; Goud et al. 2018; Goud, Alagirusamy, et al. 2019). Stolyarov et al. (2013) studied the property of different composite laminates prepared through the hybrid yarns such as direct spinning, wrap-twisting, and blending techniques. Glass fiber as the reinforcing element and polypropylene as matrix material were chosen for the development of the composite laminate. The tensile and bending strength of the laminate was compared for these three different hybrid yarn techniques. The fiber and matrix distribution inside the hybrid yarn is different due to the different combining processes.

a. **Direct spinning**: In this process, there is simultaneous spinning and blending of the reinforcing and thermoplastic fibers.
b. **Wrap twisting**: In this technique, the core reinforcing yarn is wrapped by the thermoplastic yarns. The surrounding yarn is also useful in protecting the brittle core fibers from the abrasion in the working conditions.
c. **Blending:** In this technique, the reinforcing elements and thermoplastic yarns are mixed through the compressed air jet.

The fiber distribution in the transverse direction is influenced by the hybrid yarn technology. The better the fiber matrix distribution the higher will be the impregnation quality of the hybrid yarn. Among these three technologies, the better fiber and matrix arrangement is found in the yarns made with the direct spinning process, and then in the wrap twisted yarn, the reinforcing fiber was at the core position while thermoplastic filaments were migrated to the outer surface. The non-uniform distribution of the fiber and matrix may lead to structural irregularity issues. In the second and third processes, due to the non-uniform distribution of the fiber and matrix, high temperature and pressure are required for better consolidation (Stolyarov et al. 2013; Alagirusamy et al. 2006).

11.3.1 Effect of Voids on the Mechanical Property of the Laminate

The void content of the laminate influences the mechanical property adversely. There are numerous ways of manufacturing the composite material of the same raw material. Each technique results in different void content. For the thermoplastic-based composite, the high viscosity of the resin poses challenges for better impregnation and the void is always an issue. There are various manufacturing techniques which reduce the melt flow distance of the resin to reduce the void content. The reduction in the melt flow distance results in a decrease in the void content. For reducing the melt flow distance, hybrid yarns such as powder-coated towpreg, commingled yarn technology, DREF yarn technology, micro braiding, co wrapping, etc. are useful (Chu, Ko, and Song 1992; Ramasamy, Wang, and Muzzy 1996; Bar, Das, and Alagirusamy 2017; Bar, Alagirusamy, and Das 2018; Bar, Alagirusamy, and Das 2019a; Bar, Das, and Alagirusamy 2019b; Bar, Alagirusamy, and Das 2019c; Bar et al. 2020). Ramasamy, Wang, and Muzzy (1996) studied the relationship between void content and tensile property of the composite laminate. The effect of the manufacturing technique was also studied in terms of the tensile strength as shown in Figure 11.5.

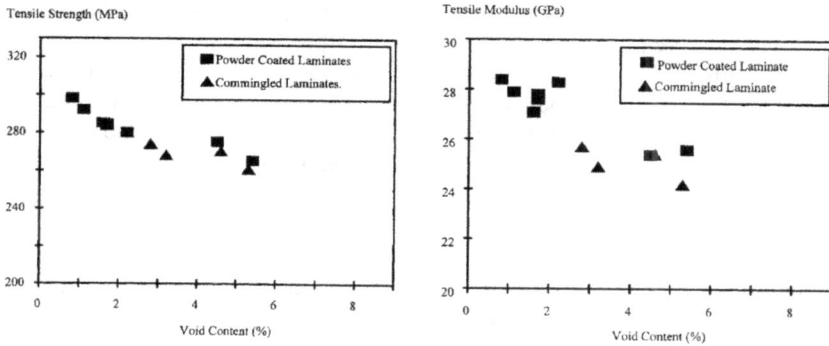

FIGURE 11.5 Voids content vs. tensile properties of the laminate. (Reproduced with permission from Ramasamy, Wang, and Muzzy 1996.)

Other mechanical properties such as compressive, flexural, and interlaminar properties were also studied and reported. The compressive strength was found to be 40–45% of the tensile strength while compressive modulus was measured to be about 70–75% of the tensile modulus. The dominant method of failure was interlaminar delamination in case of a compression test. The consolidation temperature influenced the flexural as well as the interlaminar property of the laminate significantly while there was less effect of the pressure and holding time on the property. For the same processing condition, powder-coated composites showed the superior property in comparison to the commingled towpreg-based composite. This happened due to the good impregnation of the matrix with the fiber inside the laminate and lesser void content in the case of the powder-coated composite (Chu, Ko, and Song 1992; Ramasamy, Wang, and Muzzy 1996). Long, Shanyuan, and Jianyong (2002) experimented on the composite laminate fabricated through the hot press method. The hybrid yarn was prepared through the commingling technique. A bending test was performed, and it was reported that the degree of matrix impregnation and bulkiness of the towpreg has a direct influence on the transverse flexural properties of the laminates. In this experiment, glass fiber and polypropylene filaments were used. The major finding of the work is that the structure of the commingled yarn has a direct relation with the mechanical property of the composite laminates.

11.4 EFFECT OF HYBRID YARN (TOWPREG) STRUCTURE ON MECHANICAL PROPERTIES

There are two types of processing conditions for the thermoplastic composite fabrication: one is pre-impregnation, and the other is post-impregnation. The pre-impregnation technique (also known as prepreg) fiber is wet and impregnated in one step such as in melt or solution impregnation. The prepreg is used to stack for the composite formation through hot press. But it is quite stiff and may not intertwine to form a 3D preform the room temperature. While in post-impregnation, impregnation does not occur in a single step. Here, a matrix is available in different forms such as fiber, film, powder. Fiber and matrix are blended either through physical or partial impregnation. This towpreg is flexible at room temperature and can be used in weaving any shape, either 2D or 3D according to the application. The powder impregnation is one of the famous techniques where powder penetrates even inside the filament's bundle and forms good consolidation on applying heat and press. The powder-coated towpreg can also be covered with the same material through wrapping, etc. to avoid powder loss during the weaving process. The matrix in the form of the film is used to make 2D laminates but this is not suitable for the 3D composite formation where impregnation in a through-the-thickness direction is difficult to achieve. For the 3D thermoplastic composite, the powder-coated or commingled towpreg technique are the best-suited hybrid yarn candidates. This has the capability to wet the fibers very uniformly (Kuo and Fang 2000; Goud et al. 2018; Jogur et al. 2018).

Swolfs, Gorbatikh, and Verpoest (2014) reviewed fiber hybridization in the polymer composites. The quality of the hybridization is directly linked to the quality of the composite laminates. For the complicated preform, it is very difficult to achieve uniform distribution of the matrix in the internal region. Hence, in this case, hybrid yarn is required to achieve a homogenous distribution of the matrix even at an internal interlacement point. The hybridization can be done at various levels such as at inter-layer level, intra-layer level, and yarn level as shown in Figure 11.6.

11.4.1 TENSILE PROPERTY

Lauke, Bunzel, and Schneider (1998) worked to explore the influence of hybrid yarn structure on the physical properties of the laminates. Other factors like fiber volume fraction, the orientation of the reinforcing fibers, distribution of the fiber, and matrix play an important role in controlling the behavior of the composite laminates. The presence of the discontinuous fiber leads to more stress concentration, which finally erupts as a fiber, matrix, and interface fracture and reduces the composite strength.

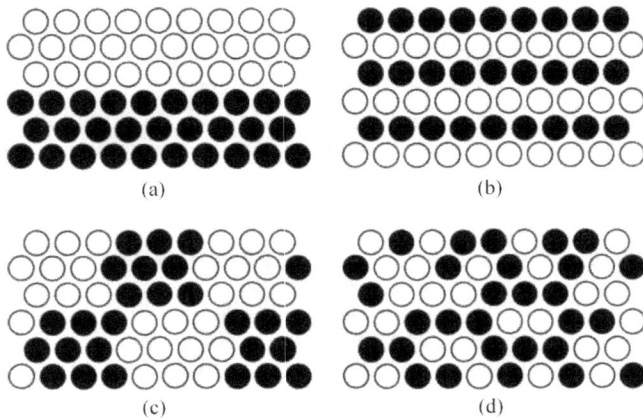

FIGURE 11.6 Degrees of hybridization at various scales. (Reproduced with permission from Swolfs, Gorbatikh, and Verpoest 2014.)

Composite laminates prepared through different hybrid yarn techniques depicted different mechanical properties. As the fiber-reinforced composites are non-isotropic material, it shows different mechanical properties in longitudinal and transverse directions. The variation of the longitudinal and transverse properties was due to the different breaking mechanisms. For the longitudinal tensile strength and modulus, the performance was found in decreasing order as SBS, FS, KEM, COM, SCH laminates respectively. There are more discontinuous yarns in the SCH composite material which resulted in different fiber orientations. Hence, SCH laminate showed the least tensile properties among these hybrid yarn techniques while in the case of commingled yarn there is waviness in the towpreg which led to the disorientation, and hence, this also showed lesser properties. In other techniques, fiber orientation was more or less unidirectional and therefore depicted better properties. The best longitudinal tensile properties were shown by the SBS hybrid yarn-based composite, while for the transverse tensile modulus and strength, the reverse trend was shown. The transverse properties were mainly dependent upon the quality of the impregnation of the fiber and matrix parts and composite microstructure. The highest transverse tensile properties were shown by the SCH and COM yarn structure composite. Hybrid yarns which have more discontinuous fibers and higher disoriented fiber arrangements showed better property in the transverse direction in comparison to the other methods, while poor property in case of FS yarn where fiber bundle is placed parallel with the matrix filaments. This happened due to large melt flow distance which resulted in poor fiber-matrix adhesion and hence, poor property in a transverse direction. The inter-laminar shear strength and tensile modulus are shown in Figure 11.7 (Rosselli and Santare 1997; Fu and Lauke 1998).

Nowadays, the demand for high specific strength material is increasing. Fiber-reinforced composite material has a high strength to weight ratio in comparison to the conventional material like metals and ceramics. The strength of the composite material mainly depends upon the reinforcing fiber, fiber orientation, and its distribution and forms of the fiber (continuous or staple) in the composite. The hybrid yarn prepared through the schappe technology where reinforcing fiber is in a staple form can never match its mechanical property with continuous fibers. This happens due

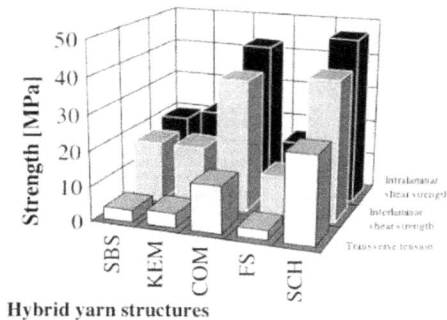

FIGURE 11.7 Properties of laminates prepared through different hybrid yarn techniques. (Reproduced with permission from Lauke, Bunzel, and Schneider 1998.)

to the disorientation and twisting of the staple fibers. In case of the friction spinning method, thermoplastic staple fiber wrapped around the filament yarn core. Due to the core/sheath structure, the impregnation of the resin inside the core element does not happen effectively. Overall, the mechanical property of the laminate prepared through the towpreg yarn is mainly dependent upon the degree of the blending of the fiber and matrix parts and also upon the multifilament damage during the production of the towpreg (Rosselli and Santare 1997; Lauke, Bunzel, and Schneider 1998; Alagirusamy et al. 2006). In another study, flat knitted fabric was produced using hybrid yarn spun through a different process such as friction spinning, air interlacing, twisting, and air texturing. The best tensile and flexural properties were found in the case of the preform prepared by the air jet texturing methods. The improvement in the mechanical properties is the result of a better degree of blending between the fiber and matrix phase (Alagirusamy et al. 2006).

Baghaei, Skrifvars, and Berglin (2013) studied the effect of the hybrid yarn structure on the mechanical property of the composite laminates. Co-wrapped hybrid yarn (as shown in Figure 11.8) was produced with hemp as reinforcing elements and PLA yarns as matrix. These hybrid yarns were produced in a different mass ratio of the fiber and matrix with different wrapping densities. The effect of the fiber volume fraction and wrapping density was studied in detail for the mechanical response of the resultant composite laminate. Since there is no loss of the matrix during the compression molding process, the laminate has the same mass fraction as in the hybrid fiber. Different composite laminates were manufactured with different fiber volume fractions. It has been observed that an increase in the fiber content in the laminate resulted in an increase in the porosity within the composite. This is due to the physical structure of the hybrid yarn and the preform. Also, the laminate made through hybrid yarn having lower wrapping density has more porosity. The reason for this behavior was due to the higher fiber volume fraction with a lower wrap density of the hybrid yarn. However, in terms of the quality of hybrid yarn, it was found that hybrid yarn through the co-wrapping technique showed high void content as well as the matrix rich and deficit zones (Miao, How, and Cheng 1994; Madsen and Lilholt 2003).

The tensile strength and modulus of the laminates made with co-wrapped hybrid yarn technique increased with an increase in the fiber volume fraction as shown in Figure 11.9. At the same time, it has been observed that for the constant fiber volume fraction the laminate made with hybrid yarns having high wrap density showed

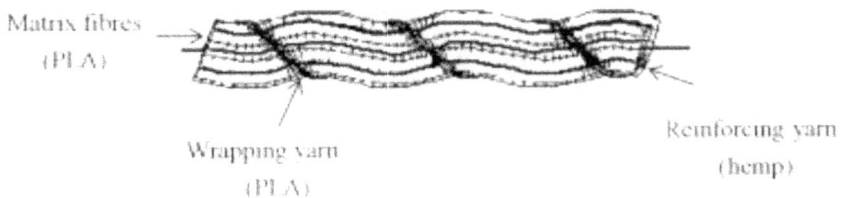

FIGURE 11.8 Co-wrapped hybrid yarn structure. (Reproduced with permission from Baghaei, Skrifvars, and Berglin 2013.)

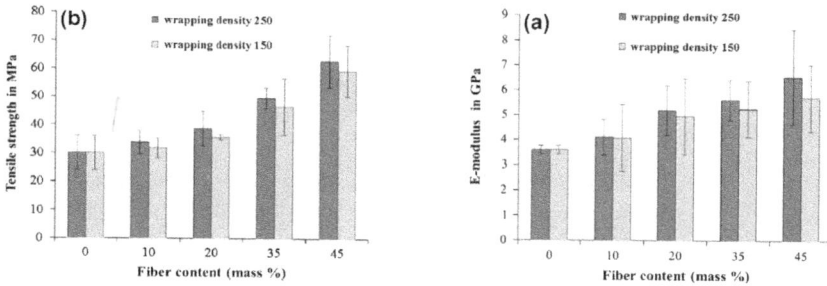

FIGURE 11.9 Influence of fiber content and wrapping density on the tensile strength and modulus of the laminate. (Reproduced with permission from Baghaei, Skrifvars, and Berglin 2013.)

higher tensile strength and the modulus. The mechanical properties of the laminate was lesser for those having lower fiber volume fraction. The increment in the tensile property can be explained based on the improved linear arrangements of the fibers with increasing wrapping density, which results in the more fiber alignments in the axial direction. At the low wrapping density, the bucking stress generated by the wrapping filament is the major reason for the yarn tortuosity. As the wrapping pitch decreases with an increase in the wrapping density, there will be less buckling of the reinforcing yarn by the wrapping filament due to the shortening of the bending span, hence improvement in the mechanical property noticed (Miao, How, and Cheng 1994; Baghaei, Skrifvars, and Berglin 2013).

Ramasamy, Wang, and Muzzy (1996) conducted an experiment on the laminate prepared through powder-coated towpregs and commingled yarn. The effect of towpreg structure was studied on the different mechanical properties of the composite materials. The powder-coated preform and commingled yarn preform were consolidated under various processing condition. The different properties such as tensile, compressive, flexural, and shear are listed at different processing conditions in Table 11.2 for both types of laminates. It has been observed that the resin has not filled all the internal gaps between the fibers inside the commingled towpreg laminates, while in the case of powder-coated composites, it was fully consolidated due to uniform matrix powder distribution throughout the tow and hence excellent fiber-matrix interface was found. This was also confirmed with SEM images as well as the better tensile properties in comparison to the commingled yarn composites. The lesser quantity of the voids in the powder-coated laminates reduced the stress concentrated weaker point, which in turn gave higher resistance against any deformation.

11.4.2 FLEXURAL PROPERTY

Long, Shanyuan, and Jianyong (2002) conducted an experiment on unidirectional glass/polypropylene composites at 50% volume fraction. The composite laminate was made with three different hybrid yarns: the first (UY1) was polypropylene filaments and glass fibers roving, the second one (UY2) was bulked glass fibers roving

TABLE 11.2

Mechanical Properties of Composite Laminates

Consolidation Conditions			Tensile Modulus (GPa)	Tensile Strength (MPa)	Compressive Modulus (Mpa)	Compressive Strength (Mpa)	Flexural Modulus (GPa)	Flexural Strength (MPa)	Shear Strength (MPa)
T (Cel.)	P (kPa)	t (min)							
Laminates from Powder-Coated Laminates									
260	345	5	26.2	282.4	18.6	124.9	16.5	422.9	65.6
260	345	30	27.8	284	19.5	128.3	16.4	444.8	69.1
260	690	5	27.9	292.3	22.5	142.5	18.5	436.1	72.3
260	690	30	28.4	298.6	21.6	138.2	22.1	432.4	75.4
232	345	5	25.6	265.5	17.2	115.2	12.2	446.6	48.9
232	345	30	25.4	275.4	18.9	118.5	12.8	456.1	52.6
232	690	5	28.3	280.1	21.2	120.3	14.4	432.1	52.3
232	690	30	27.1	285.2	20.4	111.4	15.6	461.2	58.6
245	520	18	27.6	284.5	19.6	122.3	13	439.3	61.2
Laminates from Commingled Tows									
260	345	5	23.7	260.7	14.6	111.2	15.4	446.5	58.6
260	345	30	24.9	268.1	16.5	124.4	17.5	447.9	61.2
260	690	5	25.4	270.3	18.5	130.1	18.2	434.2	70.3
260	690	30	25.7	274	17.6	128.2	21.6	452.4	75.6
245	520	18	20.2	240.6	12.1	103.2	_-	_-	_-

Source: Ramasamy, Wang, and Muzzy (1996).

and polypropylene filaments yarn, and the third one (AJCY) was air-jet commingled yarn. The bending strength and modulus of the composite made through these hybrid yarns were found in decreasing order for AJCY, UY2, and UY1 laminate respectively. This trend came out due to the differences in the mixing of the fiber and matrix constituents in the respective techniques. In the case of AJCY towpregs, there was a greater number of polypropylene filaments inside the fiber tow which reduced the impregnation distance and resulted in excellent impregnation and hence, the highest flexural properties. And the least number of PP filaments was found inside within the fiber tow in the case of UY1. The rate of the impregnation (Ux) depends upon several factors such as the viscosity of the polymer matrix, the permeability of the fibers, and the applied pressure gradient. The relationship of these factors with the rate of impregnation is given by Darcy's Law as shown in Equation 11.7 (Long, Shanyuan, and Jianyong 2002):

$$Ux = k/(h) \ x \ dp/dx \tag{11.7}$$

Where dp/dx is the pressure gradient, h is the viscosity of polymer matrix, k is the permeability of the fiber (Long, Shanyuan, and Jianyong 2002).

Klinkmuller et al. (1994) studied the flexural and shear behaviors of the composite laminates made with commingled yarn techniques at different consolidated temperatures. It was found that the transverse bending properties and shear properties were mainly governed by the matrix properties. But the higher quantity of fiber reinforcement in the composite resulted in an incomplete blending of the fiber and matrix phases which influenced the mechanical properties adversely. The probability of the unbonded zone increases with an increase in the fiber content of the composite. Therefore, the fiber-reinforced composite material with high fiber volume fraction is generally more sensitive to the processing condition for the properties in the transverse directions. The shear and flexural properties of the laminate increased with an increase in the consolidation temperature. The increase in the processing temperature results in a decrease in void content and better impregnation leading to an improvement in the physical properties of the composite laminate. Bar, Das, and Alagirusamy (2018) studied the effect of the hybrid yarn (DREF) structure on the mechanical properties of the composite. The effect of the core yarn twist and flax fiber content was studied for the flexural properties of the composite laminates. The influence of interface modification was also studied for the laminate's behavior. The bending strength and bending modulus increase with the increase in the bulkiness of the commingled yarn. An increase in the bulkiness of the towpreg enhances the velocity of the matrix impregnation into the reinforcing fibers, which results in the improvement of adhesion force between fiber and matrix (Alagirusamy et al. 2006).

11.4.3 INTERLAMINAR AND FATIGUE PROPERTY

The inter-laminar shear strength of the polymer composite mainly depends upon the quality of the fiber and matrix distribution as well as interface bonding. The commingled hybrid yarn, powder-coated towpreg, and others result in excellent interface bonding, therefore they exhibit higher inter-laminar shear strength. In case of DREF yarn and twisted yarn, fibers available in cluster forms result in poor interface and reduced inter-laminar shear strength (Alagirusamy et al. 2006). In another study, a towpreg was prepared using the commingled yarn technique, and the thermoplastic composite laminate was manufactured with woven and warp knitted structures. The fracture toughness test was carried out to study the crack propagation in three modes such as mode-I, which is simply tensile opening of the cracks; next is Mode-II, in which sliding of the layers happens; and the last one is Mode-III, where tearing shear dominates. To investigate fracture energy, a double cantilever beam (DCB) test method was used. The magnitude of fracture toughness was measured higher for the woven structure composite in comparison to the warp-knitted structure. The greater toughness for the woven laminates was attributed to the weft yarn which restricted the crack propagation and increased the energy demand for the crack propagation (Yoon and Takahashi 1993; Alagirusamy et al. 2006). On the contrary, Svensson, Shishoo, and Gilchrist (1998) reported that there was no significant difference in the fracture toughness obtained in the woven and warp knitted laminate in the mode-I test. The laminate was prepared using commingled glass/PET hybrid yarns. In this work, the fracture mechanism of the laminate was captured through the SEM images. The dominant fractures surface revealed fiber pull-out during the test. In the woven laminate, the extensive fiber pull-out was prevented by the yarns oriented in the weft direction, while more stable crack growth was found in mode-I than mode-II. In both the modes, woven laminate showed marginally higher fracture toughness than the warp-knitted laminate. The thermoplastic laminate has generally higher fracture toughness; this happens due to the extensive fiber pull-out in the thermoplastic due to the weaker fiber/matrix bonding.

There are various methods to develop hybrid yarn such as SCH, COM, SBS, KEM, FS and each has its own characteristics. The composite manufactured through these hybrid yarn techniques has different fiber distribution and orientation which influence the characteristics of the composite laminates. Lauke, Bunzel, and Schneider (1998) reported the comparison of inter-laminar and intra-laminar shear strengths of different hybrid yarn-based composites. In mode-I, critical energy release rate was determined by the DCB and SEN test while in mode-II, it was through the ENF test. In the fracture energy test, the energy was mostly consumed through fiber fracture, fiber bridging, fiber pull-out, debonding between fiber, and matrix, and also matrix related phenomena such as plastic deformation, brittle matrix fracture, etc. The critical energy rate also depends upon the length of the initial crack for the stable crack growth. The highest critical energy rate was found in the case of the COM composite laminates. When the crack started propagating inside the laminate, the disoriented fiber started debonding from the matrix but remained in contact with both the fractured surfaces and acts as a fiber bridging and consumes more energy in the crack propagation. This mechanism was also shown by the SCH composite but the peel off distance was dependent upon the fiber lengths. For the KEM and FS composite, the crack mainly propagated through fiber bundles and matrix rich zones. Although the fracture energy rate can be measured through both DCB and SEN tests, the absolute value has a considerable difference due to the different geometry and boundary conditions for the initial crack generation. However, the DCB method is more reliable to measure fracture toughness, but the SEN method is generally used for the comparison purpose of the different samples. The crack propagation mechanism is different in both DCB and SEN test methods. In the DCB test, the crack propagates through the inter-laminar delamination while in the SEN test the crack propagates through the intra-laminar fracture mechanism. The good impregnation of the resin with the fiber has a positive impact on the fracture energy rate (Davies 1990; Ye and Friedrich 1993). Yoon and Takahashi (1993) studied the inter-laminar toughness behavior of the carbon/PEEK commingled laminate. The prepreg laminate exhibited the stable crack growth while the commingled 2D laminate depicted a unstable crack growth. The performance of commingled UD laminate was found in between 2D CF/PEEK and CF/epoxy prepreg. The unidirectional composite exhibited better fracture toughness in comparison to the prepreg laminate. The main reason for this behavior was the fiber distribution in the commingled yarn-based composite laminates. The towpreg structure hurdled the straight flow of the crack growth and the fiber bridging phenomena increased the fracture toughness, while in case of 2D laminates where carbon fibers available in both the directions such as weft and warp which resulted in further higher fracture toughness.

The crack propagation was arrested by the weft yarn. Therefore, the crack became unstable and started propagating through the neighboring layers and formed another crack propagating plane, and hence, it absorbed more fracture energy (Yoon and Takahashi 1993).

For the fatigue test, commingled-based composites show fatigue sensitive behavior due to the stress concentration in the laminate caused by the waviness of the yarn and matrix-rich and fiber-rich zones (Diao, Ye, and Mai 1997; Alagirusamy et al. 2006). Dickson et al. (1985) studied the fatigue response of the carbon/PEEK laminates of different stacking sequences. The tensile fatigue behavior of the [0/90°] CF/PEEK laminate was found to be the same as that of the [0°] carbon fiber-based epoxy composite. The PEEK thermoplastic resin system formed the excellent adhesion with the fiber and inhibited the localized fiber failure and fatigue crack growth, which in turn gave improved fatigue resistance and ductility of the laminates. The fatigue response of the [+45/–45] stacked laminates was observed to be better than the [+45/–45] stacked carbon epoxy laminates due to better adhesion property of the PEEK matrix resin system. Also, the carbon/PEEK laminates showed better fatigue response than the carbon/epoxy laminates for other stacking sequences such as [0/±45/90]$_s$.

11.4.4 IMPACT PROPERTY

The impact test can be categorized as low-velocity impact and high-velocity impact based on the velocity of the projectile. This can also be categorized as drop weight impact and notched swinging pendulum impact based on the configuration of the laminate and much more. For the impact application, earlier isotropic metallic material was the first choice, but the heaviness of the metal posed the problem in the handling. The development of the composite material is the best innovation in substitution of the material in terms of strength to weight ratio. Its application is extended to the aircraft for better stiffness, cost, and ease in the manufacturing process (Cantwell and Morton 1991; Jogur et al. 2018). But the impact response of the material has some inherent problems like a localized failure. The composite material has a unique way of responding to the impact load unlike the metallic counterparts which can absorb the kinetic energy of the projectile through elastic and plastic deformation. Composite has a limited plastic deformation capacity, which results in large fracture areas with a significant reduction in the stiffness and strength of the laminates. Nowadays, composite material is being used in the manufacturing of civil and military aircraft where the consequences of the impact are very serious. Therefore, research related to the impact property of the composite material is of utmost important (Alagirusamy et al. 2006; Jogur et al. 2018).

The impact property of the fiber-reinforced composite is mainly dependent upon its constituents such as fiber, matrix, and interface bonding. The impact property of the thermoplastic composite is superior in comparison to its counterpart thermoset composite. The energy required for the fracture of a thermoplastic composite is more than that of a thermoset composite (Cantwell and Morton 1991). Several other parameters influence the impact property such as the stacking sequence, the structure of the preform, shape of the projectile, and many more (Cantwell and Morton 1991; Alagirusamy et al. 2006). Charpy and Izod are the oldest methods of impact test. By the pendulum, notch impact strength and energy absorption mechanism are used to understand. These test methods provide ease in the testing procedure and they are simplest in form with low time and cost factors. In Charpy test methodology, the specimen is fixed in the horizontal plane with simply supported arrangements, and the pendulum impactor strikes the specimen in the middle just opposite to the notch provided. The energy absorbed during the test is measured with the dial provided in the instruments. Furthermore, the information is captured through the strain gauge by instrumenting it with the impactor. This helps in finding the variation of the impact force with the time. Izod impact is the same as Charpy impact only with a slight difference in the configuration of the specimen. In this test, the specimen is fixed vertically having cantilever like support, and one end is kept free. The impactor strikes on the free end, having a notch on the same side, while the drop weight impact test is another method in which the specimen is kept in a horizontal plane and clamped property. The impactor is dropped from a particular height as per the energy level requirement. The velocity of the impactor can be calculated either through the energy equation or through the optical sensor attached just above the target plate. The impactor is well equipped, which enables it to measure force vs. time and also attached with the displacement transducer, which helps in determining energy dissipation during the impact event. This test enables a wide range of the specimen dimensions, which helps in the study of the complex damage mechanism involved in the composite laminate (Cantwell and Morton 1991; Jogur et al. 2018; Goud, Alagirusamy, et al. 2019). The composite impact property can be influenced by the structure of the towpreg and preform geometry. A complex stress is generated during the impact loading on the composite laminate. The response of the fiber-reinforced composite depends upon the load transfer occurring between the fiber and matrix. The load transfer phenomena are the function of the interface bonding between fiber and matrix which can be controlled with the manufacturing technique.

Cantwell and Morton (1991) reviewed the impact property of the composite laminate and the factors which influence the response of the laminate. The material property, such as type of fiber, matrix, and manufacturing technique, and external factors such as the shape of the impactor, boundaries condition together decide the impact property of the composite laminate (as shown in Figure 11.10).

The composite made with co-wrapped hybrid yarn having higher wrapping density showed lower impact strength than the composite having lower wrapping density. This was mainly due to the change in the fiber pull-out length. The energy absorption is proportional to the yarn pull-out length. For the longer fiber length, more energy is required to pull out during impact loading. For the higher wrapping density yarn, there was a better interfacial strength, which in turn decreased the average fiber pull-out length during the impact loading, and therefore it absorbed lesser energy than the composites made with yarns having lower wrapping density as shown in Figure 11.11 (Baghaei, Skrifvars, and Berglin 2013).

Baghaei and Skrifvars (2016) conducted a study on the laminates prepared with wrap spun hybrid yarn. The hybrid yarn was produced with hemp/PLA and hemp/lyocell/PLA through wrap spinning. The composite was fabricated using the compression

Specimen

Notch   Impactor

Weighbar tube

Inertia bar

Strain gauge —| | |—

Specimen

a

Impactor

Specimen
Notch   Moveable jaw
Fixed jaw

Strain gauges

Input bar

Yoke

b   **c**

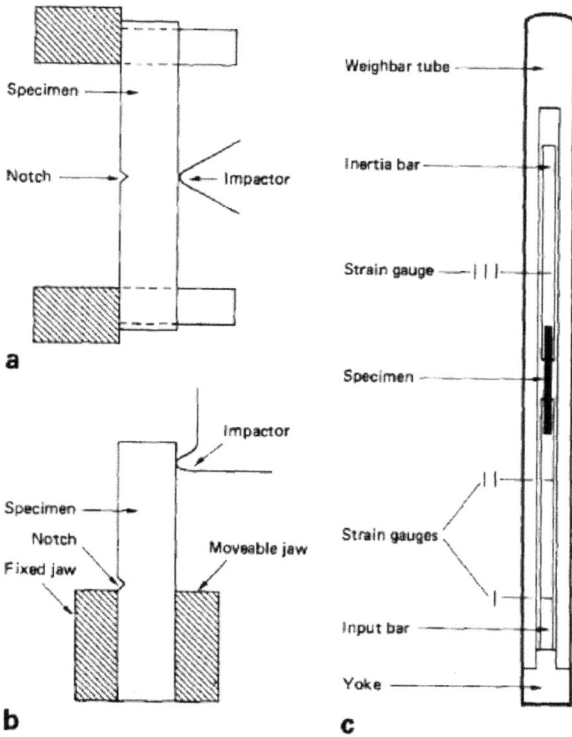

FIGURE 11.10 Various configuration for the impact test. (Reproduced with permission from Cantwell and Morton 1991.)

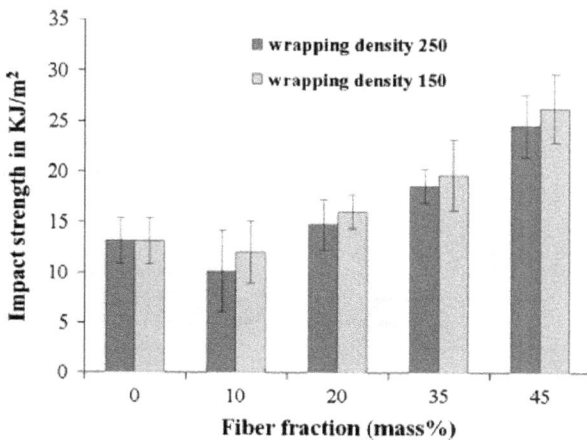

■ wrapping density 250
▨ wrapping density 150

Impact strength in KJ/m^2

Fiber fraction (mass%)

FIGURE 11.11 Impact strength of the composite laminate at different fiber volume fractions and different wrapping densities of the hybrid yarn. (Reproduced with permission from Baghaei, Skrifvars, and Berglin 2013.)

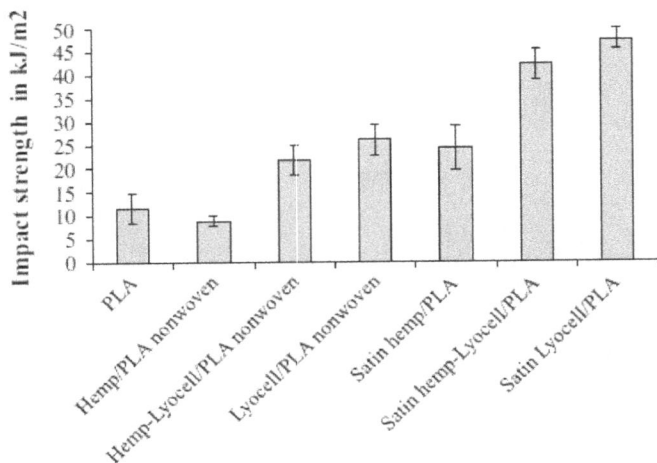

FIGURE 11.12 Impact strength of composite laminates at different preform structure. (Reproduced with permission from Baghaei and Skrifvars 2016.)

molding technique. The impact strength was measured for different fiber orientations within the laminate as shown in Figure 11.12. The maximum impact strength was found for the satin lyocell/PLA laminates in comparison to other combinations and non-woven composite laminates.

Thanomsilp and Hogg (2003) studied the penetration impact strength of composites prepared with the commingled towpreg technique. During the transverse out of plane impact test, different stresses generate at the top and bottom surfaces of the composite laminates. The complex shear, tensile, compression, and flexural stress generate in the laminates during impact loading. The development of the stresses in the composite plate depends upon the ratio of span length to the thickness of the materials as shown in Figure 11.13. The force vs. displacement and energy vs. displacement curve were plotted as shown in Figure 11.14.

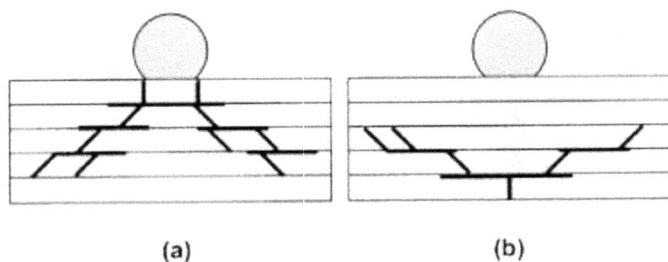

(a)   (b)

FIGURE 11.13 Illustration of crack growth (a) as a result of contact stresses and (b) as a result of flexural deformation. (Reproduced with permission from Thanomsilp and Hogg 2003.)

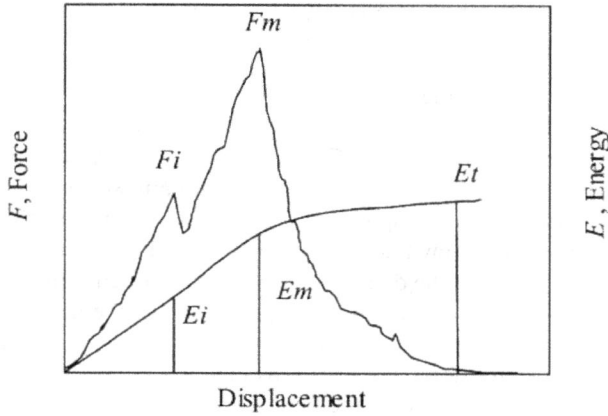

FIGURE 11.14 Graphs for the impact loading. (Reproduced with permission from Thanomsilp and Hogg 2003.)

The parameters such as resin type, lamina stacking sequence, and reinforcement architecture play a vital role in determining the energy absorption capacity. The best parameter to relate the penetration energy is the product of total fiber thickness and impactor diameter, which can cover different impact test geometries. The comparison of impact property for the carbon and glass fiber composites is shown in Figure 11.15 (Thanomsilp and Hogg 2003; Jogur et al. 2018).

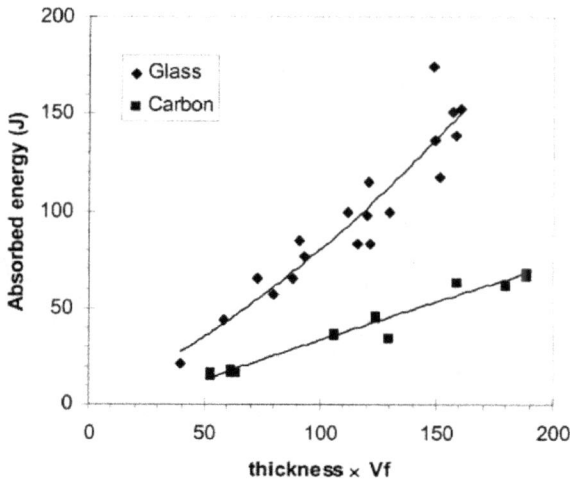

FIGURE 11.15 Energy absorption curve for the glass and carbon fiber-based composites laminates. (Reproduced with permission from Thanomsilp and Hogg 2003.)

11.5 EFFECT OF PREFORM STRUCTURE ON THE MECHANICAL PROPERTIES

The mechanical property of the composite material can be easily controlled with the help of the architecture of the preform. The behavior of the composite laminate is dependent on the structure of the reinforcement. There are various preform geometries that can be produced, such as unidirectional, plain woven, knitted, different 3D woven fabric (as shown in Figure 11.16), and many more. Each kind of reinforcement geometry has its own application and inherent properties (Alagirusamy et al. 2006). In recent times, the demand for the 3D fabric reinforced composite has increased significantly. This is due to the improvement of the mechanical property in the third direction. This is also useful in the production of the integrated hollow structures. The production of 3D preforms involves a complex weaving process where friction and abrasion between the towpregs during the weaving process is the challenge. There is a relation between fiber damage and tow size during the weaving process. For the same quantity of fiber tow, the smaller tow has a larger surface area, therefore, more abrasion during the weaving process and hence, more fiber damage occurs to the fiber available at the fiber bundle surface (Hugh et al. 1992; Kuo and Fang 2000). Traditionally, a 3D composite was prepared mainly for the thermoset composites. RTM is the most common technique for composite manufacturing where resin infuses to the desired location of the preform due to very low viscosity, while in the 3D thermoplastic composite, it is quite difficult to achieve that

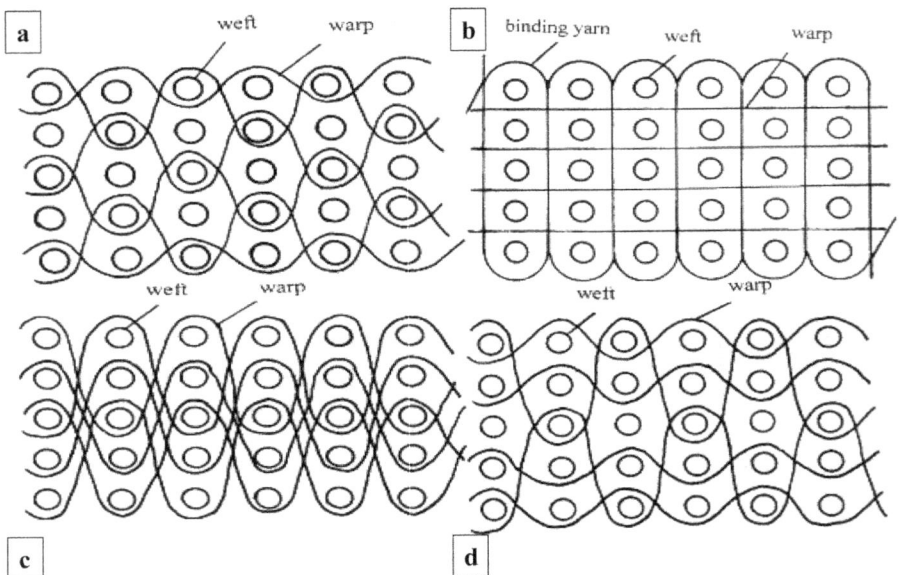

FIGURE 11.16 Illustration of orthogonal structure for layer-to-layer binding (S1), orthogonal structure (S2), angle interlock structure (S3), and modified angle interlock structure (S4), respectively. (Reproduced with permission from Gu and Zhili 2002.)

level of impregnation of the resin due to the very high viscosity of the matrix. This restricts the manufacturing of the 3D thermoplastic composite due to the poor blending of the fiber and matrix phase (Kuo and Fang 2000). But nowadays a hybrid yarn technology makes it possible to produce 3D thermoplastic composite laminates with homogenous distribution of the fiber and matrix parts, even at interlacement points.

11.5.1 Tensile Property

Gu and Zhili (2002) studied the effect of the 3D weaving geometry on the tensile property of the composites. It was found that the architecture of the reinforcement strongly influences the tensile property and dimensional stability of the composites as shown in Tables 11.3 and 11.4.

The variation in the tensile strength is observed due to the difference in the fiber volume fraction and extent of the bending of the yarns in different structures as shown in Figure 11.17. To achieve higher strength of the laminate, more straight filaments in the structure are preferred with minimum bending of the yarn (Gu and Zhili 2002; Ansar, Xinwei, and Chouwei 2011). Lee et al. (2002) studied the damage

TABLE 11.3
Tensile Strength for (a) Preform and (b) Laminate

	(a)					(b)			
No.	Tensile Strength			Average (KN)	No.	Tensile Strength			Average (KN)
S1	2.2	1.9	2	2.03	S1	6.4	6.4	6	6.27
S2	2.75	2.7	2.8	2.75	S2	6.8	7.6	8	7.47
S3	2.5	2.6	2.6	2.57	S3	5.6	5.2	5.5	5.43
S4	2.95	2.8	3	2.92	S4	6.6	7	7.6	7.07

Source: Gu and Zhili (2002).

TABLE 11.4
(a) Fiber Volume Fraction of the Composite and (b) Elongation at Break

	(a)					(b)			
No.	Fiber Volume Content			Average (%)	No.	Elongation at Break			Average (%)
S1	64.6	58.7	60.7	61.3	S1	7.8	8.06	7.72	7.82
S2	59.1	63.3	63.3	61.9	S2	6.1	6	6.09	6.06
S3	55.7	55.1	55.4	55.4	S3	4.87	5.49	5.27	5.21
S4	51.6	56.5	54.7	54.3	S4	5	5.17	5.97	5.38

Source: Gu and Zhili (2002).

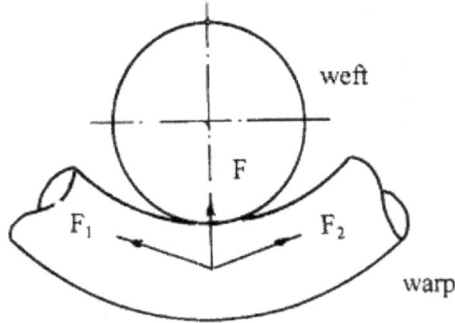

FIGURE 11.17   Force on the yarn. (Reproduced with permission from Gu and Zhili 2002.)

that occurred to the yarn during the weaving process. 3D weaving may be divided into six stages: let off, tensioning, shedding, weft insertion, beating, and take-up. The damage to the fiber during the weaving process depends upon the types of the yarns and stages involves. The yarn breakage leads to the adverse effect on the tensile properties of the composite laminates as shown in Figure 11.18.

Baghaei, Skrifvars, and Berglin (2015) studied the mechanical property of the laminate having different preform structures. The tensile property of the unidirectional hemp/PLA composite was found to be smaller than the woven fabric composite. The main reason for this behavior was the tightly woven fabric structure where all fibers were held rigidly at their place. In the woven fabric laminate, the tensile strength of the satin composite was observed to be higher than the basket 2/2 composite laminate. This behavior was attributed to the inherent fabric structure which caused weaker zones inside the laminate. There was higher crimp in the basket weave in comparison to the satin weave. This was mainly due to the presence of

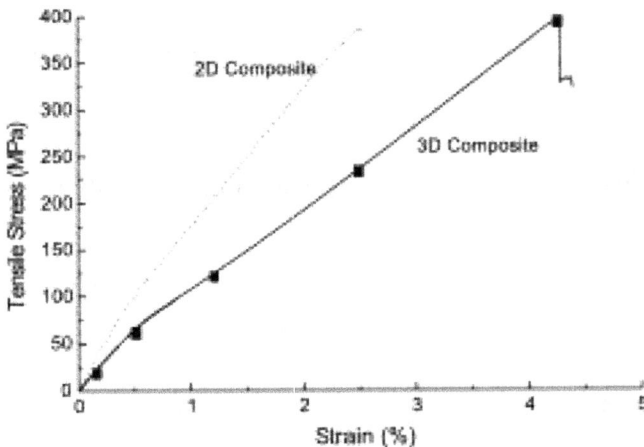

FIGURE 11.18   Tensile test response for 2D and 3D laminate. (Reproduced with permission from Lee et al. 2002.)

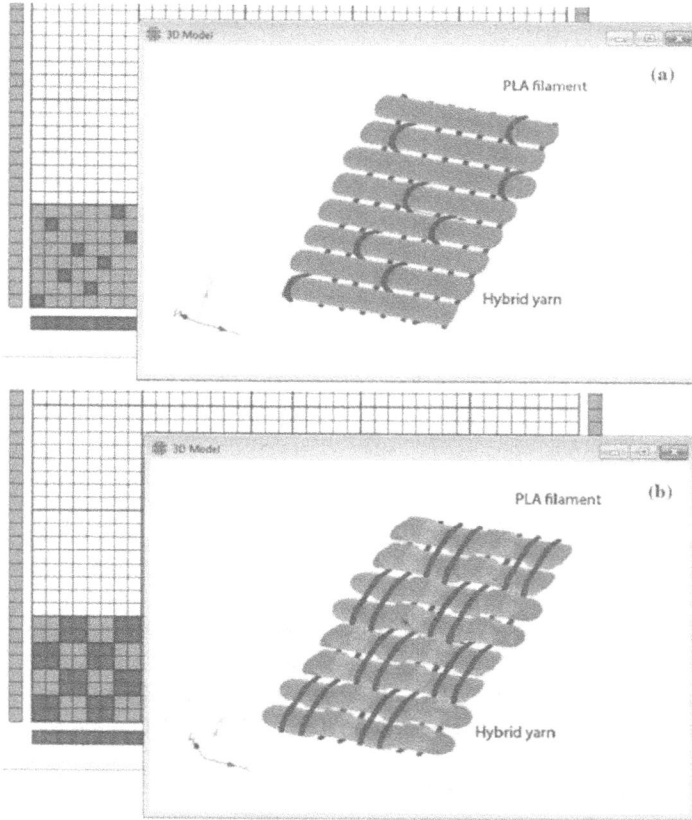

FIGURE 11.19   Illustration of (a) 8-harness satin and (b) basket weave patterns. (Reproduced with permission from Baghaei, Skrifvars, and Berglin 2015.)

more interlacement points in the basket weave. The greater crimp in the fabric led to a reduction in the tensile strength of the composite material. In the satin weave structure, each adjacent floating yarn has a different interlacement point, which facilitated uniform resin distribution, unlike in a basket weave structure where irregular resin distribution created matrix-rich and fiber-rich zones which produced stress concentration zones. This adversely affected the tensile property of the composite laminates. The matrix-rich zone acts as a weak point as it does not have any reinforcing elements. The schematic diagram for the satin and basket weave structure is shown in Figure 11.19 (Baghaei, Skrifvars, and Berglin 2013; Baghaei and Skrifvars 2016).

The same trend was found in the case of flexural and impact strengths of the laminate as shown in Figure 11.20. The satin composite showed the highest flexural strength and the modulus as well as impact strength. This happened due to the presence of the more aligned yarn in the satin composite, and there was a smaller number of stress concentration points in the same laminate. This improved the load transfer in composites during flexural and impact loading (Baghaei, Skrifvars, and Berglin 2013).

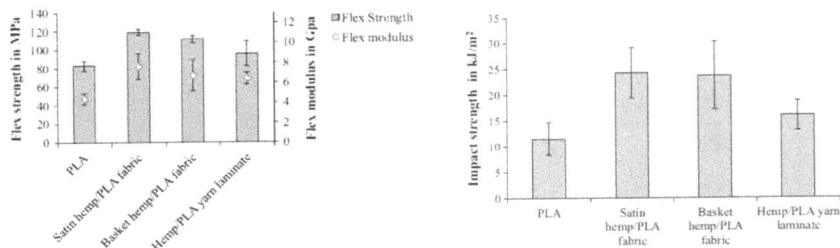

FIGURE 11.20 Flexural and impact properties of the composite laminates. (Reproduced with permission from Baghaei, Skrifvars, and Berglin 2015.)

11.5.2 FLEXURAL PROPERTY

Kuo and Fang (2000) conducted an experiment on the 3D thermoplastic composite. The main focus of the work was to explore behavior of the mechanical properties with the processing conditions and their microstructures. A powder-coated nylon/carbon yarn was used to develop 3D preform. The two types of preforms were explored: one was a 3-axis woven fabric (3A) where fiber was distributed in three orthogonal directions, and the second was a 2-step (2S) braided preform, where most of the fibers were aligned along the axial direction. The fabric specification is shown in Table 11.5. The composite was manufactured using the compression molding technique.

The load vs. deflection curve was plotted for the flexural behavior of the composites with preform geometries consolidated at different temperatures with different molding thicknesses. The load vs. deflection curve was used to describe the behavior of the composite. Two zones can be observed in Figure 11.21: one is linear, and the other is a non-linear zone. The length of the non-linear zone talks about the degree of ductility of the material. It can be observed that the thinner laminate has a lower non-linear zone and tends to behave like a brittle material. The thicker laminate has a higher length of the non-linear zone and tends to behave more like a ductile material. In general, for the uniform and brittle material, the initial slope of the curve is proportional to the cube of the thickness and the ultimate load is proportional to the square of the thickness. But the role of the thickness of the laminate is not very clear. As a composite is influenced by different factors other than the geometrical parameters due to its non-isotropic nature, a composite material is

TABLE 11.5
Fabric Details

Fabric	Number of Axial Yarns	Total Yarns in the Fabric	Width (mm)	Thickness (mm)	Pitch Length (mm)
3A	50	65	33	10.3	7.6
2S	67	81	23.2	10.1	18.9

Source: Kuo and Fang (2000).

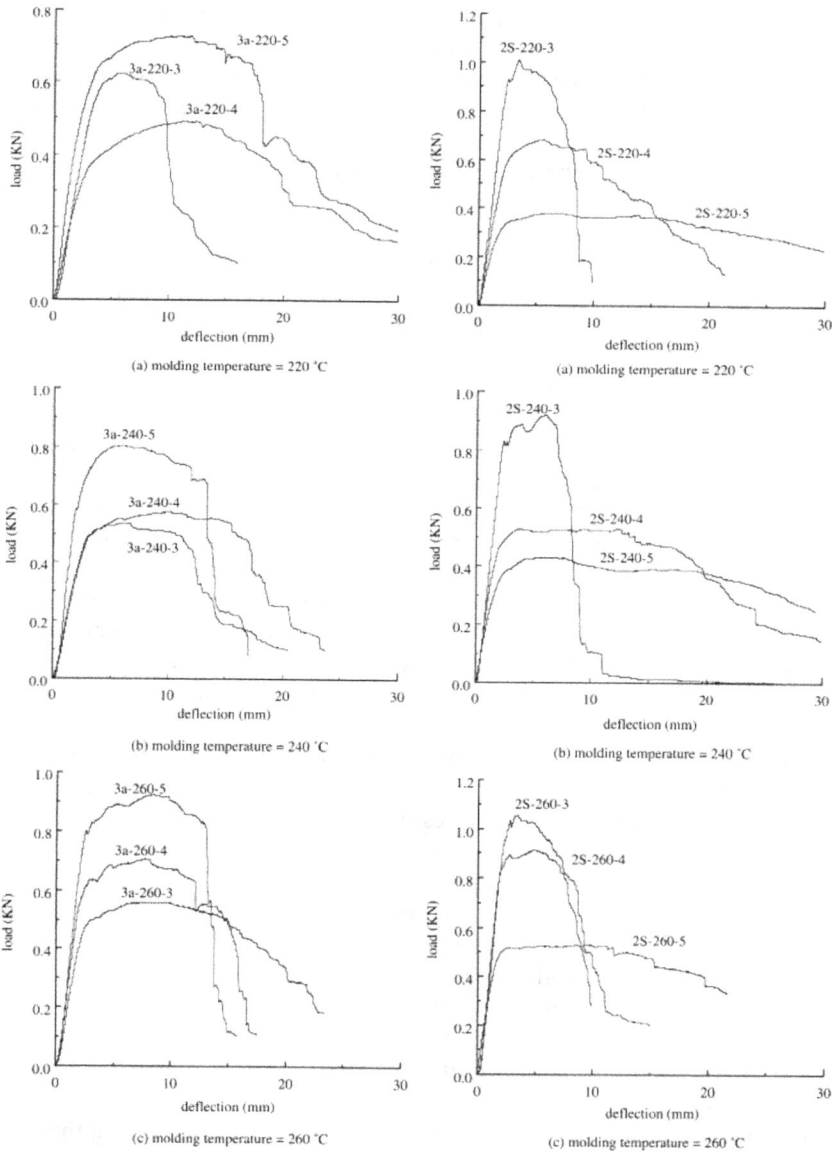

FIGURE 11.21 Flexural load vs. deflection curve for composite laminates manufactured at different consolidating temperatures. (Reproduced with permission from Kuo and Fang 2000.)

neither perfectly brittle nor ductile; therefore complex material factor comes into the picture with changes in the thickness of the laminate (Kuo and Lee 1998; Kuo and Cheng 1999; Kuo and Fang 2000).

For the 3A composite, 5 mm thick laminate has a high value of the initial slope and ultimate load for each processing temperature level, which indicates the dominancy of the moment of inertia, but the results were varying for other thickness levels. For the 2S composite, it showed more responsive behavior against the change in the thickness in comparison to the 3A composite. In the 2S composite, the thinner 3 mm laminate has the highest value of the initial slope as well as the ultimate load, which behaved more like a brittle material, and the curve dropped much earlier than the thicker ones at each temperature levels. The brittle behavior of the laminate was prominent due to the dense fiber packing. The increase in the modulus was more dominant in the 2S composite due to the high value of the fiber fraction in the axial direction. The 5 mm specimen has the lowest ultimate load and the longest non-linear portion, which indicates poor load transfer between the fiber and matrix (Byun and Chou 1996; Kuo, Ko, and Chen 1998; Kuo and Fang 2000).

The thickness of the composites is very critical to the composite modulus and damage mechanism. The composite modulus not only depends upon the fiber volume fraction but also on the bonding of the fiber and matrix. At 3 mm thickness of the laminate, the 2S composite exhibited more stiffness than the 3S composite, which was due to increased fiber volume fraction in the axial direction, while in case of the 5 mm laminate, the result was the opposite. Although the 2S laminate still has higher fiber in the axial direction, the poor bonding of the fiber and matrix parts led to the poor functional property. The laminate thickness influences the failure modes during loading. The improved bonding in the laminate causes tensile rupture and kinking of the fibers. The fiber breakage was found mainly in the buckling mode in the case of the thicker laminate whereas failure in tensile rupture mode was observed least (Byun and Chou 1996; Kuo, Ko, and Chen 1998; Kuo and Lee 1998; Kuo and Cheng 1999; Kuo and Fang 2000).

In another study, Svensson, Shishoo, and Gilchrist (1998) compared the warp-knitted laminate with a woven laminate. Warp knitted laminates were found to be slightly stiffer than the woven laminates in the warp direction, but both materials were equally strong in terms of the strength. In the weft direction woven laminate showed a better property; this was due to the presence of the glass fiber in the weft direction which, was absent in the case of the warp-knitted laminate. The same trend was found in the case of the flexural strength and modulus. Shear moduli were found to be same in both cases. Wang and Zhao (2006) studied the effect of different 3D structure, such as 3D braid (BR), 3D woven (WV) fabric, and 3D fabric with twisted yarns (TY) reinforcement on the flexural and short beam strength property of the composite. The bending and shear forces act simultaneously in a 3-point bending test. The bending load decreases with the decrease in the span length while the shear force becomes predominant. When the span length is large, the failure occurs in the bending mode, and when the span length is shorter, then the failure occurs in the inter-laminar shear mode (short beam). The ratio of the inter-laminar shear stress to the maximum bending stress is given by the beam geometry method for the linear elastic material as per Equation 11.8 (Wang and Zhao 2006).

$$\frac{\tau}{\sigma_b} = \frac{0.5}{\left(L/t\right)} \tag{11.8}$$

For the isotropic material, the failure occurs in the shear mode if $\frac{\tau}{\sigma_b}$ exceeds 0.58, as per the Von-Mises criterion. However, for the anisotropic material, the shear failure may occur even at the lower magnitude of the ratio. The shear stress measured with the maximum load in a short beam test is known as the apparent inter-laminar shear strength (ILSS). The apparent ILSS is often close to the actual one when the failure mode occurs predominantly in the shear mode; otherwise apparent ILSS will be the lesser than the actual ILSS (Wang and Zhao 2006).

11.5.3 INTER-LAMINAR AND FATIGUE PROPERTY

Diao, Ye, and Mai (1997) conducted a study on the continuous AS4 carbon fiber/ PEEK composite laminate made with the commingled yarn technique. Unidirectional fabric composites were produced having different stacking sequences, such as $[0]_{16}$, $[90]_{16}$, and $[0_2/90_2]_{2s}$. These different preforms were manufactured to investigate the response of the laminate against tension-tension fatigue loading. The fatigue life and residual strength under fatigue loading were studied for all preform structures, as shown in Figure 11.22. The composite laminate prepared using pre-impregnated tape prepreg (APC-2) has better fatigue response than the commingled composite laminate as shown in Table 11.6. The commingled laminate was found to be more sensitive to fatigue loading due to the disoriented fiber arrangement.

The S/N curve was drawn for the CF/PEEK composite with different stacking patterns as shown in Figure 11.23. The CF/PEEK commingled laminate showed quick degradation in comparison to the APC-1 laminate. This indicated that the commingled laminates are more fatigue sensitive. The residual strength degradation was plotted against number of cycles for the $[0]_{16}$ and $[0_2/90_2]_{2s}$ laminate at different fatigue stress levels in Figure 11.24 (Diao, Ye, and Mai 1997).

During the fatigue loading, there are mainly two types of a failure modes: one is splitting along the sample length and the other is transverse breakage normal to the fiber at some position. As per the experimental observation, splitting failure generally occurs during static loading while transverse failure occurs catastrophically after the

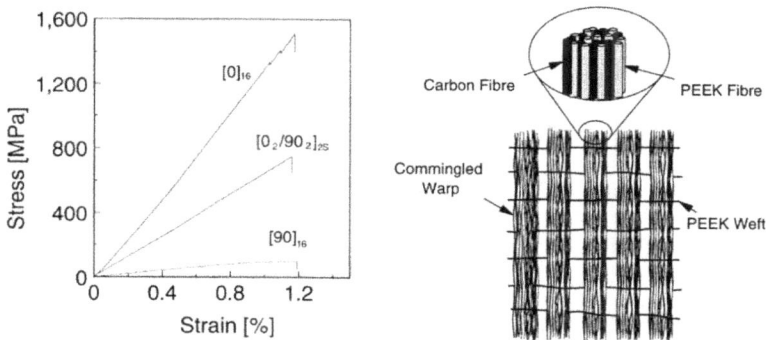

FIGURE 11.22 Stress vs. strain curve for the composite and illustration of commingled unidirectional fabric prepregs. (Reproduced with permission from Diao, Ye, and Mai 1997.)

TABLE 11.6
Mechanical Property of the Composite Laminate

Materials	Strength (MPa)		Stiffness (GPa)	
APC-1 (XAS/PEEK)	0	1256.2	0	–
	0/90	740	0/90	61
	90	–	90	–
APC-2 (AS4/PEEK)	0	2068	0	138
	0/90	1100	0/90	73
	90	86	90	10.1
AS4/Epoxy	0	2132.6	0	138.8
	0/90	932	0/90	83
	90	63.4	90	9.6
Pure PEEK Resin		100		33.7

Source: Diao, Ye, and Mai (1997).

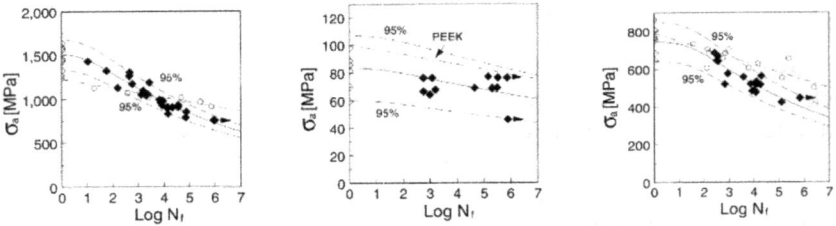

FIGURE 11.23   S/N curve for $[0]_{16}$, $[90]_{16}$ and $[0_2/90_2]_{2s}$ composite laminate respectively. (Reproduced with permission from Diao, Ye, and Mai 1997.)

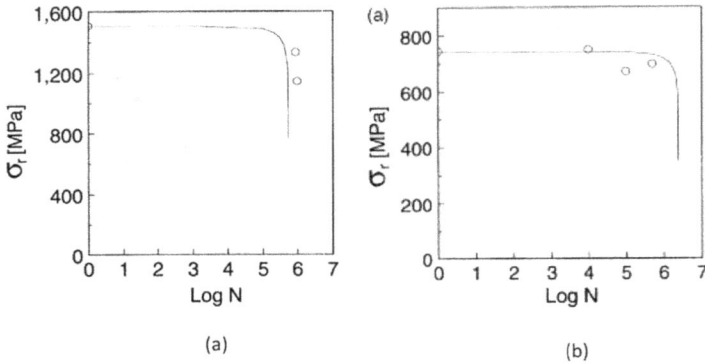

(a)

(b)

FIGURE 11.24   Residual strengths of (a) unidirectional laminate and (b) cross-ply laminate. (Reproduced with permission from Diao, Ye, and Mai 1997.)

post splitting failure. A similar failure was obtained in the unidirectional laminates $[0]_{16}$ with minimum fiber failure in the transverse directions under fatigue loading. The fiber damage is the main micro failure mode during static and fatigue loading. Fiber fails either individually or in the form of the bundles. The fiber breakage occurs when fibers are in the stretched position, while fiber bundle fracture occurs when there is a misalignment of the fiber bundle and matrix and hence, the fiber bundle fails due to the local stress concentration. Fiber bundle fractures are more prone to occur under fatigue load. The APC-1 showed better fatigue response than the CF/PEEK laminate; this was mainly due to the better fiber and matrix adhesion. The broken fiber never gets separated from the matrix parts and always is in contact giving bridging effects (Dickson et al. 1985; Diao, Ye, and Mai 1997). In the case of $[90]_{16}$ laminate, the fatigue properties of the composite were mainly governed through the matrix system. The stack sequence $[0_2/90_2]_{2s}$ laminate showed multiple damage modes. In the first stage, the damage started through transverse matrix cracking in the $90°$ plies direction. In the second stage, damage occurred through splitting into $0°$ plies. The propagation of the delamination occurred in the third stage from the tip of the crack at the interface between $0°$ and $90°$ plies, and while in the final stage, fiber breakage was observed in the $0°$ plies, which led to the catastrophic failure (Diao, Ye, and Mai 1997).

11.6   SUMMARY

In this chapter, different hybrid yarn technologies and the corresponding quality of the laminates are discussed in detail. The towpreg technique helps in achieving better impregnation quality in comparison to the conventional manufacturing techniques. The reduced melt flow distance results in low void contents, which results in superior mechanical properties of the laminates. The flexible nature of the towpreg helps in developing a complex preform structure and gives homogeneous fiber/matrix distributions on consolidation. The advancement in the thermoplastic manufacturing techniques leads to the ease in the fabrication process as well as elimination of the issues related to the inherent high viscous nature of the thermoplastic resins.

REFERENCES

Alagirusamy, R., R. Fangueiro, V. Ogale, and N. Padaki. 2006. "Hybrid Yarns and Textile Preforming for Thermoplastic Composites." *Textile Progress* 38 (4): 1–71. https://doi.org/10.1533/tepr.2006.0004

Ansar, Mahmood, Wang Xinwei, and Zhou Chouwei. 2011. "Modeling Strategies of 3D Woven Composites: A Review." *Composite Structures* 93 (8): 1947–63. https://doi.org/10.1016/j.compstruct.2011.03.010

Baghaei, Behnaz, and Mikael Skrifvars. 2016. "Characterisation of Polylactic Acid Biocomposites Made from Prepregs Composed of Woven Polylactic Acid/Hemp–Lyocell Hybrid Yarn Fabrics." *Composites Part A: Applied Science and Manufacturing* 81 (February): 139–44. https://doi.org/10.1016/j.compositesa.2015.10.042

Baghaei, Behnaz, Mikael Skrifvars, and Lena Berglin. 2013. "Manufacture and Characterisation of Thermoplastic Composites Made from PLA/Hemp Co-Wrapped Hybrid Yarn Prepregs." *Composites Part A: Applied Science and Manufacturing* 50 (July): 93–101. https://doi.org/10.1016/j.compositesa.2013.03.012

Baghaei, Behnaz, Mikael Skrifvars, and Lena Berglin. 2015. "Characterization of Thermoplastic Natural Fibre Composites Made from Woven Hybrid Yarn Prepregs with Different Weave Pattern." *Composites Part A: Applied Science and Manufacturing* 76 (September): 154–61. https://doi.org/10.1016/j.compositesa.2015.05.029

Bar, Mahadev, R. Alagirusamy, and Apurba Das. 2018a. "Properties of Flax-Polypropylene Composites Made through Hybrid Yarn and Film Stacking Methods." *Composite Structures* 197 (August): 63–71. https://doi.org/10.1016/j.compstruct.2018.04.078

Bar, Mahadev, R. Alagirusamy, and Apurba Das. 2019a. "Influence of Friction Spun Yarn and Thermally Bonded Roving Structures on the Mechanical Properties of Flax/Polypropylene Composites." *Industrial Crops and Products* 135 (September): 81–90. https://doi.org/10.1016/j.indcrop.2019.04.025

Bar, Mahadev, R. Alagirusamy, and Apurba Das. 2019b. "Development of Flax-PP Based Twist-Less Thermally Bonded Roving for Thermoplastic Composite Reinforcement." *The Journal of The Textile Institute* 110 (10): 1369–79. https://doi.org/10.1080/00405000.2019.1610997

Bar, Mahadev, R. Alagirusamy, Apurba Das, and Pierre Ouagne. 2020. "Low Velocity Impact Response of Flax/Polypropylene Hybrid Roving Based Woven Fabric Composites: Where Does It Stand with Respect to GRPC?." *Polymer Testing* 89 (September): 106565. https://doi.org/10.1016/j.polymertesting.2020.106565

Bar, Mahadev, Apurba Das, and R. Alagirusamy. 2017. "Studies on Flax-Polypropylene Based Low-Twist Hybrid Yarns for Thermoplastic Composite Reinforcement." *Journal of Reinforced Plastics and Composites* 36 (11): 818–31. https://doi.org/10.1177/0731684417693428

Bar, Mahadev, Apurba Das, and R. Alagirusamy. 2018b. "Effect of Interface on Composites Made from DREF Spun Hybrid Yarn with Low Twisted Core Flax Yarn." *Composites Part A: Applied Science and Manufacturing* 107 (April): 260–70. https://doi.org/10.1016/j.compositesa.2018.01.003

Bar, Mahadev, Apurba Das, and Ramasamy Alagirusamy. 2019c. "Influence of Flax/Polypropylene Distribution in Twistless Thermally Bonded Rovings on Their Composite Properties." *Polymer Composites* 40 (11): 4300–10. https://doi.org/10.1002/pc.25291

Boey, F.Y.C., and S.W. Lye. 1992a. "Void Reduction in Autoclave Processing of Thermoset Composites: Part 1: High Pressure Effects on Void Reduction." *Composites* 23 (4): 261–65. https://doi.org/10.1016/0010-4361(92)90186-X

Boey, F.Y.C., and S.W. Lye. 1992b. "Void Reduction in Autoclave Processing of Thermoset Composites: Part 2: Void Reduction in a Microwave Curing Process." *Composites* 23 (4): 266–70. https://doi.org/10.1016/0010-4361(92)90187-Y

Boriek, A.M., J.E. Akin, C.D. Armeniades, and M.M. El-Alem. 1990. "A Theoretical Model for Voids Distribution in Polymer Composites and Their Effect on Setting Stresses." *Probabilistic Engineering Mechanics* 5 (3): 129–37. https://doi.org/10.1016/0266-8920(90)90004-4

Byun, Joon-Hyung, and Tsu-Wei Chou. 1996. "Process-Microstructure Relationships of 2-Step and 4-Step Braided Composites." *Composites Science and Technology* 56 (3): 235–51. https://doi.org/10.1016/0266-3538(95)00112-3

Cantwell, W.J., and J. Morton. 1991. "The Impact Resistance of Composite Materials — a Review." *Composites* 22 (5): 347–62. https://doi.org/10.1016/0010-4361(91)90549-V

Centea, T., L.K. Grunenfelder, and S.R. Nutt. 2015. "A Review of Out-of-Autoclave Prepregs – Material Properties, Process Phenomena, and Manufacturing Considerations." *Composites Part A: Applied Science and Manufacturing* 70 (March): 132–54. https://doi.org/10.1016/j.compositesa.2014.09.029

Chen, Dingding, Kazuo Arakawa, and Changheng Xu. 2015. "Reduction of Void Content of Vacuum-Assisted Resin Transfer Molded Composites by Infusion Pressure Control." *Polymer Composites* 36 (9): 1629–37. https://doi.org/10.1002/pc.23071

Chu, J.N., F.K. Ko, and J.W. Song. 1992. "Time-Dependent Mechanical Properties of 3-D Braided Graphite/PEEK Composites." *SAMPE Quarterly (Society of Aerospace Material and Process Engineers);(United States)* 23 (4): 14–19.

Davies, P. 1990. "Glass/Nylon-6.6 Composites: Delamination Resistance Testing." *Composites Science and Technology* 38 (3): 211–27.

Desplentere, F., S.V. Lomov, D.L. Woerdeman, I. Verpoest, M. Wevers, and A. Bogdanovich. 2005. "Micro-CT Characterization of Variability in 3D Textile Architecture." *Composites Science and Technology* 65 (13): 1920–30. https://doi.org/10.1016/j.compscitech.2005.04.008

Diao, Xiaoxue, Lin Ye, and Yiu-Wing Mai. 1997. "Fatigue Behaviour of CF/PEEK Composite Laminates Made from Commingled Prepreg. Part I: Experimental Studies." *Composites Part A: Applied Science and Manufacturing* 28 (8): 739–47. https://doi.org/10.1016/S1359-835X(97)00025-0

Dickson, R.F., C.J. Jones, B. Harris, D.C. Leach, and D.R. Moore. 1985. "The Environmental Fatigue Behaviour of Carbon Fibre Reinforced Polyether Ether Ketone." *Journal of Materials Science* 20 (1): 60–70. https://doi.org/10.1007/BF00555899

Fu, Shao-Yun, and Bernd Lauke. 1998. "Characterization of Tensile Behaviour of Hybrid Short Glass Fibre/Calcite Particle/ABS Composites." *Composites Part A: Applied Science and Manufacturing* 29 (5–6): 575–83. https://doi.org/10.1016/S1359-835X(97)00117-6

Goud, Vijay, Ramasamy Alagirusamy, Apurba Das, and Dinesh Kalyanasundaram. 2018. "Dry Electrostatic Spray Coated Towpregs for Thermoplastic Composites." *Fibers and Polymers* 19 (2): 364–74. https://doi.org/10.1007/s12221-018-7470-7

Goud, Vijay, Ramasamy Alagirusamy, Apurba Das, and Dinesh Kalyanasundaram. 2019a. "Influence of Various Forms of Polypropylene Matrix (Fiber, Powder and Film States) on the Flexural Strength of Carbon-Polypropylene Composites." *Composites Part B: Engineering* 166 (June): 56–64. https://doi.org/10.1016/j.compositesb.2018.11.135

Goud, Vijay, Ramasamy Alagirusamy, Apurba Das, and Dinesh Kalyanasundaram. 2019b. "Box-Behnken Technique Based Multi-Parametric Optimization of Electrostatic Spray Coating in the Manufacturing of Thermoplastic Composites." *Materials and Manufacturing Processes* 34 (14): 1638–45. https://doi.org/10.1080/10426914.2019.1666991

Goud, Vijay, Dilpreet Singh, Ramasamy Alagirusamy, Apurba Das, and Dinesh Kalyanasundaram. 2020. "Investigation of the Mechanical Performance of Carbon/Polypropylene 2D and 3D Woven Composites Manufactured through Multi-Step Impregnation Processes." *Composites Part A: Applied Science and Manufacturing* 130 (March): 105733. https://doi.org/10.1016/j.compositesa.2019.105733

Gu, Huang, and Zhong Zhili. 2002. "Tensile Behavior of 3D Woven Composites by Using Different Fabric Structures." *Materials & Design* 23 (7): 671–74. https://doi.org/10.1016/S0261-3069(02)00053-5

Gupta, Mohit, Harpreet Singh, Ashraf Nawaz Khan, Puneet Mahajan, R.T. Durai Prabhakaran, and R. Alagirusamy. 2021. "An Improved Orthotropic Elasto-Plastic Damage Model for Plain Woven Composites." *Thin-Walled Structures* 162 (May): 107598. https://doi.org/10.1016/j.tws.2021.107598

Hartness, J. Timothy. 1991. "Advanced Fiber/matrix Material Systems." NASA. Langley Research Center, First NASA Advanced Composites Technology Conference, Part 2, 659–76. https://ntrs.nasa.gov/citations/19930021665

Hugh, Maylene K., Joseph M. Marchello, Janice R. Maiden, and Norman J. Johnston. 1992. "Weavability of Dry Polymer Powder Towpreg." *NASA, Langl Ey Research Cenker 1 Uncl As*, June, 175–89.

Hung, Y.Y. 1996. "Shearography for Non-Destructive Evaluation of Composite Structures." *Optics and Lasers in Engineering* 24 (2–3): 161–82. https://doi.org/10.1016/0143-8166(95)00020-8

Jogur, Ganesh, Ashraf Nawaz Khan, Apurba Das, Puneet Mahajan, and R. Alagirusamy. 2018. "Impact Properties of Thermoplastic Composites." *Textile Progress* 50 (3): 109–83. https://doi.org/10.1080/00405167.2018.1563369

Jones, T.S., D. Polansky, and H. Berger. 1988. "Radiation Inspection Methods for Composites." *NDT international* 21 (4): 277–282. https://doi.org/10.1016/0308-9126(88)90341-0

Karabutov, A.A., and N.B. Podymova. 2014. "Quantitative Analysis of the Influence of Voids and Delaminations on Acoustic Attenuation in CFRP Composites by the Laser-Ultrasonic Spectroscopy Method." *Composites Part B: Engineering* 56 (January): 238–44. https://doi.org/10.1016/j.compositesb.2013.08.040

Khan, Ashraf Nawaz, Vijay Goud, Ramasamy Alagirusamy, Puneet Mahajan, and Apurba Das. 2022 "Optimization study on wet electrostatic powder coating process to manufacture UHMWPE/LDPE towpregs." *Journal of Industrial Textiles* (January): 15280837211070995. https://doi.org/10.1177/15280837211070995

Klinkmuller, V., M.K. Um, M. Steffens, K. Friedrich, and B. -S. Kim. 1994. "A New Model for Impregnation Mechanisms in Different GF/PP Commingled Yarns." *Applied Composite Materials* 1 (5): 351–71. https://doi.org/10.1007/BF00568041

Kuo, Wen-Shyong, and Kuo-Bing Cheng. 1999. "Processing and Microstructures of 3D Woven-Fabric Composites Incorporating Solid Rods." *Composites Science and Technology* 59 (12): 1833–46. https://doi.org/10.1016/S0266-3538(99)00043-3

Kuo, Wen-Shyong, and Jiunn Fang. 2000. "Processing and Characterization of 3D Woven and Braided Thermoplastic Composites." *Composites Science and Technology* 60 (5): 643–56. https://doi.org/10.1016/S0266-3538(99)00161-X

Kuo, Wen-Shyong, Tse-Hao Ko, and Horn-I. Chen. 1998a. "Elastic Moduli and Damage Mechanisms in 3D Braided Composites Incorporating Pultruded Rods." *Composites Part A: Applied Science and Manufacturing* 29 (5–6): 681–92. https://doi.org/10.1016/S1359-835X(97)00110-3

Kuo, Wen-Shyong, and Lin-Chyuan Lee. 1998b. "Impact Response of 3-D Woven Composites Reinforced by Consolidated Rods." *Polymer Composites* 19 (2): 156–65. https://doi.org/10.1002/pc.10087

Lauke, B., U. Bunzel, and K. Schneider. 1998. "Effect of Hybrid Yarn Structure on the Delamination Behaviour of Thermoplastic Composites." *Composites Part A: Applied Science and Manufacturing* 29 (11): 1397–1409. https://doi.org/10.1016/S1359-835X(98)00059-1

Lee, L., S. Rudov-Clark, A.P. Mouritz, M.K. Bannister, and I. Herszberg. 2002. "Effect of Weaving Damage on the Tensile Properties of Three-Dimensional Woven Composites." *Composite Structures* 57 (1–4): 405–13. https://doi.org/10.1016/S0263-8223(02)00108-3

Long, Li, Wang Shanyuan, and Yu Jianyong. 2002. "Mechanical Properties of Commingled Yarn Composites." *Indian Journal of Fibre & Textile Research* 27 (September): 287–89.

Madra, Anna, Nemr El Hajj, and Malk Benzeggagh. 2014. "X-Ray Microtomography Applications for Quantitative and Qualitative Analysis of Porosity in Woven Glass Fiber Reinforced Thermoplastic." *Composites Science and Technology* 95 (May): 50–58. https://doi.org/10.1016/j.compscitech.2014.02.009

Madsen, Bo, and Hans Lilholt. 2003. "Physical and Mechanical Properties of Unidirectional Plant Fibre Composites—An Evaluation of the Influence of Porosity." *Composites Science and Technology* 63: 1265–72. https://doi.org/10.1016/S0266-3538(03)00097-6

Mehdikhani, Mahoor, Larissa Gorbatikh, Ignaas Verpoest, and Stepan V. Lomov. 2019. "Voids in Fiber-Reinforced Polymer Composites: A Review on Their Formation, Characteristics, and Effects on Mechanical Performance." *Journal of Composite Materials* 53 (12): 1579–1669. https://doi.org/10.1177/0021998318772152

Miao, Menghe, Yan Lai How, and Kwok Po Stephen Cheng. 1994. "The Role of False Twist in Wrap Spinning." *Textile Research Journal* 64 (1): 41–48. https://doi.org/10.1177/004051759406400105

Miller, A.H., N. Dodds, J.M. Hale, and A.G. Gibson. 1998. "High Speed Pultrusion of Thermoplastic Matrix Composites." *Composites Part A: Applied Science and Manufacturing* 29 (7): 773–82. https://doi.org/10.1016/S1359-835X(98)00006-2

Muralidhar, C., and N.K. Arya. 1993. "Evaluation of Defects in Axisymmetric Composite Structures by Thermography." *NDT & E International* 26 (4): 189–193. https://doi.org/10.1016/0963-8695(93)90473-8

Park, Chung Hae, Aurélie Lebel, Abdelghani Saouab, Joël Bréard, and Woo Il Lee. 2011. "Modeling and Simulation of Voids and Saturation in Liquid Composite Molding Processes." *Composites Part A: Applied Science and Manufacturing* 42 (6): 658–68. https://doi.org/10.1016/j.compositesa.2011.02.005

Podymova, N.B., and A.A. Karabutov. 2013. "Broadband Laser-Ultrasonic Spectroscopy for Quantitative Characterization of Porosity Effect on Acoustic Attenuation and Phase Velocity in CFRP Laminates." *Journal of Nondestructive Evaluation*, 33: 141–51 https://doi.org/10.1007/s10921-013-0210-z

Purslow, D. 1984. "On the Optical Assessment of the Void Content in Composite Materials." *Composites* 15 (3): 207–10. https://doi.org/10.1016/0010-4361(84)90276-3

Ramani, Karthik, Harshad Borgaonkar, and Chris Hoyle. 1995. "Experiments on Compression Moulding and Pultrusion of Thermoplastic Powder Impregnated Towpregs." *Composites Manufacturing* 6 (1): 35–43. https://doi.org/10.1016/0956-7143(95)93711-R

Ramasamy, A., Youjiang Wang, and John Muzzy. 1996. "Braided Thermoplastic Composites from Powder-Coated Towpregs. Part III: Consolidation and Mechanical Properties." *Polymer Composites* 17 (3): 515–22. https://doi.org/10.1002/pc.10641

Rosselli, F., and M.H. Santare. 1997. "Comparison of the Short Beam Shear (SBS) and Interlaminar Shear Device (ISD) Tests." *Composites Part A: Applied Science and Manufacturing* 28 (6): 587–94. https://doi.org/10.1016/S1359-835X(97)00009-2

Schell, J.S.U., M. Renggli, G.H. van Lenthe, R. Müller, and P. Ermanni. 2006. "Micro-Computed Tomography Determination of Glass Fibre Reinforced Polymer Meso-Structure." *Composites Science and Technology* 66 (13): 2016–22. https://doi.org/10.1016/j.compscitech.2006.01.003

Sisodia, S.M., S.C. Garcea, A.R. George, D.T. Fullwood, S.M. Spearing, and E.K. Gamstedt. 2016. "High-Resolution Computed Tomography in Resin Infused Woven Carbon Fibre Composites with Voids." *Composites Science and Technology* 131 (August): 12–21. https://doi.org/10.1016/j.compscitech.2016.05.010

Stolyarov, O.N., I.N. Stolyarov, T.A. Kryachkova, and P.G. Kravaev. 2013. "Hybrid Textile Yarns and Thermoplastic Composites Based on Them." *Fibre Chemistry* 45 (4): 217–20. https://doi.org/10.1007/s10692-013-9515-z

Stone, D.E.W., and B. Clarke. 1975. "Ultrasonic Attenuation as a Measure of Void Content in Carbon-Fibre Reinforced Plastics." *Non-Destructive Testing* 8 (3): 137–45. https://doi.org/10.1016/0029-1021(75)90023-7

Straumit, Ilya, Stepan V. Lomov, and Martine Wevers. 2015. "Quantification of the Internal Structure and Automatic Generation of Voxel Models of Textile Composites from X-Ray Computed Tomography Data." *Composites Part A: Applied Science and Manufacturing* 69 (February): 150–58. https://doi.org/10.1016/j.compositesa.2014.11.016

Svensson, Niklas, Roshan Shishoo, and Michael Gilchrist. 1998. "Fabrication and Mechanical Response of Commingled GF/PET Composites." *Polymer Composites* 19 (4): 360–69. https://doi.org/10.1002/pc.10109

Swolfs, Yentl, Larissa Gorbatikh, and Ignaas Verpoest. 2014. "Fibre Hybridisation in Polymer Composites: A Review." *Composites Part A: Applied Science and Manufacturing* 67 (December): 181–200. https://doi.org/10.1016/j.compositesa.2014.08.027

Thanomsilp, C., and P.J. Hogg. 2003. "Penetration Impact Resistance of Hybrid Composites Based on Commingled Yarn Fabrics." *Composites Science and Technology* 63 (3–4): 467–82. https://doi.org/10.1016/S0266-3538(02)00233-6

Wang, Youjiang, and Dongming Zhao. 2006. "Effect of Fabric Structures on the Mechanical Properties of 3-D Textile Composites." *Journal of Industrial Textiles* 35 (3): 239–56. https://doi.org/10.1177/1528083706057595

Ye, L., and K. Friedrich. 1993. "Mode I Interlaminar Fracture of Co-Mingled Yarn Based Glass/Polypropylene Composites." *Composites Science and Technology* 46 (2): 187–98. https://doi.org/10.1016/0266-3538(93)90174-F

Yoon, Hogyu, and Kiyohisa Takahashi. 1993. "Mode I Interlaminar Fracture Toughness of Commingled Carbon Fibre/PEEK Composites." *Journal of Materials Science* 28 (7): 1849–55. https://doi.org/10.1007/BF00595757

12 Towpreg-Based Thermoplastic Composites for Electromagnetic Shielding

J. Krishnasamy[a], Apurba Das[b], and R. Alagirusamy[b]
[a]PSG College of Technology
Coimbatore, India
[b]Indian Institute of Technology Delhi
New Delhi, India

CONTENTS

12.1 INTRODUCTION

In recent years, the continuous fibre thermoplastic composites have contributed more in the composites market. The unique nature of thermoplastic materials such as greater resistance to fracture, higher impact resistance, durability, lesser processing time, damping properties, greater toughness and damage tolerance have been explored by more composite manufacturing industries. However, the major challenge in processing the thermoplastic composites is their high viscosity, which makes the impregnation of continuous fibres quite difficult. This is the hindrance for commercial applications of thermoplastic materials. Researchers have identified novel manufacturing methods in which the problem of distributing high viscous polymers in the fibre bundles can be tackled. The preparation of thermoplastic flexible towpregs makes it successful by fabricating the various hybrid yarns such

DOI: 10.1201/9781003049715-12

as commingled yarns, blended conductive yarns, sheath core yarns, etc. The thermoplastic prepreg is an intermediate product to facilitate the composite processing without any limitations.

Towpregs have been developed in order to improve the ease of flow of polymer matrices during manufacturing of thermoplastic composite materials. Usually prepregs are manufactured in two ways. In the first method, direct fusion of polymer happens to obtain pre-consolidated tapes (McGregor et al., 2017). In the second method, the matrix is in the form of powder which comes in intimate contact with the fibres in the form of prepregs (Friedrich, 1999). Similar to powder coating, other methods are available for developing the thermoplastic towpreg, pre-consolidated tapes, prepregs, etc. (Nunes et al., 2003). Then the developed prepregs are made into composite materials by means of filament winding, compression molding and other suitable technologies. Hence, combing the thermoplastic material with the reinforcing material in the form of prepregs prior to composite processing leads to smooth impregnation of matrix materials in the fibre bundles. Generally, the short and long fibre reinforced composites have been developed using the towpregs for load bearing, electromagnetic shielding and other applications. In order to develop such products, high performance fibres such as glass, carbon, Kevlar, etc. and suitable matrices are used in towpreg manufacturing. In this chapter, various methods developed to obtain hybrid yarn-based composites for electromagnetic shielding applications have been reported.

12.2   SIGNIFICANCE OF TOWPREG COMPOSITES FOR ELECTROMAGNETIC SHIELDING

For shielding the electromagnetic radiation, the composite materials have made significant contributions in various fields such as the defence, domestic and aviation industries. For the shielding to be effective, the material should have good level of conductivity. Generally, textile materials are good electrical insulators. The fibres or yarns made from natural or synthetic fibres will show high electrical resistivity. However, they can be used in electromagnetic (EM) shields, if conductive or magnetic properties are added to them. Hence, attempts have been made for producing the electrically conductive fabrics. By the addition of conductive fibres, conductive filaments or by conductive coating to the textile materials, the conductive properties can be imparted. The shielding effectiveness of such materials can be assessed using the coaxial transmission line method. The other methods, such as waveguide methods and free space techniques, have also been used to assess the shielding effectiveness of material at medium and high frequencies. The shielding effectiveness of planar materials is assessed using the following formula.

$$SE = -10 \log(P_i/P_t) \qquad (12.1)$$

Where P_i is the incident power of electromagnetic waves. P_t is the transmitted power of electromagnetic waves. Figure 12.1 shows the schematic of the vector network analyser with reference and load specimens.

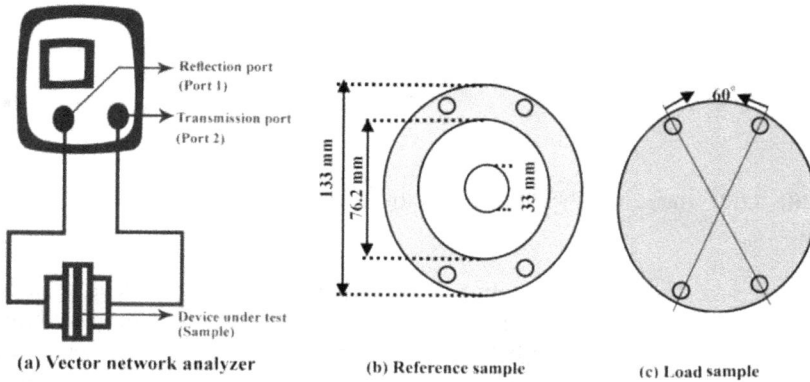

FIGURE 12.1 (a) Vector network analyzer, (b) reference and (c) load specimens. (Adapted from Krishnasamy et al., 2016.)

For assessing the shielding behaviour of planar materials, reference and load specimens are tested separately according to ASTM D4935 (1999) as observed in Figure 12.1. More details can be referred in much published literature (Geetha et al., 2009; Huang, 1995) on the assessment of shielding materials. Both discontinuous and continuous fibre composites have been adopted for shielding the electromagnetic radiation. For better shielding effectiveness and mechanical strength, continuous fibre reinforced thermoplastic composites have been preferred. As discussed earlier, the hybrid yarn-based composites provide more opportunities for tailoring the composite with specific properties. As the conductive material is to be introduced for effective shielding in textile fabrics, the hybrid yarn technique is found to be more suitable. Out of many methods, developing the flexible towpregs is the most preferred, and has many advantages such as processability, flexibility and easy tailorability. The following section explains the various towpreg materials developed for electromagnetic shielding products.

12.3 CARBON TOWPREG-BASED THERMOPLASTIC COMPOSITES

The carbon towpreg is developed by different methods such as powder coating, commingling process and other methods. The developed carbon prepreg composite can be used in shielding the electromagnetic radiation with other advantages. Several studies have been carried out on carbon-based towpregs for shielding different frequency ranges. In a study, the carbon prepreg was prepared by Zhang et al. (2020), which was composed of carbonyl iron powder-filled epoxy and the carbon nanotubes-filled epoxy used in the form of a Koch curve fractal to act as an absorbing layer. The composite was made by placing the carbon fibre prepreg in the mold along with an absorbing layer by two-step curing method, and shielding behaviour of the composite was studied. The unidirectional carbon fibre/epoxy composite prepared from carbon prepreg could be also used for electromagnetic shielding applications. In a study by Tugirumubano et al. (2018), a fabric structure was made which consisted of carbon fibre prepreg in the warp direction and wire mesh in the

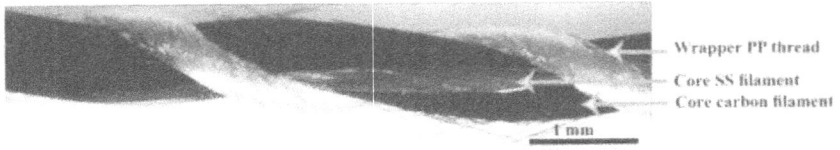

FIGURE 12.2 Image of C/SS/PP hybrid yarn (2X). (Adapted from Krishnasamy et al., 2016.)

weft direction. The two such woven fabrics were combined by means of lamination and covered with carbon fibre reinforced plastic (CFRP) plies to make bimetal carbon fibre reinforced plastic hybrid plates. Then all the layers were combined in a hot press and a composite plate was made using autoclave techniques. The hybrid composite material made by either SS/copper or SS/nickel or copper-nickel with carbon fibre reinforcement showed the maximum shielding effectiveness of 82.5 dB in the frequency range of 0.5 to 1.5 GHz. The highest shielding effectiveness (SE) of 131.6 dB was observed for SS/Cu/CFRP laminates. In addition, the mechanical strength of the composite was also higher due to presence of more carbon fibres bearing loads in that direction. In another study, Krishnasamy et al. (2016) prepared carbon/stainless steel (SS) hybrid yarns using a direct twisting machine for developing the conductive woven fabric for electromagnetic shielding applications. The image of the prepared C/SS/PP hybrid yarn is shown in Figure 12.2.

As observed in Figure 12.2, carbon was placed in the core of the yarn along with SS filament and polypropylene filament was placed in the cover of the yarn. The SE of fabrics and composites made out of C/SS/PP hybrid yarns were studied at low and C band frequency ranges. When the shielding behaviour of hybrid yarns and fabrics was compared, a significant difference is seen for woven fabric materials. The shielding effectiveness is diminished after the woven fabric is reinforced into composite materials. At low frequency range, 1/1 plain woven composite exhibited the SE of 81.4 dB, which was higher than 4 end float (76.2 dB) and 8 end float fabric composites (64 dB). This was not true for the C band frequency range, which was influenced by thread density. Figure 12.3 shows the shielding behaviour of woven fabrics and their composites in C band region (Jagatheesan et al., 2017a).

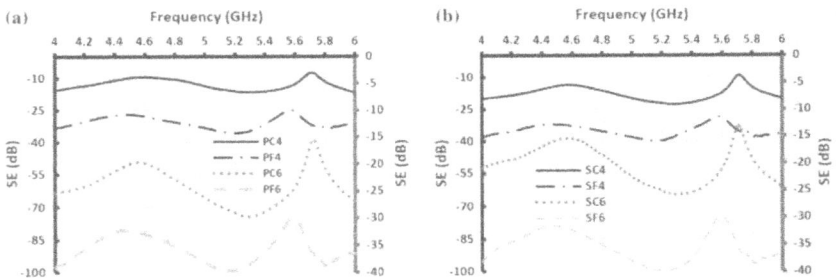

FIGURE 12.3 Shielding behaviour of (a) plain and (b) satin fabrics and their composites. (Adapted from Jagatheesan et al., 2017a.)

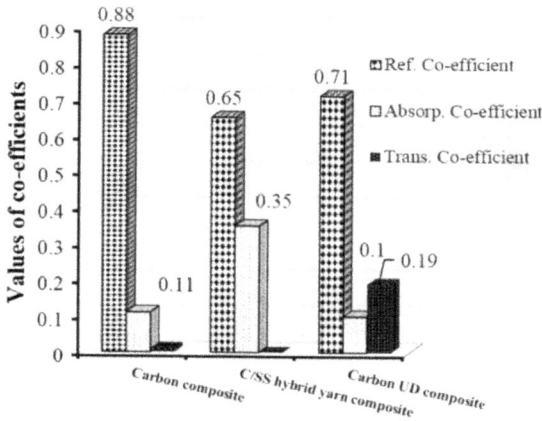

FIGURE 12.4 Reflection and absorption coefficients of fabric composites. (Adapted from Jagatheesan et al., 2017a.)

As observed in Figure 12.3, the difference in shielding effectiveness of fabrics and their composites was seen in the C band region. The same behaviour was observed for all basic fabric structures such as plain, twill and satin fabrics. At low thread density (3.93 ppcm), 8 end float fabric composites showed a higher SE than other composite types. However, at high thread density (6.3 ppcm), 1 end float fabric composite exhibited the highest SE of 29.7 dB compared to 4 end and 8 end float fabric composites (25.9 dB). Compared to fabric form, all the fabric reinforced composites showed lower shielding effectiveness at low and C band frequencies. Similarly, the shielding behaviour of both fabrics and composites were increased for the increase in thread density in both frequency ranges. The SE values were similar for 1/1 plain and 8 end fabrics due to the fibre skin effect and not by degree of yarn interlacement at high frequency. The same behaviour was seen in composite form with respect to weave structure. However, the absorption co-efficient was different for the composite when the carbon/SS hybrid yarn was introduced. The reflection and absorption co-efficients of the hybrid yarn composites are shown in Figure 12.4 (Jagatheesan et al., 2017a).

As observed in Figure 12.4, absorption co-efficient was better for C/SS hybrid yarn composites compared to other composites. Similarly, the mechanical properties of C/SS/PP hybrid yarn were tested. The work of rupture of C/SS/PP hybrid yarn was higher (6830.3 g cm) compared to pure carbon, SS and PP filaments. The tensile strength of 8 end satin fabric composites exhibited better value than 4/1 twill and plain composites. Similar to hybrid yarn woven fabrics, the shielding behaviour of hybrid yarn was also analyzed. In another study (Jagatheesan et al., 2018b), friction spun yarn was developed from carbon and SS filament yarn. The plain and rib knitted fabric were made with/without the conductive inlaid yarns for improving the fabric stability and shielding effectiveness. The microscopic images of hybrid yarn-based knitted fabrics are shown in Figure 12.5.

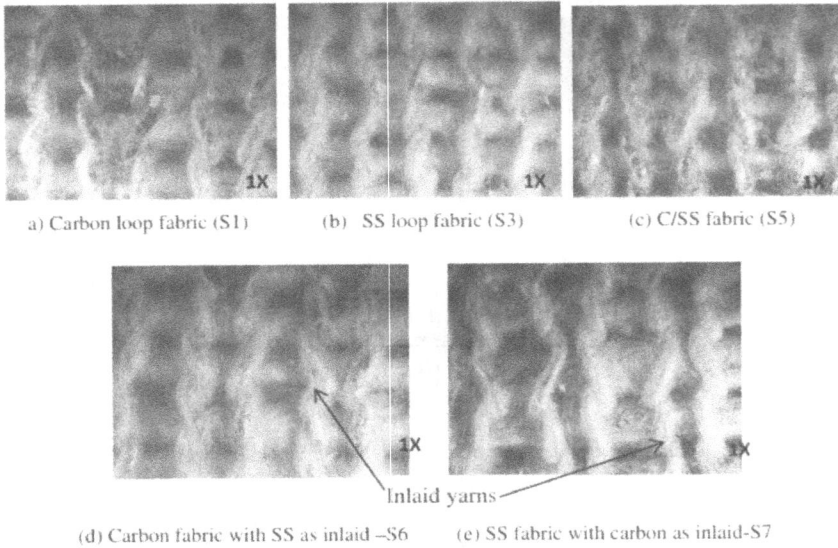

a) Carbon loop fabric (S1)   (b) SS loop fabric (S3)   (c) C/SS fabric (S5)

Inlaid yarns

(d) Carbon fabric with SS as inlaid –S6   (e) SS fabric with carbon as inlaid-S7

FIGURE 12.5 Microscopic images of surface of knitted fabric samples (a–e). (Adapted from Jagatheesan et al., 2018b.)

The conductive carbon and stainless hybrid yarns were made into knitted loops without any breakage or surface distortion (Figure 12.5). It is especially confirmed by observing the carbon loop fabric (Figure 12.5(a)) where the rupture of carbon filaments is avoided due to wrapped polypropylene fibres. Similar observation is also made for SS loop fabrics (Figure 12.5(b)). In the case of C/SS loop yarn fabric (Figure 12.5(c)), less damage is seen on the surfaces of C/SS hybrid yarn. For the inlaid fabric (Figure 12.5(d–e)), the inlaid hybrid yarns were placed in the fabric without any damages or yarn crimp. Overall, the plain knit fabric showed less resistance to knitting than rib knit fabrics. At low frequency regions (i.e. 50 MHz to 1.5 GHz), a good shielding effectiveness was observed for the plain knit fabric. However, the shielding effectiveness decreased at higher frequency (C band) range due to fibre skin effect. In the composite form, the SE was higher than in the fabric form in low and high frequencies. In particular, the composite containing conductive thread in loop configuration exhibited higher shielding effectiveness than its fabric form. However, this was not true for the fabrics and their composites containing the conductive threads in inlaid directions. Hence, the direction of fabric exposure to electromagnetic radiation played a vital role in deciding shielding effectiveness. Figure 12.6 shows the testing direction of knitted fabrics in course and wale directions.

The test revealed that conductive loop yarn fabrics and composites showed higher SE in the wale direction compared to the course direction. Hence, the fabric structure and yarn positioning are important in deciding the shielding effectiveness. Figure 12.7 shows the shielding behaviour of rib knitted fabrics in the C band region.

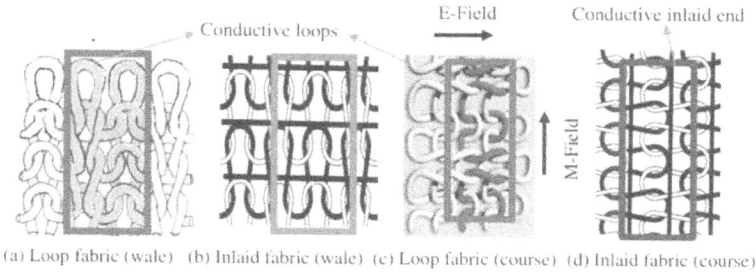

(a) Loop fabric (wale) (b) Inlaid fabric (wale) (c) Loop fabric (course) (d) Inlaid fabric (course)

FIGURE 12.6 Schematics of wale and course directions of rib-knitted fabrics. (Adapted from Jagatheesan et al., 2018b.)

From Figure 12.7(a), it was observed that when carbon hybrid yarn is present as loop yarn, it showed higher shielding effectiveness than the carbon hybrid yarn present as inlaid yarn in the course direction testing of the fabric. In addition, composites made out of carbon hybrid yarn showed higher shielding effectiveness than the fabric form containing carbon hybrid yarn. This trend is contradictory to woven fabric composites, as observed earlier. However, the case is not true for carbon inlaid fabric and the composite observed as shown in Figure 12.7(a). This is due to the presence of fewer numbers of inlaid yarns parallel to the applied E-field, which results in reduced shielding effectiveness. For the same structure, the major changes in shielding behaviour were observed when the testing direction was changed from the course to wale direction of the fabric. The carbon hybrid yarn in the loop direction of the fabric exhibited the SE of 9 dB at 6.5GHz frequency (Figure 12.6(b)). Similarly, the carbon loop yarn composite also exhibited the SE of 13 dB. In the case of inlaid yarn incorporated fabrics and their composites, higher shielding effectiveness was observed compared to other combinations as observed in Figure 12.7(b). For the increase incident frequency, the shielding effectiveness of fabrics and composites decreased in low and C band frequency range. Similarly, carbon fibre composites exhibited higher attenuation characteristics than SS fibre composites.

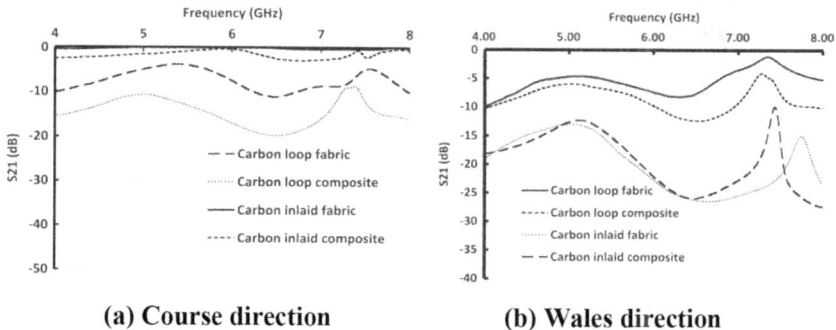

(a) Course direction

(b) Wales direction

FIGURE 12.7 Shielding behaviour of rib knitted fabrics in course (a) and wales directions (b). (Adapted from Jagatheesan et al., 2018b.)

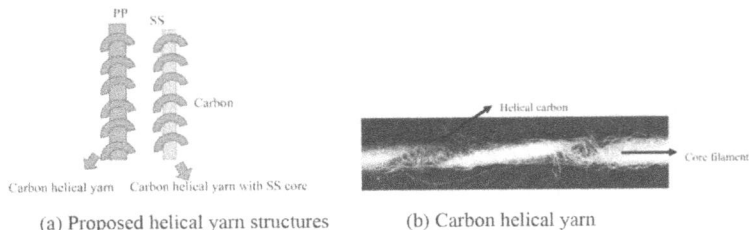

(a) Proposed helical yarn structures　　　(b) Carbon helical yarn

FIGURE 12.8 Schematic of proposed helical yarns and image of prepared helical yarns. (Adapted from Krishnasamy et al., 2018.)

In the C band region, the hybrid composites having carbon and SS filaments in both loop as well as inlaid directions alternatively showed similar shielding behaviour when tested in the wale direction. The developed hybrid yarn knitted composites could be used in aerospace components and constructing the shielded rooms.

On the other hand, a new approach was made for fabricating the conductive helical yarn-based prepreg composite. Krishnasamy et al. (2018) prepared the helical yarn using the carbon and SS multi-filaments. The polypropylene fibres were used to cover the carbon helical yarn for improving the yarn stability and to make the dry prepreg material. The SS filament was used as core material for improving the shielding effectiveness of the yarn structure. The images of the proposed and prepared carbon helical yarn are shown in Figure 12.8.

The prepared hybrid yarn, i.e. carbon helical yarn, was made into 1/1/plain woven and plain knit fabrics for developing effective shielding structures. As observed in earlier studies, the increase in thread density of the woven fabric increased the total attenuation. The shielding effectiveness of helical yarn reinforced composite showed different shielding behaviour with the absorption coefficient of 0.25. In case of helical yarn knitted fabric, a dominant EM absorption behaviour was observed for certain knitted combinations. The fabric made up of carbon helical yarn with more carbon coiling density and optimum core SS content provided higher absorption coefficients and also the overall shielding effectiveness. Similarly, the absorption co-efficients of the fabric were also increased for few types of fabrics having more number of fabric layers. When the hybrid yarn fabrics were reinforced in the polypropylene matrix, the prepared composite material showed less electromagnetic absorption behaviour compared to the plain knit fabrics. However, the overall shielding behaviour of the composite improved in the C band frequency region. For further improving the shielding efficient of helical yarn composite structure, Mn-Zn ferrite loaded PP film was fabricated and used as matrix material for developing the thermoplastic composite materials (Jagatheesan et al., 2018a). The study revealed that the increase in ferrite content improved the dielectric loss and magnetic permeability of Mn-Zn ferrite loaded polypropylene films due to increased magnetic loss. As the frequency increased from 4 to 8 GHz, the real permittivity also increased for the Mn–Zn ferrite/PP film. Hence, the conductivity and shielding effectiveness (−14.2 dB) of ferrite films also increased. Compared to pure PP/helical yarn composite, a higher absorption loss of 14.2 dB and lower reflection co-efficient of 0.345 were observed for hybrid yarn composites loaded with ferrite particles. Similarly, an increase in ferrite content (up to 45wt%) also increased the absorption loss and shielding behaviour of the composite. Hence, hybrid yarn incorporated knitted fabric could be used for developing the electromagnetic absorbing structures, and the composite form of the material could be used for reflective shielding.

12.4 STAINLESS STEEL TOWPREG-BASED THERMOPLASTIC COMPOSITES

Few research works have been done on SS-based towpregs for electromagnetic shielding applications. The conductive woven fabrics were developed using the SS/PET-based hybrid yarn and shielding behaviour of the fabric was investigated in wide frequency ranges. In a study by Das et al. (2014b), SS/polyester-based hybrid yarn was prepared for developing the woven fabric for shielding the frequency range of 300 kHz to 1.5 GHz. The effect of various woven fabric parameters such as weft density, weave pattern, proportion of conductive yarn in weft, grid size opening and number of ply yarns on shielding effectiveness were investigated. The increase in proportion of conductive yarns in weft direction increased the conducting network (by lowering the aperture ratio) and shielding effectiveness of the fabric. This was also true for the fabric having conductive thread in both warp and weft directions which increased the two way conductive network, as the fabric having lower aperture ratio showed lower fabric openness. The fabric with 1:1 weft thread ratio in particular exhibited higher SE due to the presence of self-resonance peaks of metal mesh developed by the conductive yarn. But the effect of weave structure has no significant influence on shielding effectiveness confirmed by statistical analysis. The effect of doubled yarn on shielding behaviour was investigated. A higher shielding effectiveness was observed for the fabric with doubled SS/PET yarn compared to single SS/PET yarn. The case was also true for multilayered structure for shielding the high frequency range. The fabric having maximum SS content showed the SE of 53.04 dB in the frequency range of 0.68 GHz. However, there was no significant shielding behaviour observed for the fabric below 0.5 GHz frequency range.

In another study (Das et al., 2014a), a fabric made up of both SS/PET hybrid yarn and SS multifilament yarn was investigated for shielding behaviour. The microscopic images of SS/PET hybrid yarn fabric with different thread densities of SS filaments are shown in Figure 12.9.

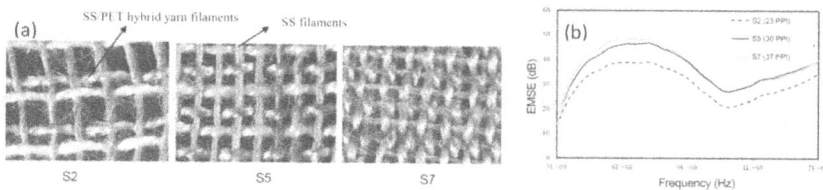

FIGURE 12.9 Microscopic images of of SS/PET fabric samples (a) and their shielding effectiveness (b) with varying SS content. (Das et al., 2014a.)

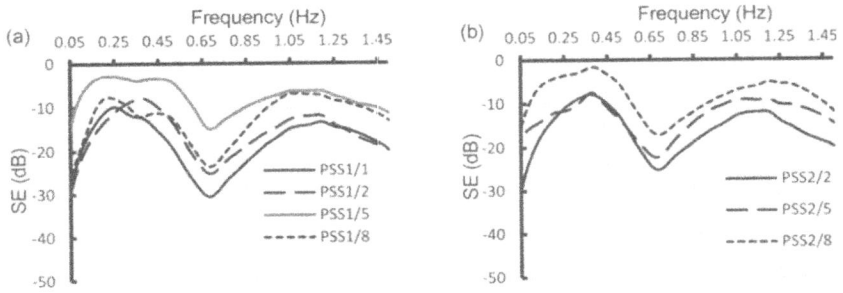

FIGURE 12.10 SE of SS fabric with varying openness in warp and weft directions; (a) warp openness (1:1) and (b) warp openness (1:2). (Adapted from Jagatheesan et al., 2017b.)

The fabric parameters such as fabric density, proportion of SS content and structure of the fabric were investigated. As seen in Figure 12.9, the increase in SS hybrid yarn density increases shielding effectiveness of fabric from 37 to 48.5 dB at the frequency of 0.62 GHz. In this study, the fabric made with SS filaments only showed better shielding effectiveness than the fabric made with SS/PET hybrid yarn. However, the case was not true for the fabric made with SS multifilament and SS/PET hybrid yarn. The hybrid fabric showed greater SE of 45 dB than the fabric made with SS filament only (40 dB). Hence SS/PET hybrid yarn in combination with SS filament yarn could be used as effective shielding. The other fabric parameters such as fabric density and weave structure have been studied. As observed in earlier studies, the increase in fabric density increased the SS content, which resulted in reduced fabric porosity and higher shielding effect. In the case of weave structure, the twill weave fabric showed higher shielding than plain weave fabric due to higher yarn float length. The shielding effectiveness of hybrid yarn woven fabric is influenced by the aperture size and aspect ratio to a great extent. Hence, a study has been carried out on the influence of aperture size and shape on the shielding behaviour of carbon and SS -based hybrid yarn woven fabric in the frequency range of 50 MHz–1.5 GHz (Jagatheesan et al., 2017b). Figure 12.10 shows the effect of grid openness on SS hybrid yarn fabric with varying fabric openness.

As observed in Figure 12.10, with decreasing weft opening, the shielding behaviour of fabrics is increased from 13.6 dB to 29.7 dB at the frequency of 0.65 GHz. The same trend is observed for warp opening of SS-based hybrid yarn fabric at different levels of opening. The shielding behaviour of carbon-based hybrid yarns as well as C/SS/PP-based hybrid yarn fabrics was also analysed. The carbon and C/SS hybrid yarn UD fabrics exhibited higher SE than SS hybrid yarns, especially at the resonance frequency of 0.66 GHz. The attenuation was higher (30–42 dB) when the hybrid yarn was incorporated in both warp and weft directions. Similarly, when the fabric openness was considered, the decreased SE of fabric was observed for the openness in weft direction despite the resonance frequency remaining the same. Further slight increase in the warp direction of the fabric did not change much of the shielding effectiveness of the fabric. It was also observed that the use of hybrid yarn improves the shielding effectiveness in the grid structures.

For the fabrics with large openness, the shielding effectiveness around 20 dB was observed in the frequency range of 50 MHz to 1.5 GHz. When the aperture ratio of the fabric was kept constant, an increase in fabric openness in warp and weft directions showed drastic reduction in attenuation level of the fabric. Similarly, the increase in fabric layers did not have much shielding effectiveness beyond the resonance frequency. In case of the fabric layering angle, multilayer fabric layered with the angle of 0°/45°/90° showed better shielding effects than other layering combinations. The prepared hybrid yarn-based grid fabrics could be suitable for developing polarization attenuators, shielding grids, etc. When the hybrid yarn is used for developing the woven-based shielding material, shielding and mechanical characteristics of thermoplastic reinforced woven composites are different from the parent fabric material. Hence, a study was conducted to analyze the shielding and mechanical performance of the fabric before and after consolidation (Jagatheesan et al., 2017a). The shielding effectiveness was carried out at the low and C band frequency region. The carbon and SS-based hybrid yarns were incorporated into woven fabrics by varying different fabric related parameters. The shielding effectiveness of fabrics was better than fabric reinforced composites irrespective of frequencies in either the low or C band frequency region.

In the case of mechanical behaviour, the fabric composites showed better mechanical strength than the fabric samples. The highest tensile strength was observed for 8 end satin fabric composites compared to 4/1 twill and satin composites. The presence of less crimp helped the satin fabric composite to have higher mechanical strength. All the composites showed higher tensile strength compared to the fabric form. Hence, the introduction of hybrid yarn improved the tensile strength and shielding effectiveness of the fabric. In the yarn form, higher work of rupture (6830.3 g cm) was observed for C/SS/PP hybrid yarn than the carbon, SS and PP filaments. In order to study the SS towpreg incorporated woven fabric, especially the effect of fabric structure, cell type weave structure was selected and the shielding effectiveness of the fabric was investigated in comparison with plain woven fabric (Krishnasamy, Jagatheesan & Das, 2020). The conductive core-sheath yarn was selected for the study in which SS/PET blended yarn as core material and polyester stable fibre are sheath materials. It was observed that the increase in core SS content in the hybrid yarn improved the shielding effectiveness of the fabric despite different fabric structure.

However, the increasing metal content did not show any improvement in SE of fabric having larger grid fabric. Similarly, the presence of conductive yarn in warp and weft directions of the fabric also improved the shielding effectiveness. As the incident frequency was increased from 0.3 GHz to 0.6 GHz, the SE of fabric was also increased due to formation of the conductive network by the fabric. Beyond the incident frequency of 0.6 GHz, the SE decreases due to fibre skin effect. Figure 12.11 shows the shielding behaviour of SS/PET hybrid yarn fabric with plain and cell type weave structure in the frequency of 50 MHz to 1.5 GHz.

The cell type fabric showed higher shielding effectiveness than the fabric having plain weave structure. The shielding effectiveness of −48 dB was observed for cell type fabric compared to plain structure (−35 dB) in the frequency of 0.7 GHz. For the same fabric, an increase in metal content (by using SS doubled yarn) improved the

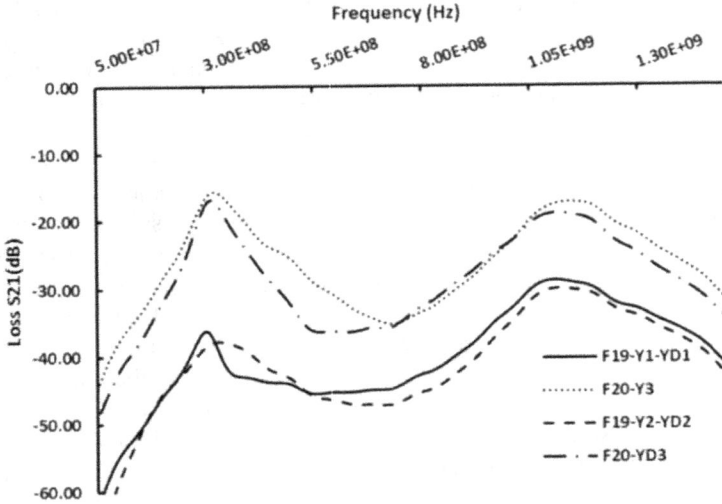

FIGURE 12.11 Comparison of shielding behaviour of plain and cell type fabrics. (Adapted from Krishnasamy, Jagatheesan & Das, 2020.)

shielding effectiveness of the fabric. The similar shielding behaviour was observed for plain fabric also. The developed hybrid yarn fabric could be used as electrostatic discharge clothing for different industries.

12.5 GLASS TOWPREG FOR THERMOPLASTIC COMPOSITES

In a study, Silveira et al. (2017) prepared the pre-impregnated glass fibre woven fabric with epoxy resin and then laminated it with carbon fibre non-woven veil metalized with Ni for developing the radar absorbing structures. The studies suggested that the epoxy resin/GF fabric prepreg could be used in the outer layer of radar absorbing structures that allows the 98.7% of electromagnetic waves through its microstructure in microwave band (8.2–12.4 GHz). The study results confirmed the low reflectivity of the material with low average attenuation values. However, the carbon fibre/Ni veil was predominantly reflective (~91%) according to this study. The developed epoxy resin/glass fibre/carbon fibre/Ni, being lightweight and low-thickness reflective multifunctional composites, could be used for buildings, communication, automation, and aerospace industries. In another study on developing the glass towpreg-based composites, Cheng et al. (2000) prepared a un-commingled yarn. The developed yarn has the copper wire and glass in the core and polypropylene in the sheath in the form of helical filaments. Using this hybrid yarn, four types of knit structure, namely double plain, double plain inlaid, 1/1 rib and 1/2 rib, were prepared to shield the electromagnetic radiation. The structure of developed un-commingled yarn is shown in Figure 12.12.

Using this un-commingled yarn, various knitted fabrics with different wale densities and course densities were produced. The advantage of developing un-commingled glass yarn is that the glass can be knitted without much damage and the lubrication

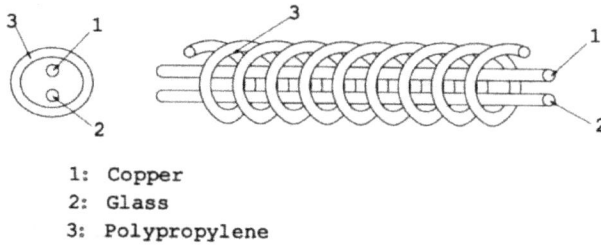

1: Copper
2: Glass
3: Polypropylene

FIGURE 12.12   The structure of the un-commingled yarn. (Adapted from Cheng et al., 2000.)

effect is brought by wrapping the PP fibres over the glass filament and copper wires. The eight fabric plies were consolidated in a compression moulding machine to produce various composite laminates. Under the action of applied temperature and pressure, the sheath PP fibres melt and impregnate through knitted glass and copper and form the conductive knitted fabric reinforced thermoplastic composite.

The effects of various fabric parameters on shielding effectiveness of the fabric composite were investigated under the frequency range of 0.3–3000 MHz. The un-commingled yarn assisted in a good impregnation of copper wires and glass fibres inside the polypropylene matrix as confirmed by microstructural examination. In addition, the amount of copper content has a strong influence of electromagnetic shielding characteristics of the composite laminates. However, the SE decreased for the increase in incident frequency for all the composite laminates. The composite laminate having the copper as inlay yarn showed higher shielding effectiveness than the other laminates. This is due to presence of copper wire in the inlaid directions, which are non-crimped in nature, which improves the shielding effectiveness. Similarly, the impact properties of conductive knitted fabric composites were also analysed. The fabric composites with thicker copper wire showed higher impact energy absorption than the composites with fine copper wire. Similarly, the laminate reinforced with higher stitch density showed better impact resistance. Hence, un-commingled yarn preparation is an effective method to prepare the flexible towpregs for developing the electromagnetic shielding composite panels.

12.6   SUMMARY AND SCOPE FOR FUTURE RESEARCH

The flexible towpregs are the choice for the composite processing industries for developing lightweight composite materials for various applications such as the automotive, aerospace, domestic and medical industries. When the towpreg is made conductive, it enlarges the scope of application of the composite product, such as electrostatic discharge and conducting materials, electromagnetic shielding applications etc. The filament towpreg can be made into composite prepreg materials by means of powder coating, blending with thermoplastic fibres, doubling with matrix filaments, preparation of sheath-core yarns, etc. The dry prepreg can be converted to fabric form by means of weaving/knitting; then it can be used as preform material for composite manufacturing.

Different methods can be adopted for introducing the conductivity in flexible towpregs such as conductive core-sheath yarns, cover yarns, wrap yarns, doubled yarns, blended yarns and helical cover yarns, which produce new flexible towpregs for electromagnetic shielding material preparation. In some approaches, over the conductive materials, thermoplastic polymers are deposited in the form of powder, fibres and yarns. Several studies have been conducted on these methods and that research is detailed in this chapter. The conductive composites developed from the conductive flexible prepregs provide improved shielding performance and mechanical properties. The reflection and absorption properties of the shielding materials could be improved by tailoring the structure of towpregs, fabric structure and layering methods. The prepreg materials also make the shielding materials free from defects, which enhances the mechanical strength suitable for aerospace and other applications. Hence, lots of research exists on preparation of flexible prepregs for developing novel electromagnetic shielding materials.

REFERENCES

ASTM D4935 -ASTM Committee D-9 on Electrical and Electronic Insulating Materials (1999). Standard Test Method for Measuring the Electromagnetic Shielding Effectiveness of Planar Materials. American Society for Testing and Materials. ASTM International, West Conshohocken, PA.

Cheng, K. B., Ramakrishna, S., & Lee, K. C. (2000). Electromagnetic shielding effectiveness of copper/glass fiber knitted fabric reinforced polypropylene composites. *Composites Part A: Applied Science and Manufacturing, 31.* 1039–1045. https://doi.org/10.1016/S1359-835X(00)00071-3

Das, A., Krishnasamy, J., Alagirusamy, R., & Basu, A. (2014a). Analysis of the electromagnetic shielding behavior of stainless steel filament and PET/SS hybrid yarn incorporated conductive woven fabrics. *Fibers and Polymers, 15*(11). https://doi.org/10.1007/s12221-014-2423-x

Das, A., Krishnasamy, J., Alagirusamy, R., & Basu, A. (2014b). Electromagnetic interference shielding effectiveness of SS/PET hybrid yarn incorporated woven fabrics. *Fibers and Polymers, 15*(1). https://doi.org/10.1007/s12221-014-0169-0

Friedrich, K. (1999). *Commingled yarns and their use for composites.* https://doi.org/10.1007/978-94-011-4421-6_12

Geetha, S., Kumar, K. K. S., Rao, C. R. K., Vijayan, M., & Trivedi, D. C. (2009). EMI shielding: Methods and materials—A review. *Journal of Applied Polymer Science, 112*(1). 2073–2086. https://doi.org/10.1002/app.29812

Huang, J. C. (1995). EMI shielding plastics. A review. *Advances in Polymer Technology, 14*(Compendex). 137. http://dx.doi.org/10.1002/adv.1995.060140205

Jagatheesan, K., Ramasamy, A., Das, A., & Basu, A. (2017a). Investigation on shielding and mechanical behavior of Carbon/Stainless steel hybrid yarn woven fabrics and their composites. *Journal of Electronic Materials, 46*(8). 5073–5088. https://doi.org/10.1007/s11664-017-5498-5

Jagatheesan, K., Ramasamy, A., Das, A., & Basu, A. (2017b). Study of aperture size and its aspect ratio of conductive hybrid yarn woven fabric on electromagnetic shielding effectiveness. *Fibers and Polymers, 18*(7). https://doi.org/10.1007/s12221-017-7044-8

Jagatheesan, K., Ramasamy, A., Das, A., & Basu, A. (2018a). Electromagnetic absorption behaviour of ferrite loaded three phase carbon fabric composites. *Smart Materials and Structures, 27*(2). https://doi.org/10.1088/1361-665X/aa9f8c

Jagatheesan, K., Ramasamy, A., Das, A., & Basu, A. (2018b). Electromagnetic shielding effectiveness of carbon/stainless steel/polypropylene hybrid yarn-based knitted fabrics and their composites. *Journal of the Textile Institute, 109*(11). 1445–1457. https://doi.org/10.1080/00405000.2018.1423883

Jagatheesan, K., & Das, A. (2020). Development of metallic core-spun yarns and hybrid conductive fabrics for electromagnetic shielding applications. *Indian Journal of Fibre and Textile Research, 45*(3). 346–351.

Krishnasamy, J., Ramasamy, A., Das, A., & Basu, A. (2016). Effect of fabric cover and pore area distribution of Carbon/Stainless Steel/Polypropylene hybrid yarn-woven fabric on electromagnetic shielding effectiveness. *Journal of Electronic Materials, 45*(6). https://doi.org/10.1007/s11664-016-4391-y

Krishnasamy, J., Ramasamy, A., Das, A., & Basu, A. (2016). Effect of fabric cover and pore area distribution of carbon/stainless steel/polypropylene hybrid yarn-woven fabric on electromagnetic shielding effectiveness. *Journal of Electronic Materials, 45*(6). 3087–3100. https://doi.org/10.1007/s11664-016-4391-y

Krishnasamy, J., Ramasamy, A., Das, A., & Basu, A. (2018). Electromagnetic absorption behaviour of carbon helical/coiled yarn woven and knitted fabrics and their composites. *Journal of Thermoplastic Composite Materials.* 1–26. https://doi.org/10.1177/0892705718759389

McGregor, O. P. L., Somashekar, A. A., Bhattacharyya, D., McGregor, O. P. L., & Duhovic, M. (2017). Pre-impregnated natural fibre-thermoplastic composite tape manufacture using a novel process. *Composites Part A: Applied Science and Manufacturing, 101.* https://doi.org/10.1016/j.compositesa.2017.05.025

Nunes, J. P., Silva, J. F., Marques, A. T., Crainic, N., & Cabral-Fonseca, S. (2003). Production of powder-coated towpregs and composites. *Journal of Thermoplastic Composite Materials, 16*(3). https://doi.org/10.1177/0892705703016003003

Silveira, D. C., Gomes, N., Rezende, M. C., & Botelho, E. C. (2017). Electromagnetic properties of multifunctional composites based on glass fiber prepreg and Ni/carbon fiber veil. *Journal of Aerospace Technology and Management, 9*(2). https://doi.org/10.5028/jatm.v9i2.657

Tugirumubano, A., Vijay, S. J., Go, S. H., Shin, H. J., Ku, K. L., & Kim, H. G. (2018). The evaluation of electromagnetic shielding properties of CFRP/metal mesh hybrid woven laminated composites. *Journal of Composite Materials, 52*(27). https://doi.org/10.1177/0021998318770511

Zhang, H., Guo, Y., Zhang, X., Wang, X., Wang, H., Shi, C., & He, F. (2020). Enhanced shielding performance of layered carbon fiber composites filled with carbonyl iron and carbon nanotubes in the Koch curve fractal method. *Molecules, 25*(4). https://doi.org/10.3390/molecules25040969

13 Flexible Towpregs and Thermoplastic Composites for Civil Engineering Applications

Nabo Kumar Barman[a], R. Alagirusamy[b], and S. S. Bhattacharya[a]
[a]The Maharaja Sayajirao University of Baroda
Gujarat, India
[b]Indian Institute of Technology
New Delhi, India

CONTENTS

DOI: 10.1201/9781003049715-13

13.1 INTRODUCTION

In civil engineering applications, structural components should provide sufficient resistance against tensile and compression stresses. Concrete offers adequate resistance under compression, and to sustain tensile forces, reinforcement components must be incorporated. Cement and steel are the prime choices for building materials. However, carbon emission, energy consumption, and environmental impact involved in manufacturing cement and steel have led researchers to explore suitable and sustainable alternatives. The alternatives for cement are supplementary cementitious materials (SCMs), which include fly ash, slag cement, metakaolin, and silica fume. The SCMs not only offer low carbon footprints but also exhibit high mechanical properties, durability, and reduce permeability to alkali ions with cost effectiveness. Similarly, steel as a reinforcement element has greater durability concerns due to exposure in concrete pore solutions which are highly alkaline (pH 12–13).

Continuous research efforts were directed for improving durability of steel reinforcements and finding non-corrosive and durable alternative reinforcements. Textile and textile-based composites have been able to overcome durability and corrosion issues encountered with steel reinforcement. Additionally, concrete reinforced with non-corrosive textiles requires lesser cover, which allow formation of lightweight structural components with reduced concrete consumption. Textile materials in different forms have been used as concrete reinforcement. The form of textile materials can be perceived as continuous and discrete in the context of reinforcement. Textile materials in short fibrous forms (discrete assemblies) of both natural and synthetic origin dispersed in the concrete as reinforcement are popularly known as fibre reinforced concrete (FRC). The FRC structures have been reported to contribute to improving strain hardening behaviour, flexural rigidity, toughness, and prevented crack propagation in concrete through fibre bridging action. However, fibre addition should be restricted to a limiting amount of 0-3% in order to retain minimum workability and prevent fibre agglomeration. The continuous textile assemblies such as unidirectional fabric, bidirectional fabric (woven, knitted, and non-crimp), nonwoven fabrics, fabric composites, fibre reinforced polymer (FRP) laminates, braided rods, and FRP rods are incorporated as reinforcement elements in concrete.

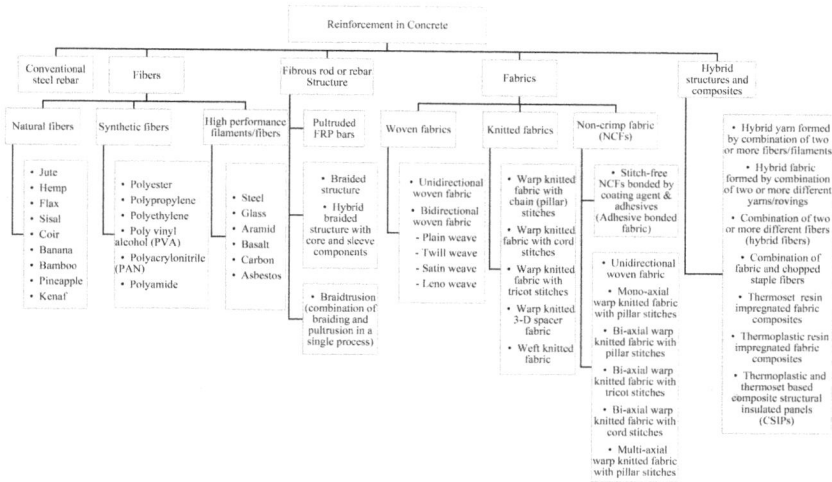

FIGURE 13.1 Classification of textiles and textile-based composites used for reinforcement in concrete structural elements.

In contrast to the discrete reinforcement, the continuous fibre reinforced structures would exhibit good structural integrity during reinforcement and hence a best exploitation of inherent properties of the reinforcement to the total structure would be possible. The different forms of textile reinforcement used in concrete reinforcement are shown in Figure 13.1. The fabric reinforcement in concrete is advantageous as customization of warp and weft threads can be made with ease to fulfil load and design requirements. Textile reinforcement is able to fulfil two major functions: (a) load bearing elements in concrete, and (b) strengthening and confinement of columns, beams mainly used as jackets for repair, rehabilitation, and restoration of structures.

Textile reinforced concrete (TRC) and textile reinforced mortar (TRM) are attracting research interest in the area of civil and structural engineering. Flexible textile assemblies as reinforcement are embedded in the concrete or mortar matrices to form TRC and TRM, respectively. TRC, recognized as TRM or fabric reinforced cementitious matrix (FRCM) comprises of two components, namely cementitious matrix and textile-based continuous structure as reinforcement component, as depicted in Figure 13.2.

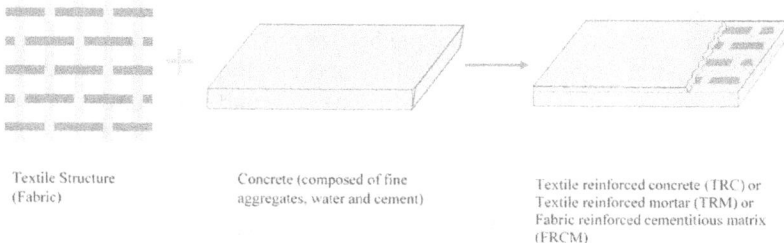

FIGURE 13.2 Schematic representation of textile reinforced concrete (TRC).

The matrix is produced from a mixture of water, cement (or in blend form with SCMs), fine aggregates (sand), and plasticizers in predefined proportions based on design requirements. Textiles are equally compatible for application in rehabilitation and repair purposes, where textiles are wrapped externally for restoring aged structures. The textiles are quite adaptive in forming complex and lightweight structural components.

Various textile materials have been used as reinforcement to serve the specific functional properties desired in structural elements. In the next sections, an attempt has been made to understand flexible towpreg (hybrid yarn) structure and its suitability in reinforcing structural elements.

13.1.1 Flexible Towpreg (Hybrid Yarn) Structure

The application of textile-based composites has seen significant rise in the last few decades. The composite materials can be easily customized by selection of suitable reinforcement and matrix components. Thermoplastic composites have now been readily accepted in various application areas such as aerospace, automobiles, wind turbines, building construction, defence and other industrial applications. They offer several advantages over thermosets in terms of high impact resistance, faster processing, longer shelf life, remolding ability, and recyclability. Towpregs are filament tows incorporated or coated with polymer matrices (Ramasamy, Wang, and Muzzy 1996). To produce flexible towpregs, the tows should be incorporated with thermoplastic polymer matrices. The concept of flexible towpregs lies in ease of material handling with flexibility during textile preforming operations.

Thermoplastic composites can be produced through following processes: (a) initially forming dry textile preforms (woven, knitted, braid, etc.) from uncoated filament tows followed by consolidating with thermoplastic matrices, (b) formation of towpregs or hybrid yarn (commingled yarn, powder coated yarn, filament wrapping) followed by textile preforming process (weaving, knitting, and braiding) and subsequent consolidation in composite structure. The former method offers challenge in terms of penetration of thermoplastic matrices, which generally have higher viscosity compared to thermosets. Another challenge is resin distribution in between filaments with closely spaced textile structure. The latter method is advantageous because the resin is incorporated in the yarn structure (towpregs) prior to the textile preforming process. Therefore, during the composite consolidation process the matrix has to cover a shorter distance (mass transfer distance) for impregnation of reinforcing filaments (Alagirusamy et al. 2006). This enables formation of composite with homogenous and controlled distribution of matrices within and in between filament bundles.

A study was carried out to examine the behavior of different yarns (basalt, polypropylene, polyester, and jute) on durability and bonding behavior (load bearing ability) when reinforced in concrete matrix. The yarn pull out behavior in terms of interfacial shear strength of basalt was found to be highest, followed by polypropylene, and was lowest for jute yarn. The accelerated ageing of yarn was investigated under NaOH and Ca(OH)$_2$ solution (10g/L) in terms of weight loss, loss in strength, and elongation behavior. The alkaline degradation in terms of loss in weight (%), strength and elongation was minimum for basalt and polypropylene yarn under alkaline conditioning, also confirmed through SEM analysis. The authors suggested use of a combination of polypropylene/polyester along with basalt in hybrid fabric as reinforcement in cement composites to improve the ductility and reduce cost (Jamshaid et al. 2018).

The high performance filament roving is composed of numerous densely packed filaments with very small inter filament spacing. The filament bundle when reinforced in a cementitious matrix offers negligible penetration of the matrix inside the bundle. Therefore, under tensile loading the reinforcing efficacy of filaments is not exploited to the actual potential. This results in recording lower tensile behavior in cement composites. Therefore, to assure all the filaments in bundle participate equally in load sharing, yarn bundles are often impregnated with polymers or resins (Kravaev et al. 2009). This enables higher utilisation of reinforcing efficacy of filament bundle and thereby provides higher resistance towards tensile forces.

Selection of reinforcement elements in concrete is most crucial, which has a profound effect on concrete durability. There are several measures to be considered in reinforcement selection such as mechanical properties (tensile, compression, flexural, and impact), thermal and insulation properties, and corrosion behaviour. However, optimization of properties based on application is of greater significance. For instance, a concrete specimen designed specifically for marine application should possess sufficient mechanical properties; however, more emphasis should be towards long term resistance against water ingress, alkaline, and corrosion resistance. Similarly, concrete structures designed for war zones and terror affected areas should ensure high protection towards impact and ballistic properties. Further, in earthquake resistance structures, the ductility behaviour of reinforcement has greater importance. Such application-based design of concrete members makes the choice of reinforcement relatively important, which should ensure long term durability during its service life. The different thermoplastic textile-based structures have been covered in this chapter along with their potential applications.

The response of the textiles on prolonged exposure in aggressive alkaline medium, high temperature, freeze thaw cycles, and UV exposure will determine the ultimate durability of the structural component. Therefore considering durability concerns, textiles produced from high performance glass, alkali resistant glass (AR glass), carbon, aramid, and basalt have gained importance in the recent times. The long-term durability effects of textile composites used in construction are discussed in the later part of this chapter.

In the context of civil engineering applications, hybrid yarn manufacturing methods, thermoplastic composite forming methods, thermoplastic fibre reinforced polymer (FRP) bars, and thermoplastic FRP laminates are also covered in this chapter.

13.2 HYBRID YARN-BASED STRUCTURES FOR CONCRETE REINFORCEMENT

There are several methods used for producing hybrid yarns such as co-wrapping, parallel winding, commingling, stretch-breaking, air-jet texturing, Kemafil technology, schappe technique, braiding, ring spinning, rotor spinning, wrap spinning, and friction spinning. A combination of two or more hybrid yarn manufacturing methods can also be used for achieving desired properties. The methods involved in hybrid yarn production have been reported in literature (Lauke, Bunzel, and Schneider 1998; Alagirusamy and Ogale 2004; Alagirusamy et al. 2006). A few relevant hybrid yarn-based structures are discussed in this section, which have good reinforcement potential in structural elements.

13.2.1 DREF-III YARN

DREF-III friction spinning is a popular technique for producing core-sheath hybrid yarn structures. The DREF spun yarn of 1590 tex was produced, which consists of basalt multifilament rovings in the core and staple polypropylene fibres as sheath components. The basalt roving/polypropylene (BR/PP) core-sheath ratio was 75.5/24.5. The authors used this process to utilise the combined properties of basalt and polypropylene. Basalt roving has good inherent alkaline resistance, heat resistance, and mechanical properties, highly desirable for concrete reinforcement. The polypropylene fibres provide high deformation and ductility behavior. The hybrid yarn sample and thermal treated hybrid yarn samples have resulted in improved tensile properties in terms of strength as well as ductility (elongation) of yarn samples. It was reported that hybrid yarn tenacity was recorded 56–95% higher as compared to parent basalt roving as depicted in Figure 13.3. The authors suggested the great potential of hybrid yarn-based reinforcement for TRC applications (Barman, Bhattacharya, and Alagirusamy 2020).

13.2.2 COMMINGLED YARN

Research was performed to study commingled hybrid yarns produced with high strength alkali resistant glass (AR-glass) filaments and polyvinyl alcohol (PVA)

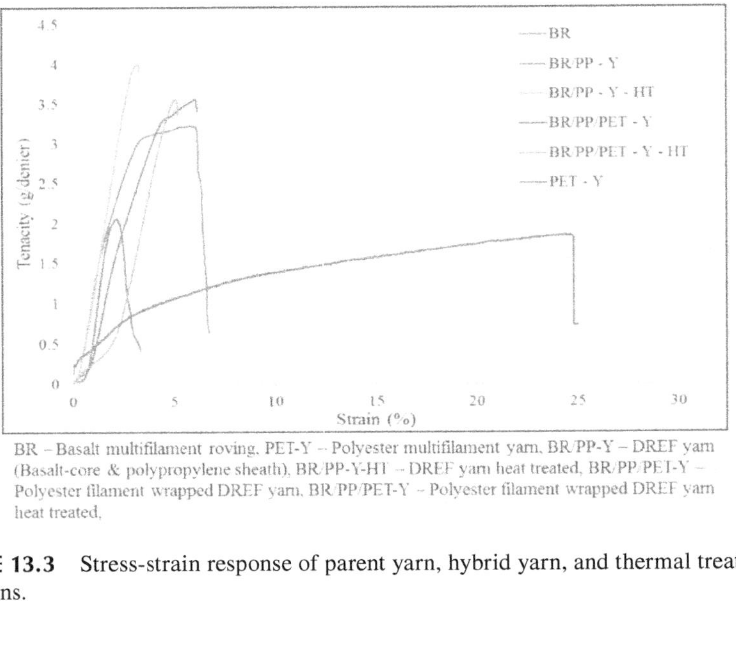

BR – Basalt multifilament roving, PET-Y – Polyester multifilament yarn, BR/PP-Y – DREF yarn (Basalt-core & polypropylene sheath), BR/PP-Y-HT – DREF yarn heat treated, BR/PP/PET-Y – Polyester filament wrapped DREF yarn, BR/PP/PET-Y – Polyester filament wrapped DREF yarn heat treated,

FIGURE 13.3 Stress-strain response of parent yarn, hybrid yarn, and thermal treated yarn specimens.

filaments (water soluble) in the proportion of 95% AR glass and 5% PVA. PVA was found to be water soluble; hence it was expected to create interstices in the yarn bundle embedded in concrete. This was intended to improve penetration of cementitious matrix in the voids created by dissolution of PVA and subsequent improvement in bonding behavior of TRC. The PVA filaments lead to improvement in concrete microstructures. In the commingling process, machine and process parameters such as nozzle type, air pressure, overfeed ratio, and yarn tension were found to have little influence on the hybrid yarn characteristics. The tensile strength of the commingled yarn was found to be 30% lower in contrast to AR glass roving. In spite of recording lower tensile strength, the commingled yarn performed better in pull-out load test of specimens embedded in concrete. Two different samples of biaxial warp knitted fabric with tricot stitches were produced: one from AR glass and another from commingled yarn. The fabric construction parameters were kept similar in both instances. The direct tensile test of the TRC components produced by reinforcing warp knitted fabrics was carried out. The results yielded 80% higher strength for the TRC components made from commingled yarn-based knitted fabrics. The SEM micrographs manifest the PVA filaments to have undergone surface etching and not complete dissolution when subjected to a alkaline cementitious matrix (Kravaev et al. 2009). In another research, behavior of spread yarn with the hybrid commingled yarn and parent AR-glass yarn were compared. The spread yarn was produced on the same set up used for commingled yarn. During spread yarn formation, an air jet ruptures the coatings on the filaments and increases the spacing between filaments within the yarn bundle while damaging the bundle as little as possible. The pull out load of the spread yarn and commingled yarn was higher by 300% and 80% respectively in contrast to the parent AR glass as shown in Figure 13.4. Moreover, the tensile strength of the TRC components produced from biaxial warp knitted fabric with spread yarn and commingled yarn was higher by 30% and 80%, respectively as compared to the TRC with AR glass-based warp knitted fabric as shown in Figure 13.5 (Janetzko et al. 2010).

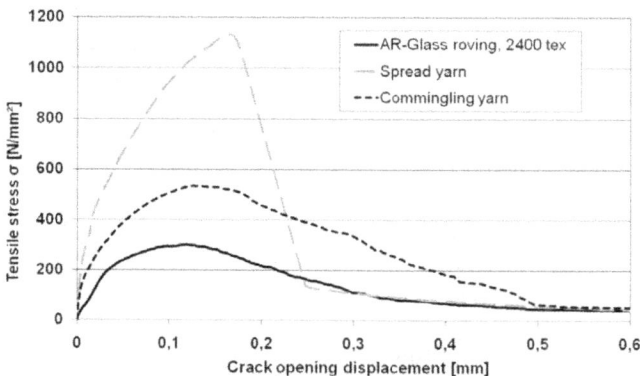

FIGURE 13.4 Yarn pull out behaviour for AR glass roving, ARG/PVA commingled yarn, and spread yarn. (Reused with permission from RILEM; Janetzko et al. 2010.)

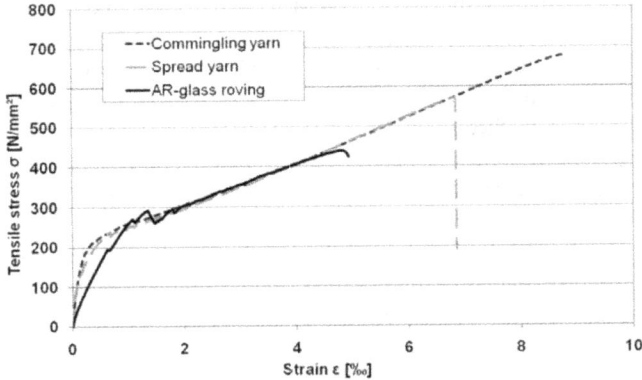

FIGURE 13.5 Stress-strain curves of TRC specimens reinforced with AR glass roving, ARG/PVA commingled yarn, and spread yarn. (Reused with permission from RILEM; Janetzko et al. 2010.)

13.2.3 Braided Composite Rod

The braiding process was effectively used for producing structural composite rods for concrete reinforcement as reported in literature (Gonilho Pereira et al. 2008; Zdraveva et al. 2010; Correia et al. 2016). The sensing and health monitoring applications of braided structures have also been reported (Rosado et al. 2012; Rana et al. 2014). The aramid braided FRP bar consolidated with pultrusion process was reported to have uniform stress-strain behaviour with large inelastic deformation. Moreover, the bond strength and ductility index were comparable to that of steel rebar (Harris, Somboonsong, and Ko 1998). The pultruded hybrid core-sheath rod structure (carbon core and glass sheath) produced through braiding process and consolidated using epoxy resin has higher interface shear strength and fatigue resistance and possess potential to replace the steel rod used in underground oil extraction applications (Li, Guo, et al. 2020; Li, Yin, Liu, et al. 2020; Li, Yin, Wang, et al. 2020). In the above discussed studies for consolidation purpose thermoset resins such as epoxy, vinylester and polyester resin were used. However, recently use of thermoplastic resin has been reported for consolidation of braided structure.

Braided hybrid structure was produced with core components composed of glass filaments and polypropylene filaments, and sheath sleeve structure was braided using polyester yarn. The structure is placed in a specially designed mold to enable uniform heat and pressure application and consolidated in an oven. They reported that the tensile properties improves and flexural strength reduces on increasing glass proportion in structure. The authors suggested the need to improve the fibre distribution and consolidation process for improved composite behaviour (Kling, Rana, and Fangueiro 2012).

In an interesting study, an investigation was carried out to examine tensile behaviour of hybrid thermoplastic composite rods produced from carbon-glass filaments (carbon core and E-glass sheath) and consolidated using a newly developed thermoplastic epoxy resin composed of difunctional epoxy resin and difunctional

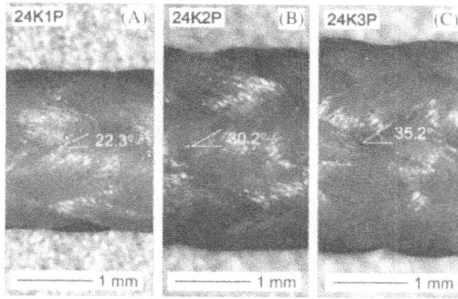

FIGURE 13.6 Outer surface showing braided glass sleeve with variable braid angles for hybrid glass-carbon thermoplastic composite rod specimens. (Reused with permission from John Wiley and Sons; Naito, Oguma, and Nagai 2020.)

phenolic compound. The sheath structure is formed using a maypole braiding setup in which 16 glass roving (16 × 12000 filaments) bundles are intertwined together to form sheath components. The core element composed of 3 bundles of 24K carbon filaments (72,000 filaments). The three different hybrid rods designated as 24K1P, 24K2P, and 24K3P having braid angles of 22.3°, 30.2°, and 35.2°, respectively are shown in Figure 13.6. All three hybrid composite rods have varying carbon/glass/matrix resin (C/G/M) proportions (in % volume fraction) – 24.6/39.8/25.5 for 24K1P, 38.3/29.6/24.5 for 24K2P, and 46.2/23.2/23.4 for 24K3P rods. The volume of void fraction (%) for 24K1P, 24K2P, and 24K3P rods were 10.2%, 7.49%, and 7.28%, respectively. The tensile strength and modulus were recorded highest at 1.84 GPa and 91 GPa, respectively for composite rod specimens-24K3P with higher carbon fibre proportion. The change in Poisson's ratio (0.39–0.45) and breaking strain (2.08–2.18) were insignificant for all the composite rod specimens. The Weibull modulus of composite rod was influenced by the proportion of carbon fibre and void content in composite. Moreover, as the braid angle is increased, the proportion of carbon fibre increases and the void content decreases, resulting in improved tensile behaviour (Naito and Oguma 2017a). The transverse compressive strength for 24K1P, 24K2P, and 24K3P rod specimens was found to be 15.26 MPa, 8.79 MPa, and 8.05 MPa, respectively. The composite rod specimens exhibited non-linear load-displacement response. The composite rods failed by forming straight lines propagating through the loading region (Naito, Nagai, and Tanaka 2019).

Further investigation was carried out to understand fatigue behaviour of carbon-glass hybrid thermoplastic composite rods. The fatigue test was carried out at 10 Hz frequency under cyclic loading at constant amplitude for 1×10^7 cycles. The test result suggested the fatigue strength of composite rods was found to be less than 30% of breaking strength, which is lower than steel rods. The authors suggested increase in proportion of carbon fibre in composite rods and reduction in void content can improve the fatigue behaviour of hybrid composite rods (Naito and Oguma 2017b). The tensile strength of composite rods are comparable to that of high strength steel rods. Similarly, the tensile modulus of composite rods are comparable to that of aluminium alloy rods, as depicted in Figure 13.7.

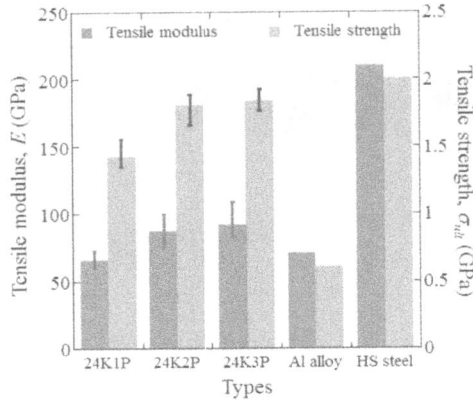

FIGURE 13.7 Tensile strength and modulus of hybrid thermoplastic composite rods compared to aluminium alloy rods and high strength steel rods. (Adapted from Naito and Oguma 2017b.)

The microscopic image of hybrid glass-carbon thermoplastic composite rod specimen is shown in Figure 13.8. The thermo-mechanical behaviour of hybrid composite rods were analyzed at different temperatures (−50°C, 0°C, 23°C, 50°C, and 80°C) using a dynamic mechanical analyzer (DMA) at 1 Hz frequency. The stress-strain response was found to be linear at all temperatures, except marginal inelastic behaviour observed at 80°C associated with resin softening. The Weibull modulus

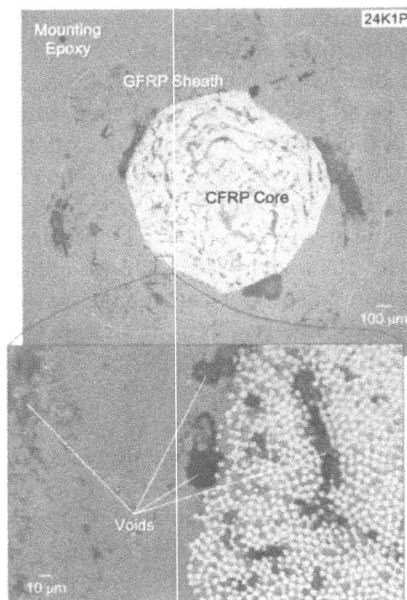

FIGURE 13.8 Microstructural image of hybrid glass-carbon thermoplastic composite rod specimen. (Reused with permission from John Wiley and Sons; Naito, Oguma, and Nagai 2020.)

is found to be greatly influenced by temperature, in addition to the proportion of carbon fibre and void content in composite rods. The temperature change influences matrix behaviour – at lower temperature uniform breaking stress and higher Weibull modulus are obtained and higher temperature results a complex behaviour with fibre pullout, and rupture causes inconsistent breaking stress and lower Weibull modulus (Naito, Oguma, and Nagai 2020).

Another study was performed in which hybrid braided structure was formed with AR glass as core and sleeve (sheath) of polypropylene. The hybrid structure is heated in a compression molding machine to form thermoplastic composites. The mechanical behaviour (compression strength and flexural strength) of the TRC reinforced with hybrid composite structure is compared with TRC reinforced with uncoated parent AR-glass yarn, epoxy coated AR glass yarn, uncoated carbon yarn, and epoxy coated carbon yarn, respectively. The flexural strength of unreinforced concrete was 10.5 MPa, whereas TRC reinforced with a hybrid composite structure, uncoated parent AR-glass yarn, epoxy coated AR glass yarn, uncoated carbon yarn, and epoxy coated carbon yarn was found to be 15.85 MPa, 14.19 MPa, 25.32 MPa, 24.78 MPa, and 40.8 MPa, respectively. The hybrid composite structure has contributed to improvement in flexural strength by 57.71% as compared to unreinforced concrete specimen. The epoxy coated specimen has improved flexural behaviour of TRC; however, higher cost of epoxy resin with tedious coating operation should also be taken in consideration. Carbon reinforcement has exhibited improved flexural behaviour as compared to AR glass reinforcement in TRC. The authors suggested that contribution of hybrid composite structure is limited as compared to epoxy coated yarn; higher strength may be achieved by performing changes in yarn production method and yarn composition (Kurban, Babaarslan, and Çağatay 2017).

13.3 THERMOPLASTIC COMPOSITES FOR STRUCTURAL APPLICATIONS

There are several methods used for manufacturing thermoplastic composites such as autoclave moulding, filament winding, thermal forming, automated tape placement, continuous consolidation, and pultrusion. These methods have already been covered in greater detail in Chapter 10. A few interesting thermoplastic composites forming methods and the advantages of thermoplastic composites are reported in this section, which have good potential for structural applications.

13.3.1 THERMOPLASTIC COMPOSITES MANUFACTURING PROCESS

The manufacturing of thermoplastic composites is rapid and hence higher production can be achieved in a short duration. Specific methods suitable for forming structural components are covered in this section.

13.3.1.1 DRIFT™ Process

This is a hot melt impregnation process in which continuous filament roving is drawn off from the creel zone and guided through tensioning devices. The extruder supplies molten polymer in the die. The filaments are passed through the die in which

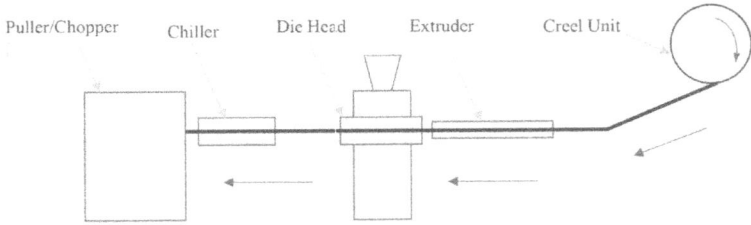

FIGURE 13.9 DRIFT™ process for manufacturing of thermoplastic tape (ribbon) shape composite structures.

the filaments are impregnated with molten polymer, and the impregnated polymer is passed through the chiller (cooling) zone for cooling the thermoplastic prepreg. The cooled prepreg can be withdrawn as continuous tape (ribbon) or can be cut into desired cut length by chopping mechanism. The line diagram for the process is shown in Figure 13.9. The reinforcement material in form of glass, aramid, carbon, and basalt filaments can be easily processed in this machine as they are able to sustain high temperature. The polymers such as polypropylene, nylon, and polyethylene can be used to form thermoplastic matrices in composites. The authors claimed that high quality composites, high molding rate, and high tensile and impact behaviour of composites can be achieved through this process (Hartness et al. 2001).

13.3.1.2 Thermoplastic Face Sheet for Thermoplastic Composite Structural Insulated Panels (CSIPs)

The thermoplastic composite structural insulated panels (CSIPs) are used for modular construction in which the core of molded expanded polystyrene, extruded polystyrene, or urethane foam is placed in a sandwich manner in between two face-sheets (Uddin and Mousa 2013). The face-sheet or laminates may be formed using thermoplastic composites; production steps of face-sheet (laminates) for CSIPs are shown in Figure 13.10.

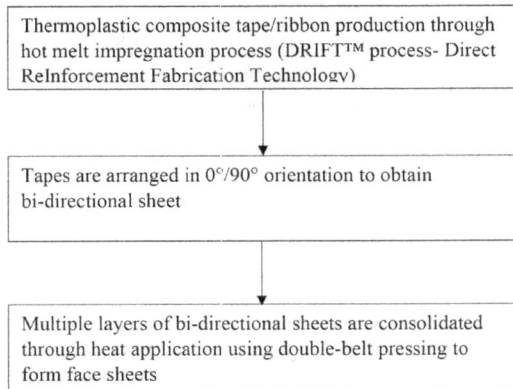

FIGURE 13.10 Production sequence for formation of thermoplastic face sheets.

13.3.1.3 Filament Winding Process

The filament winding method is commonly employed for jacketing (wrapping) the column elements which need additional strengthening from transverse forces. The filament winding technique is considered as a useful method for retrofit and restoration of column structures. In an investigation, the load carrying behaviour of E-glass/PPS and carbon/PEEK thermoplastic (high temperature resistant resin) composites tape wrapped cylinderical concrete column elements was examined (Friedrich et al. 2002). The filament winding process was used to wind the composite tapes over the cylinder specimens – precraked and undamaged, as shown in Figure 13.11. Concrete cylinder specimens of B35 grade and 100 mm diameter with compressive strength of 50 MPa and tensile strength of 4 MPa were used for the experiment. To simulate the damage of column elements, the cylinder specimens were subjected to splitting tensile tests. Under compression loading, the filament wound specimens failed in a quasi-controlled mode, whereas reference samples failed in catastrophic shear with splitting fracture into multiple pieces. The undamaged glass/PPS confined cylinders have exibited 3–5 times higher compressive stress as compared to the unconfined and undamaged cylinder specimens. A similar relationship was observed for pre-damaged cracked cylinder specimens confined with 10 layers of E-glass/PPS composite tapes, which experienced a 4.5 times higher compressive stress than the unconfined and undamaged specimens. They recommended use of glass/PPS composite for rehabilation aplications for restoring damaged columns. However, the carbon/PEEK composite confined cylinders have not exhibited significant improvement in load bearing ability. The compressive strength is influenced by fibre type, fibre proportion, and number of layers wrapped. The E-glass/PPS confined cylinder specimen with shell thickness of 3.2 mm has prevented shear failure of cylinders during compression test. The tensile strength of the composites has contributed in higher compressive strength of cylinders and restricted shear failure of specimens.

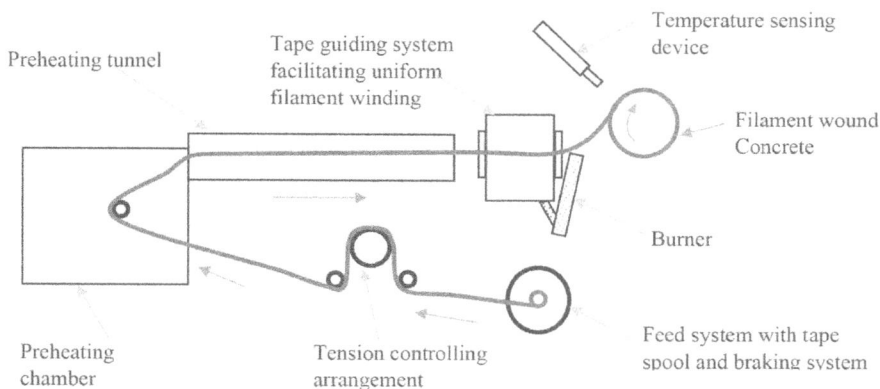

FIGURE 13.11 Schematic of filament winding process used for concrete elements.

Conservation of ancient building structures is associated with cultural, architectural, and artistic values. The use of FRP thermoset resins or adhesives when used for retrofit can effectively provide added durability and life to structure. Hybrid composites in column confinement wound with glass-Kevlar filaments and epoxy resin composite wraps showed greater ductility than those wound with glass filaments and epoxy resin composite wraps. They recommended application of hybrid composites for enhancement of performance and durability in column restoration (Liu, Tai, and Chen 2000). However, various research findings have suggested that in column confinement, contact pressure between fabric wrap/FRP and column has greater significance, rather than the adhesion (bonding) behaviour to resist the transverse forces (Cascardi et al. 2019). The application of thermoset-based FRP in restoration of historical structures is limited because of the adhesion behaviour, which is irreversible. The old thermoset FRP used in retrofit cannot be replaced by new FRP without preventing damage to the heritage structure. Therefore, newer techniques are employed to overcome the limitation of thermoset FRPs by incorporating additional elements between FRPs and columns for easy release during the replacement process. For non-adhesive filament winding use of aluminium foil (Liu, Tai, and Chen 2000), PET film, and liquid adhesion inhibitor (Cascardi, Dell'Anna, and Micelli 2018) was reported in literature. An investigation was performed to examine application of filament winding to wrap hybrid thermoplastic roving on a limestone (Lecce stone) cylinder specimen (Cascardi et al. 2019). Two different materials were selected for their experiment: (a) E-glass and polypropylene commingled flat roving (Twintex®) with width and thickness of 0.5 mm and 5 mm, respectively, (b) E-glass and polyethylene terephthalate matrix in tape form (LPET) with width and thickness of 0.8 mm and 19 mm, respectively. In filament winding machine as depicted in Figure 13.12, roving/tape was fed through unwinding of package which is passed

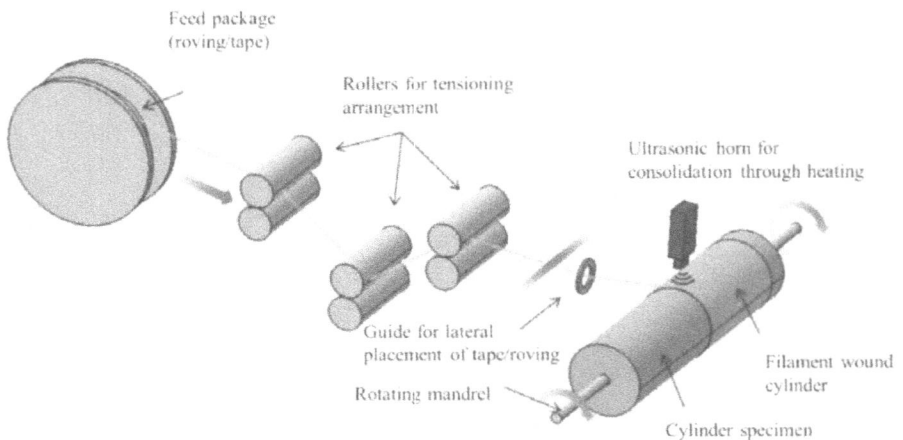

FIGURE 13.12   Filament winding process for wrapping cylindrical specimens consolidated through ultrasonic waves. (Reused with permission from Elsevier; Cascardi et al. 2019.)

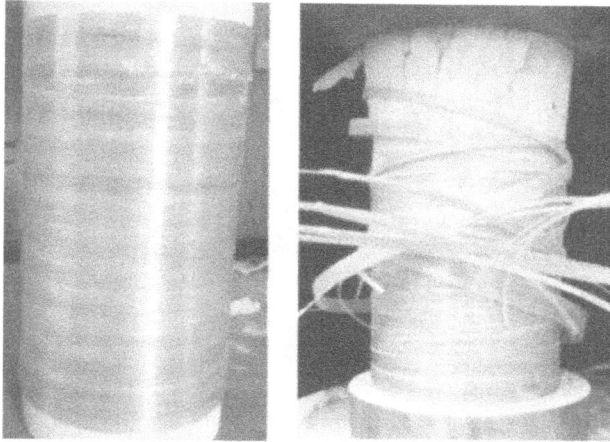

FIGURE 13.13 E-glass/polyethylene terephthalate tape filament wound cylinder specimen before test (left) and after compression test (right). (Reused with permission from Elsevier; Cascardi et al. 2019.)

through tensioning arrangement. The material is further guided through a guide which has ability for lateral positioning of roving/tape on to the rotating cylinder during winding operation. The ultrasonic titanium horn has a 20 kHz frequency and 1.5 bar horn pressure (controlled through compressed air) used for consolidation of thermoplastic wrap over cylinder specimen. The tape and roving thermoplastic material was wound over the cylinder specimen at a winding speed of 0.7 and 1.5 rad/s, respectively. The filament wound E-glass/polypropylene and E-glass/polyethylene terephthalate thermoplastic composite wraps have provided similar confinement effectiveness results (in terms of axial stress, axial strain, and lateral strain) as compared to p-phenylene-2,6-benzobisoxazole (PBO)-epoxy and carbon-epoxy confined cylinder specimens (incorporated with PET films). The easy release properties of wraps from specimens tested under compression load was observed without leaving imprints on the columns, as observed in Figure 13.13. The authors suggested use of filament winding process for restoration of heritage columns with enhanced axial properties and seismic protection.

13.3.1.4 Compression Moulding Process

This process is simple, quick, and readily used for manufacturing thermoplastic composites. Previous investigations have suggested that woven grid fabric structures have been found to be effective for concrete reinforcement. The interyarn apperture allows cement-based matrices to penetrate in the fabric structure and improve mechanical bonding with reinforcing fabric structures. In this context, development of the hybrid yarn-based woven grid was made to produce fabric composites consolidated using a compression molding machine. The fabric composites are formed

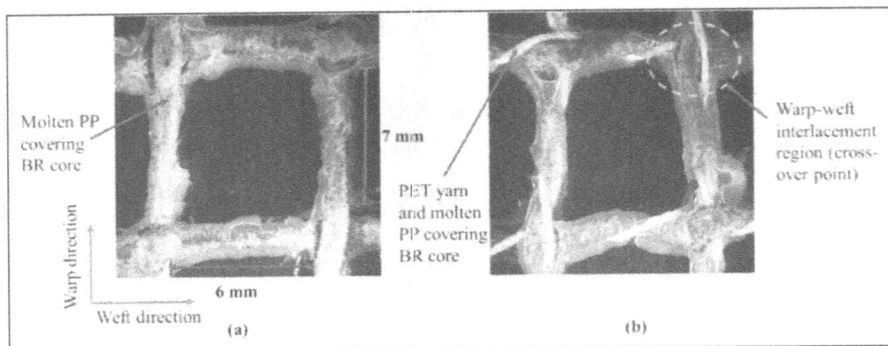

FIGURE 13.14 Microscopic image of composite samples with grid size – 6mm × 7mm. (a) BR/PP hybrid yarn-based fabric composite (left) and (b) BR/PP/PET hybrid yarn-based fabric composite (right) at 6× magnification.

at 190–200°C at 20 bar pressure and 5 minutes of curing duration. The microscopic images of the woven fabric (grid structure) composite samples are depicted in Figure 13.14), and uniform polypropylene resin distribution can be observed from the images (Barman, Bhattacharya, and Alagirusamy 2020).

13.3.1.5 Advantages of Thermoplastic Composites

There are various advantages associated with textile-based thermoplastic composites, thermoplastic FRPs, and modular thermoplastic composite panels used in building construction, as mentioned below.

- Thermoplastic composites require shorter time duration for production and curing, respectively.
- Longer shelf life for thermoplastic towpregs and composites allows bulk production and provides storage option.
- Welding of thermoplastics composites can be done with ease (e.g., using ultrasonic method), which allows feasibility to join two separate composite laminates/panels.
- Distribution of matrix throughout the composite structure is more uniform in thermoplastics as compared to the hand laying method (dependent on manpower skill), widely used for thermoset resin application in construction industries.
- Design flexibility of structural elements formed using thermoplastic composites, which is able to adapt complex architectural shape or design due to unavailability of crosslinks.
- Thermoplastics are ductile in nature, unlike thermosets. The thermosets exhibit brittle behaviour resulting in crack formation and propagation.
- Reshaping or remoulding ability of thermoplastics on application of heat makes it possible to adapt changes in shape or design even after final consolidation (composite manufacturing) based on specific requirements.

- Thermal treatment on thermoplastic composites doesn't affect resin behaviour, unlike thermosets which undergo chemical degradation.
- Prefabricated components can be easily manufactured at factories and can be transported to construction sites.
- Reduction in construction time due to availability of prefabricated thermoplastic composite panels.
- Less skilled labour is required for installation of prefabricated thermoplastic composite-based panels at construction site.
- Recycling of thermoplastic composites is possible after completion of product life, which is highly desirable considering environmental impact and sustainability.
- Monuments or heritage buildings which have cultural significance can be strengthened using thermoplastic composites as they offer ease in terms of easy replacement with new ones (reversibility) without damaging the structures due to absence of crosslinking resin.
- Thermoplastics possess superior ductile, flexural, and impact behaviour with improved fatigue resistance and recyclability (Alagirusamy et al. 2006).

The abovementioned features of thermoplastic composites are highly desirous for several high-performance applications and building structures with complex shapes and curvatures.

13.4 THERMOPLASTIC FRP BARS FOR STRUCTURAL APPLICATIONS

Fibre reinforced polymers (FRP) are composite materials with reinforcing fibres/filaments embedded in polymer matrices or resin. The combination of the components leads to superior mechanical properties. The reinforcement component is responsible for mechanical properties and the matrix component acts as a medium for inter fibre load transfer, and provides protection to the reinforcing fibres. FRP composites have a wider application network such as aerospace, automobiles, wind turbines, construction, and structural applications. The application of FRP composites in building construction generally involves high performance fibres such as glass, basalt, carbon, and aramid. The resin component may be thermoset or thermoplastic. The common thermoset resins are epoxy, unsaturated polyester, and vinylester. The popular thermoplastic resins are polypropylene, thermoplastic polyester, nylon, polyether ether ketone (PEEK), and polyether sulfone (PES).

Research was carried out to investigate the potential of glass fibre reinforced plastic (GFRP) rebar for structural application. It suggested that the GFRP rebar has the ability to replace steel rebars for applications involving prolonged resistance to corrosion, lower electro-magnetic and electrical conduction, and high specific strength (Benmokrane, Chaallal, and Masmoudi 1995). Another study reported a bending test of the reinforced concrete of the basalt fiber reinforced plastic (BFRP) reinforcing bars, which showed a linear load elongation behaviour until failure. The load carrying behaviour of BFRP rebar was higher in contrast to steel rebar.

However, BFRP rebar strengthened concrete exhibits higher deflection in contrast to steel rebar strengthened concrete and the possible reason was low modulus of BFRP rebar in contrast to steel rebar. The crack width was four times higher for BFRP rebar reinforced concrete in contrast to steel rebar strengthened concrete, and this was due to the higher deflection and deformation of BFRP rebar in concrete (Urbanski, Lapko, and Garbacz 2013; Lapko and Urbański 2015). Another study reported the response of BFRP rebar subjected to long term static load and compared with steel rebar and cable under pre-stressed condition. They observed creep behaviour was highest for steel cable. Moreover, both steel rebar and BFRP rebar have recorded lower creep (Pearson, Donchev, and Salazar 2013). Further research was carried out for BFRP rods and the result suggested that BFRP may be considered as a potential alternative to commercially available FRP rods. Rods made of BFRP showed moderate rigidity when compared to GFRP rods, but were quite deformable, and demonstrated superior tensile performance over GFRP (Quagliarini et al. 2012). The above discussed studies were based on rods consolidated using thermoset matrices. However, thermoset composite FRP bars impose limitation as they cannot be reshaped or bent once the consolidation process is completed, as resin forms chemical crosslinks. Hence, in case of thermoset bars the desired shape or bent should be predefined and incorporated during the composite forming process. Unlike steel rebars, they cannot be bent at the construction site. In case of the thermoset composite FRP bars, lap splices or joints are required for forming bent shaped structures. This results in adding cost, time, and labor cost with reduced productivity. On the contrary, thermoplastic bars can be easily remolded with application of heat into desired shapes such as shear stirrups or bendable bars for improved anchorage and flexural behaviour. The tensile strength and modulus of GFRP bars produced from thermoplastic resins are comparable with those produced from thermoset resins. The thermoplastic resin enables thermal welding of components through conduction heating or ultrasonic welding. The thermoplastic also possesses higher impact resistance and damage tolerance ability (Kocaoz, Samaranayake, and Nanni 2005).

A study was carried out to investigate behaviour of straight and bent carbonfibre reinforced plastic (CFRP) thermoplastic rebars under tensile test, pull out test, and bent bars and stirrups test. The rebar constitutes 150 rovings of 12K carbon and a thermoplastic resin. The thermoplastic CFRP rebar of 12.7 mm diameter and 380 mm length was formed using compression molding process at 4 ton loading pressure at 220°C. The molds have profile grooves which form ellipsoidal ribs in the rebar designed to improve mechanical anchorage in concrete specimens. To incorporate bent in rebars, heat treatment of 220°C for 15 minutes is carried out followed by bending operation. The tensile strength and modulus of the rebars was in the range of 710–825 MPa and 89.9–92.2 GPa. The ultimate bond strength was reported to be 3–9 MPa, which was dependent upon the embedment length of the CFRP bar inside the concrete specimen (191–445 mm). The strength of the bent specimen was found to have retained 37% of the straight CFRP rebar strength. The CFRP rebar failure was due to splitting, splitting and waviness (at failure location), and rupture as represented in Figure 13.15. The inclusion of ellipsoidal ribs in CFRP caused stress concentration in rebar, which led to premature failure.

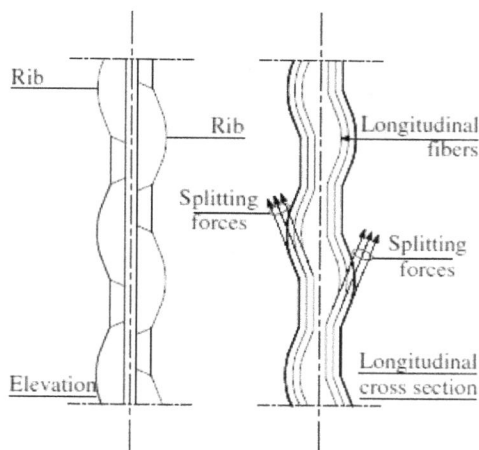

FIGURE 13.15 Schematic representation of failure in thermoplastic CRPF rebar under tension due to inclined splitting failure. (Reused with permission from Elsevier; El-Tahan, Galal, and Hoa 2013.)

Otherwise, the rebar would have achieved actual strength of 1500 MPa, according to the authors. They recommended that specimen should be kept under tension during composite formation to prevent fibre waviness formation in composites. Moreover, rebars should be symmetrically designed throughout the cross section to prevent splitting failure. They proposed composite formation through pultrusion technique followed by sand coating or filament winding, to improve composite-concrete bonding (El-Tahan, Galal, and Hoa 2013).

The thermoplastic pultruded rebar can be heated and remolded and post-formed into desired shape based on requirement. The recycling ability, higher impact resistance, and heat resistance up to 400°C are added advantages for thermoplastic composites. The thermoplastic GFRP rebar was reported to have negligible creep strain and have higher tensile strength and modulus retention when subjected to loading for 816 hours of time duration. The possible application of GFRP includes storage tanks, columns, concrete pipes, and tunnel segments (Sayed-Ahmed et al. 2017).

A detailed experimental study was carried out to investigate physical, mechanical, and durability behaviour of thermoplastic GFRP rebars (Benmokrane et al. 2021). The GFRP composite (10, 15, and 20 mm diameter) rebars were produced through pultusion process using Elium® resin (acrylic-based liquid resin) at 180–200°C and 5–20 bar pressure. The helical wraps are incorporated in the composites to improve bonding behaviour. The thermoplastic rebars were compared to equivalent vinylester-based thermoset rebars. The fibre content in thermoplastic rebar was found to be 77.7–81.1%. The water uptake for thermoplastic GFRP was reported to be 7 times higher as compared to the thermoset GFRP, which did not fulfil the ASTM and CSA standards. The rate of water diffusion was accelerated at higher temperature, unavailability of cross-links, and lower crystallinity content in resin.

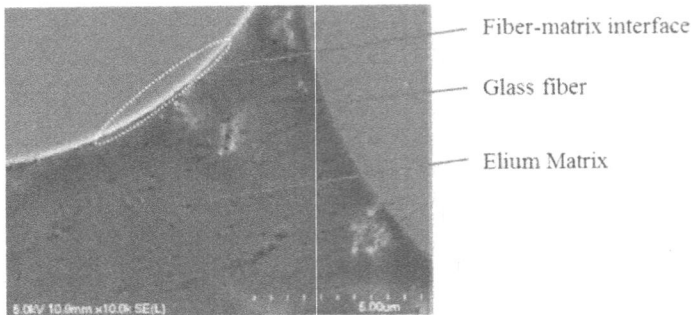

FIGURE 13.16 SEM micrographs showing fibre-matrix interface of thermoplastic GFRP rebar (15 mm diameter). (Reused with permission from Elsevier; Brahim Benmokrane et al. 2021.)

The SEM micrographs of thermoplastic GFRP indicated minimal voids within acceptable limits, consistent resin distribution, and strong fibre-matrix bonding at interface regions, comparable to thermoset composites as shown in Figure 13.16. The mechanical behaviour in terms of strength, modulus and strain, transverse shear strength, interlaminar-shear strength, and bond strength were comparable to the equaivalent thermoset GFRP and were found to be above the minimum required standards. The durability of thermoplastic GFRP was accessed through immersion of rebars in alkaline solution (pH 12.6) at 60°C for 90 days and was reported to have retained 87% of the original strength fulfilling standards. The retention in strength and modulus was 100% and 98%, respectively when subjected to alkaline solution at 40°C. The creep strain was evaluated under load equivalent to 20% and 40% of the breaking tensile strength. The 10 mm GFRP rebar has exhibited insignificant creep effect by yielding 8% creep strain after 10,000 hours of loading subjected to 40% of the breaking tensile strength, comparable to the thermoset rebars. The authors suggested need for further research on thermoplastic-based rebars and manufacturing technology for possible application of bendable rebars in the construction industry. The rebars were found to be a good alternative to the thermoset rebars with added flexibility of bending at site, higher ductlity, and impact behaviour.

Another study evaluated the performance of thermoplastic FRP rods to thermoset FRP rods for reinforcing concrete (Hokura and Miyazato 2020). The thermoplastic FRP rods consist of carbon filaments reinforced in thermoplastic epoxy resins (CFRTP) and glass filaments reinforced in thermoplastic polypropylene resins (GFRTP), while thermoset FRP rods are made up of carbon filaments reinforced in thermoset epoxy resins (CFRTS) and glass filaments reinforced in thermoset polyester resins (GFRTS). To understand the water uptake behaviour FRP rods were immersed in high alkaline environment at 60°C for a 30 day period. The weight change (%) for alkaline conditioned thermoplastic CFRTP and GFRTP rods were found to be 0.21% and 0.03%, respectively, whereas weight change (%) for alkaline conditioned thermoset CFRTS and GFRTS rods was found to be 0.46% and 0.45%, respectively. The rods were also conditioned under tap water at 20°C for 6 months and 1 year duration. A marginal change in tensile strength and modulus

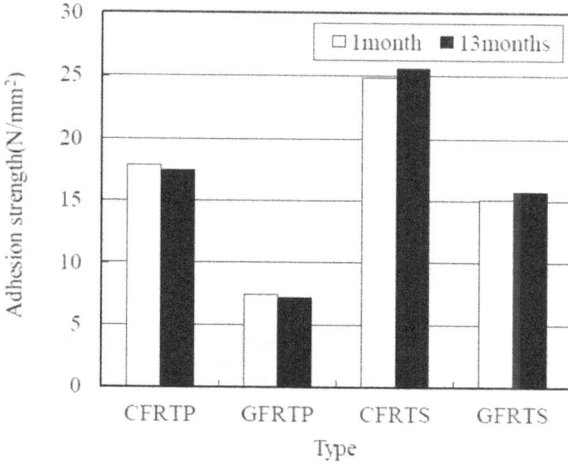

FIGURE 13.17 Adhesion strength of FRP rod reinforced concrete after 1-month and 13-month duration. (Reused with permission from Journal of the society of materials science, Japan; Hokura and Miyazato 2020.)

for thermoplastic and thermoset rods conditioned under alkaline solution and tap water was reported. The adhesion strength of the rods had not exhibited any significant change during the pull out test for FRP rod reinforced concrete specimens after 1 month and 13 months, respectively, as shown in Figure 13.17. The GFRTP rod has shown poor mechanical behaviour due to higher resin proportion of 75% in composites, which can be further improved by increasing reinforcing fibre proportion in FRP bars. The CFRTP rod has shown comparable mechanical behaviour to that of CFRTS rod and steel rebar, as shown in Figure 13.18. The thermoplastic FRP

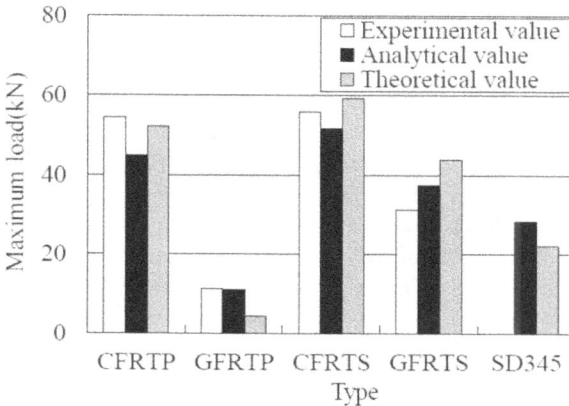

FIGURE 13.18 Maximum load of FRP rod reinforced concrete under flexural test compared with SD 345 grade steel rebar. (Reused with permission from Journal of the society of materials science, Japan; Hokura and Miyazato 2020.)

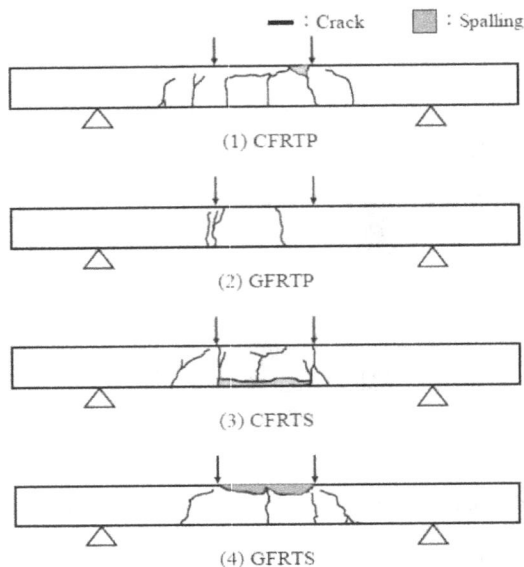

FIGURE 13.19 Failure and cracking behaviour of FRP reinforced specimens during a 4-point bending test. (Reused with permission from Journal of the society of materials science, Japan; Hokura and Miyazato 2020.)

reinforced concrete specimens in contrast to the thermoset reinforced specimens have exhibited distributed crack propagation and spalling-free failure under a four point bending test, as observed in Figure 13.19. The authors recommended use of thermoplastic FRP as a substitute to steel rebar for concrete reinforcement.

13.4.1 METHODS FOR IMPROVING SURFACE ROUGHNESS OF FRP COMPOSITE RODS

The pull out resistance is an important parameter pertaining to reinforcement component in concrete. The pull out behaviour of reinforcing yarns inside cementitious matrix is dependent on the interface behaviour of reinforcing yarn and cementitious matrix. It is desirable to achieve higher pull out resistance which will avoid slippage of reinforcement in concrete. The surface roughness or incorporation of ribbed structure in FRP rods has a profound effect in improving bonding behaviour through mechanical anchorage with concrete or cementitious matrices. Similarly, steel bars commonly used in concrete reinforcement have ribs in structures to improve bonding with cementitious matrices. Various methods such as sand-epoxy coating, filament wrapping, milling, and braiding are employed to improve the bonding behaviour of reinforcement and matrix in concrete.

Sand-epoxy coating is a common method employed for improving bond behaviour of FRP rods in which FRP rods are initially coated with epoxy resin and then finer sand particles are distributed uniformly over the epoxy coated FRP rods. The methods used for incorporating different geometrical profiles in carbon FRP bars

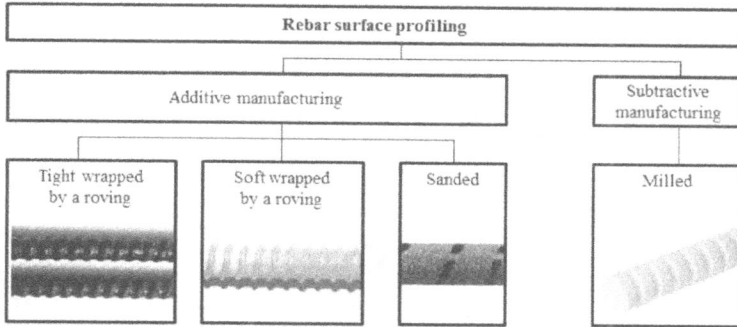

FIGURE 13.20 Profiling methods used in pultruded carbon FRP bars to improve rod bonding behaviour in concrete. (Adapted from Böhm et al. 2018.)

are shown in Figure 13.20. In the additive method, a roving or tape is wrapped over the FRP bar or sanded. In subtractive method, drilling or milling is carried out to generate desired profiles in the FRP bars (Böhm et al. 2018).

The braiding process produces a structure which has naturally non-uniform surface profile, which is indeed advantageous for improved bonding action in cementitious matrix. However, a small modification in the braiding process by introducing coarser yarn (rib yarn) enables formation of yarn structure with prominent rib profiles as shown in Figure 13.21 (Harris, Somboonsong, and Ko 1998; Fangueiro et al. 2006).

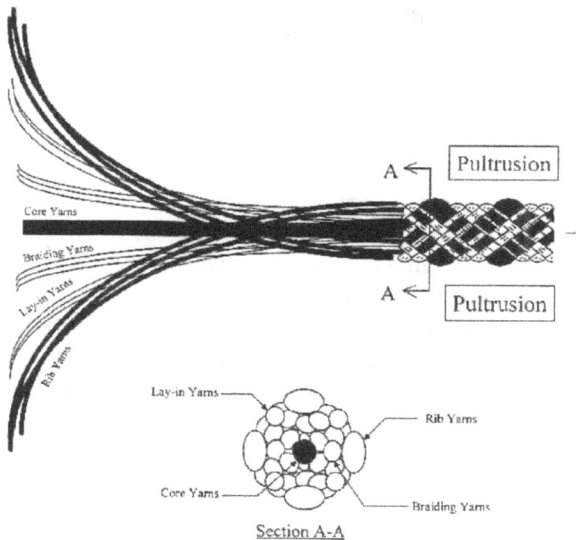

FIGURE 13.21 Rib profile in braided structure for improved mechanical anchorage. (Reused with permission from American Society of Civil Engineers; Harris, Somboonsong, and Ko 1998.)

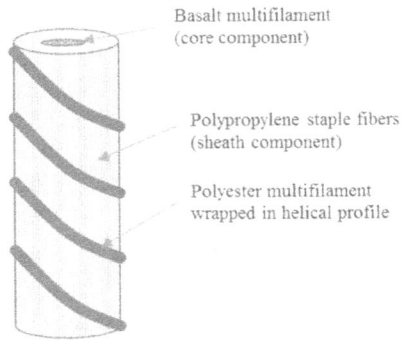

FIGURE 13.22   Schematic representation of a hybrid yarn structure with polyester filament wrapped in helical configuration.

In another study, helical wrapping of a continuous multifilament yarn is performed over the core yarn, to mimic the helical profiles present in steel rebar used as reinforcement components in concrete as illustrated in Figure 13.22. The helical configuration is intended to improve the bonding behaviour of reinforcement-matrix through mechanical anchorage (Barman, Bhattacharya, and Alagirusamy 2019).

13.4.2 Durability of Thermoplastic FRP Bars

The durability of structural members in service life is greatly influenced by the performance of reinforcing elements subjected to different environmental conditions. The long-term durability of FRP rods used in building construction is influenced by presence or exposure to moisture, temperature, salt/alkaline solution, solution concentration, thermal conditioning (freeze-thaw cycles), chemical agents, fatigue stress, ultraviolet radiation, and fire. Apart from the external factors, the material characteristics have considerable influence on durability behaviour. The understanding of chemical composition of fibre and matrix constituents, fibre-matrix interaction, and void content are of prime importance. The effect of FRP under alkaline conditions has greater importance as the pore solution in concrete specimen has a pH (12.0–13.5) associated with presence of alkaline ions. Mechanical properties of specimens are often deteriorated due to a combination of external agents such as moisture, pH, and temperature, which can readily accelerate the reaction (Micelli and Nanni 2004).

An investigation was carried out to understand mechanical and microstructural behaviour of glass-polypropylene thermoplastic composites subjected to different environmental conditions such as tap water, salt solution (3% NaCl), and freeze thaw cycles (at –18°C to +18°C for 200–800 days) for possible application in bridge decks (Robert, Roy, and Benmokrane 2010). The selected environmental conditions represent the following – tap water to emulate rain, salt solution imitate effect of dicing salt, and freeze thaw cycles to emulate harsh climatic conditions. The respective effect of tap water and salt solution on composite samples was accessed at different temperatures set at 23°C, 50°C, and 70°C for a duration of 56–184 days. The glass-polypropylene composite laminate consists of four bidirectional glass-polypropylene commingled rovings which were consolidated using roll forming techniques. The composite was formed with a fibre/matrix proportion of 60/40 (weight). The composite specimen was reported to have 8% void content, dependent on the composite fabrication process. For testing purposes, composite specimen size was kept at 20 mm × 96 mm. The water uptake by the composite specimen was found to be dependent on the temperature, type of solution (tap water, salt solution), composite consolidation process, and void content in composites. When the composite sample was immersed in tap water for a period of 184 days, the lowest water uptake was around 2.6% at 23°C and the highest at 3.75% at 70°C. The water uptake was less in case of salt solution; dissolution of NaCl in water generates a barrier preventing solution diffusion inside the composite specimen. The loss in flexural strength and Young's modulus is also influenced by type of solution (tap water, salt solution) and reaction temperature. After a period of 168 days at 70°C, the loss in flexural and elastic modulus was recorded to be 44% and 23% respectively, subjected to salt water. The loss in flexural and elastic modulus was comparatively lesser at 24% and 20% respectively, subjected to tap water. The salt solution alters the fibre-matrix interface, causing reduction in flexural behaviour. The effect of 800 freeze thaw cycles on loss in flexural and elastic modulus of moisture saturated specimens is recorded at 32% and 22%, respectively. The loss is higher when sample is saturated in moisture. The authors suggested that composite consolidation processes have a greater influence on void content, which in turn significantly affects the durability behaviour of composite specimens. They also asserted that glass-polypropylene composites have great prospects for possible applications in formation of structural components such as bridges, beams, and barriers.

Another study was carried out to understand tensile behaviour of GFRP pultruded composite bars produced from E-glass and thermoplastic polyurethane (Kocaoz, Samaranayake, and Nanni 2005). The GFRP bars have identical fibre and matrix components, fibre-matrix proportions, and diameter of 12.5 mm. Four different types of coating were used in the GFRP bars – (a) chopped E-glass fibre and polyurethane, (b) chopped carbon fibre and polyurethane, (c) pure polyurethane, and (d) Wollastonite ($CaSiO_3$) filler and polyurethane. To improve anchorage and prevent debonding of coating and core, use of expansive cementitious grout in a steel pipe and threading of FRP bars with 3 mm thread with thread density of 2 threads/cm was employed as depicted in Figure 13.23. The tensile test revealed that, GFRP composite rebar specimens failed abruptly, initiating with splitting behaviour, and followed by filament rupture. The GFRP coated with pure polyurethane has shown higher standard deviation, else the tensile strength was in the range of 979–1014 MPa. The coating and addition of filler helped to achieve improved tensile behaviour of GFRP bars. The tensile strength of GFRP bars follows Gaussian distribution and the test procedure was found reliable for estimating tensile strength of GFRP bars.

The combined effect of alkaline exposure, freeze-thaw and high temperature cycle, high relative humidity, and ultraviolet radiation on durability of CFRP and GFRP rods for possible application in concrete reinforcement was investigated (Micelli and Nanni 2004). The FRP specimens were subjected to 200 freeze-thaw cycles, 480 humidity cycles and 600 high temperature cycles. The alkaline solution was prepared with potassium hydroxide, sodium hydroxide, and calcium hydroxide

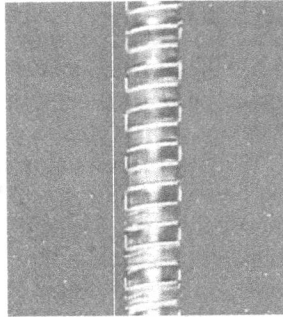

FIGURE 13.23 FRP bar threaded with 3 mm diameter thread (2 threads/cm) to improve bonding behaviour. (Reused with permission from Elsevier; Kocaoz, Samaranayake, and Nanni 2005.)

in weight proportions of 1.4%, 1%, and 0.16%, respectively in water. The result of the short beam shear test suggested that a FRP rod with E-glass polyester resin has completely lost its shear properties under immersion in alkaline solution for 42 days at 60°C. The retention in shear property was maximum at 80% for carbon epoxy FRP, followed by 70% for carbon vinylester-epoxy FRP and E-glass thermoplastic resin FRP rods. The absorption under alkaline solution was highest for E-glass polyester resin FRP rods as shown in Figure 13.24. The GFRP rod with thermoset polyester resin has exhibited poor tensile performance on exposure to accelerated aging by loss in strength by 30% and 41% when exposed to 21 days and 42 days, respectively. The resin was ineffective in preventing glass filaments from alkaline attack. The CFRP rod with epoxy and vinylester thermoset resin has performed moderately well under accelerated aging. The CFRP rod with thermoset epoxy resin and GFRP with thermoplastic resin has retained complete tensile properties under accelerated aging conditions. The fibre-matrix adhesion in the above samples was also intact, which was observed under SEM. Extensive damage was observed for E-glass polyester

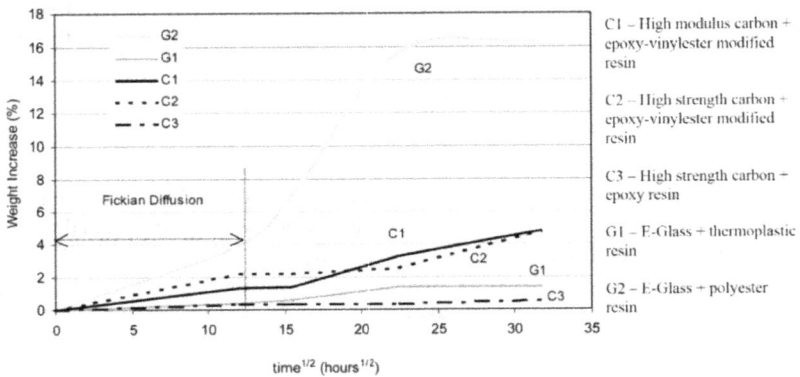

FIGURE 13.24 Absorption behaviour of different FRP rods immersed in alkaline solution. (Reused with permission from Elsevier; Micelli and Nanni 2004.)

resin FRP rods and limited chemical damage in external regions of E-glass thermoplastic resin FRP rods. No damage was observed for carbon epoxy specimens. Carbon specimens have shown superior resistance against alkali attack. The authors suggested that resin has a significant effect on determining durability of FRPs and recommended use of thermoplastic resin in GFRP for improved durability behaviour.

13.5 THERMOPLASTIC FRP COMPOSITE LAMINATES FOR STRUCTURAL APPLICATIONS

Composite materials as reinforcing elements have been able to overcome the inherent limitations of steel reinforcement in building structures. The thermoplastic FRP laminates and bars are quite effective for strengthening and rehabilitating applications. The higher specific strength, corrosion resistance, and improved fatigue tolerance of FRPs encourage its widespread application in structural applications (Kocaoz, Samaranayake, and Nanni 2005).

13.5.1 THERMOPLASTIC COMPOSITE STRUCTURAL INSULATED PANELS (CSIPs)

Modular panelised structures are increasingly used due to the advantages of ease of bulk production, cost-effectiveness, customisation possibility during prefabrication of panels, ease of assembling at construction sites, light weight, and uniform quality of components. Modular structures are well suited for disaster prone areas in which structures can be erected in a short span of time. Structural insulated panels (SIPs) are one form of modular structures which consist of foam-based core (molded expanded polystyrene, extruded polystyrene, and urethane foam) sandwiched in between face sheet or laminates. The face sheets were conventially made using cement concrete board (CCB) or oriented strand board (OSB). The application of glass/PP-based face sheet for developing thermoplastic composite structural insulated panels (CSIPs) for walls and floor panels was explored (Uddin and Mousa 2013). The face sheets consist of glass/PP laminates in 70/30 weight proportion and core was expanded polystyrene (EPS) foam. A spray of hot thermoplastic melt was used to form CSIPs by binding the laminates and foam. The conventional SIPs require 24 hours for fabrication; however, the CSIPs can be fabricated within two hours. CSIPs appear to be highly resistant to impacts, undergo marginal deterioration in strength and stiffness during flood tests, and withstand extreme windstorm (Level-5 Hurricanes).

The thermoplastic CSIP face sheets are 1.8 times lighter and 0.3 times thinner as compared to OSB face sheets. The thermoplastic CSIP structures have higher specific strength and are 11 times stiffer than OSB-based SIP structures. CSIPs also have important characteristics such as corrosion resistance, mould and termite resistance, and increased impact resistance, making them ideal for panelized constructions. It was reported that use of 3-D thermoplastic polyurethane open-core sandwich composites reinforced with GFRP is highly effective in internal separation walls. The incorporation of GFRP has improved the load bearing ability by three times as compared to structures without GFRP reinforcement (Vaidya, Uddin, and Vaidya 2007).

13.5.1.1 Flood Water Testing of CSIPs

An experimental investigation was performed to understand the behaviour of CSIPs' flexural strength when exposed to floodwater for a period of 3 days and 7 days (Mousa and Uddin 2013). The test result suggested retention of flexural strength up to 95%, 90%, and 84% for CSIPs completely immersed under flood water for a period of 3 days (dried specimen), 7 days (dried specimen), and 7 days (wet specimen), respectively. The retention in flexural strength was high as compared to conventional wood-based panels. The use of CSIPs in flood prone areas for construction of durable structures resistant to molds and termite attack was highly recommended.

13.5.1.2 Eccentric Compression Loading Behaviour of CSIPs

The behaviour of CSIPs under eccentric compression load was investigated (Mousa and Uddin 2012). The CSIP specimen demonstrated an elastic behaviour until failure associated with debonding of face sheets and core causing reduction in stiffness. The face sheet failed due to wrinkling in the compression side, which resulted in local bucking behaviour. The pull out test suggested specimen failure associated with higher out-of-plane interfacial stress overcoming core tensile strength. Irrespective of delamination behaviour, CSIPs fulfill the design criteria and limit in terms of strength and stiffness for use as possible sandwich wall panels. The core density has a profound effect on deflection behaviour and has marginal effect on face sheet compression behaviour. The core density values when kept at 3 PCF in place of 1 PCF have significantly reduced deflection by 70% and have marginal reduction in strain by 3%. The behaviour was attributed as strain values are dependent on dimensions and thicknesses of face sheet and core.

In another study, when CSIPs specimens were subjected to axial compressive load of 13.13 kN/m, the deflection was 3 mm, which is inside the maximum deflection limit of 16.25 mm specified by ACI design code (Vaidya, Uddin, and Vaidya 2010). CSIPs specimens recorded failure load of 17 kN under eccentric loading with buckling failure followed by debonding of face sheet and core, which is also higher than the minimum specified load limit of 16 kN. It was recommended to use CSIP panels for wall panel applications.

13.5.1.3 Impact Testing of CSIPs

High velocity impact behaviour was examined for CSIP materials (Vaidya, Uddin, and Vaidya 2010). CSIP panels have resisted impact energy of 66.7 N for wind missiles moving at 44.7 m/sec velocity. Debonding behaviour between face sheet and core was not observed even after a second impact test. The CSIP specimen was capable of resisting projectiles moving at 135 m/sec. They recommended CSIP for wall applications with face sheet and core thickness of 3.04 mm and 140 mm, respectively. The use of CSIP panels for wall panel applications in hurricane affected regions was found to be effective.

13.5.2 Bonding Methods for Assembling CSIPs

The composite panels are bonded using adhesive bonding, mechanical fasteners, and fusion bonding. Further investigations suggested that among the bonding methods fusion bonding via ultrasonic welding is tested to be superior compared to other methods (Uddin, Vaidya, et al. 2013). The fusion bonding method overcomes the limitations of adhesive bonding (flaw generation, lower joint strength, higher stress concentration) and mechanical fasteners (delamination and wear associated with drilling hole, thermal expansion, water ingress at joint region, longer time requirement). In ultrasonic-based fusion bonding, ultrasonic frequencies generate mechanical vibrational energy which leads to melting of polymers followed by formation of welding joints. PVC-based pultruded connectors are used to form interlocking bonding between two adjoining face sheets. The CSIP structures are 1.8 times lighter in contrast to the conventional SIPs. They recommended use of ultrasonic welding with interlocking assembly, as ultrasonic welding has exhibited highest strength during single lap shear tests, and interlocking assembly is rapid and requires less manpower.

13.5.3 Improved Composite Structural Insulated Panel (ICSIPs)

The performance of improved composite structural insulated panel (ICSIPs) was examined for structural floor and roof design applications (Liu et al. 2018). ICSIPs were developed to improve the load bearing ability and debonding behaviour of CSIPs. In ICSIPs, the face sheet on the top is replaced by recycled concrete aggregates to improve stiffness and compression resistance. The regular grade concrete (C20–C40) specimens used for testing are crushed and segregated, and 5 mm and 20 mm size aggregates are used for forming face sheets. A steel mesh is used in conjugation with recycled aggregates with grid size of 200 mm × 200 mm and 8 mm diameter steel bars. The ICSIPs' sandwich panel is composed of top recycled aggregate layers, followed by steel mesh and EPS foam and bottom layers of glass-polypropylene thermoplastic sheets, illustrated in Figure 13.25. To improve bonding between EPS foam and thermoplastic sheets, epoxy adhesive spray is used. The analysis revealed that load bearing capacity of the ICSIPs has improved by 3.46 times as compared to CSIPs. The deflection behaviour has been reduced by 6 times for ICSIPs in contrast to CSIPs. The failure of ICSIPs is through crushing of top aggregate layers, whereas the CSIPs fail through debonding of core and face sheets, followed by wrinkling of face sheets and shear failure of core elements. It was reported that presence of glass-polypropylene bottom sheets in ICSIPs has observed a significant improvement on load bearing ability and stiffness of panel components

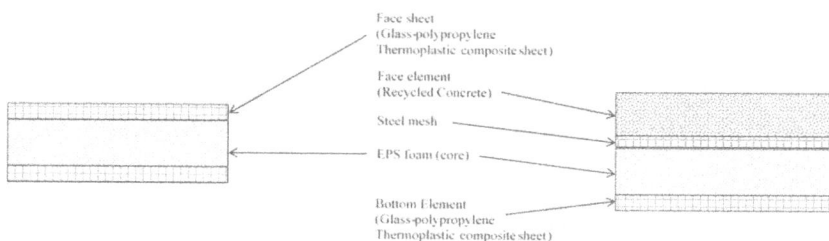

FIGURE 13.25 Schematic representation of composite-based structural panels – CSIPs (left), and modified CSIPs or ICSIPs (right).

by exploiting tensile properties of GFRP composite sheets. The ICSIPs have been recommended for application in floor and large-span roof design. They can also be used as retrofit for rehabilitation of existing concrete floors or roofs.

13.5.4 APPLICATIONS OF THERMOPLASTIC FRP LAMINATES IN BUILDING CONSTRUCTION

The unique and unmatched properties of thermoplastic FRP composites have promoted their acceptance in various sectors. A few applications specifically designed for building construction sector are discussed in this section.

13.5.4.1 Emergency Safe House Wall Application

In order to design emergency safe houses in areas where hurricanes with high wind velocity are frequent, it is advisable that walls should have good resistance against flying debris impact. An investigation was carried out to understand the effect on ballistic impact behaviour by incorporating additional polycarbonate sheets in glass/epoxy composite and graphite/epoxy composite sheets, respectively (Uddin and Vaidya 2005). The adhesion of the polycarbonate sheet to the composite laminates was carried out using an epoxy binder mixed with microballoon. It was reported that during ballistic impact testing composite laminates bonded with polycarbonate sheets have provided higher energy absorption in contrast to composite laminates without polycarbonate sheets. The effect was more prominent for glass-epoxy composites, where inclusion of polycarbonate sheet has yielded 200%–300% higher energy absorption compared to parent specimens. The polycarbonate sheet thickness has a direct influence on energy absorption and associated damage due to debonding behaviour.

13.5.4.2 Deck Composite System for Bridge Structure Application

The behaviour of glass-polypropylene composites for bridge deck slab applications as a possible replacement of timber and steel bridges was investigated (Roy et al. 2005). The composite is formed by commingled glass-polypropylene-based fabric consolidated through vacuum bag molding at 180°C. The top and bottom composite laminates have thickness of 5 mm each. The two corrugated composite laminates sandwiched in between the top and bottom laminates had a thickness of 4 mm, as depicted in Figure 13.26. The inside hollow portion is filled with polyurethane foam.

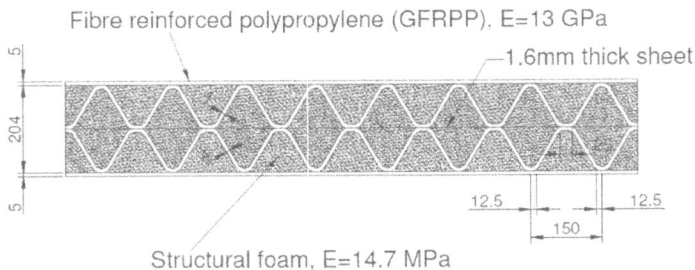

FIGURE 13.26 Double wavy corrugated sandwich structure designed for bridge deck application. (Reused with permission from Roy et al. 2005.)

Top face sheet produced by
Glass/PP tape woven structure

Sine-rib profile structure produced   Bonded together through mechanical, adhesive,
by Glass/PP tape woven structure   mechanical-adhesive combination, ultrasonic
                                          bonding methods

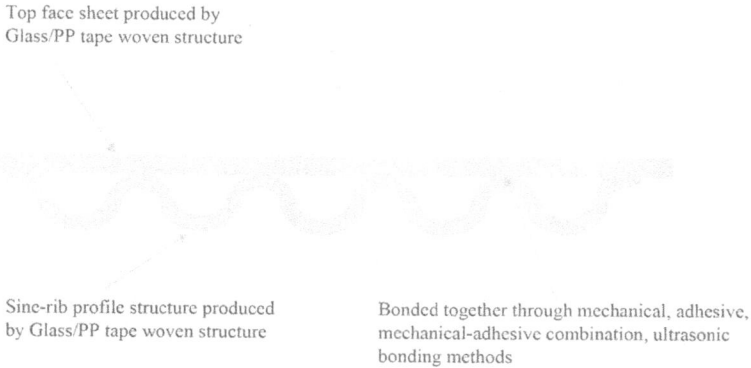

FIGURE 13.27 Woven glass-polypropylene thermoplastic deck structure with sine-rib (Hat-sine) profiles.

The double wavy corrugated sandwich structure is formed by assembling the panels using high tensile bolts and epoxy resin. The composite deck was subjected to static load of 515 kN. The slab recovered completely from the deformation upon load removal without fracture. The composite deck has an improved service life of 75 years, as compared to 15 years for timber decks. The authors recommended composite deck slab systems over timber slabs. The advantages of composite slabs are less penetration to water, improved resistance towards freeze-thaw cycles, chemical free processing unlike timber, good fatigue resistance, ease of installation and service, prolonged service life and durability, and good impact behaviour.

In another study, the design of E-glass/polypropylene composite-based bridge deck superstructures to assist highway traffic was explored (Uddin and Abro 2008). The E-glass/polypropylene (67/33 – weight proportion) tape structure was produced using the DRIFT method. The tapes are further assembled into plain woven structures. Hat-sine profile deck composite systems are represented in Figure 13.27, with variable thickness (> 1 mm) produced using thermoforming techniques at 0.345 MPa pressure. The male and female mold are designed in such as manner to deform the E-glass/polypropylene composite sheet into profile structure aided by heating. The stiffness and strength of E-glass/polypropylene deck system was compared with two other thermoset composite-based deck systems – glass/polyester-vinylester and S-glass/epoxy using finite element methods. The deck structure with nonplaner core design has adequate strength and stiffness, good interfacial stress at the junction of the face sheet, and sine rib profile structure. However, the structure has higher self-weight resulting in higher dead load/live load compared to the other two thermoset-based systems. The composite deck was recommended as an efficient and economical alternative compared to thermoset-based deck systems.

13.5.4.3 Wrapped Columns for Strengthening Bridge Structure

The ductility of column plastic hinges has greater significance during impact loading. The reinforcement in transverse direction is accountable for ductility behaviour, and resists axial compression of columns against buckling behaviour. External retrofitting is done with FRP sheets to improve ductility; mainly CFRP wraps are introduced. Thermoset composite wraps (jackets) are extensively used for wrapping and confinement of column elements. However, thermoset wrapped specimens have poor ductility and high cost. The low velocity impact and uniaxial compression behaviour of pultruded thermoplastic polypropylene confined column elements was examined. Two split shell halves of polypropylene were joined using a thermoplastic tape. They compared behaviour of polypropylene (3 mm and 6 mm thickness) confined column specimens with unconfined and thermoset (epoxy) CFRP confined specimens. The study of impact behaviour is significant considering impacts due to vehicle collision or blast. The loading behaviour was not improved with use of polypropylene jackets. However, strain values for polypropylene confined columns was three times higher compared to CFRP columns and six times as compared to unconfined columns. The transverse strain for polypropylene confined columns was reported to be 145% and 100% higher as compared to unconfined and CFRP wrapped specimens. During impact testing, the energy absorption of polypropylene confined specimens (both 3 mm and 6 mm) was reported to be higher than CFRP specimens. The energy absorption was highest for specimens confined with 3 mm thick polypropylene jackets. The polypropylene confined elements were found to exhibit ductile failure with marginal bulging in top and bottom regions of jacket in contrast to CFRP elements, which have failed through abrupt fibre rupture (catastrophic failure). However, debonding was not observed in polypropylene confined elements. They recommended use of economical polypropylene jackets for confining columns with improved ductility, higher energy absorption, resist lateral expansion, and large deformation (strain) before failure under compression loading (Uddin, Purdue, and Vaidya 2008; Uddin, Abro, et al. 2013).

13.5.4.4   Thin Shell Elements for Roof Structures

Thin shell structures are used as structural members which have high aesthetic appearance and lighter in weight which makes them suitable for long span structure. Thin shell elements resist external load primarily through axial loading. CSIPs offers quick installation, less labor, light weight, thermal insulation, fire resistance, corrosion free, and higher specific strength, and are found suitable for design of thin shell elements. The suitability of glass/polypropylene-based CSIPs for possible application as thin shell components was examined (Uddin and Du 2014). The core of CSIPs is composed of expanded polystyrene foam, which is economical, lightweight, fire-resistant, and impact-resistant. The cylindrical CSIP shell structure was compared with conventional concrete thin shells under static and dynamic loading using ANSYS software. The static response in terms of maximum von Mises stress of the of the CSIP thin shells has reported to have 31.67% lower stress as compared to concrete thin shells. Similarly, dynamic performance in terms of maximum von Mises stress of the CSIP thin shells has been reported to have 48.2% lower stress as compared to concrete thin shells under seismic loading. The CSIPs also possess ability to bear higher nonlinear buckling load (excluding self-weight) as compared to concrete thin shells. The CSIP thin shell was also almost 44 times lighter in weight compared to the equivalent concrete thin shell element. However, CSIPs have higher displacement under static and dynamic loading, which is attributed to lower elastic modulus of face sheets and core components. The limitation can be overcome by increasing stiffness and self-weight of CSIP elements in limited proportions.

Further investigation was directed to improve the stiffness of CSIPs and reduce their displacement behaviour (Du and Uddin 2017). The static analysis results suggested marginal displacement for CSIPs' folded shell structure under self-weight. Further, CSIPs' folded shells possess 5 times higher non-linear load bearing ability against bucking load compared to CSIPs' thin shell structure. Incorporation of GFRP rods further leads to improvement in structural stiffness and integrity. They recommended use of CSIPs' folded shell structure for applications in large-span roof structures.

13.6   INFLUENCE OF ADDING SHORT LENGTH (DISCRETE) FIBRES IN TRC

In an investigation, the surface of unidirectional carbon fabric was altered by incorporation of polyvinyl-alcohol (PVA) fibres (0.5% proportion by volume, and 6 mm length) to create a fluffy surface (Pakravan, Jamshidi, and Rezaei 2016). The carbon fabric and PVA fibres are bonded through epoxy resin sprayed onto surface of carbon fabric. The TRC specimens were produced using a casting (laminating) method. The 90° peel off test and three-point flexural test was carried out to understand the effect in bonding behaviour between surface-modified carbon textile and concrete specimens. The PVA incorporated carbon textile has reported improved bond strength and load-bearing ability to maximum of 43% and 56%, respectively. The addition of higher proportion of fibres can negatively influence the TRC properties. The authors suggested use of PVA fibres in limited quantity to improve the mechanical properties of TRC for potential repair and strengthening applications.

Another study was directed to understand the effect of addition of short alkali resistant (AR) glass and carbon fibres in fracture behaviour of TRC (Barhum and Mechtcherine 2012). The fibres were 6 mm length incorporated in AR-glass biaxial reinforced concrete in proportions of 0.5% and 1.0% by volume. The first crack stress was enhanced by 1.5–2 times with the addition of fibres and moderate enhancement in tensile strength. The short fibres have formed new crosslinks, which has led to enhanced bonding between textile structure and concrete matrix.

Further investigation was carried out by adding PVA staple fibres (12 mm length) with proportions of 0.5%, 1%, and 2% (by volume) incorporated in a cementitious matrix reinforced with balanced carbon woven grids (Deng, Dong, and Zhang 2020). It was reported that increase in PVA content in the matrix causes multiple cracking phenomena and lesser crack opening (spacing between cracks). The specimen failure pattern is altered from slippage to fracture due to fibre bridging and crosslinking effect, as observed in Figure 13.28. The addition of 2% PVA fibres results in improvement in tensile stress by 106%, 128%, and 87% with reinforcement ratio of 0.35%, 0.70%, and 1.05% for cementitious matrix with higher fly ash composition.

Another investigation was carried out to examine the load bearing behaviour and toughness response of concrete slab reinforced with hybrid components – basalt fabric (mesh size – 5x5 mm^2) and polypropylene fibres (45 mm length, l/d ratio-56) (Ding et al. 2020). The basalt TRC slab has a higher load bearing ability and toughness of 27% and 697%, respectively as compared to a plain concrete slab as shown

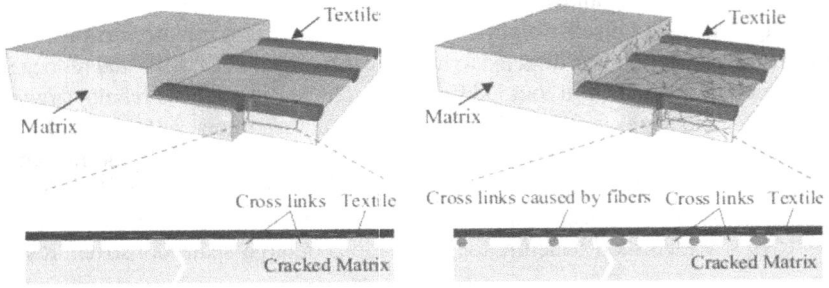

FIGURE 13.28 Textile reinforced mortar without short fibres (left) and with short fibres (right). (Reused with permission from Elsevier; Deng, Dong, and Zhang 2020.)

in Figure 13.29. The hybrid synergistic effect was observed on load bearing and toughness behaviour when basalt fabric is reinforced with polypropylene fibres. The polypropylene fibre addition in basalt TRC has resulted in ductile failure of TRC specimens with multiple cracking behaviour. The TRC composed of a single layer of basalt textile with polypropylene fibre content of 8 kg/m^3 (CFC1-8) when compared with steel reinforced concrete (RC) slab (steel content of 0.31%) has exhibited higher load bearing ability and toughness by 7% and 4.5%, respectively. The authors suggested that the hybrid basalt TRC with added polypropylene fibres has the ability and potential to replace steel mesh reinforced concrete (RC) slab.

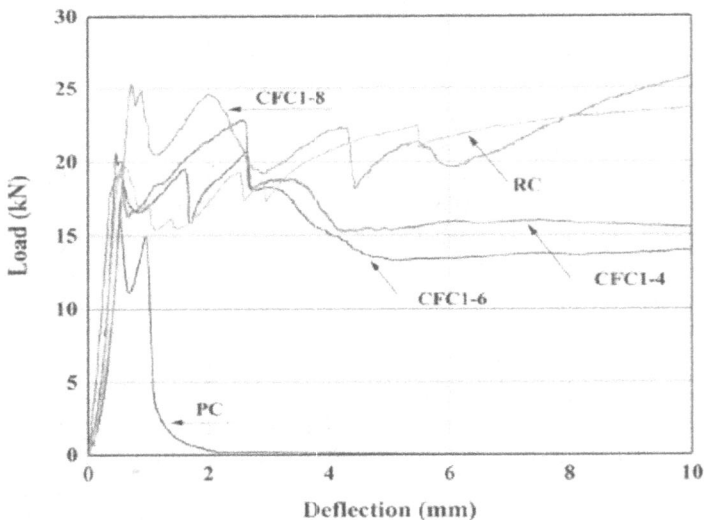

FIGURE 13.29 Load-deflection response of plain concrete slab (PC), steel reinforced concrete slab (RC), and polypropylene-basalt (hybrid) TRC slab (CFC1-4, CFC1-6, and CFC1-8 represent single layer of basalt textile with polypropylene fibre content of 4 kg/m^3, 6 kg/m^3, and 8 kg/m^3, respectively). (Reused with permission from Elsevier; Ding et al. 2020.)

13.7 SUMMARY

The application of towpregs and thermoplastic composites in building construction offers numerous advantages. The requirement of unconventional structural elements involving use of materials with unique characteristics such as light weight, low carbon footprint, durability, recyclability, remoulding, adapting complex structural shapes, reversibility and, forming bendable elements can be fulfilled with innovative towpreg-based thermoplastic composites. The important conclusions drawn from this chapter are mentioned below.

- Flexible towpregs or hybrid yarns offer high tensile, flexural, and impact behaviour with improved ductility. DREF-III yarn, commingled yarn, and braided structures provide relative uniform distribution and impregnation of resin component with reduced resin flow distance, which are capable of producing composite specimens with superior mechanical properties, highly desired for reinforcing concrete elements. Further research is required to recognize the role and potential of hybrid yarn in structural applications.
- Thermoplastic composite manufacturing processes produce composite specimens which possess uniform characteristics, improved tensile and impact behaviour, high moulding rate, rapid processing, economical process, and produce reversible structures. However, the consolidation method has a profound effect on composite behaviour. Extensive research is required to explore effects of consolidation methods on durability of thermoplastic composites used in the construction sector.
- Development of new innovative thermoplastic resin such as Elium® (acrylic-based liquid resin) and thermoplastic epoxy resin (composed of difunctional epoxy resin and difunctional phenolic compound) has resulted in improved resin impregnation behaviour in composites comparable to thermosets while retaining the unique features of thermoplastics such as recycling, remoulding, and reusing.
- Thermoplastic FRP bars allow formation of bendable bars for shear stirrups with improved anchorage and mechanical behaviour. The thermoplastic FRP bars offer comparable strength, modulus, strain, shear strength, and creep behaviour as compared to thermosets. The failure pattern is altered from rupture (catastrophic failure) to ductile and spalling-free failure. Additionally, they offer long term durability under aggressive alkaline environment. The durability of carbon FRP is better compared to glass FRP under an alkaline environment.
- Improvement of bond behaviour of thermoplastic FRP bars in cementitious matrices can be done with threading operation, filament winding, and braiding method by incorporating helical ribs in bars which are additive manufacturing processes.
- Thermoplastic CSIP laminates are thin and low cost modular construction panels which offer lightweight structures, freedom from corrosion, higher specific strength, high load bearing abilities, high impact resistance, good

seismic performance, good fatigue resistance, improved resistant towards freeze-thaw cycles and water penetration, ease of installation and service, prolonged service life resistant to termite attack, less labour requirements, ease of assembling, and rapid installation. They are found suited for applications involving construction of emergency safe houses in wars, hurricanes, or seismic prone areas; deck composite systems for bridge structure applications; thin roof structural elements; and strengthening and confinement of columns, floors, roofs, and bridge structures.

- Incorporation of thermoplastic fibres in TRC elements improved bond strength, first crack stress, toughness, load-bearing ability through fibre bridging effect, and reduced crack opening size with multiple cracking phenomena. The specimen failure pattern is altered from slippage to fracture due to fibre bridging and cross-linking effect of fibres. The hybrid synergistic effect of continuous basalt textile structures with polypropylene short fibres has the potential to replace steel mesh reinforced concrete (RC) slab.

REFERENCES

Alagirusamy, R., R. Fangueiro, V. Ogale, and N. Padaki. 2006. "Hybrid Yarns and Textile Preforming for Thermoplastic Composites." *Textile Progress* 38 (4): 1–71. https://doi.org/10.1533/tepr.2006.0004

Alagirusamy, Ramasamy, and Vinayak Ogale. 2004. "Commingled and Air Jet-Textured Hybrid Yarns for Thermoplastic Composites." *Journal of Industrial Textiles* 33 (4): 223–43. https://doi.org/10.1177/1528083704044360

Barhum, Rabea, and Viktor Mechtcherine. 2012. "Effect of Short, Dispersed Glass and Carbon Fibres on the Behaviour of Textile-Reinforced Concrete under Tensile Loading." *Engineering Fracture Mechanics* 92 (September): 56–71. https://doi.org/10.1016/j.engfracmech.2012.06.001

Barman, Nabo Kumar, S. S. Bhattacharya, and R. Alagirusamy. 2019. "Application of Textile Structures and Composites to Improve Performance and Durability of Building Structures." *Paper presented at the Second International Conference on Emerging Trends in Traditional and Technical Textiles*, Jalandhar, Punjab, India, November 1–3.

Barman, Nabo Kumar, S. S. Bhattacharya, and R. Alagirusamy. 2020. "Development of Hybrid Yarn Based Woven Textile Structures for Textile Reinforced Concrete Applications." *Paper presented at the Fifth International Conference on Industrial Textiles – Products, Applications and Prospects*, Coimbatore, Tamil Nadu, India, August 21–23.

Benmokrane, B., O. Chaallal, and R. Masmoudi. 1995. "Glass Fibre Reinforced Plastic (GFRP) Rebars for Concrete Structures." *Construction and Building Materials* 9 (6): 353–64. https://doi.org/10.1016/0950-0618(95)00048-8

Benmokrane, Brahim, Salaheldin Mousa, Khaled Mohamed, and Mahmoud Sayed-Ahmed. 2021. "Physical, Mechanical, and Durability Characteristics of Newly Developed Thermoplastic GFRP Bars for Reinforcing Concrete Structures." *Construction and Building Materials* 276 (March): 1–11. https://doi.org/10.1016/j.conbuildmat.2020.122200

Böhm, Robert, Mike Thieme, Daniel Wohlfahrt, Daniel Wolz, Benjamin Richter, and Hubert Jäger. 2018. "Reinforcement Systems for Carbon Concrete Composites Based on Low-Cost Carbon Fibers." *Fibers* 6 (3): 56. https://doi.org/10.3390/fib6030056

Cascardi, Alessio, Riccardo Dell'Anna, and Francesco Micelli. 2018. "Reversible FRP-Confinement of Heritage Masonry Columns." In *9th International Conference on Fibre-Reinforced Polymer (FRP) Composites in Civil Engineering, CICE 2018*, 35–43. Paris.

Cascardi, Alessio, Riccardo Dell'Anna, Francesco Micelli, Francesca Lionetto, Maria Antonietta Aiello, and Alfonso Maffezzoli. 2019. "Reversible Techniques for FRP-Confinement of Masonry Columns." *Construction and Building Materials* 225 (November): 415–28. https://doi.org/10.1016/j.conbuildmat.2019.07.124

Correia, Luis, Fernando Cunha, P. Subramani, and Raul Fangueiro. 2016. "Development of Hybrid Braided Composite Rods with High Ductility for Civil Engineering." In *Proceedings of the 5th International Conference on Integrity-Reliability-Failure*, edited by J. F. Silva Gomes and S. A. Meguid, 305–306. Porto/Portugal: INEGI/FEUP (2016).

Deng, Mingke, Zhifang Dong, and Cong Zhang. 2020. "Experimental Investigation on Tensile Behavior of Carbon Textile Reinforced Mortar (TRM) Added with Short Polyvinyl Alcohol (PVA) Fibers." *Construction and Building Materials* 235 (February): 1–12. https://doi.org/10.1016/j.conbuildmat.2019.117801

Ding, Yining, Qingxuan Wang, F. Pacheco-Torgal, and Yulin Zhang. 2020. "Hybrid Effect of Basalt Fiber Textile and Macro Polypropylene Fiber on Flexural Load-Bearing Capacity and Toughness of Two-Way Concrete Slabs." *Construction and Building Materials* 261 (November): 1–11. https://doi.org/10.1016/j.conbuildmat.2020.119881

Du, Wenfeng, and Nasim Uddin. 2017. "Innovative Composite Structural Insulated Panels (CSIPs) Folded Shell Structures for Large-Span Roofs." *Materials and Structures* 50 (1): 51. https://doi.org/10.1617/s11527-016-0924-3

El-Tahan, Mossab, Khaled Galal, and Van Suong Hoa. 2013. "New Thermoplastic CFRP Bendable Rebars for Reinforcing Structural Concrete Elements." *Composites Part B: Engineering* 45 (1): 1207–15. https://doi.org/10.1016/j.compositesb.2012.09.025

Fangueiro, Raul, Guilherme Sousa, Filipe Soutinho, Saíd Jalali, and Mário de Araújo. 2006. "Application of Braided Fibre Reinforced Composite Rods in Concrete Reinforcement." *Materials Science Forum* 514–516 (PART 2): 1556–60. https://doi.org/10.4028/www.scientific.net/MSF.514-516.1556

Friedrich, K., N. Glienke, J. Flöck, F. Haupert, and S. A. Paipetis. 2002. "Reinforcement of Damaged Concrete Columns by Filament Winding of Thermoplastic Composites." *Polymers and Polymer Composites* 10 (4): 273–80. https://doi.org/10.1177/096739110201000402

Gonilho Pereira, C., R. Fangueiro, S. Jalali, M. Araujo, and P. Marques. 2008. "Braided Reinforced Composite Rods for the Internal Reinforcement of Concrete." *Mechanics of Composite Materials* 44 (3): 221–30. https://doi.org/10.1007/s11029-008-9015-z

Harris, Harry G., Win Somboonsong, and Frank K. Ko. 1998. "New Ductile Hybrid FRP Reinforcing Bar for Concrete Structures." *Journal of Composites for Construction* 2 (1): 28–37. https://doi.org/10.1061/(ASCE)1090-0268(1998)2:1(28)

Hartness, Tim, George Husman, John Koenig, and Joel Dyksterhouse. 2001. "The Characterization of Low Cost Fiber Reinforced Thermoplastic Composites Produced by the DRIFT™ Process." *Composites Part A: Applied Science and Manufacturing* 32 (8): 1155–60. https://doi.org/10.1016/S1359-835X(01)00061-6

Hokura, Atsushi, and Shinichi Miyazato. 2020. "Feasibility Study on Thermoplastic FRP Rod to Rebar in Concrete." *Journal of the Society of Materials Science, Japan* 69 (4): 335–42. https://doi.org/10.2472/jsms.69.335

Jamshaid, Hafsa, Rajesh Mishra, Jiří Militký, and Muhammad Tayyab Noman. 2018. "Interfacial Performance and Durability of Textile Reinforced Concrete." *The Journal of The Textile Institute* 109 (7): 879–90. https://doi.org/10.1080/00405000.2017.1381394

Janetzko, S., P. Kravaev, T. Gries, W. Brameshuber, M. Schneider, and J. Hegger. 2010. "Textile Reinforcements with Spread and Commingled Yarn Structures." In *PRO 75: International RILEM Conference on Material Science – MATSCI*, edited by W. Brameshuber, I: 37–44. Aachen: RILEM Publications SARL.

Naito, Kimiyoshi, Chiemi Nagai, and Yoshihisa Tanaka. 2019. "Transverse Compressive Properties of Carbon/Glass Hybrid Thermoplastic Composite Rods." *Journal of Physical Science and Application* 9 (1): 25–33. https://doi.org/10.17265/2159-5348/2019.01.003

Kling, Veronika, Sohel Rana, and Raul Fangueiro. 2012. "Fibre Reinforced Thermoplastic Composite Rods." *Materials Science Forum* 730–732 (November): 331–36. https://doi.org/10.4028/www.scientific.net/MSF.730-732.331

Kocaoz, S., V.A. Samaranayake, and A. Nanni. 2005. "Tensile Characterization of Glass FRP Bars." *Composites Part B: Engineering* 36 (2): 127–34. https://doi.org/10.1016/j.compositesb.2004.05.004

Kravaev, Plamen, Steffen Janetzko, Thomas Gries, Bong-Gu Kang, Brameshuber Wolfgang, Zell Maike, and Hegger Josef. 2009. "Commingling Yarns for Reinforcement of Concrete." In *4th Colloquium on Textile Reinforced Structures (CTRS4)*, 1–13. Dresden: Technische Universität Dresden. https://tud.qucosa.de/api/qucosa%3A23805/attachment/ATT-0/

Kurban, Mutlu, Osman Babaarslan, and İsmail Hakkı Çağatay. 2017. "Hybrid Yarn Composites for Construction." In *Textiles for Advanced Applications*, edited by Bipin Kumar and Suman Thakur, 135–60. London: Intech Open. https://doi.org/10.5772/intechopen.69034

Lapko, A., and M. Urbański. 2015. "Experimental and Theoretical Analysis of Deflections of Concrete Beams Reinforced with Basalt Rebar." *Archives of Civil and Mechanical Engineering* 15 (1): 223–30. https://doi.org/10.1016/j.acme.2014.03.008

Lauke, B., U. Bunzel, and K. Schneider. 1998. "Effect of Hybrid Yarn Structure on the Delamination Behaviour of Thermoplastic Composites." *Composites Part A: Applied Science and Manufacturing* 29 (11): 1397–1409. https://doi.org/10.1016/S1359-835X(98)00059-1

Li, Chenggao, Rui Guo, Guijun Xian, and Hui Li. 2020. "Effects of Elevated Temperature, Hydraulic Pressure and Fatigue Loading on the Property Evolution of a Carbon/Glass Fiber Hybrid Rod." *Polymer Testing* 90 (July): 1–13. https://doi.org/10.1016/j.polymertesting.2020.106761

Li, Chenggao, Xiaoli Yin, Yancong Liu, Rui Guo, and Guijun Xian. 2020. "Long-Term Service Evaluation of a Pultruded Carbon/Glass Hybrid Rod Exposed to Elevated Temperature, Hydraulic Pressure and Fatigue Load Coupling." *International Journal of Fatigue* 134 (September 2019): 1–15. https://doi.org/10.1016/j.ijfatigue.2020.105480

Li, Chenggao, Xiaoli Yin, Yunjia Wang, Lei Zhang, Zhonghui Zhang, Yancong Liu, and Guijun Xian. 2020. "Mechanical Property Evolution and Service Life Prediction of Pultruded Carbon/Glass Hybrid Rod Exposed in Harsh Oil-Well Condition." *Composite Structures* 246 (March): 1–12. https://doi.org/10.1016/j.compstruct.2020.112418

Liu, Hsien-Kuang, Nyan-Hwa Tai, and Chih-Chen Chen. 2000. "Compression Strength of Concrete Columns Reinforced by Non-Adhesive Filament Wound Hybrid Composites." *Composites Part A: Applied Science and Manufacturing* 31 (3): 221–33. https://doi.org/10.1016/S1359-835X(99)00075-5

Liu, Qi, Wenfeng Du, Nasim Uddin, and Zhiyong Zhou. 2018. "Flexural Behaviors of Concrete/EPS-Foam/Glass-Fiber Composite Sandwich Panel." *Advances in Materials Science and Engineering* 2018 (October): 1–10. https://doi.org/10.1155/2018/5286757

Micelli, Francesco, and Antonio Nanni. 2004. "Durability of FRP Rods for Concrete Structures." *Construction and Building Materials* 18 (7): 491–503. https://doi.org/10.1016/j.conbuildmat.2004.04.012

Mousa, Mohammed A., and Nasim Uddin. 2012. "Structural Behavior and Modeling of Full-Scale Composite Structural Insulated Wall Panels." *Engineering Structures* 41 (August): 320–34. https://doi.org/10.1016/j.engstruct.2012.03.028

———. 2013. "Performance of Composite Structural Insulated Panels after Exposure to Floodwater." *Journal of Performance of Constructed Facilities* 27 (4): 424–36. https://doi.org/10.1061/(ASCE)CF.1943-5509.0000306

Naito, Kimiyoshi, and Hiroyuki Oguma. 2017a. "Tensile Properties of Novel Carbon/Glass Hybrid Thermoplastic Composite Rods." *Composite Structures* 161 (February): 23–31. https://doi.org/10.1016/j.compstruct.2016.11.042

————. 2017b. "Tensile Properties of Novel Carbon/Glass Hybrid Thermoplastic Composite Rods under Static and Fatigue Loading." *Matéria (Rio de Janeiro)* 22 (2). https://doi.org/10.1590/s1517-707620170002.0176

Naito, Kimiyoshi, Hiroyuki Oguma, and Chiemi Nagai. 2020. "Temperature-dependent Tensile Properties of Hybrid Carbon/Glass Thermoplastic Composite Rods." *Polymer Composites* 41 (10): 3985–95. https://doi.org/10.1002/pc.25686

Pakravan, Hamid R., Masoud Jamshidi, and Hojatolah Rezaei. 2016. "Effect of Textile Surface Treatment on the Flexural Properties of Textile-Reinforced Cementitious Composites." *Journal of Industrial Textiles* 46 (1): 116–29. https://doi.org/10.1177/1528083715576320

Pearson, Maximus, Ted Donchev, and Juan Salazar. 2013. "Long-Term Behaviour of Prestressed Basalt Fibre Reinforced Polymer Bars." *Procedia Engineering* 54 (2009): 261–69. https://doi.org/10.1016/j.proeng.2013.03.024

Quagliarini, Enrico, Francesco Monni, Stefano Lenci, and Federica Bondioli. 2012. "Tensile Characterization of Basalt Fiber Rods and Ropes: A First Contribution." *Construction and Building Materials* 34: 372–80. https://doi.org/10.1016/j.conbuildmat.2012.02.080

Ramasamy, A., Youjiang Wang, and John Muzzy. 1996. "Braided Thermoplastic Composites from Powder-Coated Towpregs. Part I: Towpreg Characterization." *Polymer Composites* 17 (3): 497–504. https://doi.org/10.1002/pc.10639

Rana, Sohel, Emilija Zdraveva, Cristiana Pereira, Raul Fangueiro, and A. Gomes Correia. 2014. "Development of Hybrid Braided Composite Rods for Reinforcement and Health Monitoring of Structures." *The Scientific World Journal* 2014: 1–9. https://doi.org/10.1155/2014/170187

Robert, Mathieu, René Roy, and Brahim Benmokrane. 2010. "Environmental Effects on Glass Fiber Reinforced Polypropylene Thermoplastic Composite Laminate for Structural Applications." *Polymer Composites* 31 (4): 604–11. https://doi.org/10.1002/pc.20834

Rosado, K. P., Sohel Rana, C. Pereira, and R. Fangueiro. 2012. "Self-Sensing Hybrid Composite Rod with Braided Reinforcement for Structural Health Monitoring." *Materials Science Forum* 730–732 (November): 379–84. https://doi.org/10.4028/www.scientific.net/MSF.730-732.379

Roy, R. J., A. S. Debaiky, H. Borazghi, and B. Benmokrane. 2005. "Glass Fibre Reinforced Polypropylene Bridge Deck Panel Design, Fabrication and Load Testing." In *33rd Annual General Conference of the Canadian Society for Civil Engineering*, 1–9. Toronto, Ontario, Canada.

Sayed-Ahmed, Mahmoud, Babak Hajimiragha, Borna Hajimiragha, Khaled Mohamed, and Brahim Benmokrane. 2017. "World's First Thermoplastic GFRP Mfx-Bars Subjected to Creep Effects." In *CDCC 2017 Fifth International Conference on Durability of FRP Composites for Construction and Rehabilitation of Structures*, 1–8. Qubec, Canada.

Uddin, N. U., and M. A. Mousa. 2013. "Innovative Fiber-Reinforced Polymer (FRP) Composites for Disaster-Resistant Buildings." In Woodhead Publishing Series in Civil and Structural Engineering, *Developments in Fiber-Reinforced Polymer (FRP) Composites for Civil Engineering*, edited by Nasim Uddin, 272–302e. Cambridge: Woodhead Publishing Limited. https://doi.org/10.1533/9780857098955.2.272

Uddin, Nasim, A. M. Abro, J. D. Purdue, and Uday Vaidya. 2013. "Thermoplastic Composites for Bridge Structures." In Woodhead Publishing Series in Civil and Structural Engineering, *Developments in Fiber-Reinforced Polymer (FRP) Composites for Civil Engineering*, edited by Nasim Uddin, 317–50e. Cambridge: Woodhead Publishing Limited. https://doi.org/10.1533/9780857098955.2.317

Uddin, Nasim, and Abdul Moeed Abro. 2008. "Design and Manufacturing of Low Cost Thermoplastic Composite Bridge Superstructures." *Engineering Structures* 30 (5): 1386–95. https://doi.org/10.1016/j.engstruct.2007.08.001

Uddin, Nasim, and Wenfeng Du. 2014. "New Thin Shells Made of Composite Structural Insulated Panels." *Journal of Reinforced Plastics and Composites* 33 (21): 1954–65. https://doi.org/10.1177/0731684414549492

Uddin, Nasim, John D. Purdue, and Uday Vaidya. 2008. "Feasibility of Thermoplastic Composite Jackets for Bridge Impact Protection." *Journal of Aerospace Engineering* 21 (4): 259–65. https://doi.org/10.1061/(ASCE)0893-1321(2008)21:4(259)

Uddin, Nasim, A. Vaidya, Uday Vaidya, and S. Pillay. 2013. "Thermoplastic Composite Structural Insulated Panels (CSIPs) for Modular Panelized Construction." In Woodhead Publishing Series in Civil and Structural Engineering, *Developments in Fiber-Reinforced Polymer (FRP) Composites for Civil Engineering*, edited by Nasim Uddin, 302–16. Cambridge: Woodhead Publishing Limited. https://doi.org/10.1533/9780857098955.2.302

Uddin, Nasim, and Uday Vaidya. 2005. "Ballistic Testing of Polymer Composites to Manufacture Emergency Safe House Shelters." *Journal of Composites for Construction* 9 (4): 369–75. https://doi.org/10.1061/(ASCE)1090-0268(2005)9:4(369)

Urbanski, Marek, Andrzej Lapko, and Andrzej Garbacz. 2013. "Investigation on Concrete Beams Reinforced with Basalt Rebars as an Effective Alternative of Conventional R/C Structures." *Procedia Engineering* 57: 1183–91. https://doi.org/10.1016/j.proeng.2013.04.149

Vaidya, A., N. Uddin, and U. Vaidya. 2010. "Structural Characterization of Composite Structural Insulated Panels for Exterior Wall Applications." *Journal of Composites for Construction* 14 (4): 464–69. https://doi.org/10.1061/(ASCE)CC.1943-5614.0000037

Vaidya, Amol, Nasim Uddin, and Uday Vaidya. 2007. "Flexural Response of Lightweight Sandwich Panels for Panelised Construction." In *International Conference on Mechanical Engineering 2007 (ICME2007)*, 1–5. Dhaka, Bangladesh. http://me.buet.ac.bd/icme/icme2007/Proceedings/PDF/ICME07-AM-68.pdf

Zdraveva, Emilijia, Cristiana Gonilho-Pereira, Raul Fangueiro, Senentxu Lanceros-Méndez, Saíd Jalali, and M. Araújo. 2010. "Multifunctional Braided Composite Rods for Civil Engineering Applications." *Advanced Materials Research* 123–125 (August): 149–52. https://doi.org/10.4028/www.scientific.net/AMR.123-125.149

14 Potential Application Areas for Thermoplastic Composites

J. Krishnasamy[a], R. Alagirusamy[b],
and G. Thilagavathi[a]
[a]PSG College of Technology
Peelamedu, Coimbatore
[b]Indian Institute of Technology Delhi
Hauz Khas, New Delhi

CONTENTS

14.1   INTRODUCTION

Metal products like buried steel tanks, piping and other metal-based fuel storage systems are generally less protective and eventually corrode and leak into the environment. Corrosion can attack the metal surface and may result in a localized hole in an unprotected tank in about 15 years. Hence, there is a need for composite material exhibiting better corrosion resistance, long life and high strength to weight ratio. In industrial applications, thermoplastic composites are gaining popularity due their merit characteristics like lighter weight, QUICKER production time, longer shelf

DOI: 10.1201/9781003049715-14

life and easier shipping compared with metals. All composite materials allow manufacturers to minimize the product weight compared to metals, which is essential
for automotive, aerospace and rail industries. By using composites, weight of the
product can be reduced by 40% to 70%. Lightweight vehicles are a big catalyst for
the composites to replace heavier metals, especially in cars and other vehicles as
automobile industries are seeking to fulfil new federal fuel efficiency requirements.
Thermoplastic composites have several advantages compared to thermoset composites, such as greater ease to process, freezer free cycles, being recyclable, etc. In aerospace industry, thermoplastic composites have been of considerable significance. For
manufacturing aerospace crafts, the contribution of carbon fibre reinforced plastic
(CFRP) is highly significant. By compounding the carbon fibre directly into the base
resin, carbon-fibre reinforced thermoplastic is made that can be easily moulded into
any form. The thermoplastic CFRP has excellent fatigue and damage resistance,
shorter processing time and low moisture absorption properties that are suitable for
aerospace applications. Among the thermoplastic composite materials, the carbon/
PPS, carbon/PEKK, carbon/PEEK, carbon/PEI, glass/PPS, glass/PEI, and glass/
PEEK are commonly used. Hence, in this chapter, the various applications of towpreg-based thermoplastic composites in aircraft, transport, automotive, marine,
medical, industrial, and other fields are reported in detail.

14.2 TOWPREG COMPOSITES FOR EASY FIBRE PLACEMENT

Carbon fibre composite materials possess internally great tensile strength and low
weight and find significant application in the areas ranging from automotive to aerospace industries. Pure roving or pre-impregnated towpregs can be used for manufacturing three-dimensional shapes using the automated fibre placement. Three types
of materials are used for composite fabrication:

- Pre-impregnated towpreg or prepreg
- Pure rovings, UD tapes or pre-confectioned tows for resin injection
- Thermoplastic tapes

During fabrication of composite product, the raw carbon tows or fibreglass rovings
are used in developing the three-dimensional parts that requires the resin injection during/after preforming the mould. When the resin is injected into the mould,
the impregnation and distribution of resin depend on the temperature and pressure
applied. The optimum viscosity and the pressure need to be applied for homogenous
fibre wetting, which requires lot of skills and challenging for 3D structures. Hence,
pure non-towpreg filament layers could not easily make the homogenous composite material during the pressure-heating application. Hence, the usage of towpregs
comes into effect. As the towpregs are already impregnated with the resin, there is
no need of injecting the resin into the finished shape prior to application of temperature and pressure. They only need to be heat treated to form crosslinking of matrix
material. Usually, curing is carried out in specialized ovens after laying up the layers

and pre-curing the materials. Towpreg-based products are demanded in automobile industries where the product is immalleable under high temperatures. In addition, it also eliminates the resin coating or injecting steps in the production process. In order to place the fibre for thick and complex structures, the automated fibre placement machines are used to avoid human error while developing huge volumes of composite products.

Offeringa (1996) has reported the various end-uses of thermoplastic composites in structural and non-structural assemblies such as access panels and doors, engine cowlings and movable wing surfaces such as elevators, flaps and ailerons. These products are made with specific thermoplastic processing technologies like thermo folding, press-forming, welding and membrane forming due to cost effective manufacturing methods, and the products developed by these techniques will meet the standards by high degree of automation. Moreover, the continuous fibre reinforced aramid/polyetherimide is used in preparing the ice-protection plate of the Dornier 328 aircraft (Hansmann and Wismar 2003). The plate edges are thermo folded and the folded flange is only 6 mm wide, which provides the fairing between the plate and the aircraft fuselage skin. Similarly, the toilet module in the new Fokker 60 is made with thermo folded panels which are commercially manufactured. The pressformed thermoplastics ribs are produced for the Dornier 328 turboprop flaps using the carbon/PEI prepreg materials (Hansmann and Wismar 2003). For structural applications, 3D woven composites have been used in large quantities to produce the woven "H-joint" and "V-joint" by Beech for the Beechcraft Starship turboprop bizjet (Smith 2001). Mouritz et al. (1999) showed the examples of demonstration structures for aircraft:

- Turbine engine thrust reversers, rotor blades and structural reinforcement
- Reinforcement and heat exchangers
- Rocket engines, nozzles and fasteners
- Mounts for engines
- T-sections for main fuselage frame structures
- Rib, cross-blade and stiffened panels with many blades
- Elements of the T- and X-shape
- Leading edges to wings

14.3   AIRCRAFT APPLICATIONS

In earlier days, aluminium was the choice for the aircraft components due to its ease of manufacturing and shaping the parts. However, due to many limitations of metal components, an alternative was needed with lightweight, high specific strength and fatigue strength properties. Nowadays, aerospace industry consumes more composite materials, up to 50% by weight of the aircraft (Brown 2014). The commercial and military aircrafts use nonstructural and structural parts made up of composite materials such as elevators, horizontal stabilizer boxes, motor castings, tail sections (Airbus), cargo liners, rudders (Boeing 767), fuselage (V22), engine nacelles, landing gear doors, wings (Prototype ATF, V22) and so on (Pilato and Michno 1995).

The carbon fibre/epoxy-based composite motor cases provide the significant weight reduction. Similarly, nozzles of exit cones and throat elements have been also been made up of carbon fibre/phenolic or carbon-carbon composites which withstand hot exhaust gases of burning propellant (Pilato and Michno 1995). Similarly, the use of 3D textile composites is also increased for structural applications. Several methods are available to develop 3D composite materials and 3D stitching technique is one of the methods by which 3D structures are made. For manufacturing the low-cost and damage-tolerant composite wings, the 3D stitching technique was used in a NASA sponsored demonstrator program. A 28-m-long sewing machine designed by Boeing was used to stitch over 25 mm thick carbon layers at a rate of over 3000 stitches per minute.

The blade stiffener flanges made with tubular braided fabrics were also stitched to the skin of the composite. Finally the resin infusion technique was used to prepare the composite panel, which showed 20% cheaper and 25% lighter properties than aluminium part (Dutton et al. 2004). Similarly, on the Beechcraft Starship, H-joint connectors made of 3D woven composites were used to join honeycomb sandwich wing frames. In this frame, the connector was necessary for the low-cost development of wings. This helps in increased stress flow at the joint and minimized the peeling stress (Karaduman et al. 2017). In recent years, thermoplastic composites have been successfully used in aircraft applications due to their superior properties. The greatest advantage of thermoplastic composites is that the bonding and riveting are eliminated, which reduces the cost and weight issues, and welding can be done with any material. The major advantages of thermoplastic prepreg tape is that it allows full automation of complex shapes with better properties and reduced cost. The thermoplastic composite made with automated fibre placements and autoclaved for full consolidation could be used fabricating fuselage of the aircraft. The aerospace grade thermoplastic tape could be made with very low porosity (< 0.5%). The artificial intelligence-supported automation could be effectively used to manufacture the aircraft composites with real-time quality management processes. In 2025, the fuselage of the airplane will be made with thermoplastic composites for a new midsized airplane (Favaloro 2018), because the thermoplastic manufacturing and processing have become more flexible in both designing and manufacturing. In addition, manufacturing the thermoplastic components into geometrically complex shapes has also been possible over the past few decades with the high tolerance up to 0.002 mm. This is most critical requirement for aerospace applications. Many established composite forming techniques could be applied, such as rapid thermoforming, press forming, autoclave processing, tape and fibre placement techniques in preparing various thermoplastic components.

The fatigue properties of thermoplastic composites are even better than metals and exhibit good resistance to deformation for larger deflections. Mostly, metal is replaced by PEEK in the aerospace industry due to its lightweight, mechanical strength, creep and fatigue resistance and ease of processing. The PEEK-based thermoplastic composite could be used in flight control, engine and aircraft interior, fuel systems and aerodynamic-related components. The polymer matrix used for aircraft applications has useful inherent properties like thermal and electrical insulation whereas metallic components require expensive secondary processing and insulation coating. For example, polyetherimide (PEI) having low thermal conductivity and high dielectric strength could be used in aircraft galley equipment. In addition, the excellent flame rating (UL94 V-O) makes it better for aircraft interior components. In fact, 30% of glass-reinforced grade of ULTEM PEI is the best replacement for aluminium due to its similar coefficient of thermal expansion. In addition to insulation properties, the thermoplastic composites also absorb the radar signals, which can be useful for stealth military aircraft applications. But for metal-based components, the electromagnetic waves are reflected, which makes it easier to detect by radar signals (compositence.com 2021). Similarly, joining the thermoplastic composite structures is very important in aerospace vehicles since it should withstand high static and fatigue loads. The various welding methods available for joining the thermoplastic polymers are reviewed and reported by da Costa et al. (2012).

14.4 TRANSPORTATION APPLICATIONS

The thermoplastic towpreg composites are used in many transportation industries such as automotive, marine and other industries. Glass towpreg-based thermoplastic composites have been used in transportation vehicle parts such as bumper beams, engine covers etc. The matrix materials like polypropylene, nylon and polyester are used as thermoplastic matrices for developing the glass fibre reinforced composite. Similarly, carbon towpreg-based composites are also used as replacements for metal materials in high temperature and high performance applications. The advantages of thermoplastic composite materials arerecyclability, superior damage tolerance and fracture toughness, and ability to produce complex shapes using the towpreg winding. Among thermoplastic composites, the automotive and transportation sector have more demands for long-fibre thermoplastic (LFT) composites. For developing beams and shells of automobile bodies, chassis and drive shafts, 3D braided composites have been used. Similarly, the composites reinforced with non-crimp multi-axial glass fabrics are used in manufacturing the car bumper bars, floor panels and door members (Pisanikovski et al. 1998).

14.5 AUTOMOTIVE VEHICLES

In the automotive vehicles, the lightweight and strong composite material should be incorporated in order to decrease the fuel consumption. The incorporation of composite material decreases the component weight by 10–75% compared to cast iron-, steel-, and aluminium-based components. As a result, the fuel consumption is decreased by 6–8% (Dufour et al. 2015). Several parts of the automotive vehicles are replaced by composite materials such as bumpers, hatch doors, cabin components, underbody panels, instrument panel carriers, battery trays, sunroof beams, door modules, seat structures, roof and trunk lids, exterior claddings, running boards, step assists, front-end carriers, lift gates and other body parts (Mallick 2010). Figure 14.1 shows the various applications of composite materials in automotive and industrial applications.

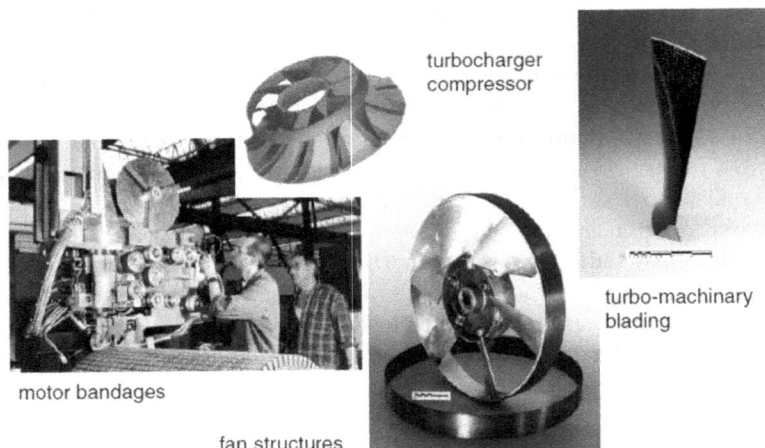

FIGURE 14.1   High performance applications of composite materials at ABB corporate research. (Reprinted with permission from Elsevier Publications; Mallick 2001.)

In a study, Turner et al. (2008) produced a front wing-fender component for automotive body panel applications. The composite was made with semi-preg systems for improved performance. The semi-preg system consisted of suface film made by thin woven glass with either side of an epoxy film and a bulk ply. The bulky ply was made of woven 3000 carbon filaments with 300 g/m² on one side and other side of woven glass with 400 g/m², and an epoxy glass microsphere core. The developed body panel exhibited good performance compared to existing stamped steel components. Usually, carbon fibre composites provide equivalent bending stiffness and 40–50% weight saving compared with steel panels (Turner et al. 2008). The composite material is also used in high performance racing cars where the weight reduction is a primary demand. Feraboli et al. (2007) fabricated the composite panels for body and integrated chassis components for the Murcièlago Roadster. The developed carbon fibre composite provided the significant weight reduction, good performance and better design flexibility.

The application of 3D textile-reinforced composites has been explored in different parts of automotive vehicles such as floor beams in trains and fast ferries, crash members in cars, containers, buses and many other parts. The 3D composites reinforced with glass-based spacer fabrics are especially used in car and truck spoilers/fairings and luggage floors of automobiles (Mouritz et al. 1999). Similarly, the oil pan can be produced by 3D woven interlock fabrics. Figure 14.2 shows the orthogonal and layer-interlock woven fibre architectures used for developing the 3D woven composites. The main difference between orthogonal and layer-interlock architectures is the pattern of the through-thickness binder yarn.

The various fibre architectures can be made by controlling the binder yarns for the through-thickness reinforcement. Different high performance yarns such as carbon, glass, Kevlar and ceramic yarns can be introduced in the 3D woven composite for producing the hybrid woven preforms that have more than one type of material.

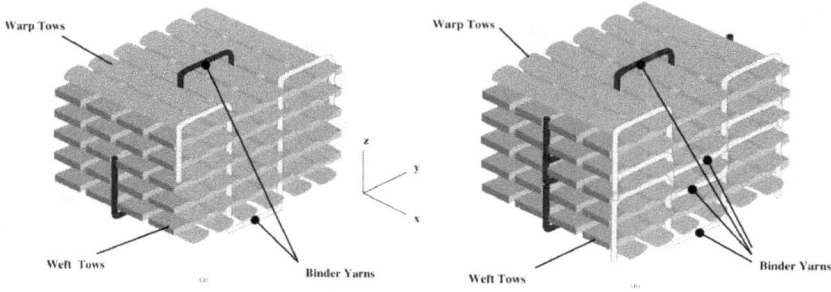

FIGURE 14.2   (a) Orthogonal and (b) layer-interlock woven fibre architectures commonly used in 3D woven composites. (Reprinted with permission from Elsevier; Mouritz et al. 1999.)

The car noses and bodies of Formula One car are made from carbon textile structures. Similarly, the braided carbon composites are used for developing the car bumpers (e.g., BMW M6, Lotus Elise) and the carbon-woven fabrics are used for fabricating the car roof of BMW M6 (Gries et al. 2008).

Hence, carbon fibre reinforced polymer composites are usually incorporated in constructing the body of the car. In addition, interior panels and headlight and tail-light assemblies are also made up of polymers and polymer matrix composites. The hybrid yarn-based composite is also used in automobile applications such as bumper beams, under-body shields, etc. due to its ease of processing, consistent quality, recyclability and specific strength. Davies et al. (2005) explored the thermoplastic matrix composites in underwater applications such as oceanography, submarine and sub-sea offshore structures. The glass/PEI and carbon/PEEK-based composite cylinders were fabricated and tested for hydrostatic pressure tests and drop weight impact tests. The developed carbon/PEEK material showed the resistance to the pressure of 90 MPa and exhibited better damage tolerance values. But the residual implosion resistance after impact was reduced due to a local buckling failure. Hence, this area needs to be focused and analyzed further for the improvement in structural damage tolerance of the composite material (Davies et al. 2005). Similarly, the use of composite material in main hull structure in underwater vehicles is reported in a project (Meng et al. 2016). The military submarines also uses the composite materials such as sonar domes and outer decks (Mouritz et al. 2001) whereas offshore industry uses large composite structures (Salama et al. 2002). In a work by Ning et al. (2009), a lightweight thermoplastic composite mass transit bus roof door was developed and prototyped to replace the existing aluminium mass transit bus roof door. The composite made with thermoplastic polyolefin (TPO) as the outer skin and a ribbed lofted glass mat thermoplastic (GMT) as inner liner is used as roof door with good structural stiffness and strength. Figure 14.3 shows the tool made from mahogany wood works as male plug with a female counterpart made of aluminium.

The developed composites roof door fitted on the bus in closed position is also shown in Figure 14.3. The composite was developed using the compression moulding technique. It can be also used for rail cars, trucks, marine and aerospace structures.

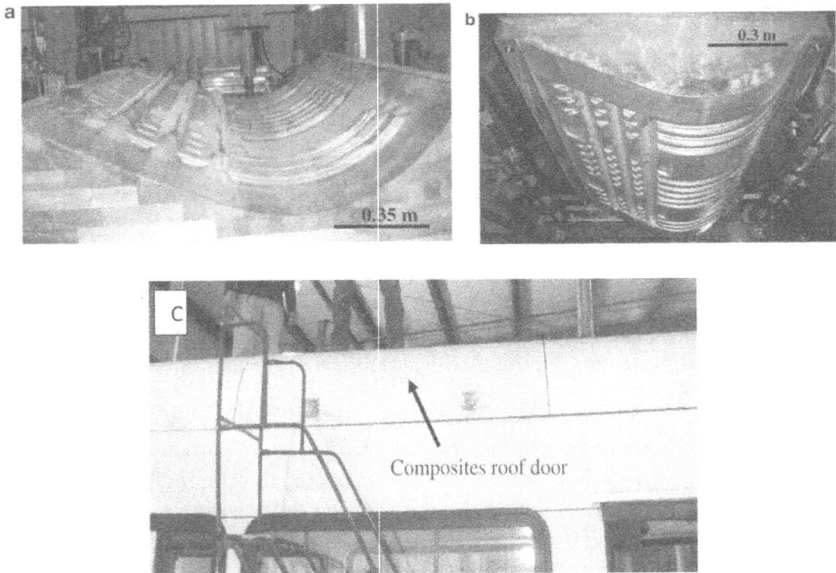

FIGURE 14.3 (a) Aluminium tool for thermoforming of TPO skin; (b) aluminium tool for thermoforming of inner SuperLite@ sheet; and (c) the composites roof door assembled on the bus (closed position). (Reprinted with permission from Elsevier; Ning et al. 2009.)

14.6 MARINE VEHICLES

In marine vehicles, polymer matrix composites are used in construction in boat hulls, boat decks, instrument panels, casings, doors, cabinets, etc. Generally, fibre glass towpregs are used along with suitable matrix materials to develop the composite products. Nowadays, carbon fibres are used in boatbuilding instead of glass fibres in the emerging trend. In addition, the composite can be also used in rail industries for preparing ceilings, platforms, interior doors, windows and front structures (Mankodi 2011). In another study, the self-healing composite structure was investigated for improving the life-cycle of thermoplastic fibre reinforced tape layered composite for onshore and offshore applications by Gupta et al. (2019). The results showed the recovery of 84% to its original strength after self-healing and exhibited long life of the structure.

14.7 MEDICAL APPLICATIONS

The attractive properties of composite materials such as high stiffness, light weight and durability find numerous applications in medicine. In cases of bone repair, composite material finds many applications. Generally, in external fixation systems of bone repair, splints, casts, braces and other external fixator systems are used. For example, traditionally, cotton-woven fabric reinforced with calcium sulfate is used as casting material. However, it has many limitations such as low success rate, high density, low stiffness and low stability in presence of water. Hence, as an alternate, glass/polyester fabric reinforced with water activated polyurethane has been applied as casting material. Similarly, the splinting material must possess lightweight, stiffness, comfortness and ability to fit into the complex contours of the limbs. For this purpose, fabrics with large openings such as leno structures, warp and weft knitted fabrics have been used, which allow easy impregnation of plaster material (Ellis 2005).

Conventionally, bone plates prepared from stainless steel and Ti alloy were used. But they have major limitations as the stiffness between the bone and plate does not match, which causes increased bone porosity and subsequently weakens the bone. In order to avoid such problems, composite bone plates have been fabricated. The stiffness of the composite plates can be tailored by selecting the proper polymer matrix and reinforcing material such as carbon fibre/epoxy or PMMA, glass fibre/epoxy, carbon fibre/polyethylene, carbon fibre/nylon, and carbon fibre/PEEK composites (Ramakrishna 2004). Similarly, stainless steel, Co-Cr and Ti alloys were used for total hip replacement (THR) and total knee replacement (TKR) in earlier days. To improve the performance of THR and TKR, the CF/PEEK and UHMWPE-based composites have been used (Evans and Gregson 1998).

The composite materials are also used in other areas such as bone cement, bone grafts, dental posts, ureter prostheses, catheters, tendons and ligaments, prosthetic limbs and other medical parts (Karaduman et al. 2017). Especially in the orthopaedic implants, the metal implants have been replaced by artificial joints and bones. For this purpose, the composite structures made by poly (D, L-lactide urethane) reinforced with polyglycolic acid showed excellent physical characteristics (Rigby et al. 1997). The composite can be formed into required shapes during surgery by applying suitable temperature for both hard and soft tissue applications. Dental implants have been made with composite materials fabricated by SiC and carbon or carbon fibre reinforced carbon (Ramakrishna et al. 2001). These materials provided special advantages over ceramics and metal alloys such as high strength with low elastic modulus and superior fatigue properties.

14.8 COMPOSITES IN CIVIL ENGINEERING APPLICATIONS

Despite better strength and stiffness of steel reinforced concrete, the problem due to steel reinforcement such as heavy weight, design limitation and corrosion could not be controlled. In the reinforced concrete structures, the use of steel as reinforcement is susceptible to corrosion, since the alkaline nature of concrete makes a passive layer on steel and protects it from corrosion. However, over a long time, the alkalinity of the steel is decreased due to penetration of outside substances. This results in less protection of steel and reduces the permanency of the structure and also the structural failure in extreme cases. In addition, civil structures today need building reinforced structures to provide stability against natural disasters such as earthquakes and hurricanes. The use of composite materials can make the concrete structures more earthquake resistant. Hence, the textile reinforced concrete materials have been explored in combination with different materials. Fibre reinforced polymeric (FRP) materials are commonly used in civil engineering applications. The fibre reinforced concrete offers lots of advantages in civil engineering applications such as low density, fabricating of complex-shaped components, low cost due to easy transportation and required aesthetic properties. Hence, textile material could be used as a replacement for steel, which provides the design freedom, durability and cost effectiveness also (Motavalli et al. 2010). The advantages of FRP composite decks are light weight, resistance to chemical and corrosion, quick construction and easy handling, high fatigue resistance and join free parts (Balaguru 1994).

Some of the textile reinforced concrete structures are bridges, pillars and road guards made with Kevlar or jute reinforced concretes. The reinforced concrete is made into any form and is extremely versatile, and it can be used for developing buildings, bridges and other structures (Rajendran et al. 2020). The glass reinforced plastics with low elastic modulus can be incorporated in building construction by the application of double curvature and folded-plate structures. The alkali-resistant glass fibres are used as reinforcement in cement-based composites for mostly non-load bearing applications. Lightweight composite panels have especially been used for partitioning and similar applications. The carbon fibre reinforced plastics (CFRP) have also been used for developing lightweight structures, including a number of bridges. But the cost is the constrain factor for vast application of fibre reinforced plastics. Different fabric forms such as woven, knitted, nonwoven, stitched and 3D fabrics are used as reinforcing structures. By converting fibrous assemblies into non-woven fabric by means of chemical, thermal and mechanical methods, different densities of fabric could be obtained ranging from $10g/cm^2$ to $100g/cm^2$ (Rajendran et al. 2020). For example, a textile reinforced pedestrian bridge is constructed in Kempten, Germany as shown in Figure 14.4. Similarly, building structures have been made up of textile reinforced concretes.

The concrete bridge consisted of four layers of woven alkali-resistant (AR) glass fibre fabrics supported by steel tendons. The span of the bridge is 8.60 m (Ehlig et al. 2012). In another study, restoration of a train station platform was made by Hering Bau in Walleshausen, Germany. To develop the structure, concrete was reinforced with two layers of carbon fibre having the panel dimensions of 2.5 × 1.35 m., and it was fixed with two stainless steel mandrels (Raupach and Morales Cruz 2016). Likewise, several composite parts have been successfully brought to usage for civil engineering applications.

FIGURE 14.4 (a) Textile reinforced concrete bridge in Kempten, Germany (Photo: Harald Michler); (b) Building consisting of TRC components. (Reprinted with permission from John Wiley and Sons; Ehlig et al. 2012.)

14.9 SPORTING GOODS INDUSTRIES

The performance of sporting goods could be tailored by incorporating composite materials. Sporting goods such as golf, tennis and bicycle racing have shown better performance when the composite material is introduced. For developing composite sporting goods, hand layup and roll wrapping have mainly been used. For tubular products such as golf shafts, sail masts, ski poles, softball bats and bicycle frame tubing, filament winding technique is preferred. The filament winding technique involves lower labor cost, design possibilities, product consistency and quality of the products (Spencer 1998). For developing sports equipment, the unsaturated polyester casting is used for tennis racquets, horizontal bars, jumping boards, fishing poles and so on. The bicycle parts made of carbon fibre composites such as frame, crank-shaft and seat bracket provide good rigidity and shock absorption behaviour (Tong 2019). The carbon fibre composite is also used in most of the sporting goods such as bicycle frames, front forks, and seat posts. Because the carbon fibre composites provide lightweight, good bending stiffness and fatigue resistance. It can be also used to produce tennis racquet frames with high strength and stiffness. The handles of the racquet frames could be made by wrapping the multiple layers of composite material over the soft core to impart vibration damping characteristics (Froes 1997). For shock absorption, glass fibre reinforced nylon wheels have been used. Similarly, carbon and Kevlar fibres reinforced with epoxy resin are used for making kayaks, skis and snowboards with soft cores and various constructions of composite layers (Van De Velde 2005).

14.9.1 Tennis Racquets

The durable composite racquets have been made using a compression moulding process with an internal bladder. Other composite making techniques such as injection moulding, filament winding, braiding and resin transfer moulding have been also used to manufacture the composite racquets. In special cases, the mixture of these processes has been used to meet the complex requirements of racquets. Filament winding is mostly used by the manufactures of racquets, since it minimizes the raw material and labour cost and improves the product consistency. In addition, design flexibility is possible by changing the strength and stiffness of nearby racquet heads, which is challenging in other processing techniques. Compared to other reinforcing materials, carbon fibre-based tennis racquets attract higher market demand in this field due to several merits such as shock absorption, design freedom, light weight and good vibration damping performance.

14.9.2 Bicycle Frames and Components

The tubes and frames of bicycles have to be manufactured with high durability and good joint designs in order to avoid stiffness mismatch between the parts. Out of different composite manufacturing methods, filament winding is the suitable technique for developing the frame tubes. As the filament winding method involves less material cost, labour and controlled quality parameters, most of the suppliers prefer this technique for developing good quality bicycle frames. Similarly, roll wrapping is also currently practiced, which requires precise cutting of prepreg plies and hand wrapping them around a metal mandrel before curing. In addition, titanium tubes of front wheel suspension systems are also reinforced with composite tubing. Similarly, wheels and spokes are also made by combining roll wrapping, hand lay-up and resin transfer molding (Spencer 1998).

14.9.3 GOLF SHAFTS

In 1400s, the first golf shaft was made from hickory wood. Later, high modulus graphite, titanium and high tensile strength steels were used to make the golf shafts. The performance of the shaft depends on the specific strength and specific stiffness of the shaft material. Hence, the materials with better specific properties have to be chosen to improve the shaft design and performance of product. The common metals such aluminium and stainless steel could not fullfil the requirements of the product, hence titanium and fibreglass have been tried. Although the specific strength and modulus of titanium are comparable to steel, the shaft cost is four to five times costlier than carbon fibre shafts, because fibre reinforced composites enhance the quality of product in lightweight shafts, oversized heads, and larger sweet spots (Berg 1999).

The initial use of carbon fibre/epoxy composites had processing difficulty because the resin system was not user friendly, hence high pressures and temperatures were applied to cure the material. In addition, complicated handling and curing requirements were not viable for many suppliers. Use of autoclave was also not affordable for many goods' component suppliers. Similarly, there were many limitations to fabricate the good quality golf shafts. In addition, fabrication of composites with low fibre areal density such as below 150 g/m^2 and low resin content such as below 32 percent was difficult. In recent years, the development of prepreg process made a breakthrough in fabricating the composite goods. Nowadays, composites with areal density of 50 g/m^2 and resin content of 25% could be made without compromising the quality of the product using the prepreg layers (Berg 1999). Hence, one can design the laminate with required fatigue strength, tip strength and surface cosmetic qualities with lower overall mass. The lightweight glass scrim materials have especially been used to improve the transverse strength of the shaft laminate. Many other materials like boron carbide and hybrid composite materials could be used in golf shaft materials.

Most commonly, graphite fibre reinforced epoxy golf shaft is used for its lightness and strength. In 1972 in the United States, CFRP (carbon fibre reinforced composite material) was used to make the golf shafts to meet the ball's flight stability requirements (Zhang 2015). Similarly, carbon fibre composite-based poles are used in pole vaulting, which exhibits lightweight, flexible yet stiff, and torsional resistant. To develop the pole shaft, layered design is used in which the outer layer has unidirectional carbon fibre/epoxy system, the middle has glass fibre web/epoxy and inner layer has wound glass fibres/epoxy. The stiffness to the shaft is provided by carbon/epoxy composite layers, whereas torsion resistance is provided by glass fibre composites (McCormick 2010). For sporting goods composites, the performance, cost and ease of processing are the critical requirements. The gold shaft developed by filament wound shafts provides better product consistency and shaft performance than the roll-wrap process. This is due to the better orientation of the plies on the mandrel that exhibits good flex and torque properties.

14.10 INDUSTRIAL APPLICATIONS

The towpreg-based composite materials are widely used in manufacturing industries due to their reduced weight, better mechanical properties, vibration damping, durability, corrosion resistance and flexural strength. In a study by Muller et al. (1994), 3D woven composite was made in the form of I-beams and used in the roof of a ski chair lift building in Germany. Since the 3D woven composite beam is lighter, it is easy to transport and lift to the steep terrain at the building site. Compared to heavy steel beams, the composite beams contributed to cost savings and improved performance. In Western countries, the manhole covers in petrol station forecourts are made up of 3D woven composite materials (Muller et al. 1994). In a study by Choi et al. (1999), CF/PEEK hybrid yarns were prepared to develop the high performance rotors for improving the failure tolerance. The commingled hybrid yarn was produced by air texturing technology due to its homogeneity and good fibre impregnation. For developing the industrial components, PPS, PEI and PEEK polymers are more suitable, can operate at high temperature and provide better wear and corrosion resistance. In addition, bushings and seals in chemical refineries and semiconductor manufacturing equipment components are also made of composite materials to improve their service life. The bicycle frames madeup of thermoplastic composites have been also widely used. Mallick (2001) reported that high-performance turbo-machinery components have been made up of thermoplastic components. Figure 14.5 shows the robotic-based thermoplastic fibre placement (TFP) in a thermoplastic overhang bandage of an ABB motor. Compared to steel wire wrapping systems, TFP process offers significant cost reduction due to reduced manufacturing times.

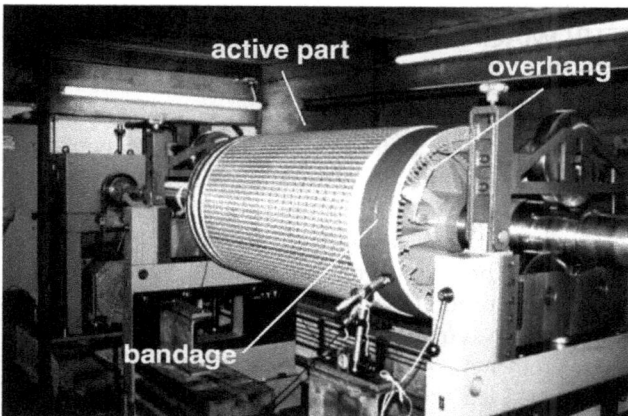

FIGURE 14.5 Motor with a thermoplastic overhang bandage from an ABB corporate. (Reprinted with permission from Elsevier; Mallick 2001.)

Similarly, high-pressure fans made by fibre-reinforced polymeric composites provide stability even at the modern impeller speeds of 270 m/s due to their light weight. In addition, the subcomponents such as inlet and back plate and blade are also manufactured using the composite materials. The rings in rotor windings of electrical machines are also made of glass/unsaturated polyester resin, which shows better properties than steel rings. Mouritz et al. (1999) stated that many engineering components could be made by composite materials such as I beams, T joints, rocket motor nozzles and rib stiffened panels. The 3D knitted fabrics could make complex shapes such as bicycle helmets and engine vanes. Moreover, 3D reinforcing material should be analyzed in detail in terms of manufacturing methods, mechanical behaviour, cost of processing and life cycle of the product, so that the potential to use various structures like woven, braided, stitched and knitted structures as reinforcement could be enhanced. For manufacturing the double-walled tanks, the wall linings for chemical storage tanks and for similar products, the distance fabrics have been used as reinforcement for manufacturing the composite materials. Generally the distance fabrics have two parallel skins of 2D fabrics connected by low density through-thickness glass yarns for imparting the good mechanical properties in thickness direction (Alagirusamy et al. 2006). In a study by Robert et al. (2010), the performance as well as the environmental effects of using glass fibre reinforced polypropylene thermoplastic composite laminate in the civil structural applications were investigated. The durability of PP/glass composites depends on the quality of consolidation. The PP/glass composites have potential usage in prefabricated structures such as bridge decks, beams and barriers. The thermoplastic composites can be used in many different industries such as automotive, aerospace or sports. Moreover, the natural fibre reinforced composites (NFRC) have been also used as a substitute for glass or carbon reinforced polymer composites. The other possible application areas are automobile interior linings (roof, rear wall, side panel lining), shipping pallets, construction products, furniture and household products (Sepe et al. 2018).

14.11 ELECTROMAGNETIC SHIELDING APPLICATIONS

The composite board incorporated with conductive materials is used for electromagnetic shielding applications. The conductive fibre reinforced polymeric composite materials provide the electromagnetic shielding effect with other mechanical properties. In addition, the filler loaded particles board has been also used for shielding. The wide range of frequencies could be shielded by designing the composite with suitable reinforcement and fillers. In a study, the authors have confirmed that the composite could provide the reflection co-efficients of 0.8 and above when the textile fibres are incorporated into the fabric (Krishnasamy et al. 2018). To arrest the radiation, lead was traditionally used due to its cost and radiation shielding capability. But the problems associated with health and environmental concerns are high. Hence, traditional lead-based materials are now considered to be toxic and banned in many countries. The new composite materials developed with lead and other lighter materials provide sufficient attenuation. In addition, they are lighter than traditional lead and provide same lead equivalency protection levels. It also showed no signs of leakage or corrosion even after long time disposal (Globe 2021). Lead-free composites can be also fabricated using eco-friendly and durable materials with low density, high impact strengths and other required properties.

14.12 SUMMARY AND FUTURE SCOPE OF COMPOSITE MATERIALS

Advanced yarn formation techniques such as commingled yarn, hybrid yarn, etc. can be used to develop fabric reinforcement along with matrix forming polymers. Compared to conventional composite manufacturing, prepreg-based composite manufacturing offers lots of advantages such as ease of processing, good fibre wetting, uniform resin distribution, etc. The composites with tailorable properties in any direction can be imparted by using the textile reinforcement. The textile reinforcements like UD, 2D woven fabric, biaxial fabric, triaxial fabric, multiaxial fabric, braided structure, 2D and 3D knitted fabric, spacer fabric, co-we-knit, multilayer fabric, stitched fabric and 3D woven fabric can be used to develop composite laminates with desirable physical and mechanical properties. Similarly, high performance fibres can be used in one direction or blended with other material for meeting the required stiffness and strength for the composite materials. To improve the toughness of the composite material, 3-phase composite material could be made, which is suitable for many applications such as aerospace, automobile, sports and medical industries. To improve the shielding behaviour of composite materials, multilayer fabric can be incorporated as a reinforcement. This helps in shielding the wide range of frequencies by selecting the suitable fibres in different layers.

REFERENCES

Alagirusamy, R., Fangueiro, R., Ogale, V., & Padaki, N. (2006). Hybrid yarns and textile preforming for thermoplastic composites. *Textile Progress*, *38*(4), 1–68. https://doi.org/10.1533/tepr.2006.0004

Balaguru, P. (1994). Fiber-reinforced-plastic (FRP) reinforcement for concrete structures: Properties and applications. *Cement and Concrete Composites*, *16*(1). https://doi.org/10.1016/0958-9465(94)90032-9

Berg, J. S. (1999). Composite material advances in the golf industry. In *ICCM12 Proceedings* (pp. 1–8). Europe.

Brown. (2014). The use of composites in aircraft construction. https://vandaair.com/2014/04/14/the-use-of-composites-in-aircraft-construction/

Choi, B.-D., Diestel, O., & Offermann, P. (1999). Commingled CF/PEEK hybrid yarns for use in textile reinforced high performance rotors. In *12th International Conference on Composite Materials (ICCM)* (pp. 796–806). Paris.

compositence.com. (2021). Towpreg composites for easy fiber placement. https://compositence.com/s/en/fiber-placement/innovation/towpreg/index.html. Accessed 11 February 2020.

da Costa, A. P., Botelho, E. C., Costa, M. L., Narita, N. E., & Tarpani, J. R. (2012). A review of welding technologies for thermoplastic composites in aerospace applications. *Journal of Aerospace Technology and Management*, 255–265. https://www.researchgate.net/publication/259558052_A_Review_of_Welding_Technologies_for_Thermoplastic_Composites_in_Aerospace_Applications

Davies, P., Riou, L., Mazeas, F., & Warnier, P. (2005). Thermoplastic composite cylinders for underwater applications. *Journal of Thermoplastic Composite Materials*, *18*(5). https://doi.org/10.1177/0892705705054397

Dufour, C., Pineau, P., Wang, P., Soulat, D., & Boussu, F. (2015). Three-dimensional textiles in the automotive industry. In *Advances in 3D Textiles* (pp. 265–291). UK: Woodhead Publishing, https://doi.org/10.1016/b978-1-78242-214-3.00010-3

Dutton, S., Kelly, D., & Baker, A. (2004). *Composite Materials for Aircraft Structures.* Reston, VA: American Institute of Aeronautics and Astronautics. https://doi.org/10.2514/4.861680

Ehlig, D., Schladitz, F., Frenzel, M., & Curbach, M. (2012). Textilbeton: Ausgeführte projekte im Überblick. *Beton- und Stahlbetonbau, 107*(11), 777–785. https://doi.org/10.1002/best.201200034

Ellis, J. G. (2005). Textile reinforced composites in medicine. In *Design and Manufacture of Textile Composites* (pp. 436–443). UK: Woodhead Publishing. https://doi.org/10.1533/9781845690823.436

Evans, S. L., & Gregson, P. J. (1998). Composite technology in load-bearing orthopaedic implants. *Biomaterials, 19*(15), 1329–1342. https://doi.org/10.1016/S0142-9612(97)00217-2

Favaloro, M. (2018). Thermoplastic composites in aerospace – The future looks bright. https://www.compositesworld.com/articles/thermoplastic-composites-in-aerospace-past-present-and-future. Accessed 11 February 2021.

Feraboli, P., Masini, A., & Bonfatti, A. (2007). Advanced composites for the body and chassis of a production high performance car. *International Journal of Vehicle Design, 44*(3–4), 233–246. https://doi.org/10.1504/IJVD.2007.013641

Froes, F. H. (1997). Is the use of advanced materials in sports equipment unethical?. *Journal of the Minerals, Metals and Materials Society, 49*(2), 15–19. https://doi.org/10.1007/BF02915473

Globe. (2021). Globe composite blog. *3 Different Types of Radiation Shielding Materials.* https://www.globecomposite.com/blog/3-different-types-of-radiation-shielding-materials

Gries, T., Stueve, J., Grundmann, T., & Veit, D. (2008). Textile structures for load-bearing applications in automobiles. In *Textile Advances in the Automotive Industry* (pp. 301–319). Cambridge: Woodhead Publishing Limited. https://doi.org/10.1533/9781845695040.4.301

Gupta, R., Huo, D., White, M., Jha, V., Stenning, G. B. G., & Pancholi, K. (2019). Novel method of healing the fibre reinforced thermoplastic composite: A potential model for offshore applications. *Composites Communications, 16.* https://doi.org/10.1016/j.coco.2019.08.014

Hansmann, H., & Wismar, H. (2003). Compendium, composites ASM handbook/extraction polyester resins. ASM handbook/extraction. https://studylib.net/doc/10324251/

Muller, J., Zulliger, A., & Dorn, M. (1994). Economic production of composite beams with 3D fabric tapes. *Textile Month,* 9(9–13).

Karaduman, N. S., Karaduman, Y., Ozdemir, H., & Ozdemir, G. (2017). Textile reinforced structural composites for advanced applications. In *Textiles for Advanced Applications* (pp. 87–131). UK: IntechOpen Publishing. https://doi.org/10.5772/intechopen.68245

Krishnasamy, J., Ramasamy, A., Das, A., & Basu, A. (2018). Electromagnetic absorption behaviour of carbon helical/coiled yarn woven and knitted fabrics and their composites. *Journal of Thermoplastic Composite Materials,* 1–26. https://doi.org/10.1177/0892705718759389

Mallick, P. K. (2010). *Materials, Design and Manufacturing for Lightweight Vehicles* (pp. 1–376). UK: Woodhead Publishing. https://doi.org/10.1533/9781845697822

Mallick, V. (2001). Thermoplastic composite based processing technologies for high performance turbomachinery components. *Composites - Part A: Applied Science and Manufacturing, 32*(8), 1167–1173. https://doi.org/10.1016/S1359-835X(01)00064-1

Mankodi, H. R. (2011). Developments in hybrid yarns. In *Specialist Yarn and Fabric Structures: Developments and Applications* (pp. 23–56). UK: Woodhead Publishing. https://doi.org/10.1533/9780857093936.21

McCormick, M. (2010). Soaring to new heights: The evolution of pole vaulting and pole materials. *Illumin, 12,* 1–5.

Meng, L., Lin, Y., Zheng, R., Xu, H., Gu, H., & Jia, Q. (2016). Mechanical design and implementation of a modular autonomous underwater vehicle. *Jiqiren/Robot, 38*(4), 395–401. https://doi.org/10.13973/j.cnki.robot.2016.0395

Pilato, L. A., & Michno, M. J. (1995). *Advanced composite materials. Polymer International* (pp. 215–216). Heidelberg: Springer. https://doi.org/10.1002/pi.1995.210380216

Motavalli, M., Czaderski, C., Schumacher, A., & Gsell, D. (2010). Fibre reinforced polymer composite materials for building and construction. In *Textiles, Polymers and Composites for Buildings* (pp. 69–128). Cambridge: Woodhead Publishing Limited. https://doi.org/10.1533/9780845699994.1.69

Mouritz, A. P., Bannister, M. K., Falzon, P. J., & Leong, K. H. (1999). Review of applications for advanced three-dimensional fibre textile composites. *Composites Part A: Applied Science and Manufacturing, 30*(12), 1445–1461. https://doi.org/10.1016/S1359-835X(99)00034-2

Mouritz, A. P., Gellert, E., Burchill, P., & Challis, K. (2001). Review of advanced composite structures for naval ships and submarines. *Composite Structures, 53*(1), 21–42. https://doi.org/10.1016/S0263-8223(00)00175-6

Ning, H., Pillay, S., & Vaidya, U. K. (2009). Design and development of thermoplastic composite roof door for mass transit bus. *Materials and Design, 30*(4), 983–991. https://doi.org/10.1016/j.matdes.2008.06.066

Offringa, A. R. (1996). Thermoplastic composites – Rapid processing applications. *Composites Part A: Applied Science and Manufacturing, 27*(4), 329–336. https://doi.org/10.1016/1359-835X(95)00048-7

Pisanikovski, T., Wahlberg, J., Andersson, C. H., Eng, K., Zah, W., & Baeten, S., et al. (1998). Thermoplastic non crimped composites from split film co-knitted fabrics. *Asian Textile Journal, 7*(8), 61–65.

Rajendran, Sathyakumar, Nallusami, R., Chakravarthy, D., Dharan, B., & Shasvanth (2020). Textile fibre reinforced concrete. *International Research Journal of Engineering and Technology (IRJET), 7*(3), 1417–1424.

Ramakrishna, S. (2004). An Introduction to Biocomposites. Series on Biomaterials and Bioengineering (Vol. 1). UK: Imperial College Press.

Ramakrishna, S., Mayer, J., Wintermantel, E., & Leong, K. W. (2001). Biomedical applications of polymer-composite materials: A review. *Composites Science and Technology, 61*(9), 1189–1224. https://doi.org/10.1016/S0266-3538(00)00241-4

Raupach, M., & Morales Cruz, C. (2016). Textile-reinforced concrete: Selected case studies. In *Textile Fibre Composites in Civil Engineering* (pp. 295–299). Duxford: Elsevier. https://doi.org/10.1016/B978-1-78242-446-8.00013-6

Rigby, A. J., Anand, S. C., & Horrocks, A. R. (1997). Textile materials for medical and healthcare applications. *Journal of the Textile Institute, 88*(3), 83–93. https://doi.org/10.1080/00405009708658589

Robert, M., Roy, R., & Benmokrane, B. (2010). Environmental effects on glass fiber reinforced polypropylene thermoplastic composite laminate for structural applications. *Polymer Composites, 31*(4), 604–611. https://doi.org/10.1002/pc.20834

Salama, M. M., Storhaug, T., & Spencer, B. (2002). Recent developments of composites in the oil/gas industry. *SAMPE Journal, 38*(1). 30–38.

Sepe, R., Bollino, F., Boccarusso, L., & Caputo, F. (2018). Influence of chemical treatments on mechanical properties of hemp fiber reinforced composites. *Composites Part B: Engineering, 133*, 210–217. https://doi.org/10.1016/j.compositesb.2017.09.030

Smith, H. (2001). U-99 uninhabited tactical aircraft preliminary structural design. *Aircraft Engineering and Aerospace Technology, 73*(1), 31–35. https://doi.org/10.1108/00022660110367051

Spencer, B. E. (1998). Composites in the sporting goods industry. In *Handbook of Composites* (pp. 1045–1052). Dordrecht: Springer. https://doi.org/10.1007/978-1-4615-6389-1_50

Tong, Y. (2019). Application of New Materials in Sports Equipment. In *IOP Conference Series: Materials Science and Engineering, 493.* https://doi.org/10.1088/1757-899X/493/1/012112

Turner, T. A., Harper, L. T., Warrior, N. A., & Rudd, C. D. (2008). Low-cost carbon-fibre-based automotive body panel systems: A performance and manufacturing cost comparison. *Proceedings of the Institution of Mechanical Engineers, Part D: Journal of Automobile Engineering, 222*(1), 53–63. https://doi.org/10.1243/09544070JAUTO406

Van De Velde, K. (2005). *Textile Composites in Sports Products. Design and Manufacture of Textile Composites.* Cambridge: Woodhead Publishing Limited. https://doi.org/10.1533/9781845690823.444

Zhang, L. (2015). The application of composite fiber materials in sports equipment. In *5th International Conference on Education, Management, Information and Medicine (EMIM 2015)* (pp. 450–453). https://doi.org/10.2991/emim-15.2015.88

15 Future Trends in Towpreg-Based Thermoplastic Composites

Pierre Ouagne and Mahadev Bar
Université de Toulouse
Tarbes, France

CONTENTS

15.1 INTRODUCTION

Over the past few decades, thermoplastic composites have gained significant interest in the consumer industry. This tendency holds to some intrinsic properties that offer some advantages to the thermoplastic matrices over the thermoset one viz.: no transportation restriction, long-term storage at room temperature, low toxicity during processing, low processing time, ability to weld and their potential of being

DOI: 10.1201/9781003049715-15

reprocessed/recycled owing to their thermal reversibility (Fortea-Verdejo et al. 2017; John and Thomas 2008). However, the thermoplastic polymers have some drawbacks originating from their high melt viscosity. The high melt viscosity of the thermoplastics hinders the resin impregnation into the reinforcing textile structures. This leads to non-uniform fibre-resin distribution in the composite structure and further development of a composite with voids and inferior mechanical properties (Alagirusamy et al. 2006). The use of flexible towpreg during composite manufacturing can overcome the above drawbacks of the thermoplastic composites (Bar, Das, and Alagirusamy 2019). Previous chapters of the present book discuss the manufacturing of different flexible towpregs, their characteristics and their influence on thermoplastic composite properties. However, the development of thermoplastic composites with improved performance for global applications is an ongoing, continuous process and the researchers are working towards it. The present research in flexible towpregs for thermoplastic composites is focusing on the following areas which are discussed in detail in the present chapter:

- Natural fibres from non-conventional resources for composite reinforcement
- Fancy yarn for composite
- Composites with improved interphase
- 3D-printed composites
- Environmental issues and recyclability

15.2 NATURAL FIBRES FROM NON-CONVENTIONAL RESOURCES FOR COMPOSITE REINFORCEMENT

At present, the natural fibres are gaining attention of the composite researchers and manufacturers as composite reinforcing material. This is mainly due to the advantages of the natural fibres over their synthetic counterparts. The advantages of the natural fibres as composite reinforcements have been mentioned in much recent literature (Dong 2018; Peças et al. 2018). Due to these advantages, the natural fibre composites have found wide applications in the areas of automobile, construction, packaging, households etc. Along with composite reinforcement, the natural fibres are also used for textile, rope and sack manufacturing. The consumption of natural fibre in these areas is growing exponentially every year. However, the production of conventional natural fibres such as flax, jute, sisal etc. is not increasing at the same rate. In fact, with growing world population, the cultivation land used for fibre-crop production is decreasing day-by-day. Hence to maintain the natural fibre supply chain, the researchers are trying to extract fibres from the non-conventional plant resources such as linseed flax, hemp, banana etc. The challenges related to fibre extraction from these non-conventional plant resources such as hemp and linseed flax are discussed below.

15.2.1 HEMP FIBRE EXTRACTION

Cannabis, the plant behind the hemp fibre, is believed to find its origins in central Asia, probably between the Caspian Sea and the south of Lake Baikal (Li 1974).

Historic evidence shows that the cannabis plant was cultivated primarily to grow fibres in China at about 4000 BC. The fibres were used for the production of textile fabrics, ropes and fishing nets as well as for the production of paper. The traditional extraction of hemp involves water retting, which requires large amounts of manual labour. This water-retting process also has a high environmental impact as it involves high water consumption and high oxygen demand (BOD) of the used water. Hence water retting is prohibited in most of the advanced countries in the world. At present in Europe, hemp stems are mainly processed using hammer mill (Bag et al. 2011), which produces short length fibres with poor mechanical properties suitable for the production of papers and bio-composites for non-load bearing applications (González-García et al. 2010; Yan et al. 2013). In order to avoid fibre damage during fibre extraction, the due retting process is adopted as an alternate. In due retting, after harvesting the hemp stems are left under the open sky where the retting is carried out by the fungi relying on the rain water and air humidity. This process is environmentally friendly and economical but is strictly dependent on weather conditions. An increase in precipitations and relative humidity may lead to an increase in moisture content in the stems and could then make their collection difficult or impossible, due to rotting. Grégoire et al. (2020) have extracted long and fine hemp fibres from due retted hemp stem using a lab-scale scutching and hackling fibre extraction device as shown in Figure 15.1.

This device has three main sections: (i) breaking section, (ii) scutching section, and (iii) hackling section. The first section of this machine breaks the wooden part of the hemp stems and allows a first extraction of the shives and the dust. The material obtained is then automatically transported to the scutching unit which consists of two turbines rotating in opposite direction to each other. Their role is to beat the

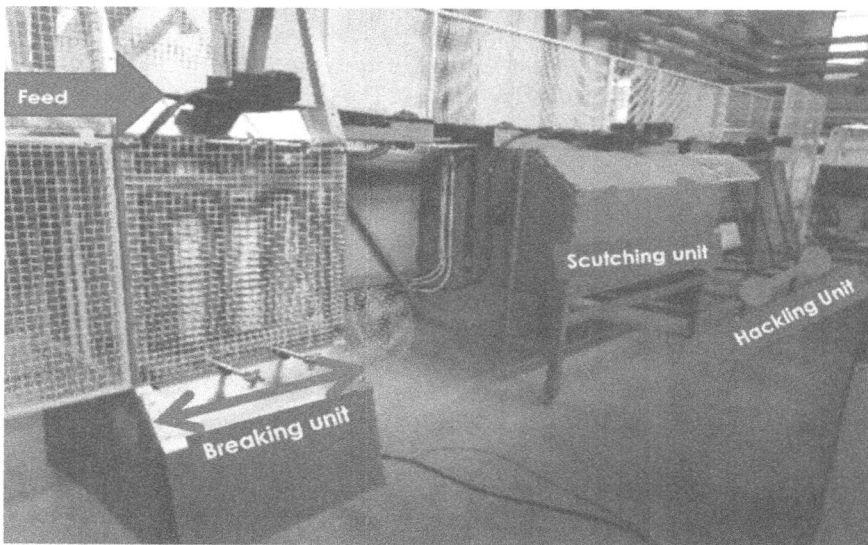

FIGURE 15.1   Lab-scale scutching and hackling fibre extraction device.

fibre and shives mixture with the objective to remove the remaining shives from the fibres. Finally, the fibres are subjected to a progressive hackling stage to align the fibre and also to reduce the technical fibre diameter. Long and fine hemp fibres having sufficient strength for load-bearing composite applications can be extracted through the above process. The combined action of the three sections of the above-mentioned machine performs the same job as that of conventional flax scutching and hacking machine. Thus, the due retted hemp stems can be processed using industrial flax scutching and hacking machineries. The hemp stems are longer than the flax stems and morphology of both stems are not similar. Hence, the hemp stems need to be cut into pieces and the machine parameters should be adjusted to process retted hemp stems in conventional flax scutching and hacking machinery.

15.2.2 Linseed Flax Fibre Extraction

Flax is a dual-purpose crop plant generally cultivated for the fibre and seed purpose. The flax that is grown for fibre purposes is known as textile flax while the other one is known as linseed flax. Although the linseed flax is grown for oil seed purpose, the plant is rich in good quality fibrous materials. The height of the linseed flax plant is comparatively smaller than the textile flax and after harvesting the seed the linseed flax stems are left randomly in the field. Thus, extraction of fibre from linseed flax plants using conventional flax fibre extraction machines is not possible. Pillin et al. (2011) have studied the mechanical properties of the elementary linseed flax fibre of different varieties and compared them with the mechanical properties of the same variety of textile flax fibres. They have reported that the mechanical properties of the linseed flax fibres are comparable to that of textile flax fibres. Thus, the oleaginous flax fibres could be considered as good candidates for substitution to glass fibres in polymer matrix reinforcements. Ouagne et al. (2017) have extracted fibres from linseed flax stems in two ways, i.e. using a cadet machine and using a lab-scale scutching and hackling fibre extraction device (shown in Figure 15.1). A schematic diagram of the cadet machine is shown in Figure 15.2. The fibre yield in the cadet is higher than that of lab-scale scutching and hackling fibre extraction devices. However, the fibre extracted using the cadet has short fibre length and it contains lot of shives, which disturbs the further processing of the fibre. The short fibre length of the fibres extracted using cadets restricted their utilization in the low and medium load bearing applications only. On the other hand, the scutching and hackling fibre extraction

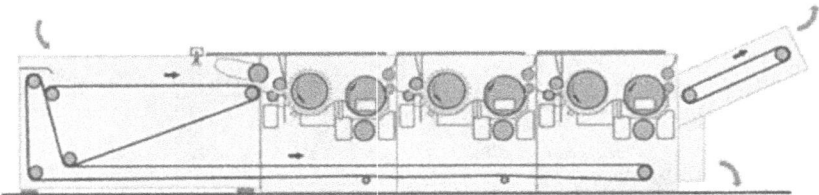

FIGURE 15.2 Schematic diagram of the cadet machine. (Reproduced with permission from Grégoire et al. 2020.)

device can extract long and fine linseed flax fibres, but this is a lab scale process not suitable for large scale production. Thus, the extraction of long, fine and cleaned fibres from linseed flax stems is still a challenging task.

15.3 FANCY YARN FOR THERMOPLASTIC COMPOSITE REINFORCEMENT

The main problem of the thermoplastic composite is the poor fibre-matrix distribution in the composite structure. This non-uniform fibre-resin distribution develops a composite with voids and inferior mechanical properties. The use of hybrid yarn during composite manufacturing improves the fibre-matrix distribution in the composite by lowering the effective resin flow distance. Hybrid yarns are widely used for composite manufacturing both in yarn and in fabric form. Bar et al. (2020) have observed that use of hybrid yarn-based woven fabric during composite manufacturing exhibits uniform fibre-resin distribution in the composite structure, but it left some void at the yarn crossover point of the reinforced fabric (as shown in Figure 15.3). Some matrix materials in film form were added after each woven preform layer to eliminate the above-mentioned macro-void from the hybrid yarn-based woven fabric composite. This additional matrix film layer enhances the effective matrix content in the composite and ultimately reduces the mechanical properties of the resultant composite. Instead of adding an extra matrix-film layer, if one could add some extra matrix at the yarn cross-over point of the woven fabric, that can remove the micro-void from the composite without sacrificing its effective fibre content. Although this kind of study has not been reported in the literature, the authors believe that the use of hybrid fancy yarns during composite manufacturing can solve this purpose.

Fancy yarns are a type of yarn that is made with a distinctive irregular profile or construction that differs from basic single and folded yarns. The primary objective of the fancy yarn is to enhance the aesthetics of the end product with respect to visual and textural properties. Fancy yarns essentially give fashion touches to the fabric

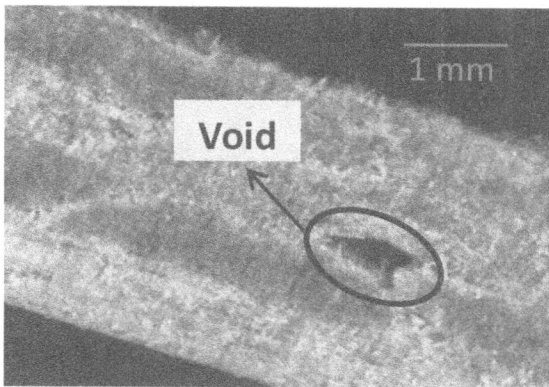

FIGURE 15.3 Hybrid yarn-based woven fabric compressed composite with macro-void. (Reproduced with permission from Bar et al. 2020.)

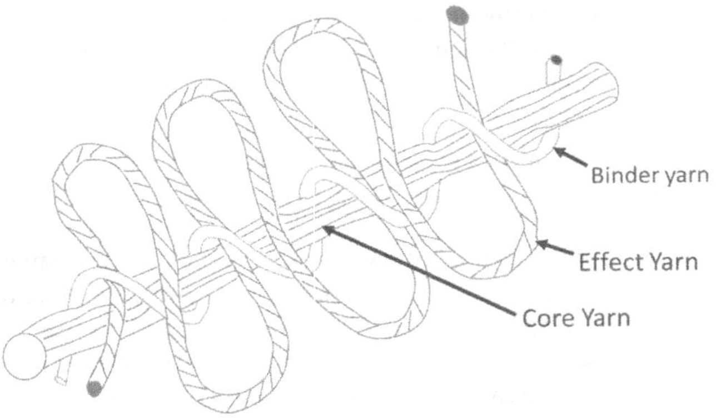

FIGURE 15.4 Basic components of a fancy yarn.

but this type of yarn could be very useful for enhancing composites' mechanical property. Fancy yarns are of different types and various names are used to describe the different fancy yarn effects. If one carefully looks at the fancy yarn structures, it would show that the various fancy yarn effects mainly consist of eight basic profiles. The basic fancy yarn profiles are spiral, gimp, slub, knop, loop, cover, chenille and snarl (Bellwood 1978). The above-mentioned fancy yarn profiles are generally made up of three basic yarn component which are known as core component, binder and effect component. The names of the yarn components reflect their function in the final fancy yarn. The buckling of the effect component around the core produces the aesthetic effect, and the twisting of the binder around the assembly locks the feature in place. The basic yarn components of fancy yarn are shown in Figure 15.4. The above-mentioned effects on the fancy yarn structure are developed mainly in two stages of fancy yarn manufacturing, i.e. in the profile twisting stage and binding stage. In the profile twisting stage the effect and core components are generally involved while the binder yarn is introduced in the binding stage to stabilize the structure. Different ply-twisting and spinning techniques are generally involved in the fancy yarn manufacturing process. Among these techniques, ring spinning, hollow spindle spinning, DREF spinning, REPCO spinning and fancy yarn twisting are well known for fancy yarn manufacturing.

There are different types of fancy yarn available. Among those, the fancy yarn having effect component in its structure will be more suitable for manufacturing fabric preform for thermoplastic composite. Some of these fancy yarn structures are shown in Figure 15.5. The effect component of the fancy yarns for thermoplastic composites has to be essentially thermoplastic. The arrangement of the effect component of the fancy yarn has to be made in such a way that on its weaving it should remain in the yarn crossover point. Then during composite manufacturing, this thermoplastic effect component will be melted and will be uniformly distributed in the yarn crossover point of the woven textile preform-based thermoplastic composite. This will reduce the void content of the resulted composite and will ultimately develop a composite with improved mechanical properties.

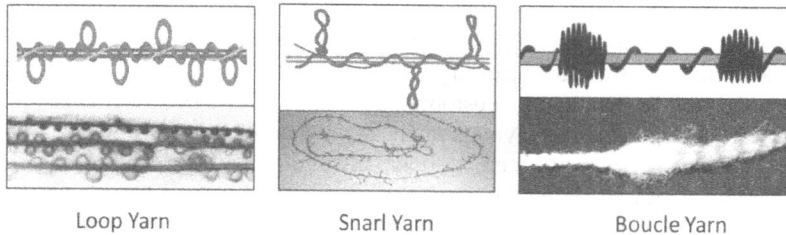

Loop Yarn   Snarl Yarn   Boucle Yarn

FIGURE 15.5   Some fancy yarn structures having effect components. (Adapted from https://www.textiletoday.com.bd/basic-idea-fancy-yarn.)

15.4   COMPOSITES WITH IMPROVED INTERPHASE

The composite interface plays an important role in determining the mechanical properties of the composite as the stress transfer from the matrix to the fibre takes place through this phase. In different fibre reinforced polymer composite systems, it is observed that the reinforcing fibre and the polymer matrices are not compatible, which leads to the development of an imperfect composite interface and ultimately a composite with inferior mechanical properties. To overcome this problem, the fibre surfaces are modified generally in three ways, i.e. through chemical treatment, physical treatment and enzyme treatment (de Farias et al. 2017; Mohanty, Khan Mubarak, and Hinrichsen 2000; Xie et al. 2010). The treatment of fibre with alkali, maleic anhydride, silane coupling agent etc. are some examples of fibre surface modification through chemical treatment approach. Fibre surface modification through plasma treatment, corona treatment etc. are the examples of some physical treatment approaches. In case of enzyme treatment approach of fibre surface modification, an enzyme is involved. At present, the enzyme treatment approach for fibre surface modification is becoming more popular day by day due to the environment friendliness of the process. However, the enzymes are substrate specific and need more time to complete the process. Fibre surfaces modified through physical treatment approaches show significant improvement in composite's interfacial bonding, but these processes consume lots of energy and involve high initial investment. Hence, the physical treatment approach of the fibre surface modification is not so popular in the small-scale industries. The chemical treatment approach is the most popular form of fibre surface modification among the composite researchers and manufactures due to ease of the process, ease of material handling etc. However, most of the chemicals used for fibre surface modification in order to improve the composite interface are hazardous and are not environmentally friendly. Hence the research is going on with an aim to find eco-friendly chemical treatments for fibre surface modification in order to improve the composite interface.

15.5   3D-PRINTED COMPOSITES

3D printing is a process of manufacturing materials which functions as a layer-by-layer addition approach to make objects according to 3D model data. This process was first described by Charles Hull in 1986 (Hull 1986). Since then, the 3D printing technology popularly known as additive manufacturing technology has become a legitimate method for functional parts and product manufacturing. It is mainly due to the advantages of the 3D printing techniques such as minimal waste generation, low cost of the production, customized product geometry and ease of material change. The thermoplastic polymers such as polycarbonate, polyamide, polylactic acid etc. and the thermosetting epoxy polymers can be used as printing material in the 3D printing system (Wang 2017). 3D printing is assisting different industries in many ways. Depending on the selection of the materials, the 3D-printed polymers have found application in the areas of automobile, aerospace, medical fields, art fields, architectural industries etc. (Kroll and Artzi 2011; Wang et al. 2017; Wong and Hernandez 2012). There are several techniques of 3D printing available. Among them: fused deposition modelling (FDM), selective laser sintering (SLS), powder-liquid 3D-printing (PLP), stereolithography (SLA) and robocasting are commonly used process for 3D printing (Wang et al. 2017). The schematic diagram explaining all the above-mentioned 3D-printing techniques is shown in Figure 15.6. Every 3D printing technique, considering the material processed, processing conditions and product functionality, has some advantages and drawbacks, which are discussed below.

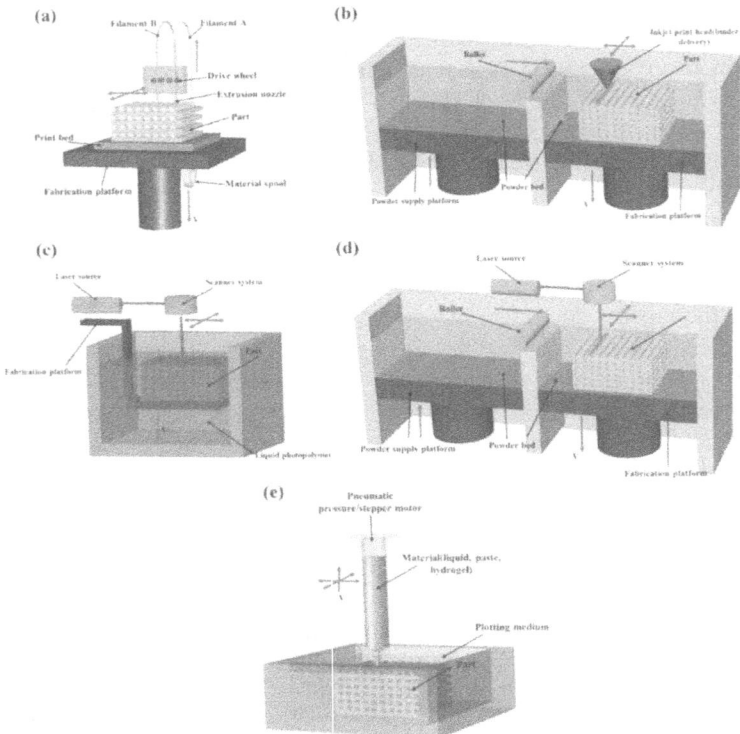

FIGURE 15.6 Schematic representation of various 3D printing techniques: (a) fused deposition modelling, (b) powder-liquid 3D printing, (c) stereolithography, (d) selective laser sintering, and (e) robocasting. (Reproduced with permission from Wang et al. 2017.)

15.5.1 FUSED DEPOSITION MODELLING (FDM)

FDM is the most common method used for 3D printing of the low melting temperature thermoplastic polymers. In this process, the polymer filaments of particular diameters are fed to the printing head of the printer, where it melts and extrudes on the base platform layer-by-layer through the nozzle until it solidifies into the final part. This is a simple, low-cost, rapid prototyping process that could print different materials simultaneously for producing multifunctional parts. However, this method could not control the uniform dispersion of the filler materials during printing.

15.5.2 SELECTIVE LASER SINTERING (SLS)

Stereolithography is the technique in which UV-laser light is used to cure the photopolymer. This process is the first and earliest technique of AM technology based on liquid photopolymers. This process is suitable for thermosetting polymers such as epoxy, acrylic etc. The quality of the SLS-printed products depends on different parameters such as laser exposure time, intensity of the exposure, printing resolution etc. However, the SLA is an expensive, relatively slow process which can handle a limited range of materials.

15.5.3 POWDER-LIQUID 3D-PRINTING (PLP)

Powder-liquid 3D printing is also known as drop-on-powder printing or 3-dimensional printing (3DP). This technique is based on powder-liquid processing and was developed at the Massachusetts Institute of Technology (MIT) in 1993. In this process, the polymeric powders are initially spread on the building platform and then the binding liquid is pumped and deposited dropwise on the powder layer at desired places by moving the inkjet print-head in the X-Y direction. Once the desired pattern is formed the platform lowers and the next layer of powder is spread. This cycle is repeated and at the end the unbounded powders are removed to get the final product. This technique is very flexible to different materials, and it functions at room temperature. The quality of the final product is controlled by the powder size, binder viscosity, binder deposition speed and the interaction between binder and powder. The contamination of powder is the major issue of this system.

15.5.4 STEREOLITHOGRAPHY (SLA)

Stereolithography also works in a similar principle to that of the PLP technique. In this method, instead of a liquid binder, a laser beam is used to fuse the layer. Hence, the photopolymers which are curable by the laser beam can be considered for the present process. In this process, the powder and the high-power laser beam interact in a controlled environment which results in fusion of powder in the interaction region. After every layer-fusion operation, the piston-operated platform is lowered, and a new powder layer is spread on the operation bed until the final product is ready. The laser power, scanning speed and powder size are the main factors that influence the SLA product quality. The main advantages of this technique are high resolution and excellent quality. It is a support-free technique because the support is provided by unbounded powder present around the fabricated part. However, like other 3D printing techniques, the SLA technique also has some drawbacks. For instance, this method is time-consuming, costly, and has some issues with porosity, especially when the powder does not melt and fuse properly.

15.5.5 ROBOCASTING

The robocasting process is also known as direct ink writing or 3D-plotting process. This process involves the extrusion of viscous polymer liquid from a pressurized orifice. In this technique, the printing platform is stationary while the orifice head moves in all three directions to create a layer-by-layer print product. The quality of the final robocasted product depends on the material viscosity and deposition speed. This technique has a vast range of material flexibility. Ceramics, solutions, suspensions and pastes of materials can be processed in this method.

The 3D-printed polymer products which are available at present are mainly used as conceptual prototypes rather than functional components. It is mainly due to the poor mechanical properties of the 3D-printed composites which makes them unacceptable for load bearing composite application. To overcome this problem different fillers are incorporated in the 3D-printed composite structures. The reinforcements for 3D-printed polymer composites are of two types, namely particle filler and fibrous fillers. Based on their dimensions, particles are of two types: micro particles and nano particles. The micro-sized particle reinforced polymers can be printed through FDM, SLS and SLA techniques respectively. The problem of poor tensile strength and low storage modulus of the 3D-printed polymers can be improved by the addition of the glass breads. Hwang et al. (2015) have employed FDM techniques to study the thermos-mechanical properties of the metal/polymer composite. Copper and iron particles having average particle size 24 μm and 43 μm respectively were used in the above-mentioned study where the filler concentrations were varied between 10% and 80% respectively. Kalsoom et al. (2016) have manufactured a composite for heat sink using the SLA technique. The composite material was prepared by adding 30% (w/v) micro-diamond particles with acrylate resins. The results demonstrated that addition of diamond particles improved the heat transfer rates of the composite heat sink compared to pure polymer heat sink.

At present, the nano materials are finding wide applications in the areas of automobile, electronics, medicine, sports etc. Incorporation of different nano particles such as graphene, carbon nanotubes, nano clays etc. in polymer matrix systems improves the mechanical, thermal and the electrical properties of the resultant composites. Lin et al. (2015) studied the mechanical properties of graphene oxide/photopolymer composites using SLA technique. They have observed that the addition of 0.2% of graphene oxide in photopolymer resulted in a 62.2% and 12.8% increase in tensile strength and ductility of the printed composite. 3D printing techniques also allow different fractions of nanoparticles to be delivered to different regions during fabrication, which provides optimized functional results. Fibrous materials as composite reinforcement have already drawn the attention of the composite researchers and manufacturers, which is mainly due to the low density and other advantages of the fibrous materials. Recently, the researchers have reinforced the polymer matrix with fibrous materials through 3D printing techniques, mainly through FDM and direct printing methods. In the FDM process, the reinforcing fibres and polymers are mixed and passed through extruders to manufacture filaments. These filaments are utilized for 3D printing of desired geometries. Carneiro, Silva, and Gomes (2015) have fabricated glass fibre–reinforced polypropylene composites through FDM techniques. They have reported that the 3D-printed glass-PP composites exhibit 40% higher tensile strength and 30% higher tensile modulus than that of pure PP. However, the process of 3D-printing of the fibre reinforced polymers has some drawbacks which are discussed below.

- The reinforcement of continuous fibres offers a higher strength to the polymer composites than that of short-fibre reinforced composites. However, the 3D-printing of the continuous fibre reinforced composites is a real challenge.
- 3D-printed polymer composites have high void content which deteriorates the mechanical properties of the final composites.
- Controlling the fibre orientation in short fibre reinforced 3D-printed polymer composites is very difficult.
- Fibre content plays an important role in determining the mechanical properties of the printed polymer composite. The high fibre volume fraction makes the fabrication of 3D-printed composite very difficult.
- All polymers are not suitable for the 3D printing process. The selection of materials depends on the type of printing process and on the process parameters.

15.6 ENVIRONMENTAL ISSUES AND RECYCLABILITY

The demand of fibre reinforced polymers due to their low density and durable characteristics is increasing every year. It is expected that in the next 20 years, the annual demand of the fibre reinforced polymers will be doubled and it will exceed 600 million tons. These polymer composites are a cause for concern at the end of their life-cycle as they end up in landfills and contribute to different forms of environmental pollution. Recycling of the end-of-life composite products and composite manufacturing process wastes can overcome the above-mentioned problems. At present, glass, metals, plastics, and many other engineering products are recycled to a great extent. However, the recycling of composites is not so common so far. It is mainly due to the inherent heterogeneous nature of the composite constituents, leading to poor material recyclability (Job 2010). Because of these challenges, most of the recycling activities for composite materials are limited to down-recycling such as energy or fuel recovery with little materials recovery such as reinforcement fibres. Each engineering recycling process involves a chain of operations. Failure in any step of this chain implies that the recycling process cannot be completed. Figure 15.7 illustrates the chain of operations for composite recycling.

The majority of the fibre reinforced polymers are comprised of thermosetting polymers. Reprocessing of thermosetting resin requires chemical digestion to the monomer and then polymerizing to produce a new thermoset product. This strategy of recycling thermoset composites is not economical. Although the market share of thermoplastic composites is much lower than the thermoset composites, the former

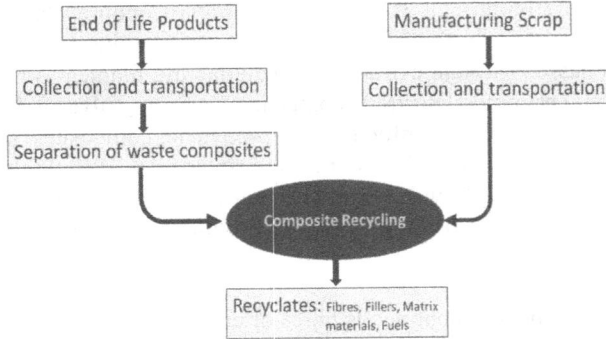

FIGURE 15.7   Chain of operations for composite recycling.

one is advantageous over the later one in terms of their recyclability. Recycling of composite materials is generally carried out in three ways: mechanical recycling, chemical recycling, and thermal recycling. Different techniques of composite recycling are summarized in Figure 15.8.

15.6.1 MECHANICAL RECYCLING

This process involves shredding and grinding followed by screening to separate fibre-rich and resin-rich fractions for re-use. Thermoplastic shredded composites can be remoulded further while the shredded thermoset composites can be used as filler material. The method is very energy-intensive and the recyclates have relatively low quality.

15.6.2 THERMAL PROCESSING

This process uses high temperatures to decompose the resin and subsequently separates the reinforcing fibres and fillers. Clean fibres or inorganic fillers are recovered

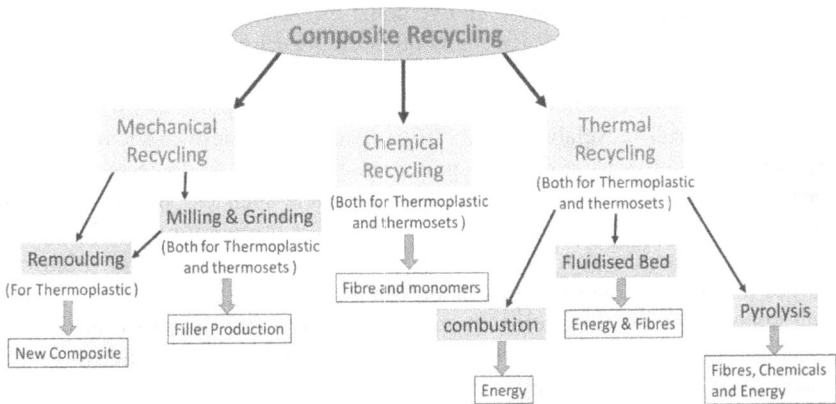

FIGURE 15.8   Different approaches for polymer composite recycling.

and the thermal energy can be produced through pyrolysis, gasification or combustion of the waste composites. However, the thermal processing deteriorates the quality of the recovered fibres or filler materials to a varying extent.

15.6.3 CHEMICAL RECYCLING

This process aims at chemical depolymerization or removal of the matrix and liberation of fibres for further recycling by using an organic or inorganic solvent.

The above-mentioned processes can recycle both the thermoplastic and thermoset composites in an effective way, but the development of a recycled composite with improved mechanical properties is the main challenge for the material scientists and manufacturing and application industry. Besides the researchers are also trying to find a low-cost, efficient recycling technology to separate and recycle composite materials. Along with the above problems, the companies involved in composite recycling have some more constraints which are mentioned below (Gopalraj and Kärki 2020; Yang et al. 2012):

- Non-availability of the composite scrap for the economical scale of recycling operations.
- The quality of the recycled composites is not comparable with the existing composites available in the markets.
- The overall cost and new environmental burdens in the recycling technology.
- Profitability and sustainable operation of the recycling business.

The green composites do not require any recycling steps as they are biodegradable and environment-friendly in nature. However, the self-life of these composites is comparatively low and they are not suitable for outdoor applications. The mechanical properties of the green composites are also comparatively lower than the non-biodegradable composites. The above-mentioned drawbacks limit the usages of green composites in non-load bearing or low load bearing indoor applications only. The research to overcome the above-mentioned drawbacks of the green composites is ongoing.

15.7 SUMMARY

Over the last few decades, a resurgence of interest in thermoplastic composites has been observed. It is mainly due to recyclability, lightweight, and ease of manufacturing and other advantages of the thermoplastic composites. However, the thermoplastic composites have some drawbacks due to their high melt viscosity which leads to formation of high amount of voids in composites resulting in inferior mechanical properties. Use of flexible towpregs during composite manufacturing overcame the above-mentioned drawbacks of the thermoplastic composites. Hence the mechanical properties of the towpreg-based thermoplastic composites are superior to that of film stacked or other conventional composites. The development of thermoplastic composites with improved performance for global applications is an ongoing continuous process. The present chapter highlights the ongoing developments and the future research scopes in the area of flexible towpreg-based thermoplastic composites.

REFERENCES

Alagirusamy, R., R. Fangueiro, V. Ogale, and N. Padaki. 2006. Hybrid yarns and textile preforming for thermoplastic composites. *Textile Progress* 38(4): 1–71. doi.org/10.1533/tepr.2006.0004

Bag, R., J. Beaugrand, P. Dole, and B. Kurek. 2011. Viscoelastic properties of woody hemp core. *Holzforschung* 65 (2): 239–247. doi.org/10.1515/hf.2010.111

Bar, M., R. Alagirusamy, A. Das, and P. Ouagne. 2020. Low velocity impact response of flax/polypropylene hybrid roving based woven fabric composites: Where does it stand with respect to GRPC?. *Polymer Testing* 89: 106565. doi.org/10.1016/j.polymertesting.2020.106565

Bar, M., A. Das, and R. Alagirusamy. 2019. Influence of flax/polypropylene distribution in twistless thermally bonded rovings on their composite properties. *Polymer Composites* 40(11): 4300–4310. doi.org/10.1002/pc.25291

Bellwood, L. 1978. Novelty yarns for speciality fabrics, *Textile Industries* 63–68.

Carneiro, O. S., A. F. Silva, and R. Gomes. 2015. Fused deposition modeling with polypropylene. *Materials and Design* 83: 768–776. doi.org/10.1016/j.matdes.2015.06.053

de Farias, J. G. G., R. C. Cavalcante, B. R. Canabarro, H. M. Viana, S. Scholz, and R. A. Simão. 2017. Surface lignin removal on coir fibers by plasma treatment for improved adhesion in thermoplastic starch composites. *Carbohydrate Polymers* 165, 429–436. doi.org/10.1016/j.carbpol.2017.02.042

Dong, C. 2018. Review of natural fibre-reinforced hybrid composites. *Journal of Reinforced Plastics and Composites* 37(5): 331–348. doi.org/10.1177/0731684417745368

Fortea-Verdejo, M., E. Bumbaris, C. Burgstaller, A. Bismarck, and K. Y. Lee. 2017. Plant fibre-reinforced polymers: where do we stand in terms of tensile properties?. *International Materials Reviews* 62(8): 1–24. doi.org/10.1080/09506608.2016.1271089

Grégoire, M., B. Barthod-Malat, L. Labonne, P. Evon, E. De Luycker, and P. Ouagne. 2020. Investigation of the potential of hemp fibre straws harvested using a combine machine for the production of technical load-bearing textiles. *Industrial Crops and Products* 145: 111988. doi.org/10.1016/j.indcrop.2019.111988

González-García, S., A. Hospido, G. Feijoo, and M. T. Moreira. 2010. Life cycle assessment of raw materials for non-wood pulp mills: Hemp and flax. *Resources, Conservation and Recycling* 54(11): 923–930. doi.org/10.1016/j.resconrec.2010.01.011

Gopalraj, S. K., and T. Kärki. 2020. A review on the recycling of waste carbon fibre/glass fibre-reinforced composites: fibre recovery, properties and life-cycle analysis. *SN Applied Sciences* 2(3): 1–21. doi.org/10.1007/s42452-020-2195-4

Job, S. 2010. Composite recycling—summary of recent research and development. Materials KTN Report.

John, M. J., and S. Thomas. 2008. Biofibres and biocomposites. *Carbohydrate Polymers* 71: 343–364. doi.org/10.1016/j.carbpol.2007.05.040

Hull, C. W. 1986. *Apparatus for production of three-dimensional objects by stereolithography.* Google Patents.

Hwang, S., E. I. Reyes, K. S. Moon, R. C. Rumpf, and N. S. Kim. 2015. Thermo-mechanical characterization of metal/polymer composite filaments and printing parameter study for fused deposition modeling in the 3D printing process. *Journal of Electronic Materials* 44(3): 771–777. doi.org/10.1007/s11664-014-3425-6

Kalsoom, U., A. Peristyy, P. N. Nesterenko, and B. Paull. 2016. A 3D printable diamond polymer composite: a novel material for fabrication of low cost thermally conducting devices. *RSC advances* 6(44): 38140–38147. doi.org/10.1039/C6RA05261D

Kroll. E, and D. Artzi. 2011. Enhancing aerospace engineering students' learning with 3D printing wind-tunnel models. *Rapid Prototyping Journal.* doi.org/10.1108/13552541111156522

Li, H. L. 1974. An archaeological and historical account of cannabis in China. *Economic Botany.* 28(4): 437–448.

Lin, D., S. Jin, F. Zhang, C. Wang, Y. Wang, C. Zhou, and G. J. Cheng. 2015. 3D stereolithography printing of graphene oxide reinforced complex architectures. *Nanotechnology* 26(43): 434003. doi:10.1088/0957-4484/26/43/434003

Mohanty, A. K., A. Khan Mubarak, and G. Hinrichsen. 2000. Surface modification of jute and its influence on performance of biodegradable jute-fabric/biopol composites. *Composite Science and Technology* 60(11): 15–24. doi.org/10.1016/S0266-3538(00)00012-9

Peças, P., H. Carvalho, H. Salman, and M. Leite. 2018. Natural fibre composites and their applications: a review. *Journal of Composites Science* 2(4): 66. doi.org/10.3390/jcs2040066

Pillin, I., A. Kervoelen, A. Bourmaud, J. Goimard, N. Montrelay, and C. Baley. 2011. Could oleaginous flax fibers be used as reinforcement for polymers?. *Industrial Crops and Products* 34(3): 1556–1563. doi.org/10.1016/j.indcrop.2011.05.016

Ouagne, P., B. Barthod-Malat, P. Evon, L. Labonne, and V. Placet. 2017. Fibre extraction from oleaginous flax for technical textile applications: influence of pre-processing parameters on fibre extraction yield, size distribution and mechanical properties. *Procedia Engineering* 200: 213–220. doi.org/10.1016/j.proeng.2017.07.031

Wang, X., M. Jiang, Z. Zhou, J. Gou, and D. Hui. 2017. 3D printing of polymer matrix composites: A review and prospective. *Composites Part B: Engineering* 110: 442–458. doi.org/10.1016/j.compositesb.2016.11.034

Wong, K. V., and A. Hernandez. 2012. A review of additive manufacturing. *International scholarly research notices* doi:10.5402/2012/208760

Xie, Y., C. A. S. Hill, Z. Xiao, H. Militz, and C. Mai. 2010. Silane coupling agents used for natural fiber/polymer composites: a review. *Composites Part A: Applied Science and Manufacturing* 41: 806–819. doi.org/10.1016/j.compositesa.2010.03.005

Yan, Z. L., H. Wang, K. T. Lau, S. Pather, J. C. Zhang, G. Lin, and Y. Ding. 2013. Reinforcement of polypropylene with hemp fibres. *Composites Part B: Engineering* 46: 221–226. doi.org/10.1016/j.compositesb.2012.09.027

Yang, Y., R. Boom, B. Irion, D. J. van Heerden, P. Kuiper, and H. de Wit. 2012. Recycling of composite materials. *Chemical Engineering and Processing: Process Intensification* 51: 53–68. doi.org/10.1016/j.cep.2011.09.007

Index

A

Absorbed energy, 103
Acetylation, 26–27
Acoustically fluidized bed, 48–49
    electrostatic spray coating *vs.,* 54
Additive manufacturing technology, 422;
    see also 3D printing
Adhesive bonding, 73–74
Aircraft applications, thermoplastic composites
    in, 399–401
Air-jet associated commingling, 114–116
    nozzle design and settings, effect of, 120–123
    process parameters, effect of, 119–120
    raw materials, effect of, 118–119
    structure and properties, factors affecting,
        118–123
Air jet texturing, 43
Alkali treatment, 26
Amylopectin, 11
Amylose, 11
Angle interlock (AI) structure, 183
Anionic polyamide 6
    liquid moulding process for, 21
Anisotropic viscous flow, constitutive relation
    for, 237–239
Autoclave molding process, 279–281
    advantages, 281
    limitations, 281
    machine and process description, 279–281
Autohesion, 221
Automated fiber placement (AFP), 287–288
Automated tape placement (ATP), 287–291
    advantages, 291
    limitations, 291
    machine and process description, 288–291
Automated tow placement process, 20
Automotive vehicles, thermoplastic composites
    in, 401–404

B

Barber pole, 37
Basalt fiber reinforced plastic (BFRP) rebars,
    373–374
Bi-axial knitted structures, 185–186
Bi-dimensional (2D) fabric, 16
Bio-commingling, 77, 86
Bis(2-hydroxyethyl) terephthalate, 7
Bladder/diaphragm forming, 274–275
Blending, 74, 312

degree of mixing, 75–76
    indices of, 75
Blending index, 125
Blend irregularity, 76–77
Braided fabrics, 16
Braiding, 42–43, 53
    composite rod for concrete reinforcement,
        364–367
    flexible preforms based on, 189–192
British Pat. No. 1,380,004, 73

C

Cantilever beam bending test, 52
Capillary pressure, 229
Carbon fibre reinforced plastic (CFRP), 344
    thermoplastic rebars, 374–375
Carbon filaments reinforced in thermoplastic
    epoxy resins (CFRTP), 376–377
Carbon filaments reinforced in thermoset epoxy
    resins (CFRTS), 376–377
Carbon/stainless steel (SS) hybrid yarns,
    344–349
Carbon towpreg-based thermoplastic composites,
    343–349
Carmen-Kozeny constant, 203
Chemical recycling, 427
Circumferential winding pattern, 286
Civil engineering applications, 358–360
    composite structural insulated panels for, *see*
        Composite structural insulated panels
        (CSIP)
    flexible towpreg structure, 360–361
    FRP bars for, 373–383
    FRP laminates in, *see* FRP laminates
    hybrid yarn-based structures for concrete
        reinforcement, 361–367
    thermoplastic composites for, 367–373,
        405–406
Coalescence, degree of, 217–220
COMFIL®, 117
Commingled yarn (COM), 124, 125, 126–127,
    168, 304–310, 362–364
    mechanical properties of, 128–131
    moulding of, 127–128
    and voids distributions in laminate, 304–310
Commingling, 43–44, 114–118
    air-jet associated, 114–116, 118–123
    online, 116–118
    powder coating *vs.,* 52–54
    quality of mixing in, 166